Recent Advances on the Modular Organization of the Cortex

Manuel F. Casanova • Ioan Opris
Editors

Recent Advances on the Modular Organization of the Cortex

Springer

Editors
Manuel F. Casanova
Department of Psychiatry
University of Louisville
Louisville, KY
USA

Ioan Opris
Department of Physiology & Pharmacology
Wake Forest University School of Medicine
Winston-Salem, NC
USA

ISBN 978-94-017-9899-0 ISBN 978-94-017-9900-3 (eBook)
DOI 10.1007/978-94-017-9900-3

Library of Congress Control Number: 2015943586

Springer Dordrecht Heidelberg New York London

Printed on acid-free paper

Springer Science+Business Media B.V. Dordrecht is part of Springer Science+Business Media (www.springer.com)

Preface

The cortical minicolumn is a continuing source of research and debate more than half a century after it was identified as a basic component of brain organization. The cortical minicolumn is a sophisticated local network of neural cells that contain within it the elements for redundancy and plasticity that account for its success as an information processing unit. Although it is sometimes compared to subcortical nuclei, the minicolumn is a distinctive form of module that has evolved specifically in the neocortex. It unifies the horizontal and vertical components of cortex within the same columnar space. Minicolumns are highly repetitive, even clone-like, units. However, they display considerable heterogeneity between areas and species, and even within a given macrocolumn. It is this heterogeneity and resultant "competition" in-between modules that provide for the temporal dynamics of information transfer that characterize the most basic unit of the cortical hierarchy. The significance of this cortical module is underscored by recent studies showing its participation as an elemental unit in the emergence of higher cognitive functions and probable participation as a core neuropathological feature of several mental disorders.

This book arose out of a series of lectures that took place in a conference regarding cortical modularity in Louisville, Kentucky. The goals of the conference were to develop a focused group with a coherent view on how to identify cortical modules and their basic components, and to bring new investigators and research techniques into the field. The lectures expounded on the anatomical basis of the cortical minicolumn and introduced the necessary concepts towards understanding the role of this structure in comparative neuroanatomy, neuropathology, neurophysiology and anthropology. As in the conference, the goal of this book is to provide a cohesive forum on the latest anatomical approaches to understand the basic components of cortical modularity and their involvement in different pathological states. The topics will span a wide spectrum of resolution, that is, from minicolumns and their parcellation into different components (e.g., apical dendritic bundles) all the way to macrocolumns and networks of the same. The book includes internationally recognized leaders and talented new investigators. By bringing together these researchers, the book will help coalesce apparently disparate subjects under the umbrella of cortical modularity.

The book is aimed at the neuroscience community from students to well-established investigators. We hope that the book will help promote the field of cortical modularity and in doing so help bridge the gap between theory and practice. We would like to express our appreciation to all of the contributing authors who have helped to make this a useful and high-quality publication. We would also like to thank the publication team at Springer for their efficient services.

Louisville, KY, USA Manuel F. Casanova
Winston-Salem, NC, USA Ioan Opris

Contents

Chapter 1
Introduction

Manuel F. Casanova and Ioan Opris

A selective history of developmental and mechanical constrains on brain maturation and evolution includes both micro- and macroscopic theories. The general idea regarding the possible existence of overarching laws had its beginning in the early nineteenth century in the work of two prominent scientists; the French zoologist Étienne Geoffroy Saint-Hilaire (1772–1844) and the comparative anatomist George Cuvier (1769–1832). Their multiple debates in 1830 at the Royal Academy of Sciences in Paris examined whether animal structures could be explained by either function (Cuvier) or by morphological laws (Geoffroy). The question was summarized by the zoologist and historian of science E. R. Russell: "Is function the mechanical result of form, or is form merely the manifestation of function or activity? What is the essence of life – organization or activity?" (Russell 1916) The view espoused by Geoffrey, later known as the "doctrine of unity of composition", argued that function was dependent on structure and that an archetype of a basic structural plan (*Bauplan* or body map) accounted for homologies across different animal phyla. The word homology was coined only after the Geoffroy-Cuvier debate by Owen to define "the same organ in different animals under every variety of form and function" (Medina 2007). Although, at the time, Geoffroy was judged to be on the losing side of the debate, modern discoveries of evolutionary conserved developmental control genes seemingly support his account of a construction plan that is shared by all bilateral animals (Hirth and Reichert 2007).

Geoffroy's ideas engendered followers primarily in the persons of Robert Edmond Grant and Étienne Serres. The doctrine of unity of composition of all

---

M.F. Casanova, M.D. (✉)
Department of Psychiatry, University of Louisville, Louisville, KY, USA
e-mail: manuel.casanova@louisville.edu

I. Opris, Ph.D.
Department of Physiology and Pharmacology, Wake Forest University Health Sciences, Winston-Salem, NC 27103, USA
e-mail: ioanopris.phd@gmail.com

M.F. Casanova, I. Opris (eds.), *Recent Advances on the Modular Organization of the Cortex*, DOI 10.1007/978-94-017-9900-3_1

vertebrates proposed by Geoffroy thus gave rise, in a grander scheme, to a link between ontogeny (i.e., development of an organism from embryo to adult) and phylogeny (i.e., the evolutionary history of a species). Ernst Haeckel (1834–1919), the eminent German biologist and philosopher, summed up previous versions of this idea and proposed that ontogeny repeated forms of the ancestors, a view now called 'strong' recapitulation. This way of thinking has fallen into disrepute, in part, due to the fact that Haeckel committed various acts of academic fraud in order to support his theory, e.g., forging diagrams and passing an incomplete dog embryo as that of a human. Although the idea is imprecise and has been refuted in its original form, a 'weaker' form is accepted. Indeed, embryonic stages of different species resemble each other more closely than later stages of development (Brüne 2008).

Almost 100 years after the Cuvier-Geoffroy debate, Cornelius Ubbo Ariëns Kappers (1877–1946) introduced the concept of neurobiotaxis (Gk. *νευρον*, nerve + *βίος*, life + *τάξις*, arrangement) as the major law governing vertebrate brain development (Ariëns Kappers 1921). Neurobiotaxis theorizes that nerve cells migrate during development in the direction from which they receive stimuli. This point of view was discredited by Sperry (Sperry 1943) whose work on regeneration of the retinotectal pathways gave rise to the influential "chemoaffinity hypothesis" (Meyer 1998). The most famous antecedent for chemotaxis is the neurotrophic hypothesis of Ramón y Cajal used in explaining the guidance of growth cones towards their final targets (De Castro et al. 2007). Sperry's hypothesis suggested the existence of unique molecular addresses ("nametags") and precise connections within the central nervous system. However, proof for a synaptic nametag that matched axonal growth to their targets remained poor and was later superseded by other theories.

Possibly the most influential developmental theory is that of chemical gradients (Linden 2007). This framework evolved from a multidisciplinary perspective that included embryology, computer sciences, and genetics. As with the chemoaffinity hypothesis, major limitations to this mechanism have been noted. Most significant among the objections is the fact that diffusion by itself is not an efficient patterning mechanism. As a zygote divides into a multicellular organism diffusing morphogens face the barriers of multiple cell membranes. The chemoaffinity hypothesis also leaves unaccounted the role of experience-dependent refinement of synaptic connections (Cline 2003).

References to chemical gradients persist till present with various modifications, e.g., the role of neuronal activity during development and the counterbalancing of signals by different chemical mediators (Turing 1952; Linden 2007). It is now known, for example, that gradients of Wnt3 molecules, counterbalanced by WPhrinB1-EphB, control the structuring of elements required to convey spatial information from the eye to its terminal cortical fields (Schmitt et al. 2006). Similarly, opposing gradients of EphA and ephrin-A control the dispersion of clonally related neurons as they assemble into cortical columns (Torii et al. 2009). Computer models emphasize that the crucial conditions for pattern formation are local self-enhancement and long range inhibition (Gierer 1981). The fact that physical-chemical process could be reduced to the forces of attraction and repulsion was probably first theorized during the nineteenth century by du Bois Reymond (Sulloway 1979). Competition lies at the heart of pattern formation. If one competing force is too strong, form disappears into featureless homogeneity. Patterns emerge when the size of the field becomes larger than the range of the activator.

Another influential idea regarding developmental constrains is the rule or law of associative learning espoused by Donald Hebb (1949). His theory about the operation of the brain is often paraphrased as: "Neurons that fire together wire together." Hebbian learning appears limited to the stabilization of existing synapses within cell assemblies. The resulting pattern of neurons connected through conjoint activation provides for memory traces called engrams. A major limitation of Hebbian precepts and some of the previously mentioned theories is that they apply to the microscopic realm and are foreign to both macroscopic and evolutionary considerations.

Broad-scale developmental and mechanical constrain theories regarding the brain have been the purview of anthropologists. Finlay and Darlington noted that schedules of neurogenesis and regional birth dates across species are fairly preserved (Finlay and Darlington 1995). This observation led to the argument that allometric expansion of regions (i.e., the relation between the size of any organism and any of its parts) follows the order in which they were generated. Finlay and Darlington therefore suggested that later born regions become disproportionately larger with increasing brain size. Alternatively, Sven Ebbesson (1980, 1984) proposed parcellation as the major principle of brain evolution: "[N]ervous systems become more complex; not by one system invading another, but by a process of parcellation that involves the selective loss of connections of the newly formed daughter aggregates and subsystems" (Ebbesson 1980, p. 213). A different perspective was publicized by Terrence Deacon as a displacement hypothesis wherein "the correlations between structural neogenesis, functional specialization and size changes in brain evolution are explained by a theory of competitive displacement of neural connections by others during development under the biasing influences of differential allometry, cell death or axon-target affinity changes" (Deacon 1990). Deacon's displacement hypothesis provides the notion that larger anatomical regions are better connected (Striedter 2005). It should be clearly noted that many of the previously divulged theories have been criticized or have, as of yet, to be proven. Thus, Barton (2001) disputed Finlay and Darlington's (1995) assertions by suggesting a significant evolutionary correlation in the size of interconnected brain regions (Katz and Lasek 1978). Furthermore, Ebbeson's original idea has fallen out of favor with the recognition that brains evolve not only by losing connections but also by establishing novel ones (Striedter 2007).

Less known but of equal importance is the work of the renowned embryologist Erich Blechschmidt who first formulated the underlying formative processes molding human development as force fields (Blechschmidt 2004). Blechschmidt recognized a variety of biodynamic metabolic fields giving rise to the differential growth of tissue. His overall views on essential physical dynamics enabled him to discern a remarkably close interrelationship between external form (morphogenesis) and internal structure (tectogenesis) for each developing organ. The scientific evidence for his views is detailed in almost 200,000 serial sections and 64 enlarged reconstructions of the human embryo at the Museum of the University of Göttingen.

Blechschmidt's proposal of a possible relationship between an external force or appearance and an internal circuitry has also been used in explaining the construction of organs in terms of modular arrangements. In this regard modules are functionally specialized regions of the brain that are domain specific for different cognitive process. In electrophysiology the term makes reference to a group of neurons within the cerebral cortex that have nearly identical receptive fields and encode similar features. Anatomical connections within and between modules constrain their functional interactions while negative feedback provides the capacity for self-regulation.

Serb and Oakley famously said that, "Modules are the distinct processes or units that act cooperatively in an organism, but have a degree of dissociability due to their internal integration. These modules are spatiotemporally bounded with discrete origins, histories, and deaths through evolutionary time and can be treated as individuals". It is the aim of this book to bring together observations in regards to cortical modules, from different research perspectives, that can generalize to other fields of science. We will emphasize the role of the smallest module capable of information processing: the minicolumn.

Neurons are not reiterative elements of the brain. They vary from each other in terms of size, shape, connectivity, location, and function. Neuroanatomists classify these cells with various appellations, e.g., GABAergic, DOPAminergic, Cajal-Retzius cell, etc. These names describe differences in type not in degree. Single neurons do not provide for higher cognitive processes. A great gap remains between neuronal pathology and the psychiatric characteristics of a given disorder. One gains insight into the workings of the brain only by looking outside of the neuron and into minicolumns. In this regard the cell minicolumn is a self-organizing ecosystem of neurons and their projections where connectivity within the module is stronger than connectivity between the nodules (Casanova 2004). Indeed, it is by looking at the interaction between minicolumns that the cerebral cortex provides for the emergence of higher cognitive functions (Opris et al. 2013).

In summary, modernist insights suggest an underlying fundamental reality to life's processes. The worldview of Émile Durkheim reflected the unbounded existence of social facts independent of individuals. The fundamental cognitive orientation for humans according to Sigmund Freud was the unconscious. In similar fashion, the multiplicity of, and confusion with, "overarching laws" in neuroscience entail a lingering expectation of finding a physical principle, a worldview sort of speaking, that can explain the anatomical complexity of the brain (Nieuwenhuys et al. 1998). Within this book the arguments will revolve not around the certainty but rather the degree to which developmental, aging, and evolutionary changes are constrained by the function of nested modular arrangements within the cerebral cortex.

References

Ariëns Kappers CU (1921) On structural laws in the nervous system: the principles of neurobiotaxis. Brain 44:125–149

Barton RA (2001) The coordinated structure of mosaic brain evolution. Behav Brain Sci 24:281–282

Blechschmidt E (2004) The ontogenetic basis of human anatomy: a biodynamic approach to development from conception to birth. North Atlantic Books, Berkeley

Brüne M (2008) Textbook of evolutionary psychiatry: the origins of psychopathology. Oxford University Press, Oxford

Casanova MF (2004) Intracortical circuitry: one of psychiatry's missing assumptions. Eur Arch Psychiatry Clin Neurosci 254:148–151

Cline H (2003) Sperry and Hebb: oil and vinegar? Trends Neurosci 26:655–661

De Castro F, López-Mascaraque L, De Carlos JA (2007) Cajal: lessons on brain development. Brain Res Rev 55:481–489

Deacon TW (1990) Rethinking mammalian brain evolution. Am Zool 30:629–705

Ebbesson SOE (1980) The parcellation theory and its relation to interspecific variability in brain organization, evolutionary and ontogenetic development, and neuronal plasticity. Cell Tissue Res 213:179–212

Ebbesson SOE (1984) Evolution and ontogeny of neural circuits. Behav Brain Sci 7:321–331

Finlay BL, Darlington RB (1995) Linked regularities in the development and evolution of mammalian brains. Science 268:1578–1584

Gierer A (1981) Generation of biological patterns and form: some physical, mathematical, and logical aspects. Prog Biophys Mol Biol 37:1–47

Hebb DO (1949) The organization of behavior: a neuropsychological theory. Wiley, New York

Hirth F, Reichert H (2007) Basic nervous system types: one or many? In: Striedter GF, Rubenstein JLR (eds) Evolution of nervous systems: a comprehensive reference, volume 1: theories, development, invertebrates. Elsevier Academic, Amsterdam, pp 55–72

Katz MJ, Lasek RJ (1978) Evolution of the nervous system: role of ontogenetic mechanisms in the evolution of matching populations. Proc Natl Acad Sci U S A 75:1349–1352

Linden DJ (2007) The accidental mind. Belknap, Cambridge, MA

Medina L (2007) Field homologies. In: Striedter GF, Rubenstein JLR (eds) Evolution of nervous systems: a comprehensive reference, volume 1: theories, development, invertebrates. Elsevier Academic, Amsterdam, pp 73–87

Meyer RL (1998) Roger Sperry and his chemoaffinity hypothesis. Neuropsychologia 36:957–980

Nieuwenhuys R, Ten Donkelaar HJ, Nicholson C (1998) The central nervous system of vertebrates. Springer, Berlin

Opris I, Santos L, Gerhardt GA, Song D, Berger TW, Hampson RE, Deadwyler SA (2013) Prefrontal cortical microcircuits bind perception to executive control. Sci Rep 3:2285

Russell ES (1916) Form and function: a contribution to the history of animal morphology. J. Murray, London

Schmitt AM, Shi J, Wolf AM, Lu C-C, King LA, Zou Y (2006) Wnt–Ryk signalling mediates medial–lateral retinotectal topographic mapping. Nature 439:31–37

Sperry RW (1943) Effect of 180 degree rotation of the retinal field on visuomotor coordination. J Exp Zool 92:263–279

Striedter GF (2005) Principles of brain evolution. Sinauer Associates, Sunderland

Striedter GF (2007) A history of ideas in evolutionary neuroscience. In: Striedter GF, Rubenstein JLR (eds) Evolution of nervous systems: a comprehensive reference, volume 1: theories, development, invertebrates. Elsevier Academic, Amsterdam, pp 1–5

Sulloway FJ (1979) Freud, biologist of the mind: beyond the psychoanalytic legend. Burnett Books, London

Torii M, Hashimoto-Torii K, Levitt P, Rakic P (2009) Integration of neuronal clones in the radial cortical columns by EphA and ephrin-A signalling. Nature 461:524–528

Turing AM (1952) The chemical basis of morphogenesis. Philos Trans R Soc Lond B Biol Sci 237:37–72

Chapter 2
Vernon Benjamin Mountcastle

Manuel F. Casanova

Abstract The vertical organization of the cortex was first described by several neuroanatomists who used the principle in order to parcellate the brain into different regions. Mountcastle validated the concept of the minicolumn as a functional unit of information processing. His discovery, set against strong opposition from the neuroscience community, became a paradigm shift to future studies on sensory function of the cerebral cortex, which have all been based on his columnar model. In effect, over the years, the concept of the minicolumn has become a way of integrating disparate elements of neuroanatomy and physiology into an element of computation that permeates all of the neocortex. More recently the study of the minicolumn in both health and disease has promoted new concepts in regards to cognition and the neuropathology of psychiatric conditions.

Keywords Mountcastle • Minicolumns • Microcolumns • Columns • Cerebral cortex • Electrophysiology

> [To] Vernon Mountcastle, whose discovery of columns in the somatosensory cortex was surely the single most important contribution to the understanding of the cerebral cortex since Cajal (Hubel 1981, p. 37)

2.1 Personal History

Vernon B. Mountcastle, Jr., was born on July 15, 1918 in Shelbyville, Kentucky into a family of farmers of Scottish descent on both sides. His father was a railroad contractor and his mother a teacher. No member of his family had attained a university education before his own generation. His mother treated him and his siblings as students asking them to repeat their school lessons to her. By the time he was 4 years of age, Mountcastle could read and write.

M.F. Casanova, M.D. (✉)
Department of Psychiatry, University of Louisville, Louisville, KY, USA
e-mail: manuel.casanova@louisville.edu

M.F. Casanova, I. Opris (eds.), *Recent Advances on the Modular Organization of the Cortex*, DOI 10.1007/978-94-017-9900-3_2

At 3 years of age Mountcastle moved with his family to Roanoke, Virginia where he attended primary and secondary school. His precocious reading and writing abilities enabled him to skip the first two grades of elementary school education. During childhood he started playing tennis, a sport he continued to play until he was 80 years. In 1935 Mountcastle enrolled in Roanoke College in Salem and in 3 years earned a B.S. in Chemistry with honors. During his college years Mountcastle lived at home and had to commute every single day to a different town. This was in the midst of the Great Depression and Mountcastle always felt lucky he had been able to attend college.

Primarily influenced by one of his chemistry professors, who studied at Johns Hopkins, Mountcastle applied and was accepted at Johns Hopkins Medical School. At that institution, Mountcastle is quoted as saying that he "...had the most fantastic experience. I felt that I was welcomed into a society of scholars. For example we never got any grades. I learned later that there was a very detailed grading of everything. But you were never told. And that produced a fantastic atmosphere. You never felt that you were competing with another student; you were competing with the subject" (Venere 1998).

Mountcastle earned his M.D. from Johns Hopkins in 1942 and interned in Surgery before serving in the U.S. Naval Amphibious Forces for 3 years with two campaigns in the European Theater: Anzio, Italy and Normandy, France. After the war ended, he married Nancy Clayton Pierpont in September, 1945. From his marriage he had three children: Vernon B. Mountcastle III in 1947, Anne Clayton Mountcastle in 1948, and George Earle Pierpont Mountcastle in 1949 (deceased in 1969). He enjoys having six grandchildren and one great grandchild. He considers the time raising his family as the happiest years of his life.

2.2 Academic Career

Scientists during the twentieth century assigned body functions to areas of the cortex from observations of region-specific deficits after brain injuries. The legacy of Broca, Wernicke, and Jackson was later on complemented by cytoarchitectural studies that used parameters for cell morphology, density and lamination in order to parcellate the cerebral cortex into an orderly arrangement. Despite the prominent reputation of its proponents the idea of "one" brain region having "one" brain function was never fully accepted and soon faced serious challenges. The British Army neurologist Gordon Holmes (1876–1963) performed experiments on the somatotopic mapping of brain injured patients during World War I that suggested how specific functions could occupy relatively large regions of the cortex. Furthermore, some complex perceptual functions such as spatial localization of an object were not based on a specific area of the brain. This idea received substantial support from the American psychologist and behaviorist Karl Spencer Lashley (1890–1958) whose ablation experiment in rats demonstrated that the amount of cortical tissue removed had a more significant effect on maze learning than the specific locus of a lesion. Lashley's results lead to the promulgation of the well-known concepts of mass action and equipotentiality. In parallel to these early efforts on brain localization research was spurred by technological needs of World War I.

The signature injury of World War I (1914–1918) was shell shock. The steel helmet was introduced only at the beginning of 1916. Soldiers engaged in static trench warfare without adequate protection were subjected to a barrage of artillery and mortar attacks (Jones et al. 2007). Research into neurotrauma was the harbinger for improved electrical components that would allow mapping of activity within the cortex. Gerard, Marshall, and Saul were the first to apply microelectrodes to the sensory cortex of monkeys to measure the slow-wave potential evoked by the activation of peripheral somatosensory receptors. Sometime in the mid-nineteen-thirties the Physiology Department at the Johns Hopkins School of Medicine developed a program to explore the possibilities afforded by this technique. Under the guidance of Philip Bard (1898–1977), a cadre of dynamic young researchers, including Wade Marshall and Clinton Woolsey, systematically mapped the somatotopic representation of the body of a monkey onto its cortical surface. Although the concept of the homunculus derived from human experiments by Penfield and colleagues gained the attention of the public press the contemporaneous efforts of the Hopkins investigators led to the description of "animalculi" in numerous mammalian species. It thus seemed quite evident that somatotopic representations were an organizing principle for the central nervous system in general and not for the human primate brain alone.

The technique of recording electrical activity from single neurons with an extracellular electrode (single-unit recording) was popularized by Gilbert Ling at the University of Illinois. Ling had done his postdoc at Johns Hopkins Medical School but moved to join Professor Gerard when the latter founded a large research laboratory at the University of Illinois. Soon, studies were been done and published in both invertebrate and vertebrate species as well as in recordings of the brainstem nuclei and thalamus. At the end of World War II Stephen Kuffler (1913–1980) moved from Australia to Hopkins and used the single-unit recording technique to investigate the receptive fields of retinal ganglion cells in the cat. His famous discovery of their center-surround properties stands as one of the classics in neuroscience.

During World War II Mountcastle served 3 years as an officer in the U.S. Naval Medical Corp. After his military duties were over in 1946 he joined the Department of Physiology at the Johns Hopkins School Medicine as a postdoctoral fellow and he never left. He was a contemporary of Stephen Kuffler and to some of his fellows (Hubel and Wiesel). Mountcastle would eventually rise through the academic ranks to become a full professor and hold the position of Director of the Department of Physiology for 17 years (1964–1980).

Shortly after World War II Mountcastle was able to experience developments in equipment and techniques that opened the door to research opportunities previously unimaginable. Most notably among these innovations were the Horsley-Clarke head-holder and the Davies isolation chamber. The use of the Horsley-Clarke head-holder was based on the reproducible relationships between landmarks of the skull and anatomical structures within the brain. The cranial fixation points allowed for a stereotactic coordinate system allowing the accurate localization of subcortical structures in small animals. The Davies isolation chamber dampened the vibrations from vascular pulsations and physiologic tremor allowing for accurate microelectrode recordings.

In addition to the technological innovations the Hopkins environment enjoyed the synergism brought about by the close collaboration with people from different field. John Chubbuck, an engineer from the Applied Physics Laboratory, took care in helping design and manufacture some of the complex setups required for stimulating and recording from the brains of animals. From the standpoint of anatomy, the introduction of new neurohistologic techniques allowed the reconstruction of the microelectrode path of penetration and the identification of specific neurons subjected to recordings. Some of the neurohistologic advancements were procured by Mountcastle's colleague, Jerzy Rose.

Mountcastle's initial work on electrophysiology was done under the supervision of Elwood Henneman (1915–1996). Together they mapped the tactile representation of the body as a "figurine" in the ventrolateral thalamus of cats and monkeys (Mountcastle and Henneman 1949, 1952). The close collaboration was brought to an end when Henneman departed in 1955 to establish a department of Neurobiology at Harvard University.

All of the previous developments set the stage for Vernon Mountcastle's discovery of the functional cortical columns in the somatosensory cortex of cats and monkeys. His studies were based on recordings from over 2,300 neurons for which, in many cases, he collected the latency and spiking frequencies. These responses were analyzed in terms of their receptive field and modality, i.e., the nature of the external stimuli.

A funny story in regard to Mountcastle's awe-impressive cumulus of work was provided by Hubel in his Nobel lecture of 1981:

> . . .we [Hubel and Wiesel] had gone to a lecture by Vernon (this was a few years after his discovery of cortical columns) in which he had amazed us by reporting on the results of recording from some 900 somatosensory cortical cells, for those days an astronomical number. We knew we could never catch up, so we catapulted ourselves to respectability by calling our first cell No. 2000 and numbering subsequent ones from there. When Vernon visited our circus tent we were in the middle of a S-unit recording, cell Nos. 3007, 3008, and 3009. We made sure that we mentioned their identification numbers. All three cells had the same receptive-field orientation but neither Vernon nor we realized, then want that implied (Hubel 1981, p. 27)

Although Mountcastle had no *a priori* expectation of a columnar organization to the cortex, careful observation revealed that the receptive fields of cells in successive layers of the cortex largely overlapped. Mountcastle's results on the columnar organization of the cerebral cortex were described in two papers both published in 1957 (Mountcastle 1957; Mountcastle et al. 1957). The columnarity principle was reproduced and expanded to show a gradient of modality representation from Brodmann areas 3 to 1 to 2 in a series of publications coauthored with Tom Powell (Mountcastle and Powell 1959a, b; Powell and Mountcastle 1959a, b).

Recognition of the columnar organizing principle, guided future investigations so as to acquire a record of the angle at which the microelectrode penetrated the cortex. The idea of a columnar organization was soon realized after early experiments were made wherein penetration at 45° to the surface demonstrated a change in modality as the microelectrodes transversed different layers of the cortex (Mountcastle 1957).

Mountcastle's view on the columnar organization of the cortex went far beyond its putative role in regards to information processing. In fact the columnar research led him to describe several physiological principles of cortical function. Mountcastle was the first person to describe how the neurons within each column responded to one specific submodality and dermatomal locale. The columns extended throughout the cortex from layers II through VI and each cell responded with approximately the same latency to stimulation, at least early in the response. Within proprioceptive columns, Mountcastle described how cells in adjacent columns exhibited activation to alternating flexion and extension of joints. Because activity in one cell led to the inactivation of the cell in the opposing column, Mountcastle inferred the mechanism of reciprocal lateral inhibition. Finally, Mountcastle described a pattern of surround inhibition where stimulation of areas in the peripheral receptive field actually inhibited cell activity.

All of Mountcastle's discoveries were done in a field that was antagonistic to his way of thinking and provided mounting criticism. Among his two publications in 1957 the columnarity principle was elaborated upon in the one having Mountcastle as the sole author. This was done at the request of his coauthors Drs. Davis and Berman, both of whom believed that columnarity was a radical departure from accepted tenets! In fact, Mountcastle was subjected to a great deal of criticism, even by his good friend Jerzy Rose. Back then the perspective of classical anatomists permeated neurosciences and cytoarchitecture was primarily about laminae, not columns.

Before Mountcastle only a few neuroanatomists had paid attention to the columnar organization of the cortex. Von Economo commented on this structural arrangement in his description of the auditory cortex. However, it was Lorente de Nó who first posited a vertical arrangement of neurons that guided the flow of information in and out of the cortex. According to Lorente de Nó: "All the elements of the cortex are represented in it, and therefore it may be called an elementary unit, in which, theoretically, the whole process of transmission of impulses from the afferent fiber to the efferent axon may be accomplished" (Lorente de Nó 1938).

The idea of columnarity as a primordial element of circuitry and anatomical organization, was antagonized to such an extent that in different occasions Mountcastle was compelled to quote the work of Roger Sperry with subpial cortical dicing in order to sustain his own observations (Sperry et al. 1955). In effect, Sperry had shown that the insertion of numerous wires or insulating mica plates perpendicular to the striatal cortex of monkeys had no effect on visual function or motor coordination. These experiments demonstrated that perception in the visual cortex depended on vertically oriented afferent and efferent connections.

The surgical procedure of subpial cortical dicing is presently pursued in patients with Landau Kleffner syndrome (infantile acquired aphasia). In this condition children lose their language skills due to seizures in either Broca or Wernicke's area. Multiple parallel incisions perpendicular to the surface of the cortex limit the generalization of the seizure without disturbing the flow of information. The incisions do not perturb the vertical flow of information through minicolumns, only lateral connections across the same appear affected.

Given the initial antagonism to Mountcastle's ideas and their present acceptance it is of importance to note the words of Zeki:" When confronted with a difficult problem which goes against their way of thinking, scientists often begin by shutting their eyes to the evidence and pretending that it does not exist. The next stage consists of accepting the evidence but pretending that it is not important or that it can be adequately explained by the known facts. The third and final stage consists of admitting the evidence and its significance, but pretending that it has all been said before" (Zeki 1993).

2.3 Mountcastle's Legacy

Mountcastle single unit recording studies provided the basis for subsequent studies by Hubel and Wiesel thus fostering an outburst of work into perceptual physiology. Mountcastle was the first person to clearly enunciate and show based on his own work the column as an independent unit of function. Hubel and Wiesel further elucidated how information could be transformed across orientation-specific steps into higher levels of perceptual complexity. David Hubel in his Nobel Prize acceptance speech graciously accorded Mountcastle the distinction his work deserved: "[the] discovery of columns in the somatosensory cortex was surely the single most important contribution to the understanding of cerebral cortex since Cajal" (Hubel 1981).

Mountcastle's efforts involved academic administration as much as they did research. For several decades he was able to steer the academic future of neuroscience within the Johns Hopkins Medical School. As the Director of the Department of Physiology he devoted considerable time to recruiting promising scientist in fields outside his own. He was instrumental in establishing the first Department of Clinical Neurology at that institution for which he acquired 2 endowed chairs: Guy McKhann and Richard Johnson. Later on he had the opportunity to retain Solomon Snyder at Hopkins by establishing the first Department of Neuroscience. In 1988, Mountcastle proposed the creation of the Zanvyl Krieger Mind/Brain Institute which was then established as part of the Department of Neuroscience with Guy McKhann as its first director. It is also noteworthy that Mountcastle was a founding member of the Society for Neurosciences for which he served as its first president (having run against Seymour Ketty in the election).

Mountcastle always felt great pride in mentoring a large number of fellows all of whom became his friends and many of whom went to occupy academic positions of great distinction. His first postdoctoral fellow was Edward Perl who arrived at Hopkins in 1950. In all, the total number of postdoctoral fellows during Mountcastle's career was 48, these included in alphabetical order: Carlos Acuna, Richard Andersen, Sven Anderson, Pradep Atluri, Frank Baker, Alvin Berman, James Campbell, Giancarlo Carii, Mirko Carreras, Ian Darian-Smith, John Downer, Charles Duffy, Robert Dykes, Solomon Erulkar, Apostolos Georgopoulos, Edward Glaser, Gundez Gucer, Thomas Harrington, Clinton Harrison, Juhani Hyvarinen, Kenneth Johnson, Cecil Kidd, Hans Kornhuber, Robert LaMotte, James Lane, Randall Long, James Lynch, Michael Merzenich, Mark Molliver, Brad Motter, Hiroshi Nakahama, Edwardo Oswaldo-Cruz, Edward Perl, Gian Poggio, Thomas Powell, Barbara Renkin, Rodolfo Romo, Sten Skoglund, Michael Steinmetz, Tadaaki Sumi, William Talbot, James Taylor, and Thomas Yin.

Mountcastle retired completely from academic duties in November 2005, at the age of 87. During his long career he won many accolades including the Schmitt Prize and Medal from MIT, the U.S. National Medal of Science, and the Fyssen Foundation Prize from France. In 1978 he was awarded the Louisa Gross Horwits Prize from Columbia University and in 1983 the Albert Lasker Award for Basic Medical Research. He holds 6 honorary degrees. He is a member of the Royal Society, *Academie des Sciences*, and the National Academy of Sciences. From the latter he received the United States National Medal of Science in 1986 for his lifetime of work in the neurosciences. Despite all of these awards Mountcastle remained a humble individual deferential to his modest upbringing in Virginia. He always kept in touch with his childhood friends and specially enjoyed those years spent raising his family. He was always the first person to credit his success to his mentors and colleagues.

It seems fitting to conclude this biographical introduction to Vernon Mountcastle with a paragraph from my introduction to *Neocortical Modularity and the Cell Minicolumn*: "Mountcastle is an outsized character that marched to his own beat. Never the person he "ought" to be, he was never boxed into preconceived ideas. This character trait was of immense help throughout his research career: Having no shoulders to stand on, he set his own goals and built upon his strengths. He aspired to an objective truth, the Holy Grail of neurosciences. In his research he overcame narrowness of scientific vision and became an heir to Socrates. While others went for easily accessible and exploitable prizes, Mountcastle claimed no glory, no acclaim. Personally he treated others by the categorical imperative: with respect and dignity as ends themselves. He is a unique individual: We can't exchange Mountcastle for someone else and have an equal" (Casanova 2005, p. xii).

References

Casanova MF (2005) Introduction. In: Neocortical modularity and the cell minicolumn. Nova Biomedical Books, New York, p xii

Hubel DH (1981) Evolution of ideas on the primary visual cortex, 1955–1978: a biased historical account. Nobel lecture, 8 December 1981. http://www.nobelprize.org/nobel_prizes/medicine/laureates/1981/hubel-lecture.pdf

Jones E, Fear NT, Wessely S (2007) Shell shock and mild traumatic brain injury: a historical review. Am J Psychiatr 164:1641–1645

Lorente de Nó R (1938) The cerebral cortex: architecture, intracortical connections, motor projections. In: Fulton JF (ed) Physiology of the nervous system. Oxford University Press, London, pp 274–301

Mountcastle VB (1957) Modality and topographic properties of single neurons of cat's somatic sensory cortex. J Neurophysiol 20:408–434

Mountcastle VB, Henneman E (1949) Pattern of tactile representation in thalamus of cat. J Neurophysiol 12:85–100

Mountcastle VB, Henneman E (1952) The representation of tactile sensibility in the thalamus of the monkey. J Comp Neurol 97:409–440

Mountcastle VB, Powell TPS (1959a) Central nervous mechanisms subserving position sense and kinesthesia. Bull Johns Hopkins Hosp 105:173–200

Mountcastle VB, Powell TPS (1959b) Neural mechanisms subserving cutaneous sensibility, with special reference to the role of afferent inhibition in sensory perception and discrimination. Bull Johns Hopkins Hosp 105:201–232

Mountcastle VB, Davies PW, Berman AL (1957) response properties of neurons of cat's somatic sensory cortex to peripheral stimuli. J Neurophysiol 20:374–407

Powell TPS, Mountcastle VB (1959a) The cytoarchitecture of the postcentral gyrus of the monkey *Macaca mulatta*. Bull Johns Hopkins Hosp 105:108–131

Powell TPS, Mountcastle VB (1959b) Some aspects of the functional organization of the cortex of the postcentral gyrus of the monkey: a comparison of findings obtained in a single unit analysis with cytoarchitecture. Bull Johns Hopkins Hosp 105:133–162

Sperry RW, Miner N, Myers RE (1955) Visual pattern perception following sub-pial slicing and tantalum wire implantations in the visual cortex. J Comp Physiol Psychol 48:50–80

Venere E (1998) Vernon Mountcastle: head honcho. Gazette (Johns Hopkins) 37:31

Zeki SA (1993) Vision of the brain. Blackwell, Oxford, p 87 (Quoted by Mountcastle VB (1998) Perceptual neuroscience: the cerebral cortex. Harvard University Press, Cambridge, MA)

Chapter 3
Modular Organization of the Prefrontal Cortex: The Legacy of Patricia Goldman-Rakic

Lynn D. Selemon

Abstract The prefrontal cortex (PFC) is the brain area that distinguishes us as uniquely human. It guides decision making based on past experience, allowing us to break away from reflexively responding to the sensory conditions on hand at any moment. Sequential planning, cognitive flexibility and rational thinking all rely on the PFC. When Patricia Goldman-Rakic began her research career in the 1970s, the PFC was a *terra incognita*. Her work dispelled the notion that higher cognitive function is beyond the reach of the scientific method and revealed the basic neurobiology governing prefrontal executive function. Goldman-Rakic started her journey by probing the behavior disrupted by lesions in the PFC. An early foray into anatomic tract tracing led to the discovery of columns of afferent terminal labeling in the PFC. From this foothold, she began a relentless quest to understand the modular organization of prefrontal architecture and how the vertical arrangement of functionally related neurons translates into the mechanistic underpinnings of working memory. She characterized neuronal activity in the PFC that forms the essence of spatial mnemonic capacity. Her work teased apart the intricacies of local neurocircuitry in the PFC and examined modulation of this circuitry by monoaminergic neurotransmitters, especially dopamine. Goldman-Rakic's studies of executive function extended from the regional level in distinguishing the dorsal and ventral domains that process spatial and object information, respectively, to the subcellular level in mapping the precise distribution of dopamine receptors. Her life's work stands today as the foundation for our contemporary understanding of prefrontal cortical function.

Keywords Working memory • Dopamine • Pyramidal cells • Interneurons • Executive function • Cognition

---

L.D. Selemon (✉)
Department of Neurobiology, Yale School of Medicine, Yale University, PO Box 208001, New Haven, CT 06520-8001, USA
e-mail: lynn.selemon@yale.edu

M.F. Casanova, I. Opris (eds.), *Recent Advances on the Modular Organization of the Cortex*, DOI 10.1007/978-94-017-9900-3_3

Patricia Goldman-Rakic's career spanned decades and encompassed a wide range of research topics. Her studies touched on nearly every corner of the brain from the cerebral mantle to the brain stem and delved into the workings of many different transmitter systems, including glutamate, the monoamines, and neuropeptides. She investigated how basic brain mechanisms might be altered in diseased states like schizophrenia and examined the neurophysiologic action of antipsychotic drugs used to treat schizophrenia. However, there was never any question about where her real passion lay, that is in understanding the basic neurobiology of working memory function in the prefrontal cortex (PFC). Therefore, this review focuses on the centerpiece of her research: elucidating the neuroanatomic and neurophysiologic substrates of working memory function, their columnar organization, the manner in which dopamine (DA) afferent input modulates cognitive function in the PFC, and parcellation of the PFC into separate processing domains. The narrow focus by no means diminishes the importance or impact of her many satellite studies but rather underscores Dr. Goldman-Rakic's conviction that characterizing the basic neurobiology of the PFC is essential for progress in understanding altered states induced by drugs or disease.

Goldman-Rakic was at the pinnacle of her career when her life tragically was cut short more than a decade ago in July, 2003. Her untimely passing is an enormous loss to the field of neurobiology. She left a number of studies in midstream. As a result, publication of her work continued for 5 years after her passing, allowing us a tantalizing glimpse of the direction it was taking, though we will never know what how much farther she could have advanced our understanding of the PFC. This review represents a retrospective look at Patricia Goldman-Rakic's seminal findings in the PFC that is intended to refresh her work in the minds of her contemporaries and to introduce her findings to a new generation of neurobiologists and clinicians.

3.1 Columns in the Dorsolateral Prefrontal Cortex

Patricia Goldman-Rakic's interest in the dorsolateral PFC (dlPFC) began with studies of the behavioral consequences of prefrontal lobectomy in nonhuman primates. Building on work by her mentors, H. Enger Rosvold and Mortimer Mishkin, Goldman-Rakic further distinguished the functional domains of the orbitofrontal cortex and the dlPFC and the capacity for recovery of function following early brain lesions (Goldman and Rosvold 1970; Goldman et al. 1970a, b, 1971, 1974; Goldman 1971, 1976). She found that the dlPFC, and in particular the cortex surrounding the principal sulcus, plays a critical role in mediating the working memory component of the spatial Delayed Response (DR) task (Goldman and Rosvold 1970). Working memory, the ability to hold information in the mental sketchpad is the essence of executive functioning of the PFC because this processing mechanism allows internalized representations of the world, rather than immediate sensory stimulation, to guide behavior (Baddeley and Hitch 1974; Goldman-Rakic 1987).

In the mid-1970s, Goldman-Rakic trained in Walle Nauta's laboratory in order to master the then new neuroanatomic tracing method utilizing tritiated amino acids anterogradely transported from the site of injection to label terminal fields throughout the brain. They made a rather astonishing finding, that cortical terminal fields in diverse cortical areas including the dlPFC exhibit a columnar patterning (Goldman and Nauta 1977). For example, injection in the cortex surrounding the principal sulcus produces wide bands of terminal labeling in the contralateral principal sulcal cortex that extend throughout all six cortical layers with unlabeled bands interspersed (Fig. 3.1). Prior to this discovery, cortico-cortical columnar terminal labeling had been described in the sensory cortex (Jones et al. 1975; Shanks et al. 1975), and patterned thalamo-cortical labeling had been found in the visual cortex, corresponding to the vertically orientated ocular dominance columns identified by physiologic recordings (Hubel and Wiesel 1972). From these early anatomic studies, Goldman-Rakic recognized that the PFC exhibits the same modular organization as sensory cortices and therefore that higher cognitive functioning could be interrogated with neurophysiologic and neuroanatomic methods comparable to those used to discern the basis of perception. With this insight, Goldman-Rakic began what would become a life-long pursuit to eludicate the cellular basis of higher cognitive function as governed by the PFC. She chose to study the PFC in the rhesus macaque monkey because development, expansion, and specification of the cortical mantle in monkeys more closely parallels that of humans compared to other species (Rakic 1988).

Goldman-Rakic and her collaborators established that columnar patterning represents an overriding organizational principal of the PFC. Columnar terminal labeling is a feature of both callosal and long tract intrahemispheric connections to the PFC, as for example those from the posterior parietal cortex (Schwartz and Goldman-Rakic 1984; Cavada and Goldman-Rakic 1989). Moreover, columnar labeling in cortical areas and patchy, patterned terminal distributions in subcortical areas characterize a widespread network of projections originating in the PFC (Goldman-Rakic et al. 1984; Selemon and Goldman-Rakic 1985, 1988), as well as subcortical afferent terminal labeling in the PFC from the mediodorsal thalamus (Giguere and Goldman-Rakic 1988). Similarly, intrinsic circuitry of the PFC exhibits a distinct modular organization. Small injections of a retrogradely transported tracer into each of the six cortical layers label a vertically oriented column of neurons that extend from the site of injection to the pial surface above and to the white matter below (Kritzer and Goldman-Rakic 1995). Of particular interest, injections in layer IIIc produce a tangential pattern that extends as far as 7 mm laterally and consists of clusters of distant layer IIIc neurons, predominantly pyramidal neurons, interspersed with unlabeled territories (Fig. 3.2; Kritzer and Goldman-Rakic 1995). This patterning indicates that a single column in the PFC receives input from multiple distant columns that are interdigitated with columns that do not send input to that same column. In parallel with these findings, Lewis and colleagues established that pyramidal neurons give rise to a disjunctive

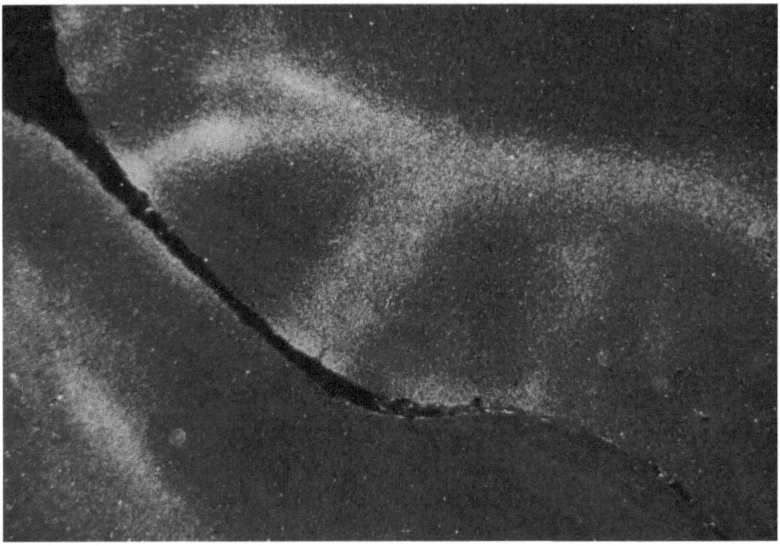

Fig. 3.1 Darkfield photomicrograph illustrating columns of terminal labeling in the principal sulcal cortex following an injection of tritiated amino acids in the principal sulcal cortex of the opposite hemisphere (From Goldman and Nauta 1977. Reproduced with permission of Elsevier BV in the format reuse in a book/textbook via Copyright Clearance Center)

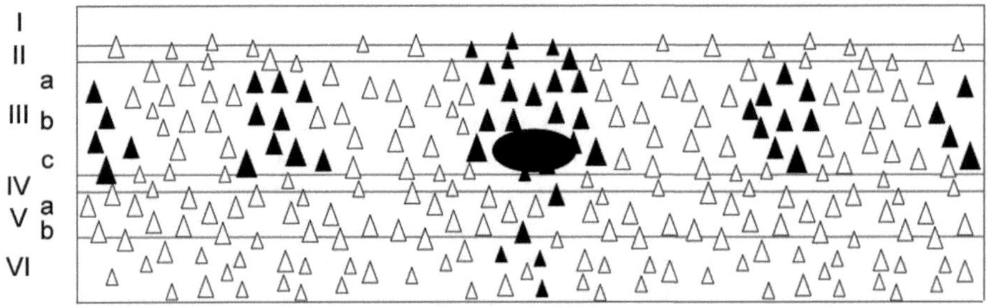

Fig. 3.2 Schematic diagram of labeled neurons in the PFC following an injection into layer IIIc of the retrogradely transported tracer, cholera toxin (B-subunit) conjugated to colloidal gold. Labeled neurons exhibit a disjunctive patterning that includes a vertical column surrounding the injection site and multiple clusters of neurons located at spaced intervals from the injection site (Figure generously provided by Dr. Mary F. Kritzer)

intralaminar projection and that these distant pyramidal terminals synapse on the dendritic spines of other pyramidal neurons forming excitatory contacts (Levitt et al. 1993; Pucak et al. 1996; Melchitzky et al. 1998). Thus, recurrent excitatory input from one pyramidal cell to another in layer III is an important feature of interconnectivity between distant columns.

3.2 Prefrontal Cortical Activity During a Spatial Working Memory Task

Single unit recording had established that neurons in the prefrontal cortex are active during spatial working memory tasks and that a substantial proportion of active neurons fire during the delay period, i.e. when the cue is no longer present but before a response is required (Fuster and Alexander 1971; Kubota and Niki 1971; Fuster 1973; Niki 1974a, b; Fuster et al. 1982; Joseph and Barone 1987). Goldman-Rakic and colleagues re-examined neuronal activity in the dlPFC using an oculomotor version of the spatial DR task. These studies revealed several properties that had not been recognized, among them the precise spatial tuning of many neurons in the principal sulcal cortex of the dlPFC (Fig. 3.3; Funahashi et al. 1989, 1990). A large proportion of neurons exhibit enhanced firing in the cue, delay, or response phases of the task, with some active in more than one phase (Funahashi et al. 1989, 1990). Interestingly, the majority of cue (97 %) and delay (79 %) period neurons show directionality such that activity is either significantly enhanced or inhibited when the cue appears in one location of visual space while unaltered when stimuli appears in other quadrants of the visual axis (Funahashi et al. 1989, 1990). In some instances, units that are fire more frequently in response to a particular direction of space are inhibited by a stimulus appearing in the opposite direction, i.e. when the two stimuli are 180° apart (Funahashi et al. 1989). Moreover, most of the neurons are tuned to the visual space of the contralateral hemisphere such that lesions of the cortex in around the principal sulcus result in failure to perform oculomotor DR when cues are located in the contralateral hemifield (Funahashi et al. 1989, 1990, 1993). Importantly, spatially discriminative neurons in the dlPFC are selectively activated when *memory* of the spatial location is required, thus distinguishing them from neurons in visual cortex that are activated by stimuli in specific retinotopic locations but only when the stimulus is present (Kojima and Goldman-Rakic 1984). In sum, these findings established that sustained firing of neurons in the dlPFC during the delay period represents the electrophysiologic basis for retaining a mental trace of spatial information until the proper response is required and that neurons in the dlPFC have preferred retinotopic locations for the memory trace, i.e., "memory fields."

3.3 Intrinsic Neurocircuitry of the Dorsolateral Prefrontal Cortex

Goldman-Rakic and colleagues learned more about how persistent firing of delay period neurons is related to cellular networks and the modular organization of prefrontal cortex by recording simultaneously from pairs of neurons in the dlPFC. One focus of their studies was elucidating the interaction between pyramidal cells, identified electrophysiologically by their regular spiking (RS) pattern, and

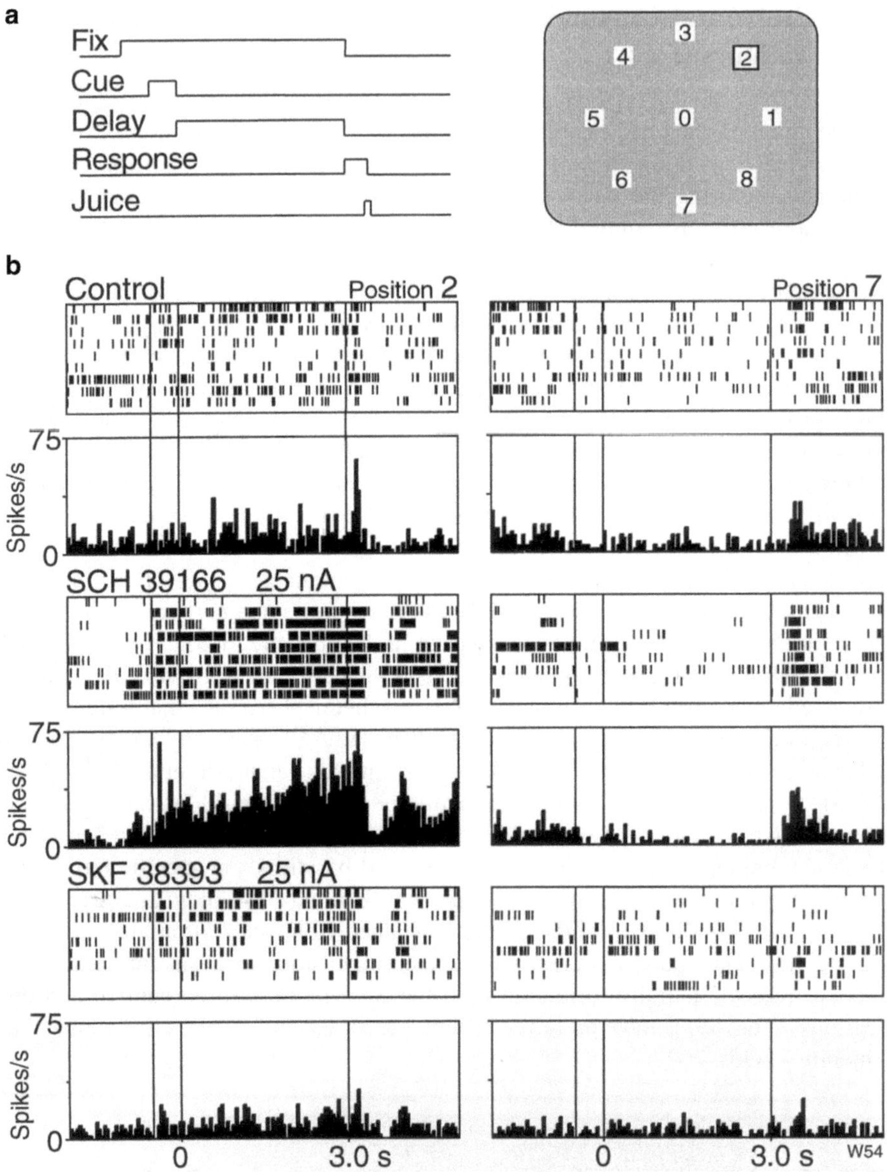

Fig. 3.3 Neuronal activity of a single unit during the oculomotor DR task in control conditions and after treatment with D1R acting drugs. (**a**) A schematic diagram (*left*) shows the sequence of events occurring during the task. Note that rasters and histograms directly below are aligned to these task epochs. During the task, the monkey fixates on the center position (0) of the screen (*right*) while one of the eight target positions is lit briefly (cue period). The target light is extinguished for a period of 3–6 s (delay period) during which time the monkey must remember the location of the previously lit target. The central fixation point is then extinguished as a signal for the monkey to make a saccade in the direction of the remembered target location (response period) to receive a juice reward. (**b**) The control condition shows that this unit is weakly activated during the delay period by a target in position 2 and not responsive to a target in position 7. Iontophoresis of the D1R antagonist SCH 39166 enhances the delay period activation of this

populations of interneurons characterized by a fast spiking (FS) pattern of activity. FS neurons have been identified as the parvalbumin-positive basket and chandelier subtypes of interneurons (Kawaguchi 1995). They found that FS neurons are also spatially tuned to a restrictive area of visual space. Pairs of RS-RS neurons in close proximity (<400 μm) are not only spatially tuned to the same visual quadrant but also respond alike with increased or decreased firing rate to the preferred visual location (Wilson et al. 1994). In contrast, RS-FS pairs have inverted patterns of activity such that excitation of activity in a RS neuron is accompanied by inhibition in the FS pair, or vice versa (Wilson et al. 1994). The opposite patterns of activation and inhibition in FS-RS pairs suggested that interneurons inhibit pyramidal neurons with opponent tuning. Further study confirmed that closely aligned neurons are isodirectionally tuned and showed that shared responsiveness to a particular spatial location of neurons within a column is due to activation by the same afferent stimulation (Rao et al. 1999). Such neurons with shared spatial directionality form "microcolumns" within the larger column. Notably, in this 1999 study, the majority of FS-RS pairs, which were in even closer proximity to each other than in the earlier study, showed isodirectional tuning rather than opponent tuning. Together these studies indicate that interneurons may be involved in two inhibitory processes: iso-directional inhibition and cross-directional inhibition (Fig. 3.4; Rao et al. 1999). Iso-directional tuning functions locally to allow sharpening of tuning to a specific spatial direction via inhibition between neighboring microcolumns that are tuned to similar, though not identical, locations of visual space. Cross-directional inhibition suppresses activity between adjacent columns that are tuned to opposite quadrants of space.

Additional features of columnar organization in the PFC were revealed by paired recordings of pyramidal neurons located in supragranular layers. For example, neurons that share directional tuning to a particular quadrant of visual space and are active during the same task epoch, i.e. cue, delay, or response, show the highest level of cross correlation activity, essentially a measure of time-linked co-activation (Constantinidis et al. 2001). Co-activation of neurons thus is a key component that defines local microcircuits of neurons that code a specific spatial location during a particular temporal epoch of oculomotor DR. These studies also revealed that neurons tuned to all angles of visual space are encountered in an expanse of 200–300 μm, suggesting that this width constitutes the limits of a single column comprised of microcolumns representing each individual direction in space. Perhaps surprisingly, cross correlation activity is even higher for pairs of interneurons identified by their fast spiking properties, indicating that tightly correlated activity is a particularly important property of local networks (Constantinidis and Goldman-Rakic 2001).

Fig. 3.3 (continued) unit to the preferred location (2) while suppressing activation to the non-preferred location (7). These effects are reversed by the partial D1 agonist SKF 38393. (From Goldman-Rakic 1999a. Reproduced with permission of New York Academy of Sciences in the format reprint in a book via Copyright Clearance Center)

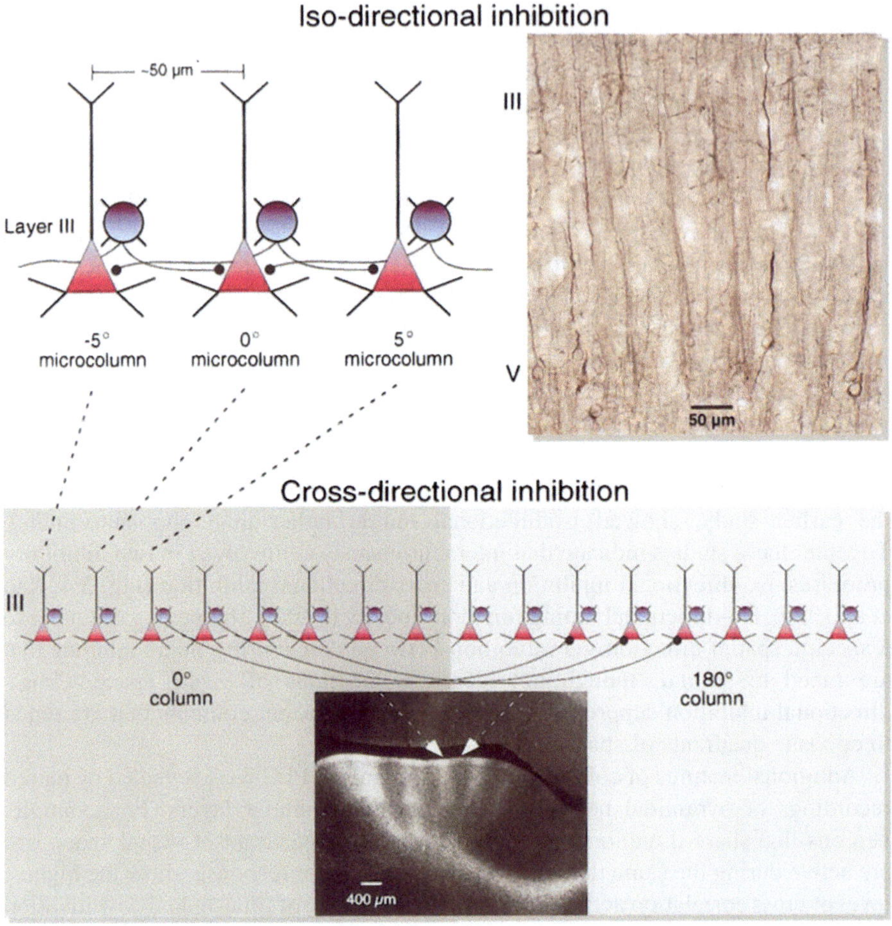

Fig. 3.4 Schematic illustration of iso-directional (*top*) and cross-directional (*bottom*) inhibition. (*Top left*) Interneurons (*round* somas) make short distance connections with pyramidal cells (*triangular* somas) in neighboring microcolumns that are tuned to nearly the same region of visual space. (*Top right*) 5-HT_{2A}-like immunoreactivity in the PFC showing prominent bundles of apical dendrites presumably from pyramidal neurons in single microcolumns. (*Bottom*) Interneurons make long distance connections with pyramidal cells that reside further away and are tuned to the opposite visual quadrant (From Rao et al. 1999. Reproduced with permission of American Physiological Society [etc.] in the format reuse in a book/textbook via Copyright Clearance Center)

3.4 Multiple Roles for Interneurons in Local Networks

The aforementioned findings show that interneuron populations play a critical role in spatial mnemonic function. Indeed, application of a γ-aminobutyric acid subtype A ($GABA_A$) receptor inhibitor alters spatial tuning of PFC neurons by eroding and broadening the tuning of some units and unmasking tuning in others

(Rao et al. 2000). Interneurons are a heterogeneous population with respect to morphology (Lund and Lewis 1993) and electrophysiologic properties (Kawaguchi 1995); therefore, it stands to reason that distinct populations of interneurons play different roles in integrating information in the PFC. In slice preparations of the ferret PFC, Krimer and Goldman-Rakic (2001) found evidence for this diversity of function in interneuron populations. They used dual whole-cell voltage recording to examine the relationship between pyramidal cells and closely aligned interneurons of three categories (local arbor, medium arbor, and wide arbor) and intracellular filling to examine the synaptic contacts between recorded pairs of pyramidal neurons and interneurons. Previous studies had established that local and wide arbor neurons are parvalbumin-positive GABA neurons whereas medium arbor interneurons use cholecystokinin as a transmitter (Lund and Lewis 1993). Pyramidal neurons reliably excite local arbor interneurons, including chandelier neurons, via a single synaptic contact whereas progressively greater numbers of synaptic contacts link pyramidal neurons with medium and wide arbor interneurons, and activation at multiple synapses is required for depolarization of these interneurons subtypes (Krimer and Goldman-Rakic 2001). Notably, axon length is also longer in wider arbor cells, as local arbor interneurons have axons extending only 300 μm tangentially whereas medium and wide arbor interneurons extend 600 and 900 μm lateral to the cell soma (Krimer and Goldman-Rakic 2001). Because interneurons are thought to give rise to the interposed inhibitory synapse between pyramidal cells in the PFC, different classes of interneurons are in prime position to mediate intracolumnar and intercolumnar inhibition. Local arbor interneurons, receiving excitatory input from nearby pyramidal cells, may suppress activity in pyramidal cells belonging to adjacent microcolumns with spatial field tuning that is similar but not exactly the same as that of the original pyramidal cell, i.e. iso-directional inhibition. In contrast, medium and wide arbor neurons, receiving input from nearby pyramidal cells, could mediate suppression at more distant pyramidal cells that have intermediate and opponent spatial tuning. In particular, wide arbor interneurons with the longest axons are capable of integration between columns, i.e. cross-directional inhibition.

Similar analysis of the reverse connectivity between adjacent interneurons and pyramidal neurons also uncovered cell type specificity. FS neurons with basket cell morphology predominantly innervate the soma and proximal dendrites of pyramidal cells whereas non-FS interneurons exhibiting bitufted, double bouquet, or Martinotti cell morphologies contact the distal dendrites of pyramidal cells (Gao et al. 2003). The importance of this connectional distinction will be discussed in the context of dopaminergic modulation of neural circuitry in the next section.

3.5 Dopaminergic Modulation of Prefrontal Network Activity

Early studies by Goldman-Rakic and colleagues established that dopamine is a key player in modulating neuronal activity in the PFC. Dopamine is present in higher levels in the PFC than in more posterior cortices of the primate brain (Brown and Goldman 1977; Brown et al. 1979), and depletion of dopamine in the PFC produces a cognitive deficit nearly as severe as ablation of the same region (Brozoski et al. 1979). Deciphering how dopamine exerts this powerful control over PFC activity became a major focus of Goldman-Rakic's studies and one that has profound implications for understanding cognitive dysfunction in diseases like schizophrenia and Parkinson's disease (Goldman-Rakic et al. 2000, 2004).

Neuroanatomic studies from the Goldman-Rakic laboratory established the precise patterning of dopamine innervation of the prefrontal cortex, the distribution of receptor subtypes and synaptic contacts. Dopamine-immunoreactive fibers in frontal cortex innervate all cortical areas although differences in density of fibers across medio-lateral and rostro-caudal axes are apparent (Williams and Goldman-Rakic 1993). In most of the frontal cortex, including the dlPFC, dopamine fibers are densely packed in two broad bands encompassing (1) layer I, II, and upper III and (2) layer V-VI, with the middle layers less densely innervated (Williams and Goldman-Rakic 1993). The origin of the mesocortical dopamine projection in the nonhuman primate, arising from all three dopaminergic midbrain groups A8-A10, is more widespread than that of rodent species, underscoring parallels between the phylogenetic expansion of the dopaminergic system and the enlargement and differentiation of the primate prefrontal cortex (Williams and Goldman-Rakic 1998).

Ultrastructural analysis of dopaminergic terminals revealed that dopaminergic boutons form symmetric (inhibitory) contacts with the soma, shafts, and dendritic spines of pyramidal cells (Goldman-Rakic et al. 1989). Most notably, dopamine terminations on dendritic spines constitute one third of a triadic arrangement with the other components identified as the postsynaptic spine and an afferent asymmetric (excitatory) terminal contacting the same postsynaptic spine (Fig. 3.5; Goldman-Rakic et al. 1989). This triadic arrangement suggests that dopamine is poised to modulate excitatory input although the exact origin of this afferent excitatory stimulation is not known; it could emanate from distant cortical areas, the thalamus, or perhaps recurrent axons terminals of local pyramidal neurons. Further analysis revealed that dopaminergic appositions are concentrated on distal dendrites and, in concordance with earlier findings of dopamine fiber distributions, that pyramidal neurons in layer II/superficial III are most heavily laden with dopamine contacts (Krimer et al. 1997). However, dopamine innervation is not confined to pyramidal neurons as dopaminergic boutons also form symmetric contacts with interneurons in the PFC (Sesack et al. 1995; Krimer et al. 1997).

The dopamine D1 receptor (D1R) is the preponderant receptor subtype in the PFC, with a density 10–20 times greater than that of the D2R (Lidow et al. 1991).

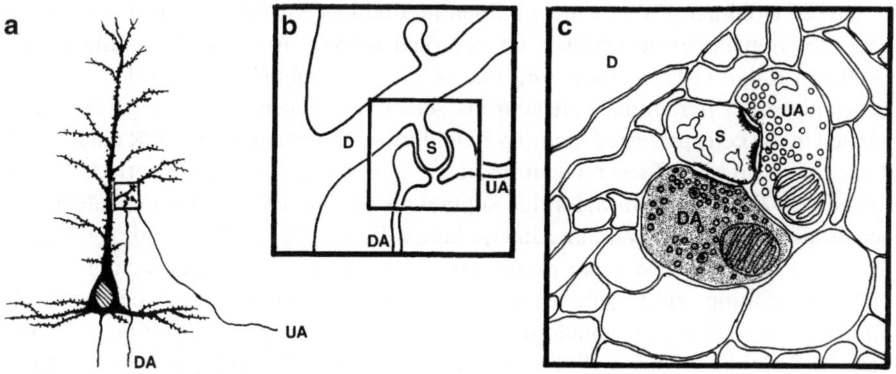

Fig. 3.5 Drawings illustrating the triadic arrangement of dopamine afferents (DA) with dendritic spines and unidentified excitatory afferents (UA). (**a**) Pyramidal neurons receive dopaminergic afferents predominantly on distal dendritic spines. (**b**) Dopamine-containing afferents contact a dendritic spine that is also contacted by an afferent that lacks dopamine immunoreactivity. (**c**) At the ultrastructural level, the dopamine-positive terminal forms a symmetrical (inhibitory) synapse with the spine. In contrast, the dopamine-negative terminal is excitatory as evidenced by the presence of an asymmetric synapse on the spine. (From Goldman-Rakic 1995. Reproduced with permission of Cell Press in the format reuse in a book/textbook via Copyright Clearance Center)

D1R is distributed most densely in superficial layers (I, II and superficial III) and predominantly on spines of pyramidal cells, suggesting that this is the receptor subtype that mediates dopaminergic modulation of excitatory transmission (Goldman-Rakic et al. 1990; Lidow et al. 1991; Smiley et al. 1994; Bergson et al. 1995). In contrast, the D2R is largely concentrated in layer V (Goldman-Rakic et al. 1990). Notably, D1R is found in perisynaptic and extrasynaptic regions of the membrane and, despite the triadic arrangement described previously, D1R-positive spines rarely are opposed by axons having features of dopamine boutons (Smiley et al. 1994). Both of these characteristics suggest that volumetric neurotransmission of dopamine at sites distant to the synapse may represent an alternative mode of action for dopamine to impact neural circuitry in the PFC. Although D1R is most abundantly found on the dendritic spines and shafts of pyramidal neurons, this receptor subtype has also been localized to excitatory axon terminals (Smiley et al. 1994; Paspalas and Goldman-Rakic 2005). Moreover, D1R is found in parvalbumin-positive (FS type) interneurons, both in the dendritic and terminal compartments (Muly et al. 1998). Interestingly, D1R-positive axon terminals only contact pyramidal cell spines; they do not synapse with parvalbumin-positive interneurons (Paspalas and Goldman-Rakic 2005). Considered together, these findings suggest that D1R is present in multiple elements of the local neuronal network that governs spatial mnemonic behavior in the PFC, yet there is an elegant specificity to the targeting of dopamine that preferentially modulates excitatory input onto pyramidal cell spines.

One of Goldman-Rakic's most influential findings emanated from studies of the action of dopaminergic agonists on neuronal activity in the PFC. Williams and Goldman-Rakic (1995) discovered that *in vivo* iontophoresis of a selective D1R antagonist onto pyramidal cell neurons selectively accentuates firing for targets located in the cells preferred memory field during the delay period. D1R antagonists do not have similar effects on neurons active during cue or response period, and the selective D2R antagonist raclopride suppresses activity of neurons non-selectively across all task epochs (Williams and Goldman-Rakic 1995). These findings suggest that delay period neurons are preferentially targeted by dopamine acting via the D1R. In addition, enhancement of delay period activity is dose dependent with higher doses of the D1R antagonist diminishing the delay activity (Fig. 3.3; Williams and Goldman-Rakic 1995). This dose sensitive relationship is best described as an inverted "u" such that too little or too much dopamine D1R stimulation has detrimental effects on delay period firing of neurons during the oculomotor DR. In concert with these findings, local injection of D1R antagonists into the PFC induces errors and increases latency in performance on the oculomotor DR in a dose dependent fashion (Sawaguchi and Goldman-Rakic 1991), and systemic injection of D1R acting compounds modulate working memory performance of monkeys on a manual DR task (Arnsten et al. 1994).

Exploration of the mechanisms by which dopamine may exert control over delay firing in PFC neurons revealed multiple roles for D1R in network activity. For example, in deep layer V, dopamine acts via presynaptic mechanisms to decrease the lateral excitation of recurrent excitatory contacts between local pyramidal cells thought to be important in maintaining delay period firing (Gao et al. 2001). In addition, findings by the Pittsburgh group suggest that in layer III dopamine increases excitability of pyramids through the D1R and acts via a combined D1R and D2R mechanism to suppress excitatory input from distant afferent sources, e.g. the thalamus or other cortical areas (Henze et al. 2000; Urban et al. 2002). These properties further support a role for dopamine in modulating excitatory input onto pyramidal dendritic spines in the PFC. Although decreased excitatory neurotransmission at the pyramidal spine seems at odds with activation and maintenance of sustained firing during the delay period, it must be remembered that not all spines and not all excitatory terminals are D1R-containing and therefore specificity in the modulation of excitatory input onto pyramidal cells may curtail input from irrelevant stimuli and consequently accentuate the signal to noise ratio for relevant stimuli during a memory related task.

Dopamine also influences network activity by acting on interneurons. In contrast to its modulatory role at the pyramidal to pyramidal synapse, dopamine does not decrease excitatory potency at excitatory contacts between pyramidal cells and interneurons (Gao and Goldman-Rakic 2003). Instead, dopamine directly excites the majority of interneurons (Zhou and Hablitz 1999; Gorelova et al. 2002; Gao and Goldman-Rakic 2003). In this regard, dopamine's selectivity in presynaptic site of action corresponds with the anatomic specificity of D1R distribution, i.e. located in excitatory terminals in apposition to pyramidal cells spines but not those contacting interneurons (Paspalas and Goldman-Rakic 2005). The exquisite specificity of dopamine's action also extends to modulation of interneuron inhibition of pyramidal cells. Dopamine acts via the D1R to facilitate inhibition at FS interneuron (basket cell) synapses that are known to innervate the perisomatic region of pyramidal cells while diminishing inhibitory neurotransmission of non-FS interneuron (double bouquet, tufted and Martinotti cells) that contact the distal dendrites of pyramids (Gao et al. 2003). The multitude of dopamine actions in the PFC, including dampening excitatory transmission between pyramidal cells and both facilitating and curtailing interneuron inhibition of pyramidal cells, attests to the complexity of dopaminergic modulation of the mnemonic-related network activity in the PFC. Certainly, a full understanding of the modulatory role of dopamine on PFC network circuitry remains beyond our grasp. However, what is clear is that optimal dopamine stimulation enhances relevant input for a PFC neuron's memory field and optimizes delay period firing for the target location.

3.6 Spatial and Object Working Memory Domains

Goldman-Rakic's work centered on the dlPFC, specifically the cortex surrounding the principal sulcus; however, studies of the connections and the physiologic properties of more ventrally located cortex in the PFC led to a remarkable discovery: the ventral PFC is specialized for the memory of "what" in contrast to the more dorsal cortex that remembers "where." Goldman-Rakic and colleagues found that cortical neurons located on the ventral convexity of the PFC are responsive to stimuli hitting the foveal region of the retina and that they are tuned to the color and form of objects (Wilson et al. 1993). Some neurons in the ventral PFC exhibit even greater specificity by responding only to faces (Wilson et al. 1993; O'Scalaidhe et al. 1997, 1999). Importantly, during the spatial oculomotor DR, a task in which the location of the target must be remembered, neurons in the ventral PFC are not active during the delay period; however, when an object must be remembered as a signal for the correct saccadic response, these same neurons in the ventral PFC exhibit sustained firing to bridge the gap between cue and response periods (Fig. 3.6, Wilson et al. 1993).

The segregation of working memory function into dorsal and ventral domains that respectively mediate memory for spatial location and object features fits perfectly with the known connectivity of dorsal and ventral PFC. A wealth of data have established that discrete features of visual information coming from the retina are processed separately in the visual system (Livingston and Hubel 1988). Moreover, visual information leaving the primary visual cortex to reach areas of higher cortical processing are also segregated. Visual attributes related to form and color are channeled from the visual cortex into a "ventral stream" that relays information through the inferotemporal cortex to the ventral PFC while spatial location is conveyed through the "dorsal stream" via the posterior parietal cortex to the dlPFC (Macko et al. 1982; Barbas 1988; Cavada and Goldman-Rakic 1989; Morel and Bullier 1990; Baizer et al. 1991; Webster et al. 1994). The partitioning of

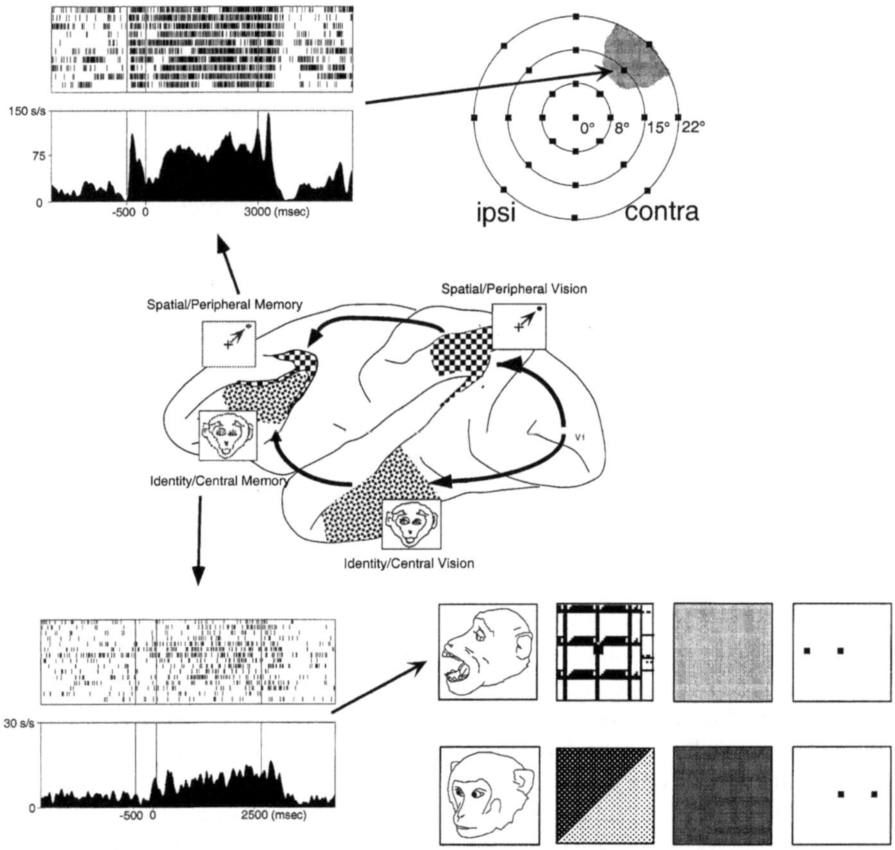

Fig. 3.6 Schematic representation of separate visual streams that feed into spatial and object domains in the dorsal and ventral PFC. (*Top*) Rasters and histograms showing enhanced activity of a delay period neuron in the dlPFC to the preferred target location. (*Bottom*) Rasters and histograms illustrating the response of a neuron to a monkey face in the ventral PFC. (From Goldman-Rakic 1999b. Reproduced with permission of Elsevier Inc. in the format reuse in a book/textbook via Copyright Clearance Center)

cortex into spatial and object domains is also supported by lesion studies in the non-human primate that show a dissociation of spatial and non-spatial deficits by dorsal and ventral site (e.g., Mishkin and Manning 1978; Passingham 1975; Levy and Goldman-Rakic 2000). It should be noted that this division of labor for the PFC has not been universally accepted with some arguing that different regions of the PFC mediate higher versus lower levels of cognitive processing (Owen et al. 1996; Petrides et al. 2002). To some degree discrepancies in location of data collection may account for conflicting evidence on this topic (for a review, see Romanski 2004).

Parcellation of the PFC with respect to the location or identity of the memorandum is not limited to the visual modality. Auditory afferent input conveying location and feature identity are also segregated in the PFC: the rostral principal sulcal cortex and surrounding cortices, including the inferior convexity, receive information about auditory features while the caudal principal sulcal and dorsal periarcuate cortices are innervated by auditory projections that convey spatial location (Romanski et al. 1999a, b). The electrophysiologic specificity in responding to stimuli again mirrors the anatomic duality. Neurons in the dlPFC have auditory responsive properties and have been shown to be sensitive to sound source location (Azuma and Suzuki 1984; Kikuchi-Yorioka and Sawaguchi 2000) whereas neurons that respond to complex features of sound, e.g., species-specific vocalizations, are located in the ventral PFC adjacent to the visual object domain (Romanski and Goldman-Rakic 2002; Romanski et al. 2005). Some evidence also suggests that somatosensory information may contribute to memory for objects in the ventral PFC (Romo et al. 1999). These data further support Goldman-Rakic's (1996, 1999b) concept of separate domains for memory of "what" and "where" that may be even further modularized into subdomains that process information from diverse sensory streams.

3.7 Summary

Perhaps the most extraordinary aspect of Goldman-Rakic's work is the breadth and depth of her studies on the PFC. She examined the entire hierarchy of PFC organization extending from the "macromodular" level of organization represented by these large, functionally distinct domains down to the microcolumnar structure devoted to representation of a single aspect of a memorandum, e.g. spatial location in the dlPFC. Her theories of prefrontal function represent the synthesis of work from many different methodologic approaches, encompassing the fields of neuroanatomy, neurophysiology, neuropharmacology, and behavior. Patricia Goldman-Rakic's diverse studies laid the foundation for contemporary understanding of prefrontal structure and function. Although she is clearly missed, the many young scientists whom she mentored continue her legacy as they pursue a multitude of avenues of research relating to the PFC.

References

Arnsten AF, Cai JX, Murphy BL, Goldman-Rakic PS (1994) Dopamine D1 receptor mechanisms in the cognitive performance of young adult and aged monkeys. Psychopharmacology 116:143–151

Azuma M, Suzuki H (1984) Properties and distribution of auditory neurons in the dorsolateral prefrontal cortex of the alert monkey. Brain Res 298:343–346

Baddeley AD, Hitch GJ (1974) Working memory. In: Bower GA (ed) Recent advances in learning and motivation, vol 8. Academic, New York, pp 47–90

Baizer JS, Ungerleider LG, Desimone R (1991) Organization of visual inputs to the inferior temporal and posterior parietal cortex in macaques. J Neurosci 11:168–190
Barbas H (1988) Anatomic organization of basoventral and mediodorsal visual recipient prefrontal regions in the rhesus monkey. J Comp Neurol 276:313–342
Bergson C, Mrzljak L, Smiley JF, Pappy M, Levenson R, Goldman-Rakic PS (1995) Regional, cellular, and subcellular variations in the distribution of D1 and D5 dopamine receptors in primate brain. J Neurosci 15:7821–7836
Brown RM, Goldman PS (1977) Catecholamines in neocortex of rhesus monkeys: regional distribution and ontogenetic development. Brain Res 124:576–580
Brown RM, Crane AM, Goldman PS (1979) Regional distribution of monoamines in the cerebral cortex and subcortical structures of the rhesus monkey: concentrations and *in vivo* synthesis rates. Brain Res 168:133–150
Brozoski TJ, Brown RM, Rosvold HE, Goldman PS (1979) Cognitive deficit caused by regional depletion of dopamine in prefrontal cortex of rhesus monkey. Science 205:929–932
Cavada C, Goldman-Rakic PS (1989) Posterior parietal cortex in rhesus monkey: II. Evidence for segregated corticocortical networks linking sensory and limbic areas with the frontal lobe. J Comp Neurol 287:422–445
Constantinidis C, Goldman-Rakic PS (2001) Correlated discharges among putative pyramidal neurons and interneurons in the primate prefrontal cortex. J Neurophysiol 88:3487–3497
Constantinidis C, Franowicz MN, Goldman-Rakic PS (2001) Coding specificity in cortical microcircuits: a multiple-electrode analysis of primate prefrontal cortex. J Neurosci 21:3646–3655
Funahashi S, Bruce CJ, Goldman-Rakic PS (1989) Mnemonic coding of visual space in the monkey's dorsolateral prefrontal cortex. J Neurophysiol 61:331–349
Funahashi S, Bruce CJ, Goldman-Rakic PS (1990) Visuospatial coding in primate prefrontal neurons revealed by oculomotor paradigms. J Neurophysiol 63:814–831
Funahashi S, Bruce CJ, Goldman-Rakic PS (1993) Dorsolateral prefrontal lesions and oculomotor delayed response performance: evidence for mnemonic "scotomas". J Neurosci 13:1479–1497
Fuster JM (1973) Unit activity in prefrontal cortex during delayed-response performance: neuronal correlates of transient memory. J Neurophysiol 36:61–78
Fuster JM, Alexander GE (1971) Neuron activity related to short-term memory. Science 173:652–654
Fuster JM, Bauer RH, Jervey JP (1982) Cellular discharge in the dorsolateral prefrontal cortex of the monkey in cognitive tasks. Exp Neurol 77:679–694
Gao W-J, Goldman-Rakic PS (2003) Selective modulation of excitatory and inhibitory microcircuits by dopamine. Proc Natl Acad Sci 100:2836–2841
Gao W-J, Krimer LS, Goldman-Rakic PS (2001) Presynaptic regulation of recurrent excitation by D1 receptors in prefrontal circuits. Proc Natl Acad Sci U S A 98:295–300
Gao W-J, Wang Y, Goldman-Rakic PS (2003) Dopamine modulation of perisomatic and peridendritic inhibition in prefrontal cortex. J Neurosci 23:1622–1630
Giguere M, Goldman-Rakic PS (1988) Mediodorsal nucleus: areal, laminar, and tangential distribution of afferents and efferents in the frontal lobe of rhesus monkeys. J Comp Neurol 277:195–213
Goldman PS (1971) Functional development of the prefrontal cortex in early life and the problem of neuronal plasticity. Exp Neurol 32:366–387
Goldman PS (1976) The role of experience in recovery of function following orbital prefrontal lesions in infant monkeys. Neuropsychologia 14:401–412
Goldman PS, Nauta WJH (1977) Columnar distribution of cortico-cortical fibers in the frontal association, limbic, and motor cortex of the developing rhesus monkey. Brain Res 122:393–413
Goldman PS, Rosvold HE (1970) Localization of function within the dorsolateral prefrontal cortex of the rhesus monkey. Exp Neurol 27:291–304
Goldman PS, Rosvold HE, Mishkin M (1970a) Evidence for behavioral impairment following prefrontal lobectomy in the infant monkey. J Comp Physiol Psychol 70:454–463

Goldman PS, Rosvold HE, Mishkin M (1970b) Selective sparing of function following prefrontal lobectomy in infant monkeys. Selective sparing of function following prefrontal lobectomy in infant monkeys. Exp Neurol 29:221–226
Goldman PS, Rosvold HE, Vest B, Galkin TW (1971) Analysis of the delayed-alternation deficit produced by dorsolateral prefrontal lesions in the rhesus monkey. J Comp Physiol Psychol 77:212–220
Goldman PS, Crawford HT, Stokes LP, Galkin TW, Rosvold HE (1974) Sex-dependent behavioral effects of cerebral cortical lesions in the developing rhesus monkey. Science 186:540–542
Goldman-Rakic PS (1987) Circuitry of the prefrontal cortex and the regulation of behavior by representational knowledge. In: Plum F, Mountcastle V (eds) Handbook of physiology, vol 5. American Physiological Society, Bethesda, pp 373–417
Goldman-Rakic PS (1995) Cellular basis of working memory. Neuron 14:477–485
Goldman-Rakic PS (1996) Regional and cellular fractionation of working memory. Proc Natl Acad Sci U S A 93:13473–13480
Goldman-Rakic PS (1999a) The "psychic" neuron of the cerebral cortex. Ann N Y Acad Sci 868:13–26
Goldman-Rakic PS (1999b) The physiological approach: functional architecture of working memory and disordered cognition in schizophrenia. Biol Psychiatry 46:650–661
Goldman-Rakic PS, Selemon LD, Schwartz ML (1984) Dual pathways connecting the dorsolateral prefrontal cortex with the hippocampal formation and parahippocampal cortex in the rhesus monkey. Neuroscience 12:719–743
Goldman-Rakic PS, Leranth C, Williams SM, Mons N, Geffard M (1989) Dopamine synaptic complex with pyramidal neurons in primate cerebral cortex. Proc Natl Acad Sci U S A 86:9015–9019
Goldman-Rakic PS, Lidow MS, Gallager DW (1990) Overlap of dopaminergic, adrenergic, and serotoninergic receptors and complementarity of their subtypes in primate prefrontal cortex. J Neurosci 10:2125–2138
Goldman-Rakic PS, Muly EC III, Williams GV (2000) D1 receptors in prefrontal cells and circuits. Brain Res Rev 31:295–301
Goldman-Rakic PS, Castner SA, Svensson TH, Siever LJ, Williams GV (2004) Targeting the dopamine D1 receptor in schizophrenia: insights for cognitive dysfunction. Psychopharmacology 174:3–16
Gorelova N, Seamans JK, Yang CR (2002) Mechanisms of dopamine activation of fast-spiking interneurons that exert inhibition in rat prefrontal cortex. J Neurophysiol 88:3150–3166
Henze DA, Gonzalez-Burgos GR, Urban NN, Lewis DA, Barrionuevo G (2000) Dopamine increases excitability of pyramidal neuron in primate prefrontal cortex. J Neurophysiol 84:2799–2809
Hubel DH, Wiesel TN (1972) Laminar and columnar distribution of geniculo-cortical fibers in the macaque monkey. J Comp Neurol 146:421–450
Jones EG, Burton H, Porter R (1975) Commissural and cortico-cortical "columns" in the somatic sensory cortex of primates. Science 190:572–574
Joseph JP, Barone P (1987) Prefrontal unit activity during a delayed oculomotor task in the monkey. Exp Brain Res 67:460–468
Kawaguchi Y (1995) Physiologic subgroups of nonpyramidal cells with specific morphological characteristics in layer II/III of the rat frontal cortex. J Neurosci 15:2638–2655
Kikuchi-Yorioka Y, Sawaguchi T (2000) Parallel visuospatial and audiospatial working memory processes in the monkey dorsolateral prefrontal cortex. Nat Neurosci 3:1075–1076
Kojima S, Goldman-Rakic PS (1984) Functional analysis of spatially discriminative neurons in prefrontal cortex of rhesus monkey. Brain Res 291:229–240
Krimer LS, Goldman-Rakic PS (2001) Prefrontal microcircuits: membrane properties and excitatory input of local, medium, and wide arbor interneurons. J Neurosci 21:3788–3796
Krimer LS, Jakab RL, Goldman-Rakic PS (1997) Quantitative three-dimensional analysis of the catecholaminergic innervation of identified neurons in the macaque prefrontal cortex. J Neurosci 17:7450–7461

Kritzer MF, Goldman-Rakic PS (1995) Intrinsic circuit organization of the major layers and sublayers of the dorsolateral prefrontal cortex in the rhesus monkey. J Comp Neurol 359:131–143
Kubota K, Niki H (1971) Prefrontal cortical unit activity and delayed alternation performance in monkeys. J Neurophysiol 34:337–347
Levitt JB, Lewis DA, Yoshioka T, Lund JS (1993) Topography of pyramidal neuron intrinsic connections in macaque monkey prefrontal cortex (areas 9 and 46). J Comp Neurol 338:360–376
Levy R, Goldman-Rakic PS (2000) Segregation of working memory functions within the dorsolateral prefrontal cortex. Exp Brain Res 133:23–32
Lidow MS, Goldman-Rakic PS, Gallager DW, Rakic P (1991) Distribution of dopaminergic receptors in the primate cerebral cortex: quantitative autoradiographic analysis using [3H] raclopride, [3H]spiperone and [3H]SCH23390. Neuroscience 40:657–671
Livingston M, Hubel D (1988) Segregation of form, color, movement, and depth: anatomy, physiology, and perception. Science 240:740–749
Lund JS, Lewis DA (1993) Local circuit neurons of developing and mature macaque prefrontal cortex. Golgi and immunocytochemical characteristics. J Comp Neurol 328:282–312
Macko KA, Jarvis CD, Kennedy C, Miyaoka M, Shinohara M, Sokoloff L, Mishkin M (1982) Mapping the primate visual system with $^{2\text{-}14}$C-deoxyglucose. Science 218:394–397
Melchitzky DS, Sesack SR, Pucak ML, Lewis DA (1998) Synaptic targets of pyramidal neurons providing intrinsic horizontal connections in monkey prefrontal cortex. J Comp Neurol 390:211–224
Mishkin M, Manning FJ (1978) Non-spatial memory after selective prefrontal lesions in monkeys. Brain Res 143:313–323
Morel A, Bullier J (1990) Anatomical segregation of two cortical visual pathways in the macaque monkey. Vis Neurosci 4:555–578
Muly EC III, Szigeti K, Goldman-Rakic PS (1998) D1 receptor in interneurons of macaque prefrontal cortex: distribution and subcellular localization. J Neurosci 18:10553–10565
Niki H (1974a) Differential activity of prefrontal units during right and left delayed response trials. Brain Res 70:346–349
Niki H (1974b) Prefrontal unit activity during delayed alternation in the monkey. I. Relation to direction of response. Brain Res 68:185–196
O'Scalaidhe SP, Wilson FAW, Goldman-Rakic PS (1997) Areal segregation of face-processing neurons in prefrontal cortex. Science 278:1135–1138
O'Scalaidhe SP, Wilson FAW, Goldman-Rakic PS (1999) Face-selective neurons during passive viewing and working memory performance of rhesus monkeys: evidence for intrinsic specialization of neuronal coding. Cereb Cortex 9:459–475
Owen AM, Evans AC, Petrides M (1996) Evidence for a two-stage model of spatial working memory processing within the lateral frontal cortex: a positron emission tomography study. Cereb Cortex 6:31–38
Paspalas CD, Goldman-Rakic PS (2005) Presynaptic D1 dopamine receptors in primate prefrontal cortex: target-specific expression in the glutamatergic synapse. J Neurosci 25:1260–1267
Passingham R (1975) Delayed matching after selective prefrontal lesions in monkeys (*Macaca mulatta*). Brain Res 92:89–102
Petrides M, Alivisatos B, Frey S (2002) Differential activation of the human orbital, mid-ventrolateral, and mid-dorsolateral prefrontal cortex during the processing of visual stimuli. Proc Natl Acad Sci U S A 99:5649–5654
Pucak ML, Levitt JB, Lund JS, Lewis DA (1996) Patterns of intrinsic and associational circuitry in monkey prefrontal cortex. J Comp Neurol 376:614–630
Rakic P (1988) Specification of cerebral cortical areas. Science 241:170–176
Rao SG, Williams GV, Goldman-Rakic PS (1999) Isodirectional tuning of adjacent interneurons and pyramidal cells during working memory: evidence for microcolumnar organization in PFC. J Neurophysiol 81:1903–1916

Rao SG, Williams GV, Goldman-Rakic PS (2000) Destruction and creation of spatial tuning by disinhibition: $GABA_A$ blockade of prefrontal cortical neurons engaged by working memory. J Neurosci 20:485–494
Romanski LM (2004) Domain specificity in the primate prefrontal cortex. Cogn Affect Behav Neurosci 4:421–429
Romanski LM, Goldman-Rakic PS (2002) An auditory domain in the primate prefrontal cortex. Nat Neurosci 5:15–16
Romanski LM, Bates JF, Goldman-Rakic PS (1999a) Auditory belt and parabelt projections to the prefrontal cortex in the rhesus monkey. J Comp Neurol 403:141–157
Romanski LM, Tian B, Fritz J, Mishkin M, Goldman-Rakic PS, Rauschecker JP (1999b) Dual streams of auditory afferents target multiple domains in the primate prefrontal cortex. Nat Neurosci 2:1131–1136
Romanski LM, Averbeck BB, Diltz M (2005) Neural representation of vocalizations in the primate ventrolateral prefrontal cortex. J Neurophysiol 93:734–747
Romo R, Brody CD, Hernandez A, Lemus L (1999) Neuronal correlates of parametric working memory in the prefrontal cortex. Nature 399:470–473
Sawaguchi T, Goldman-Rakic PS (1991) D1 dopamine receptors in prefrontal cortex: involvement in working memory. Science 251:947–950
Schwartz ML, Goldman-Rakic PS (1984) Callosal and intrahemispheric connectivity of the prefrontal association cortex in rhesus monkey: relation between intraparietal and principal sulcal cortex. J Comp Neurol 226:403–420
Selemon LD, Goldman-Rakic PS (1985) Longitudinal topography and interdigitation of corticostriatal projections in the rhesus monkey. J Neurosci 5:776–794
Selemon LD, Goldman-Rakic PS (1988) Common cortical and subcortical targets of the dorsolateral prefrontal and posterior parietal cortices in the rhesus monkey: evidence for a distributed neural network subserving spatially guided behavior. J Neurosci 8:4049–4068
Sesack SR, Snyder CL, Lewis DA (1995) Axon terminals immunolabeled for dopamine or tyrosine hydroxylase synapse on GABA-immunoreactive dendrites in rat and monkey cortex. J Comp Neurol 363:264–280
Shanks MF, Rockel AJ, Powell TPS (1975) The commissural fiber connections of the primary somatic sensory cortex. Brain Res 98:166–171
Smiley JF, Levey AI, Ciliax BJ, Goldman-Rakic PS (1994) D1 dopamine receptor immunoreactivity in human and monkey cerebral cortex: predominant and extrasynaptic localization in dendritic spines. Proc Natl Acad Sci U S A 91:5720–5724
Urban NN, Gonzalez-Burgos G, Henze DA, Lewis DA, Barrionuevo G (2002) Selective reduction by dopamine of excitatory synaptic inputs to pyramidal neurons in primate prefrontal cortex. J Physiol 593:707–712
Webster MJ, Bachevalier J, Ungerleider LG (1994) Connections of inferior temporal areas TEO and TE with parietal and frontal cortex in macaque monkeys. Cereb Cortex 4:470–483
Williams GV, Goldman-Rakic PS (1995) Modulation of memory fields by dopamine D1 receptors in prefrontal cortex. Nature 376:572–575
Williams SM, Goldman-Rakic PS (1993) Characterization of the dopaminergic innervation of the primate frontal cortex using a dopamine-specific antibody. Cereb Cortex 3:199–222
Williams SM, Goldman-Rakic PS (1998) Widespread origin of the primate mesofrontal dopamine system. Cereb Cortex 8:321–345
Wilson FAW, O'Scalaidhe SP, Goldman-Rakic PS (1993) Dissociation of object and spatial processing domains in primate prefrontal cortex. Science 260:1955–1958
Wilson FAW, O'Scalaidhe SP, Goldman-Rakic PS (1994) Functional synergism between putative γ-aminobutyrate-containing neurons and pyramidal neurons in prefrontal cortex. Proc Natl Acad Sci U S A 91:4009–4013
Zhou F-M, Hablitz JJ (1999) Dopamine modulation of membrane and synaptic properties of interneurons in rat cerebral cortex. J Neurophysiol 81:967–976

Chapter 4
The Types of Functional and Structural Subdivisions of Cortical Areas

Jon H. Kaas and Pooja Balaram

Abstract The term "cortical column" has been used to identify a number of different types of subdivisions of cortical areas. Here we describe and discuss several of these types of modular subdivisions in cortex, and suggest criteria for distinguishing these types. Minicolumns are narrow, vertical arrays of densely interconnected neurons that cross all cortical layers, and may represent a basic computational unit that is common to all cortical areas across mammalian species. Other types of columns are more variable across cortical areas and mammalian species, and have different developmental and evolutionary functions and histories. Classical columns are so named because they correspond to the column type first described by Mountcastle (J Neurophysiol 20:408–434, PMID 13439410, 1957), and consist of larger alternating patches of cortex that span all cortical layers, and contain neurons that have different response properties from those of neurons in adjoining columns. Several types of classical columns have been identified, and evidence for more types will likely emerge with further investigation. Classical columns divide cortical areas into sets of functionally specialized modules. A third type of cortical subdivision, unbounded columns, collectively represent a continuously varied stimulus dimension, such as the orientation of a bar or line, across a cortical area. Thus, vertical rows of neurons across a patch of cortex representing a small portion of the visual field will vary continuously in the orientation of a stimulus bar that best activates them, and there are no obvious borders between rows of neurons that prefer one orientation and adjoining rows that prefer a slightly different orientation. Other continuously variable stimuli such as color are likely to be represented by unbounded columns as well. A fourth type of cortical column consists of adjoining blocks of neurons that represent different parts of the receptor surface. The ocular dominance columns in primary visual cortex constitute a well-known example, where blocks of neurons are activated preferentially by one eye or the other. The banded representation of digits in somatosensory cortex of monkeys and raccoons provide another example. Lastly, cortical areas are sometimes divided into specialized regions or domains that are distinct from one another and do not repeat across the cortical surface. Such domains appear to exist in motor and

J.H. Kaas (✉) • P. Balaram
Department of Psychology, Vanderbilt University, 301 Wilson Hall 111 21st Ave S, Nashville, TN 37240, USA
e-mail: jon.h.kaas@vanderbilt.edu

M.F. Casanova, I. Opris (eds.), *Recent Advances on the Modular Organization of the Cortex*, DOI 10.1007/978-94-017-9900-3_4

premotor cortex, posterior parietal cortex, and temporal visual cortex of primates. Finally, we address questions of how columns emerge in development, and how mechanisms for their development evolved. Clearly there are genetic factors, especially for those that control activity-dependent neural mechanisms of synapse formation and selection.

Keywords Columns • Minicolumns • Module • Barrels • Blob • Hypercolumn

4.1 Introduction

Early investigators that conducted comparative studies of neocortex in mammals, such as Brodmann (1909), focused on deducing the locations and numbers of large functional subdivisions of neocortex, the cortical areas, from subtle differences in cortical architecture. As the size of neocortex varies greatly across mammals, from a thin cap over the forebrain in small insect-eating tenrecs (Krubitzer et al. 1997) to a dominating 80 % of the total brain volume in humans (Azevedo et al. 2009), the sizes of cortical areas vary from under a few mm^2 to over 3,200 mm^2, the size of primary visual cortex in humans (Kaas 2013). The number of cortical areas also varies considerably, from an estimated 20 or so in small mammals to an estimated 200 or more in humans (Kaas 2008, 2012b; Van Essen et al. 2012). More recently, studies have turned to describing subdivisions within individual cortical areas, variably termed columns, modules, and domains (Kaas 2012a). Mountcastle (1957) introduced the concept of a patchwork of physiologically defined subdivisions of somatosensory areas of cortex, the cortical columns, which extended through the depth of the cortical sheet. Goldman and Nauta (1977) are known for considering the patchy patterns of cortical connections within frontal, motor, and limbic cortex as evidence of modular subdivisions. More recently, array sequencing and molecular labeling have been used to delineate cortical areas (for example, Belgard et al. 2011; Bernard et al. 2012; Zilles et al. 2005; Mashiko et al. 2012). As cortical areas vary greatly in size, however, smaller areas could be misinterpreted as modules, so it is important to have distinguishing criteria when discussing functional divisions within the cerebral cortex (Kaas 1990).

Cortical areas, according to Brodmann (1909), are the "organs of the brain". Brodmann (1909) used structural differences across the laminar appearance of cortex in Nissl-stained brain sections to delimit cortical areas, and did so for a great range of species. To this day, the Nissl method is widely used to define cortical boundaries in most species. However, when using this architectonic method alone, many uncertainties and outright errors are likely, and other methods of revealing identifying features are now available. Additionally, the congruence of multiple identifying features provide much more compelling evidence than any single feature alone (Kaas 1982). Well defined cortical areas often contain a systematic and complete representation of a sensory receptor surface, such as the contralateral visual hemifield, contralateral body surface, or the full range of the strip of auditory hair cells in a cochlea. As such representations have topographic patterns of cortical and subcortical connections with other representations of the same modality, such patterns of connections are another defining feature of cortical sensory representations, and of other cortical areas as well. Cortical areas are also often histologically distinct, not only in the Nissl and myelin preparations of yesteryear, but also in the vast array of immunological and histochemical preparations that are available today (e.g. Zilles et al. 2005; Wong and Kaas 2009). Importantly, the agreement of a range of histological preparations on the location of areal borders provides some of the most powerful evidence for defining cortical areas. Of course, the response properties of neuronal populations in areas should also differ, as should the behavioral consequences of damaging or inactivating regions within a cortical area.

Cortical modules are functional subdivisions within areas, and as such they make areas more complex, and may be mistaken for areas on their own. In addition, clearly related species may or may not have similar modules within a specific area, but distantly related mammals may have comparable modules. Thus, similar modules may exist because they were already present in a common ancestor and are homologous features of an area, or they may have been independently derived, and thus constitute convergent features of evolution. Such differences in origin raise issues of genetic, developmental, and environmental factors in the evolution of modules. Given that there are also several types of modular subdivisions within cortical areas, it is useful to consider the characteristics of each type. Here, several types of cortical modules are considered: minicolumns, the classic columns of Mountcastle's description, unbounded columns, modules representing separated parts of sensory surfaces, and larger, functionally defined subdivisions or areas we call domains.

This review is an expanded version of an earlier review that focused on the evolution of types of columns (Kaas 2012a). An earlier review of Mountcastle (1997) is an important source of information and concepts. Other reviews that provide valuable opinion and theory include those of Towe (1975), Eccles (1981), Szentágothai (1975), Purves et al. (1992), Buxhoeveden and Casanova (2002), da Costa and Martin (2010), Rockland (2010), and Molnár (2013). Some of the cortical areas containing columns that are discussed are presented in Fig. 4.1.

4.2 Minicolumns

Discussions of the features and functions of minicolumns (Buxhoeveden and Casanova 2002) are a special topic of this volume, so they are only mentioned briefly here for the sake of completeness. While definitions may vary somewhat, minicolumns are narrow, vertical arrays of neurons that cross all cortical layers. Neurons are densely connected within the arrays (e.g. Lachica et al. 1992; Yu et al. 2009), and have common response properties, while still differing according to the varied functions and connections of each layer. In some cortical locations, the

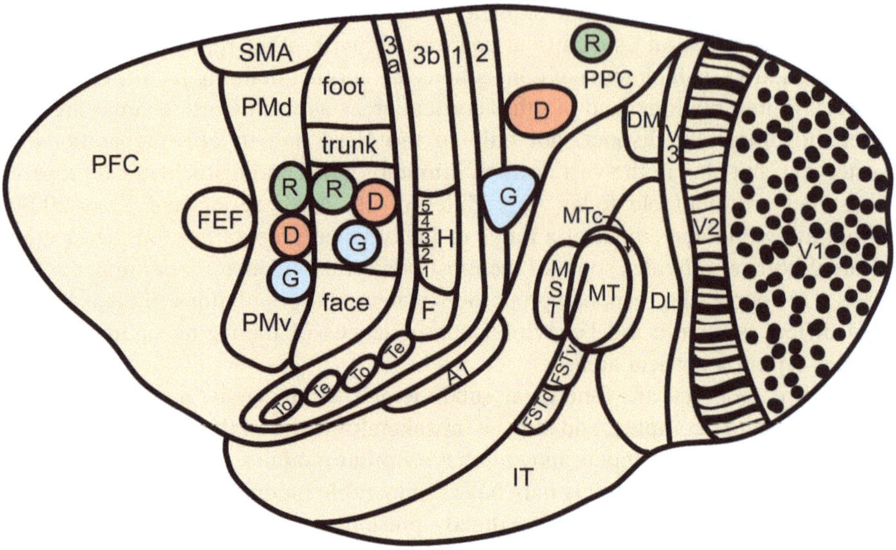

Fig. 4.1 Some of the cortical areas of a squirrel monkey brain that contain columns discussed in this review. Classical columns concerned with either rapidly or slowly adapting afferents are found in the hand (*H*) representation in somatosensory area *3b* (digit territories are numbered). Subdivisions of are *3b* (primary somatosensory cortex) for each digit (1–5) or teeth (*Te*) or tongue (*To*) are another type of modular organization. Other somatosensory areas include *3a*, *1*, and *2* of Brodmann (1909). Classical columns have also been well described in primary (*V1*) and secondary (*V2*) visual areas, and in other visual areas (see text), which include the third visual area (*V3*), the dorsomedial visual area (*DM*), the dorsolateral visual area (*DL*), the middle temporal visual area (*MT*), the *MT* crescent (*MTc*), the middle superior temporal area (*MST*) and dorsal and ventral divisions of the fundal area of the superior temporal sulcus (*FSTd* and *FSTv*). Inferior temporal cortex (*IT*) is a higher order visual region. Primary auditory cortex (*A1*) may have band-like modules for differently processing information from the two ears. Posterior parietal cortex (*PPC*), primary motor cortex (*M1*), and dorsal and ventral premotor cortex (*PMd* and *PMv*) have larger subdivisions, called domains, related to functionally important complex movements (*R*-reach, *D*-defensive movement, *G*-grasp), as well as a looking domain (not shown) that relates to the frontal eye field (*FEF*). IT cortex has regions selective for faces or objects. Anatomical studies support that many or all other cortical areas have classical or other types of modular columns, including prefrontal cortex (*PFC*)

cell bodies within one minicolumn are somewhat separated from adjacent minicolumns by neuropil (e.g. Buxhoeveden et al. 2001), but in other locations, such separations may not be apparent. The dominance of vertical connections within a minicolumn suggests that they represent a computational unit that is basic to all cortical areas and functions (Lorente de Nó 1949). With the advent of recordings from single cortical neurons with penetrating microelectrodes, it became apparent that vertical arrays of neurons everywhere across the thickness of cortex have some response properties that unites them within a common processing circuit. Local lateral connections also closely tie neurons in one minicolumn with those in adjoining or nearby minicolumns (Krieger et al. 2007). Developmentally, minicolumns appear to reflect the progenitor cells of the ventricular zone that stem clonally and migrate to form the ontogenetic column (Rakic 2007). Once developed, minicolumns vary in width, and are wider in the large human brain compared to similar minicolumns in monkeys (Buxhoeveden et al. 2000). On subsequent pages of this review, we discuss functional evidence for minicolumns in primary motor cortex as parts of larger functional regions called domains.

4.3 Classical Columns

Somewhat differently, Mountcastle (1957) and Powell and Mountcastle (1959) described larger patches of somatosensory cortex that appeared to be columnar in form, extending across the depth of cortex, but wider than minicolumns, varying around 0.5 mm in diameter. These investigators used microelectrode penetrations that coursed tangential to the surface of somatosensory cortex in cats and monkeys to record short sequences of neurons that responded to light touch on the skin or more intense pressure on the skin, and related these responses to presumed differences in activation from superficial skin receptors or deep receptors in muscles and joints. These findings and interpretations led to the widespread search for similar evidence in other cortical areas for neurons that are grouped in patches of cortex extending column-like from surface to white matter, are linked by responding to a common type of input, and alternate with other patches that responded to a different type of input. This interest in modular subdivisions resulted in further evidence for the Mountcastle type of column, but also for other types of columns, all revealed by differences in neuronal activity, anatomical structure, or both. Here, it is useful to distinguish one type of column, the classical column, as columnar patches of cortex that respond to one category of input, while insulated from each other by patches of cortex that respond to a different category of input. The patches might vary in size and shape, but they should involve neurons across cortical layers, and repeat within a cortical area many times. Typically, two or more types of such a column would represent small parts of the same receptor surface, but respond differently due to activation by separate classes of peripheral receptors, or by afferents from functionally different classes of computational outcomes in central nervous system processing.

While some investigators continued to characterize neurons in primary somatosensory cortex of cats and monkeys as responsive to deep or superficial receptors on the body, evidence from other studies suggest that somatosensory cortex is activated throughout by superficial cutaneous receptor inputs, and that alternating columns related to either deep or superficial receptors do not exist in primary somatosensory cortex (e.g. Dykes et al. 1980; Favorov and Diamond 1990; Pons et al. 1987; Sur et al. 1981; Krubitzer et al. 2004). The reason for this discrepancy is unclear, but one possibility is that varying anesthetic levels raised or lowered stimulus threshold levels, leading to cutaneous activation that required more intense stimulation, which was classified as related to deep receptors. It now appears that deep receptor inputs from muscle spindles and joint receptors are relayed over a separate pathway to the ventroposterior superior nucleus of the thalamus to areas 3a and 2 of somatosensory cortex of primates (Cusick et al. 1985; Kaas 2011). Alternations of deep receptor and cutaneous receptor activation territories have been reported for both areas 3a and 2 (McKenna et al. 1982). However, most neurons in both areas are likely activated by proprioceptive inputs from the thalamus and tactile inputs from area 3b and 1 (Pons et al. 1985; Krubitzer et al. 2004).

Nevertheless, another type of modular organization has been proposed for primary somatosensory cortex, area 3b, of primates. Inputs from the skin involve two major types of afferents, the slowly adapting (SA) type I and the rapidly adapting (RA) type I afferents, related to Merkel's disk and Meissner corpuscle receptors, respectively. These two types of afferents activate separate groups of neurons in the dorsal column nuclei and the ventroposterior nucleus (Kaas 2011), although the modular organization of these groups at each level is yet unknown. However, in area 3b of monkeys, separate groups of cortical layers and neurons appear to be activated by either SA or RA inputs, thus involving columns of neurons that process either SA or RA information (Sur et al. 1981, 1984; Friedman et al. 2004). As these alternating types of columns (Fig. 4.2) represent overlapping distributions of SA and RA receptors in the skin, and are differently responsive to tactile stimulation, they can be regarded as columns of the classical type. Whether they exist or not in primary somatosensory cortex of all primates, or any nonprimate mammals, is yet uncertain.

The barrels of somatosensory cortex in some rodents may be considered as another example of classical columns. The 'barrel' is an unusual term given to one unit of a set of repeated structures, approximately shaped as such, in the primary somatosensory cortex of rats, mice, and a few other rodents (Woolsey and Van der Loos 1970) and some marsupials (Waite et al. 1998). Each barrel can be identified by a dense packing of neurons and higher levels of the metabolic enzyme, cytochrome oxidase (CO) (Fig. 4.3), and represents or is isomorphic with a single whisker on the face, such that neurons in a barrel are driven mainly by the movement of its associated whisker, although adjacent whisker movements may reduce the overall response (see Ebner and Kaas 2015). By themselves, barrels would not constitute a classical column as the same types of input activate each barrel, they simply derive from different whiskers. However, the barrels are separated from each other by narrow septal regions, which get different inputs from the thalamus, have different cortical connections, and contain neurons that have different response properties. For example, neurons in barrels are best activated by touch on a single whisker, while neurons in septa are responsive to many whiskers. Thus, the whisker region of the face is represented once within the barrels and again within the septal regions around the barrels. While the barrels and septal regions are of different proportions, neuron packing densities, and shapes, they do constitute parts of separate, but interacting, processing streams for information from the large sensory whiskers of the face (Alloway 2008). However, barrels vary in size and number of neurons according to the whisker they represent (Meyer et al. 2013).

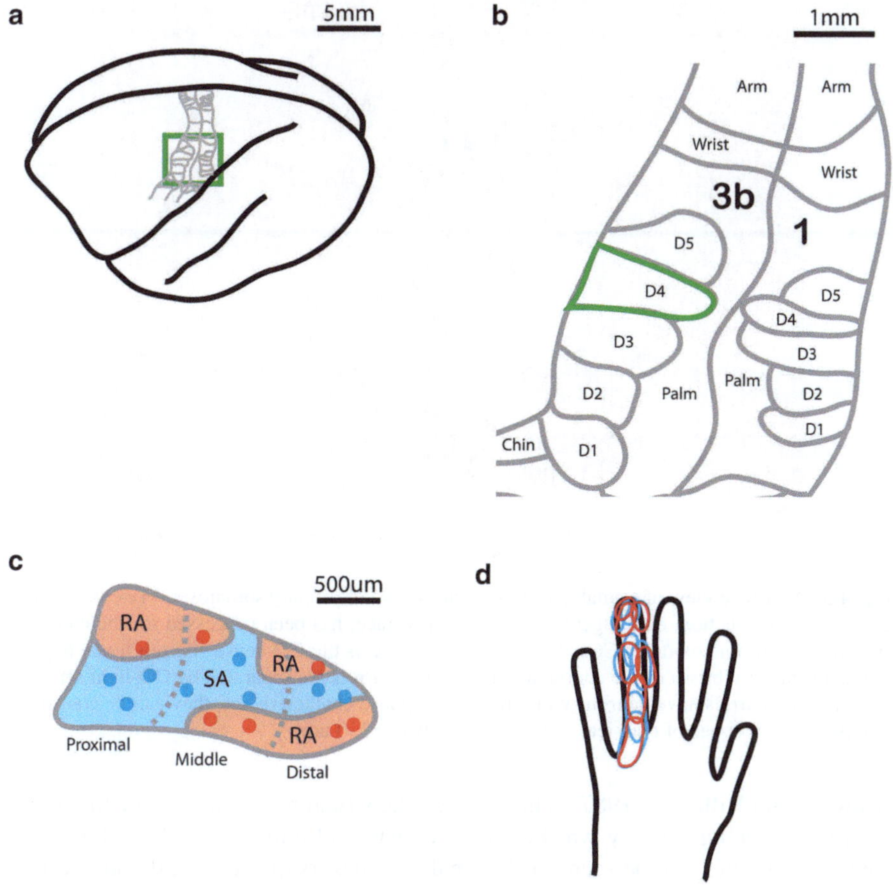

Fig. 4.2 Evidence for classical columns in somatosensory cortex of owl monkeys. (**a**). A dorsolateral view of the owl monkey brain with the primary somatosensory area, *S1* proper, which contains areas *3b* and *1*, outlined. The region within the square is shown at higher magnification below. (**b**). The somatotopic organization of areas *3b* and *1* in somatosensory cortex of owl monkeys. In the hand representation of area *3b*, the digits are represented from the thumb (*D1*) to the little finger (*D5*) in a lateral to medial sequence, and again in area *1*. (**c**). An enlarged view of the representation of digit *4* (*D4*) in area *3b* shows the separate territories that are responsive to either rapidly adapting (*RA*) or slowly adapting (*SA*) cutaneous afferents subserving digit *4*. The *dots* indicate the locations of electrode penetrations that sampled neurons in the cortex. (**d**). The glabrous surface of the monkey hand with the locations of receptive fields from *SA* or *RA* neurons recorded within the cortical representation of *D4* in area *3b*. Note that the digit is represented twice, once in *SA* territory and once in *RA* territory. Based on the results of Sur et al. 1981, 1984

Thus, they do not form an array of identical modules, but are variable and computationally specialized for the different functions of whiskers according to position in the face. In addition to the well-known barrels, other parts of the body are represented in isolated patches of cortex in primary somatosensory cortex of rats

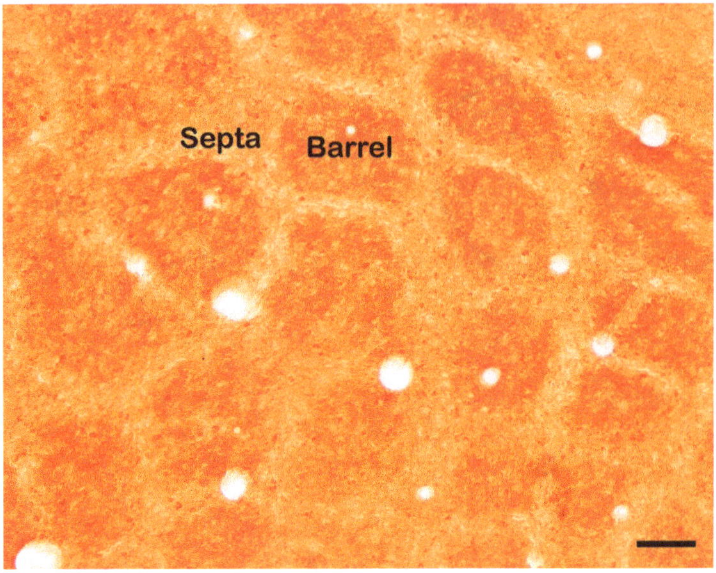

Fig. 4.3 A surface view of a small part of the barrel field of primary somatosensory cortex, S1, of a rat. The brain section, cut tangential to the brain surface, has been processed for the metabolic enzyme cytochrome oxidase (CO). This preparation reveals the CO-dense "barrels", one for each of the mystacial vibrissa of the contralateral face. The barrels are separated by CO-light septa. As the septa and barrels have neurons with different response characteristics, they can be considered to represent two types of classical columns. Scale bar is 250 um

(Dawson and Killackey 1987). Thus, barrel-like structures are histologically apparent for other long sensory whiskers on the head and other parts of the body, and smaller structures can be seen for the smaller whiskers of the buccal pad. Narrow, cell poor septa separate these "barrels", and other groups of neurons that represent digits and glabrous pads of the forepaw and hindpaw. Whether these narrow septal regions constitute a second set of processing modules is uncertain, and there is no evidence that the cortex representing the fur bearing skin of the body is subdivided by septa as well.

It is important to recognize that narrow, cell-poor septal regions separating the representations of distant body parts are commonly observed in primary somatosensory cortex of mammals, and sometimes in other cortical areas as well (Kaas and Catania 2002). Most notably, the unusual star nosed mole, with 11 fleshy appendages extending from the face around the nostril, has the representation of each appendage or ray separated from those for adjacent rays in primary somatosensory cortex by a narrow septum. In a similar manner, the re-representation of rays in the second somatosensory area S2 is divided by septa as well (Catania and Kaas 1995). Even the digits of the forepaw are partly separated by septa in S2. These narrow septa have few neurons, but are favored by corpus callosum connections. Thus, septal regions could be considered as a second set of modules. Finally, similarly narrow septa separate the representation of digits in the primary somatosensory cortex of monkeys (Jain et al. 1998; Qi et al. 2011; Liao et al. 2013), and other patches of cortex for the teeth and for the tongue are separated by cell-poor regions as well (Kaas et al. 2006; Cerkevich et al. 2013).

A third type of modular organization exists in the large amount of cortex devoted to the representation of the bill of the duck-billed platypus (Krubitzer 1995). The ancestors of the aquatic platypus evolved a new receptor on the bill, one extremely sensitive to weak electric currents. Thus, electroreception became one of the sensory inputs from the skin of the bill, allowing them to detect the weak currents generated by muscle contractions of their underwater prey. As there wasn't a system already present for electroreception when electroreceptors evolved, this new modality was mediated by the somatosensory system, although in a segregated pathway. Thus, information from the electroreceptors was segregated from tactile inputs in the trigeminal complex, the ventroposterior nucleus, and in somatosensory cortex (Krubitzer 1995). In the huge representation of the bill, occupying as much as one third of neocortex, interdigitated bands of S1 respond to touch and electroreception. Although the shapes of the modules are band-like, rather than columnar, they conform to the definition of classical columns.

Visual cortex also has versions of classical columns. As a well known example, the subdivisions of layer 3 in primary visual cortex of primates into small cytochrome oxidase (CO)- dense patches surrounded by CO-light cortex is well known (Casagrande and Kaas 1994). These patches are found in all primates (Preuss and Kaas 1996), but not in the close relations of primates (tree shrews, rodents, and lagomorphs). Surprisingly, similar structures evolved independently in cats and ferrets (Wong-Riley 1979; Murphy et al. 1995; White et al. 1999). In primates, the dark CO-dense puffs (Carroll and Wong-riley 1984) or blobs (Horton and Hubel 1981) in layer 3 are a reflection of dense, focused terminations of inputs from the koniocellular layers of the lateral geniculate nucleus, the so-called third visual pathway that, in diurnal primates, is dominated by chromatic signals from S (blue) cones (Shostak et al. 2002). Hence, neurons in blob columns are especially involved in color vision, while neurons in the interblob regions are more involved in form vision. A problem for this general interpretation is that nocturnal owl monkeys and prosimian galagos, without blue cones or color vision, have blobs. Thus, it may be more correct to conclude that blobs are part of a pathway mediating contrast, whether color or brightness (Allman and Zucker 1990). This may also be the case for the independently evolved "W-cell" pathway to blobs in cats and ferrets (LeVay and Gilbert 1976), which lack color vision as well (Van Hooser 2007).

In primates, the layer 3 blobs (Fig. 4.3) are not only CO-dense, but can be distinguished as myelin-dense patches in the otherwise even distribution of myelinated fibers in layer 3 (Rockoff et al. 2014). Additionally, blobs correspond to patches of dense labeling of the neurotransmitter transporter VGLUT2, which is found in the terminations of lateral geniculate axons. Inputs from the magnocellular and parvocellular geniculate layers form a continuous distribution in layer 4, and inputs from the koniocellular geniculate layers form the blob pattern in layer 3 (Wong and Kaas 2010; Balaram et al. 2011; Garcia-Marin et al. 2013; Bryant et al. 2012). Because of their vertical connections within cortex, the blob columns

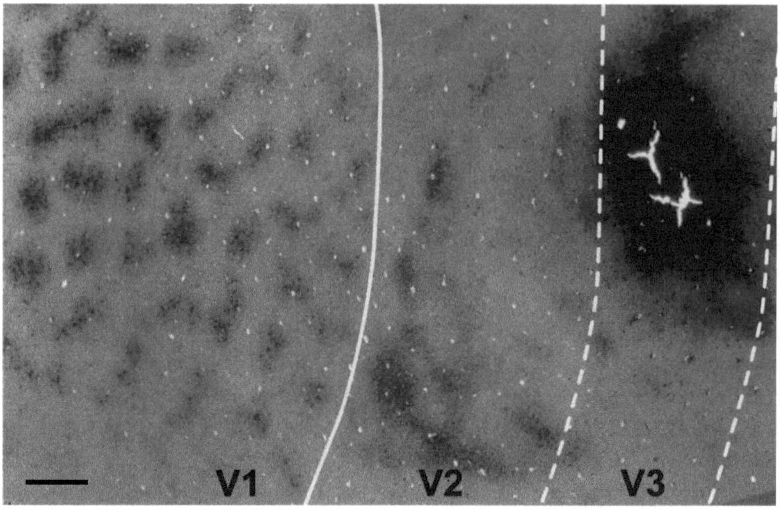

Fig. 4.4 An example of a pattern of cortical connections that provides evidence for a modular organization of primary visual cortex, or *V1*. The injection of a retrograde and anterograde neural tracer, cholera toxin subunit B, CTB, into dorsal *V3* (*solid dark oval*) of an owl monkey labeled a regular array of patches of cortex in primary visual cortex, *V1*, and a less uniform array in the second visual area, *V2*. Cortex was flattened, cut, and processed parallel to the pial layer to provide a "surface" view of the three cortical visual areas and their regions of label. The patches in *V1* likely correspond to the cytochrome oxidase (CO) blobs of *V1*, while the patches in *V2* likely mark the CO-dark thin stripes. The selective labeling of blobs in *V1* suggests that *V3* has a modular organization, with one set of modules connecting to the *V1* blobs. Similar results have been published by Krubitzer and Kaas 1993; Beck and Kaas 1998; and Angelucci et al. 2002

and the surrounding interblob columns involve all cortical layers. As functional components of partly segregated systems, blobs and interblobs have differences in inputs, connections within V1, callosal connections, and connections with extrastriate cortex (Casagrande and Kaas 1994). Thus, the blobs and interblobs of V1 can be considered as alternating classes of classical columns (Fig. 4.4).

Classical columns are also found in other visual areas of primates. For example, the second visual area, V2, of most primates is characterized by a repeating series of CO-dense thin stripes and a similar series of CO-dense thick stripes that are separated by CO-pale interstripes (Fig. 4.5) (Casagrande and Kaas 1994). Thus, there are twice as many CO-pale interstripes as CO-dense thin and thick stripes. Overall, there are about 30 sets of stripes in V2 of macaque monkeys (Olavarria and Van Essen 1997) and likely more in V2 of humans. Early studies focused on deriving the connections and functional properties of thin, thick, and pale stripes, assuming three types of modules (DeYoe and Van Essen 1985; Livingstone and Hubel 1984, 1987; Shipp and Zeki 1985; Sincich and Horton 2002). More recent evidence indicates that the interstripes are of two types, differing in connections from V1 (Federer et al. 2013), and thereby having different functional properties (Xu et al. 2004). As the visual hemifield would need to be separately represented in each of the four sets of stripes, V2 contains four fragmented visual representations (Roe and T'so 1995), with a global visuotopic organization of the lower visual quadrant through the upper visual quadrant, arranged mediolaterally across the cortical surface. Thus, a set of four adjacent stripes – a CO-dense thick stripe, a CO-dense thin stripe, and two CO-pale interstripes – would combine to form a single hypercolumn of the sort described by Hubel and Wiesel (1972, 1977). Optical imaging in V2 of owl monkeys (Xu et al. 2004) revealed that the medial interstripes in each set are highly selective for stimulus orientation, while the lateral interstripes in each set are not. The CO-dense thick stripes, also sensitive to stimulus orientation, receive inputs from layer 3C of the magnocellular pathway in V1, and project to the orientation sensitive middle temporal area (MT), while the CO-dense thin stripes are sensitive to luminance changes (Kaskan et al. 2009) and to color (Hubel and Livingstone 1987) in those primates with color vision. Although stripe shaped, and crossing the width of V2, the four functionally and anatomically different stripes of V2 qualify as classical columns.

Less is known about the understudied area V3 (Kaas and Lyon 2001). However, optical imaging experiments in owl monkeys, where much of dorsal V3 is exposed on the brain surface, have revealed the existence of a pattern of orientation selective stripes across V3 that are wider than those in V2 (Xu et al. 2004; Kaskan et al. 2009). Parts of these orientation stripes were sensitive to luminance change or sensitive to binocular disparity. In addition, a band-like or patchy pattern of CO-dense cortex has been observed in V3. Connection patterns also suggest the existence of modular subregions with inputs from blob or interblob divisions of V1 (Fig. 4.4). Thus, functionally distinct classical columns appear to exist in V3.

The middle temporal visual area, MT (Fig. 4.1), is a widely recognized visual area of primates that is easily distinguished by its systematic representation of the contralateral visual field and dense myelination (Allman and Kaas 1971). Optical imaging experiments in owl monkeys and prosimian galagos, where MT is exposed on the brain surface and available for imaging, reveal that MT has orientation sensitive neurons throughout, and these are grouped according to their preferred stimulus orientation (Malonek et al. 1994; Xu et al. 2004, 2006). In addition, patches of neurons that are selective for orientation are subdivided into smaller regions that are selective for motion, either in one direction orthogonal to the orientation preference or the opposite direction orthogonal to the orientation preference (Kaskan et al. 2010). This evidence suggests that MT has two types of alternating modules, where orientation selective neurons preferring movement in one direction are grouped together and adjacent to modules preferring the opposite direction of movement. Other evidence is limited (Born and Bradley 2005), but the connection pattern of MT with the adjoining dorsal division of FST is patchy (Kaas and Morel 1993), suggesting a pattern of alternating patches of cortex, with one set of patches projecting to FST and the other not. Similarly, the belt-like MTc area surrounding much of MT is characterized by a series of CO-dense patches of cortex separated by CO-light cortex, and also contains patchy connections with the ventral division of FST (Kaas and Morel 1993). Thus, MTc appears to be subdivided by

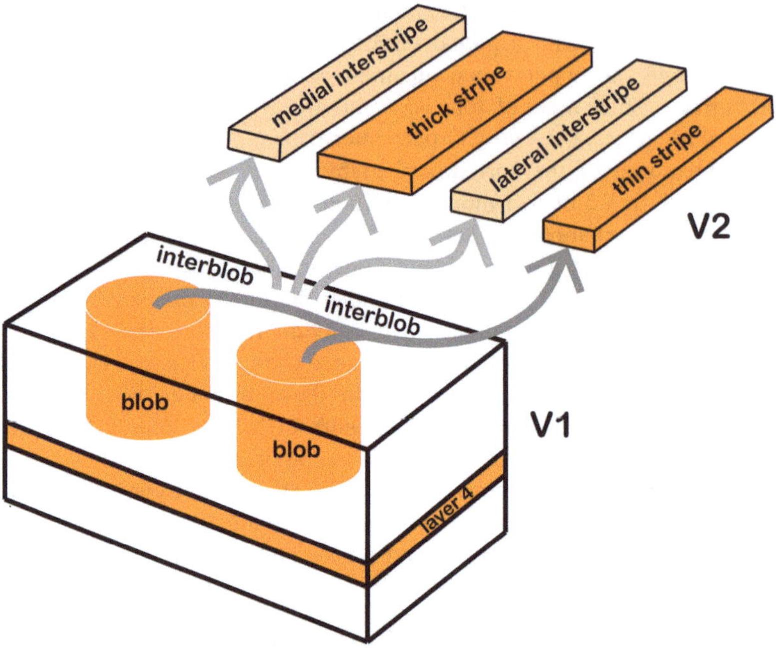

Fig. 4.5 The blob and interblob modules of *V1* in primates differ in functional properties, thalamic inputs, and (as shown here) projections to band-like modular subdivisions of *V2*. The projections from *V1* blobs carry information abut stimulus contrast and color to *V2* thin stripes, while projections from *V1* interblob regions carry information about stimulus orientation from magnocellular and predominantly parvocellular geniculate inputs to the thick stripes and interstripes, respectively. See text for details

alternating patches and interpatches of cortex that are likely functionally distinct, and can be classified as classical columns.

Another well studied visual area, the dorsolateral (DL) visual area (Allman and Kaas 1974) or V4 (Zeki 1973) is located between V3 and MTc (Fig. 4.1). DL/V4 represents the contralateral lower to upper visual hemifield in a dorsoventral cortical sequence (Allman and Kaas 1974; Gattas et al. 1988), and receives a matching visuotopic pattern of input from V2 (Stepniewska et al. 2005a). Recent optical imaging results from macaques provide evidence that modular regions in V4 alternate to represent either stimulus color or orientation (Tanigawa et al. 2010) and color regions are organized to represent dimensions of hue, lightness, and saturation (Li et al. 2014). The color and orientation modules appear to fulfill the requirements of classical columns, while the complex arrangement of hue, lightness, and saturation representations deserves further study. In addition, the possible arrangement of separate modules for faces or objects in IT of temporal cortex (Fig. 4.1) in macaques suggests that this cortex may have classical columns as well (Tsao et al. 2003; Moeller et al. 2008; Tsao et al. 2008).

In conclusion, as more is learned about the internal organizations of cortical areas, evidence for classical columns has accumulated. Most is now known about the well studied visual areas, and very little about auditory areas, although bands in primary auditory cortex of cats that are excited by both ears (E-E bands) or excited by one and inhibited by the other (E-I bands) have been described (Imig and Adrian 1977; Middlebrooks et al. 1980). The patchy cortical connections from single locations in frontal cortex (Goldman and Nauta 1977) and elsewhere, as well as the modular structure of some cortical regions (Manger et al. 1998), suggest that many more examples of classical columns will emerge, and that they are present in many cortical areas.

4.4 Columns That Spatially Represent a Stimulus Dimension – Unbounded Columns

A number of visual areas, V1, V2, V3, MT, and possibly DM, have regions of cortex that systematically represent stimulus orientation for a given region of visual space. Similar adjoining arrays represent adjoining visual spaces. The hypercolumn of Hubel and Wiesel (1977) is a well-known attempt to portray the continuous change of orientation preference that is revealed by electrode penetrations that cross the vertical arrays of neurons in cortex. In this well known method, the "boundaries" between vertical arrays of neurons with slightly different orientation preferences reveal progressive changes in preferred orientation until the full set of 180° of preferred orientation occurs for stimuli in a given place in the visual field. Other such arrays represent stimulus orientation for other spaces in the visual field, resulting in a large number of hypercolumns for V1 of larger primates. Because the changes in orientation preference across cortex appear to be largely gradual and continuous, the illustrated borders of the so-called orientation columns are arbitrary, and they are usually portrayed as about the width of a minicolumn. If they are to be considered columns, they need to be defined in a way that makes their borders less subjective. As it is, the portrayed orientation columns do not conform to the definition of a classical column, because they are not separated by columns with another class of inputs. However, a full set of orientation columns has at least some definable boundaries. One full set of orientation selective regions constitutes a type of column known as a pinwheel, where orientation preferences continually change through 180° around a point or singularity in cortex (Blasdel 1992). Thus, orientation preferences change sharply across that critical point. Other sorts of discontinuities occur in orientation maps, and these help define the borders of orientation "hypercolumns", but without some type of separating column with different inputs, orientation columns may not exist as classical columns, but only as a continuously mapped response feature of cortical neurons, much like visual space or auditory tones would be represented in areas without classical columns.

Interestingly, while orientation hypercolumns and the systematic representation of orientation is a feature of V1 and other cortical areas in all primates, and in V1 of tree shrews (Bosking et al. 1997), which are a close relative of primates, they do not exist in V1 of rodents (Van Hooser et al. 2005), which are only slightly more distant relatives of primates (Kaas 2012a, b). However, orientation selective columns have been independently evolved (see Kaschube et al. 2010) in V1 and V2 of carnivores, including cats (Bonhoeffer and Grinvald 1991; Crair et al. 1997) and ferrets (Rao et al. 1997; White et al. 2001).

4.5 Ocular Dominance Columns and Other Modules Representing Separated Parts of Sensory Surfaces

Sensory surfaces may be discontinuous or have disruptions that can be reflected in cortical representations. For example, most mammals relay separate representations of the binocular visual hemifield for each eye from the dorsal lateral geniculate nucleus (Kaas et al. 1972) to a visuotopic representation in primary visual cortex, but they do this in different ways. In some mammals, the relay of the segregated right or left eye inputs to primary visual cortex is often merged in layer 4 of V1, while being distinct at the single neuron level, or at the sublaminar level as in tree shrews (Fitzpatrick 1996). Alternatively, the inputs from one eye or the other are often segregated in a patch and surround, or a band-like pattern in layer 4 of V1 (Fig. 4.6), and each input may be the dominant driving source of activity in other layers, giving rise to the concept of ocular dominance columns (Hubel and Wiesel 1968). Present evidence suggests that some form of patchy to band-like segregation of neuron groups by ocular dominance is common to many mammals, including carnivores (Löwel and Singer 1987; Anderson et al. 1988; Issa et al. 1999), but especially in primates. However, the form of ocular dominance columns is highly variable, even within V1, and sometimes across individuals of a species (Horton and Hocking 1996; Horton and Adams 2005). Very pronounced ocular dominance columns are revealed by the sharp segregation of geniculate inputs related to one eye or the other in layer 4 of V1 in Old World monkeys, apes and humans (Horton and Hedle-Whyte 1984; see Florence and Kaas 1992, for review). For most of V1, these segregated inputs take the form of meandering bands for right or left eye inputs that merge and part in various ways. In this large part of V1, the inputs from each eye are equal, but in the part of V1 devoted to peripheral vision, the input from the ipsilateral eye is reduced and this results in a dot-like pattern of ipsilateral eye dominance surrounded by a continuous field of contralateral eye dominance (Tanaka 1991). Of course, these segregated inputs from each eye selectively activate the neurons in bands of cells immediately above and below them, so the ocular dominance bands extend from white matter to cortical surface. This is clearly revealed by similar patterns of expression of activity-dependent genes when input from one eye is blocked with tetrodotoxin (TTX; Takahata et al. 2009). The

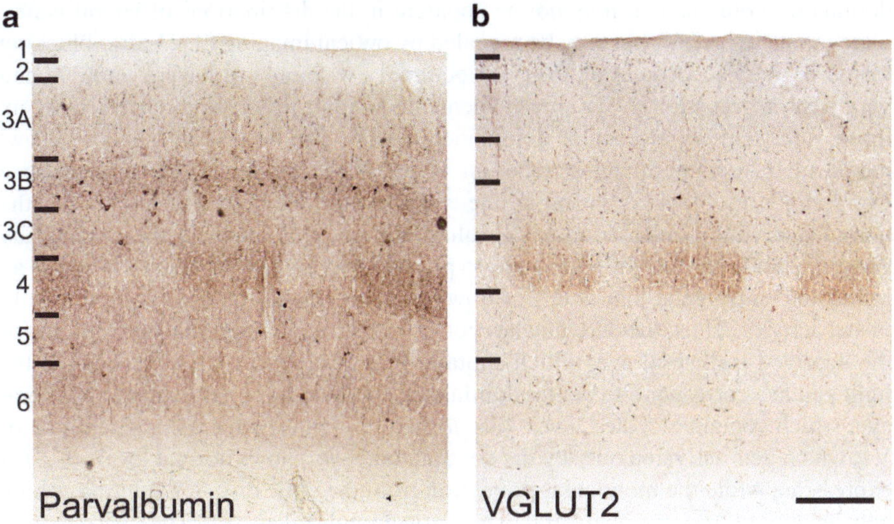

Fig. 4.6 Ocular dominance territories are revealed in V1 of a baboon with a longstanding visual deficit in one eye, via immunolabeling for parvalbumin and vesicular glutamate transporter 2 (*VGLUT2*). Both markers are robustly expressed in the terminals of LGN neurons that synapse in layer 4 of V1. Thus, the dark parvalbumin- and *VGLUT2*-positive patches in layer 4 correspond to active LGN terminations from the intact eye, while the light patches correspond to inactive terminations from the deprived eye. Scale bar is 1 mm

alteration in gene expression also reveals a previously undetected feature of ocular dominance columns. In layer 4, neurons at the border of an activated column adjacent to a deactivated column form a "border strip" where neurons are activated at a higher level than neurons elsewhere in the activated column. This is because they receive most or all of their activating input from the intact eye, while roughly half of their inhibitory surround inputs are from the deactivated eye. For most neurons along the border of a column, inhibition comes from the adjoining neurons of the activated and the deactivated column, while neurons in central locations in a column receive most of their inhibition from within the column itself (Takahata et al. 2009).

In New World monkeys and prosimian primates, the expression of ocular dominance columns is highly variable. Some of the larger New World monkeys, such as cebus and spider monkeys, have ocular dominance columns that are just as pronounced, and of the same band-like form, as in Old World monkeys (see Florence and Kaas 1992 for review). In contrast, New World squirrel monkeys have an individually variable pattern, ranging from ocular dominance patches of various sizes to a salt and pepper representation with no obvious dominance across a group of neurons (Livingstone 1996; Adams and Horton 2003). Some marmosets may have highly overlapping distributions of inputs from the two eyes in V1 as adults, but partly separated inputs early in development or after rearing with monocular deprivation, while other mature marmosets have patent ocular dominance columns that may not be apparent in the distributions of lateral geniculate inputs to layer 4, but may be revealed by optical imaging of V1 after blocking vision in one eye (Roe et al. 2005). Nocturnal owl monkeys have a variable and very weak segregation of eye specific geniculate inputs to V1 (Kaas et al. 1976), but these otherwise cryptic ocular dominance patches can be revealed in V1 of owl monkeys by optical imaging (Kaskan et al. 2007) or activity-dependent gene expression after blocking vision in one eye (Takahata et al. 2014). In Fig. 4.7, the normally cryptic ocular dominance columns in an owl monkey are revealed by visually depriving one eye for a short period (1–3 h) and labeling for *c-fos*, an immediate early gene that is up- or downregulated by V1 neurons in response to visual activity. Thus, the dark patches correspond to ocular dominance columns for the deprived ipsilateral eye, which contain little or no *c-fos* expression, while the light patches correspond to ocular dominance columns for the normal contralateral eye, which contain elevated levels of *c-fos* expression. The monocular segment of V1, which gets activated only by the contralateral eye, shows normal levels of *c-fos* expression, while the monocular projection from the optic disc of the retina, which gets input only from the ipsilateral eye, is downregulated as well. The expression of activity-dependent genes reveals that owl monkeys, as for macaques, have border-strip neurons with higher activity along the layer 4 margins of activated columns with deactivated columns. In cats, the patchy ocular dominance columns extend into V2, and this may also occur in owl monkeys (Takahata et al. 2014). In cats, callosal connections preferentially originate in ocular dominance columns for the ipsilateral eye, except for the border zone between V1 and V2, where they originate from ocular dominance columns from the contralateral eye instead (Olavarria 2001).

In summary, the existence of ocular dominance columns in V1 is not likely to be universal across mammals, but likely more common than presently understood, as the optical imaging and gene expression approaches are more sensitive than anatomical methods for revealing ocular dominance differences. The patterns of neurons favored by one eye or the other vary across species, and sometimes within species. The most pronounced columns are band-like rather than patch-surround, and this band-like pattern has evolved independently in Old World monkeys and some New World monkeys. The ocular dominance columns are not classical columns, however, because their modular alternation is by retina of origin rather than by functional class of input.

In the somatosensory cortex of many mammals, parts of the receptor surface, the skin, are represented in discrete blocks of tissue separated by narrow cell-poor septal regions. The "barrels" in the cortex of mice and rats are a well-known example (Fig. 4.2) (Woolsey and Van der Loos 1970). Other examples include the cortical representation of the rays of the nose in star nosed moles (Catania and Kaas 1995), and the modular representation of the digits of the forepaw in moles (Catania and Kaas 1995), raccoons (Welker and Seidenstein 1959) and monkeys (Jain et al. 1998; Qi and Kaas 2004; Liao et al. 2013). These discrete clusters and bands of neurons for a particular input are separated from each other, and isolated from other similar inputs, by narrow-cell poor septa that occasionally contain

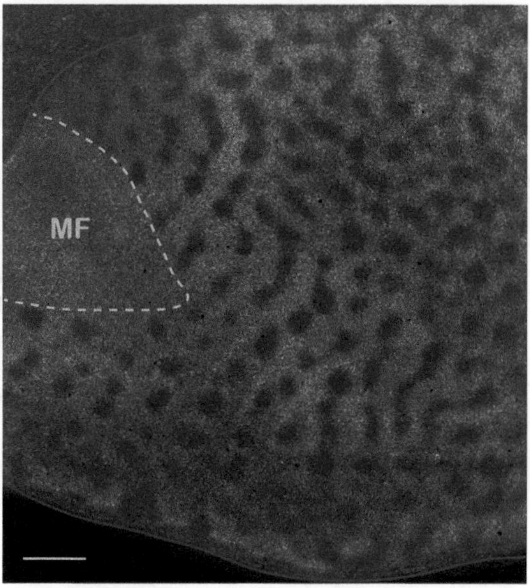

Fig. 4.7 A surface view of part of V1 in an owl monkey reveals the ocular dominance columns related to the ipsilateral eye (*dark patches*) and the contralateral eye (*light patches*), after visually depriving the ipsilateral eye for 1–3 h. Note the dominance of the contralateral eye in terms of cortical territory, and the wedge of continuous contralateral eye territory corresponding to the monocular visual field (*MF*) of the contralateral eye. Scale bar is 1 mm

enough neurons to form another type of processing module, as in the case of the barrel and septal systems in somatosensory cortex of rodents (Woolsey and Van der Loos 1970). In other cases, it may be better to consider the dividing septa as an extracellular isolating mechanism which separates groups of cells that have different activation patterns because they receive inputs from separated regions of the receptor sheet (Kaas and Catania 2002).

4.6 Larger Functional Divisions of Cortical Areas

Larger cortical areas are sometimes divided into functionally specialized subregions, or domains, that do not repeat across cortex and are of several unique types. The clearest example of this type of module comes from primary motor cortex of primates. This area has long been known to contain an overall representation of the contralateral body (e.g. Leyton and Sherrington 1917), in that brief sequences of electrical pulses via surface electrodes evoke movements from the foot, trunk, forearm, hand, face and tongue in a mediolateral sequence across precentral cortex, the location of the primary area M1. However, when the representation was considered in more detail, with microelectrode stimulation sites in the deeper output layers of M1, the fine-grain analysis revealed a mosaic or fractured organization, where any type of movement evoked from a specific site could also be evoked from other sites, which were separated by sites for different movements altogether (Gould et al. 1986; for review see Schieber 2001). Thus, clusters of sites, or sets, included those for different digit movements, as well as those for wrist, elbow, and shoulder movements, and other sets had clusters that evoked many of the same movements (Fig. 4.8). A possible logic for this repetition appeared when Graziano and coworkers (e.g. Graziano et al. 2002, 2005) reported that longer trains of electrical pulses, of about 0.5 s each, evoked purposeful sequences of movements from different parts of the forelimb-hand region of M1, such as reach to grasp, hand to mouth, and arm lifts to protect the head. In this way, M1 was subdivided into sets of putative minicolumns, which together could mediate a complex movement pattern that corresponded to motor primitives (Flash and Hochner 2005; Mussa-Ivaldi and Bizzi 2000) or ethologically relevant behaviors (Graziano 2010). We have found such functional divisions in M1 and in premotor cortex of New World monkeys, macaques, and prosimian galagos (Stepniewska et al. 2005b; Gharbawie et al. 2011a, b; Stepniewska et al. 2014; Kaas et al. 2011). Other groups are currently contributing to the evidence that M1 is subdivided into functional domains (e.g. Adelsberger et al. 2014; Cooke et al. 2012) Our experiments and those of Cooke and colleagues (2003) indicate that similar domains that are functionally matched to the premotor and M1 domains exist in posterior parietal

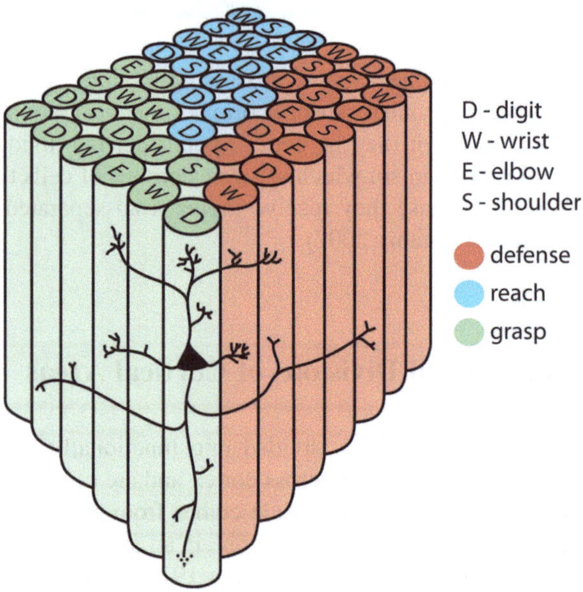

Fig. 4.8 The proposed relationship of minicolumn arrangements to the larger functional divisions we call domains in the hand-forearm segment of primary motor cortex of monkeys. Brief trains of electrical pulses at near threshold levels via microelectrodes at any specific location, that is in one of the minicolumn-like modules, produces a brief, single movement of a digit (*D*), wrist (*W*), elbow (*E*), or shoulder (*S*), in a somatotopy that is fractured to produce a mosaic of radial columns of neurons related to one movement or another. The movements are locally mixed. We suggest that the columns relate to each other via intrinsic connections within domains to mediate more complex movements over short periods (0.5 s) of time that differ according to the domain (grasp, reach, and defense shown here). Connections between domains allow them to interact to prevent a competing complex movement, or a combination of complex movements

cortex (PPC), in the region of the intraparietal sulcus. Thus, domains in PPC activate matching domains in premotor and motor cortex, with M1 providing the critical cortical output for motor commands (Stepniewska et al. 2014).

As another example, the visual cortex of the junction of occipital with temporal cortex and the adjoining inferior temporal cortex (IT in Fig. 4.1) of monkeys and humans is known to contain a mosaic of functionally specialized regions, or domains, that are activated by viewing faces, objects, bodies, or scenes (Grill-Spector and Malach 2004; Tsao et al. 2008; Tanaka 2003; Pinsk et al. 2005; Haxby et al. 2000). Thus, domain-like regions in temporal cortex for different, visually complex stimuli separate each other in temporal cortex, and such domains may repeat. For example, as many as six face selective domains exist in the temporal cortex of monkeys (Freiwald and Tsao 2010). The domains are interconnected, and seem to represent successive steps in the processing of face information, from preferring specific views of faces to being responsive across viewpoints. Other areas of frontal and prefrontal cortex might also be subdivided into domains specialized for complex functions as well (Goldman and Nauta 1977; Nelissen et al. 2005).

4.7 How Do Columns Emerge in Brain Development, and Are They Functionally Important?

The prevailing opinion is that columns of the classical type, as well as those related to disjunctions in the receptor sheet, are the result of correlation-based competition for synaptic space (e.g. Kaas 1990; Kaas and Catania 2002; Katz and Crowley 2002; Shatz 1996; Erzurumlu and Kind 2001; Purves et al. 1992). In regard to the cortical barrel field of rats and mice, Van der Loos and Dörfl (1978) asked the question "Does the skin tell the somatosensory cortex how to construct a map of the periphery?", and the answer is clearly yes. However, the activity-dependent molecular mechanisms of synapse formation and selection are under genetic control, as mutations that alter aspects of sensory signaling can alter or abolish the formation of barrels in the cortex of rats (Erzurumlu and Kind 2001). In addition, experimentally induced activity-dependent competition can induce the formation of columns in structures where they never normally exist (Merlin et al. 2013). For example, the surgical transplantation of an extra eye on the head of a developing frog introduces a competition for synaptic space between inputs from the added eye and the normal eye in the contralateral optic tectum. This results in the inputs dividing up the tectum into a series of alternating bands, or ocular dominance columns, for one eye or the other (Constantine-Paton and Law 1978). Thus, the potential for columns always existed, but obviously for other functions than forming columns. Such observations on the variability of columns across species, their absence or modification after some mutations, and their formation after experimental manipulations, all raise questions about the functional importance of columns. As one example, the great variability in ocular dominance columns, from highly distinct bands to barely visible patches, across individual squirrel monkeys suggests no more than weak selection for the genetic control of ocular column structure in this primate (Krubitzer and Kaas 2005). Such reasoning has led some investigators to suggest that cortical columns only reflect developmental mechanisms and provide no special advantage in neural processing (Purves et al. 1992; Horton and Adams 2005). Others have suggested that columns provide an economy of connections, as neurons that interact are placed closer together (Hubel and Wiesel 1963). According to this view, cortical processing would require longer and metabolically more costly connections in the well-developed primary visual cortex of squirrels, as they do not have orientation selective pinwheels (Van Hooser et al. 2005), than in tree shrews with a similarly developed V1 but with orientation selective pinwheels (Bosking et al. 1997). However, the processing of stimulus orientation information seems to be adequate in both mammals. Yet, the observations that squirrels and tree shrews appear to have similar visual abilities, or that mutant rats without cortical barrels still process somatosensory information does not argue that the orientation pinwheels or barrels provide no benefit, whether sensory or metabolic.

While it does seem possible that the separation of inputs from disjunctive body parts such as whiskers and digits into bands or barrels of cells separated by narrow cell-poor septa has no function other than to isolate groups of neurons with different inputs, such segregations may have provided opportunities for classical columns to emerge. Structures that emerge for one purpose in evolution are often repurposed for something else (Gould and Lewontin 1979). Thus, the narrow, cell-poor separating septa in a sensory representation could be invaded by late-developing cortical inputs, and thereby transform septal regions into a distinct processing unit, as appears to have happened with the barrel and septal subsystems in somatosensory cortex of rats (Alloway 2008).

References

Adams DL, Horton JC (2003) Capricious expression of cortical columns in the primate brain. Nat Neurosci 6:113–114. PMID 12536211

Adelsberger H, Zainos A, Alvarez M et al (2014) Local domains of motor cortical activity revealed by fiber-optic calcium recordings in behaving nonhuman primates. Proc Natl Acad Sci U S A 111:463–468. PMID 24344287

Allman JM, Kaas JH (1971) A representation of the visual field in the caudal third of the middle temporal gyrus of the owl monkey (Aotus trivirgatus). Brain Res 31:85–105. PMID 4998922

Allman JM, Kaas JH (1974) A crescent-shaped cortical visual area surrounding the middle temporal area (MT) in the owl monkey (Aotus trivirgatus). Brain Res 81:199–213. PMID 4215542

Allman J, Zucker S (1990) Cytochrome oxidase and functional coding in primate striate cortex: a hypothesis. Cold Spring Harb Symp Quant Biol 55:979–982. PMID 1966774

Alloway KD (2008) Information processing streams in rodent barrel cortex: the differential functions of barrel and septal circuits. Cereb Cortex 18:979–989. PMID 17702950

Anderson PA, Olavarria J, Van Sluyters RC (1988) The overall pattern of ocular dominance bands in cat visual cortex. J Neurosci 8:2183–2200. PMID 3385494
Angelucci A, Levitt JB, Walton EJ et al (2002) Circuits for local and global signal integration in primary visual cortex. J Neurosci 22:8633–8646. PMID 12351737
Azevedo FA, Carvalho LR, Grinberg LT et al (2009) Equal numbers of neuronal and nonneuronal cells make the human brain an isometrically scaled-up primate brain. J Comp Neurol 513:532–541. PMID 19226510
Balaram P, Takahata T, Kaas JH (2011) VGLUT2 mRNA and protein expression in the visual thalamus and midbrain of prosimian galagos (Otolemur garnetti). Eye Brain 3:5–15. PMID 22984342
Beck PD, Kaas JH (1998) Cortical connections of the dorsomedial visual area in New World owl monkeys *(Aotus trivirgatus)* and squirrel monkeys *(Saimiri sciureus)*. J Comp Neurol 400:18–34. PMID 9762864
Belgard TG, Marques AC, Oliver PL et al (2011) A transcriptomic atlas of mouse neocortical layers. Neuron 71:605–616. PMID 21867878
Bernard A, Lubbers LS, Tanis KQ et al (2012) Transcriptional architecture of the primate neocortex. Neuron 73:1083–1099. PMID 22445337
Blasdel GG (1992) Orientation selectivity, preference, and continuity in monkey striate cortex. J Neurosci 12:3139–3161. PMID 1322982
Bonhoeffer T, Grinvald A (1991) Iso-orientation domains in cat visual cortex are arranged in pinwheel-like patterns. Nature 353:429–431. PMID 1896085
Born RT, Bradley DC (2005) Structure and function of visual area MT. Annu Rev Neurosci 28:157–189. PMID 16022593
Bosking WH, Zhang Y, Schofield B, Fitzpatrick D (1997) Orientation selectivity and the arrangement of horizontal connections in tree shrew striate cortex. J Neurosci 17:2112–2127. PMID 9045738
Brodmann K (1909) Vergleichende Lokalisationslehre der Gro hirnrinde. Verlag von Johann Ambrosius Barth, Leipzig
Bryant KL, Suwyn C, Reding KM et al (2012) Evidence for ape and human specializations in geniculostriate projections from VGLUT2 immunohistochemistry. Brain Behav Evol 80:210–221. PMID 22889767
Buxhoeveden DP, Casanova MF (2002) The minicolumn hypothesis in neuroscience. Brain 125:935–951. PMID 11960884
Buxhoeveden DP, Switala AE, Roy E, Casanova MF (2000) Quantitative analysis of cell columns in the cerebral cortex. J Neurosci Methods 97:7–17. PMID 10771070
Buxhoeveden DP, Switala AE, Litaker M et al (2001) Morphological differences between minicolumns in human and nonhuman primate cortex. Am J Phys Anthropol 115:361–371. PMID 11471134
Carroll EW, Wong-Riley M (1984) Quantitative light and electron microscopic analysis of cytochrome-oxidase rich zones in the striate cortex of the squirrel monkey. J Comp Neurol 1:1–17. PMID 6321561
Casagrande VA, Kaas JH (1994) The afferent, intrinsic, and efferent connections of primary visual cortex in primates. In: Peters A, Rockland KS (eds) Cerebral cortex: primary visual cortex of primates, vol 10., pp 201–259
Catania KC, Kaas JH (1995) Organization of the somatosensory cortex of the star-nosed mole. J Comp Neurol 351:549–567. PMID 7721983
Cerkevich CM, Qi HX, Kaas JH (2013) Thalamic input to representations of the teeth, tongue, and face in somatosensory area 3b of macaque monkeys. J Comp Neurol 521:3768–3790. PMID 23873330
Constantine-Paton M, Law MI (1978) Eye-specific termination bands in tecta of three-eyed frogs. Science 202:639–641. PMID 309179
Cooke DF, Taylor CS, Moore T, Graziano MS (2003) Complex movements evoked by microstimulation of the ventral intraparietal area. Proc Natl Acad Sci 100:6163–6168. PMID 12719522

Cooke DF, Padberg J, Zahner T, Krubitzer L (2012) The functional organization and cortical connections of motor cortex in squirrels. Cereb Cortex 22:1959–1978. PMID 22021916

Crair MC, Ruthazer ES, Gillespie DC, Stryker MP (1997) Ocular dominance peaks at pinwheel center singularities of the orientation map in cat visual cortex. J Neurophysiol 77:3381–3385. PMID 11135259

Cusick CG, Steindler DA, Kaas JH (1985) Corticocortical and collateral thalamocortical connections of postcentral somatosensory cortical areas in squirrel monkeys: a double-labeling study with radiolabeled wheatgerm agglutinin and wheatgerm agglutinin conjugated to horseradish peroxidase. Somatosens Res 3:1–31. PMID 4070891

Da Costa NM, Martin KA (2010) Whose cortical column would that be? Front Neuroanat. Ecollection 2010. PMID 20640245

Dawson DR, Killackey HP (1987) The organization and mutability of the forepaw and hindpaw representations in the somatosensory cortex of the neonatal rat. J Comp Neurol 256:246–256. PMID 3558880

DeYoe EA, Van Essen DC (1985) Segregation of efferent connections and receptive field properties in visual area V2 of the macaque. Nature 317:58–61. PMID 2413132

Dykes RW, Rasmusson DD, Hoeltzell PB (1980) Organization of primary somatosensory cortex in the cat. J Neurophysiol 43:1527–1546. PMID 7411175

Ebner FF, Kaas JH (2015) Somatosensory system. In: Paxinos G (ed) The rat nervous system, vol 4. Elsevier Sciences, Academic Press, Oxford, pp 673–699

Eccles JC (1981) The modular operation of the cerebral neocortex considered as the material basis of mental events. Neuroscience 6:1839–1855. PMID 7301112

Erzurumlu RS, Kind PC (2001) Neural activity: sculptor of 'barrels' in the neocortex. Trends Neurosci 24:589–595. PMID 11576673

Favorov OV, Diamond ME (1990) Demonstration of discrete place-defined columns – segregates – in the cat S1. J Comp Neurol 298:97–112. PMID 2212100

Federer F, Williams D, Ichida JM et al (2013) Two projection streams from macaque V1 to the pale cytochrome oxidase stripes of V2. J Neurosci 33:11530–11539. PMID 23843523

Fitzpatrick D (1996) The functional organization of local circuits in visual cortex: insights from the study of tree shrew striate cortex. Cereb Cortex 6:329–341. PMID 8670661

Flash T, Hochner B (2005) Motor primitives in vertebrates and invertebrates. Curr Opin Neurobiol 15:660–666. PMID 16275056

Florence SL, Kaas JH (1992) Ocular dominance columns in area 17 of Old World macaque and talapoin monkeys: complete reconstructions and quantitative analyses. Vis Neurosci 8:449–452. PMID 1375096

Freiwald WA, Tsao DY (2010) Functional compartmentalization and viewpoint generalization within the macaque face-processing system. Science 330:845–851. PMID 21051642

Friedman RM, Chen LM, Roe AW (2004) Modality maps within primate somatosensory cortex. Proc Natl Acad Sci 101:12724–12729. PMID 15308779

Garcia-Marin V, Ahmed TH, Afzal YC, Hawken MJ (2013) Distribution of the vesicular glutamate transporter 2 (VGLUT2) in the primary visual cortex of the macaque and human. J Comp Neurol 521:130–151. PMID 22684983

Gattas SR, Sousa AP, Gross CG (1988) Visuotopic organization and extent of V3 and V4 of the macaque. J Neurosci 8:1831–1845. PMID 3385477

Gharbawie OA, Stepniewska I, Kaas JH (2011a) Cortical connections of functional zones in posterior parietal cortex and frontal cortex motor regions in new world monkeys. Cereb Cortex 21:1981–2002. PMID 21263034

Gharbawie OA, Stepniewska I, Qi H, Kaas JH (2011b) Multiple parietal-frontal pathways mediate grasping in macaque monkeys. J Neurosci 31:11660–11677. PMID 21832196

Goldman PS, Nauta WJH (1977) Columnar distribution of cortico-cortical fibers in the frontal association, limbic, and motor cortex of the developing rhesus monkey. Brain Res 122:393–413. PMID 402978

Gould SJ, Lewontin RC (1979) The spandrels of San Marco and the Panglossian paradigm: a critique of the adaptationist programme. Proc R Soc Lond B Biol Sci 205:581–598. PMID 42062

Gould HJ 3rd, Cusick CG, Pons TP, Kaas JH (1986) The relationship of corpus callosum connections to electrical stimulation maps of motor, supplementary motor, and the frontal eye fields in owl monkeys. J Comp Neurol 247:297–325. PMID 3722441

Graziano MSA (2010) Ethologically relevant movements mapped on the motor cortex. In: Platt ML, Ghazantos AA (eds) Primate neuroethology. Oxford University Press, Oxford, pp 454–470

Graziano MS, Taylor CS, Moore T (2002) Complex movements evoked by microstimulation of precentral cortex. Neuron 34:841–851. PMID 12062029

Graziano MS, Aflalo TN, Cooke DF (2005) Arm movements evoked by electrical stimulation in the motor cortex of monkeys. J Neurophysiol 94:4209–4223. PMID 16120657

Grill-spector K, Malach R (2004) The human visual cortex. Annu Rev Neurosci 27:649–677. PMID 15217346

Haxby JV, Hoffman EA, Gobbini MI (2000) The distributed human neural system for face perception. Trends Cogn Sci 4:223–233. PMID 10827445

Horton JC, Adams DL (2005) The cortical column: a structure without a function. Philos Trans R Soc Lond Ser B Biol Sci 360:837–862. PMID 15937015

Horton JC, Hedley-Whyte ET (1984) Mapping of cytochrome oxidase patches and ocular dominance columns in human visual cortex. Philos Trans R Soc Lond Ser B Biol Sci 304:255–272. PMID 6142485

Horton JC, Hocking DR (1996) Intrinsic variability of ocular dominance column periodicity in normal macaque monkeys. J Neurosci 16:7228–7239. PMID 8929431

Horton JC, Hubel DH (1981) Regular patchy distribution of cytochrome oxidase staining in primary visual cortex of macaque monkey. Nature 292:762–764. PMID 6267472

Hubel DH, Livingstone MS (1987) Segregation of form, color, and stereopsis in primate area 18. J Neurosci 7:3378–3415, PMID 2824714

Hubel DH, Wiesel TN (1963) Shape and arrangement of columns in cat's striate cortex. J Physiol 165:559–568. PMID 13955384

Hubel DH, Wiesel TN (1968) Receptive fields and functional architecture of monkey striate cortex. J Physiol 195:215–243. PMID 4966457

Hubel DH, Wiesel TN (1972) Laminar and columnar distribution of geniculo-cortical fibers in the macaque monkey. J Comp Neurol 146:421–450. PMID 4117368

Hubel DH, Wiesel TN (1977) Ferrier lecture. Functional architecture of macaque monkey visual cortex. Proc R Soc Lond B Biol Sci 198:1–59. PMID 20635

Imig TJ, Adrián HO (1977) Binaural columns in the primary field (A1) of cat auditory cortex. Brain Res 138:241–257. PMID 589474

Issa NP, Trachtenberg JT, Chapman B et al (1999) The critical period for ocular dominance plasticity in the Ferret's visual cortex. J Neurosci 19:6965–6978. PMID 10436053

Jain N, Catania KC, Kaas JH (1998) A histologically visible representation of the fingers and palm in primate area 3b and its immutability following long-term deafferentations. Cereb Cortex 8:227–236. PMID 9617917

Kaas JH (1982) Why do sensory systems have so many subdivisions? In: Neff WD (ed) Contributions to sensory physiology, vol 7. Academic, New York, pp 201–240

Kaas JH (1990) Processing areas and modules in the sensory-perceptual cortex. In: Edelman GM, Gall WE, Cowan WM (eds) Signal and source: local and global order in perceptual maps. Wiley, New York, pp 67–82

Kaas JH (2008) The evolution of the complex sensory and motor systems of the human brain. Brain Res Bull 75:384–390. PMID 18331903

Kaas JH (2011) Somatosensory system. In: Mai JK, Paxinos G (eds) The human nervous system, vol 3. Elsevier, London, pp 1064–1099

Kaas JH (2012a) Evolution of columns, modules, and domains in the neocortex of primates. Proc Natl Acad Sci 109(Suppl 1):10655–10660. PMID 22723351

Kaas JH (2012b) The evolution of neocortex in primates. Prog Brain Res 195:91–102. PMID 22230624

Kaas JH (2013) The evolution of brains from early mammals to humans. Wiley Interdiscip Rev Cogn Sci 4:33–45. PMID 23529256

Kaas JH, Catania KC (2002) How do features of sensory representations develop? Bioessays 24:334–343. PMID 11948619

Kaas JH, Lyon DC (2001) Visual cortex organization in primates: theories of V3 and adjoining visual areas. Prog Brain Res 134:285–295. PMID 11702549

Kaas JH, Morel A (1993) Connections of visual areas of the upper temporal lobe of owl monkeys: the MT crescent and dorsal and ventral subdivisions of FST. J Neurosci 13:534–546. PMID 8381166

Kaas JH, Guillery RW, Allman JM (1972) Some principles of organization in the dorsal lateral geniculate nucleus. Brain Behav Evol 6:253–299. PMID 4196831

Kaas JH, Lin CS, Casagrande VA (1976) The relay of ipsilateral and contralateral retinal input from the lateral geniculate nucleus to striate cortex in the owl monkey: a transneuronal transport study. Brain Res 106:371–378

Kaas JH, Qi HX, Iyengar S (2006) Cortical network for representing the teeth and tongue in primates. Anat Rec A Discov Mol Cell Evol Biol 288:182–190. PMID 16411246

Kaas JH, Gharbawie OA, Stepniewska I (2011) The organization and evolution of dorsal stream multisensory motor pathways in primates. Front Neuroanat. eCollection 2011. PMID 21716641

Kaschube M, Schnabel M, Löwel S et al (2010) Universality in the evolution of orientation columns in the visual cortex. Science 330:1113–1116. PMID 21051599

Kaskan PM, Lu HD, Dillenburger BC et al (2007) Intrinsic-signal optical imaging reveals cryptic ocular dominance columns in primary visual cortex of New World owl monkeys. Front Neurosci 1:67–75. PMID 18974855

Kaskan PM, Lue HD, Dillenburger BC, Kaas JH, Roe AW (2009) The organization of orientation-selective, luminance-change and binocular-preference domains in the second (V2) and third (V3) visual areas of New World owl monkeys as revealed by intrinsic signal optical imaging. Cereb Cortex 19:1394–1407. PMID 18842661

Kaskan PM, Dillenburger BC, Lu HD et al (2010) Orientation and direction-of-motion response in the middle temporal visual area (MT) of New World owl monkeys as revealed by intrinsic signal optical imaging. Front Neuroanat. Ecollection 2010. PMID 20661299

Katz LC, Crowley JC (2002) Development of cortical circuits: lessons from ocular dominance columns. Nat Rev Neurosci 3:34–42. PMID 11823803

Krieger P, Kuner T, Sakmann B (2007) Synaptic connections between layer 5B pyramidal neurons in mouse somatosensory cortex are independent of apical dendrite bundling. J Neurosci 27:11473–11482. PMID 17959790

Krubitzer L (1995) The organization of neocortex in mammals: are species differences really so different? Trends Neurosci 18:408–417. PMID 7482807

Krubitzer L, Kaas JH (1993) The dorsomedial visual area (DM) of owl monkeys: connections, myeloarchitecture, and homologies in other primates. J Comp Neurol 33:497–528

Krubitzer L, Kaas JH (2005) The evolution of the neocortex in mammals: how is phenotypic diversity generated? Curr Opin Neurobiol 15:444–453. PMID 16026978

Krubitzer L, Manger P, Pettigrew J, Calford M (1995) Organization of somatosensory cortex in monotremes: in search of the prototypical plan. J Comp Neurol 351:261–306. PMID 7699113

Krubitzer L, Künzle H, Kaas JH (1997) Organization of sensory cortex in a Madagascan insectivore, the tenrec (Echinops telfairi). J Comp Neurol 379(3):399–414

Krubitzer L, Huffman KJ, Disbrow E, Recanzone G (2004) Organization of area 3a in macaque monkeys: contributions to the cortical phenotype. J Comp Neurol 471:97–111. PMID 14983479

Lachica EA, Beck PD, Casagrande VA (1992) Parallel pathways in macaque monkey striate cortex: anatomically defined columns in layer III. Proc Natl Acad Sci 89:3566–3570, PMID 1314392

LeVay S, Gilbert CD (1976) Laminar patterns of geniculocortical projection in the cat. Brain Res 113:1–19. PMID 953720

Leyton ASF, Sherrington CS (1917) Observations on the excitable cortex of the chimpanzee, orang-utan, and gorilla. Exp Physiol 11:135–222

Li M, Liu F, Juusola M, Tang S (2014) Perceptual color map in macaque visual area V4. J Neurosci 34:202–217. PMID 24381282

Liao CC, Gharbawie OA, Qi HX, Kaas JH (2013) Cortical connections to single digit representations in area 3b of somatosensory cortex in squirrel monkeys and prosimian galagos. J Comp Neurol 521:3768–3790. PMID 23749740

Livingstone MS (1996) Ocular dominance columns in New World monkeys. J Neurosci 16:2086–2096. PMID 8604053

Livingstone MS, Hubel DH (1984) Anatomy and physiology of a color system in the primate visual cortex. J Neurosci 4:309–356. PMID 6198495

Livingstone MS, Hubel DH (1987) Connections between layer 4B of area 17 and the thick cytochrome oxidase stripes of area 18 in the squirrel monkey. J Neurosci 7:3371–3377. PMID 2824713

Lorente de Nó R (1949) Physiology of the nervous system. In: Fulton DS (ed) The cerebral cortex: architecture, intracortical connections, motor projections. Oxford University Press, London

Löwel S, Singer W (1987) The pattern of ocular dominance columns in flat-mounts of the cat visual cortex. Exp Brain Res 68:661–666. PMID 3691734

Malonek D, Tootell RB, Grinvald A (1994) Optical imaging reveals the functional architecture of neurons processing shape and motion in owl monkey area MT. Proc Biol Sci 258:109–119. PMID 7838851

Manger P, Sum M, Szymanski M, Ridgway SH, Krubitzer L (1998) Modular subdivisions of dolphin insular cortex: does evolutionary history repeat itself. J Cogn Neurosci 10:153–166. PMID 9555104

Mashiko H, Yoshida AC, Kikuchi SS et al (2012) Comparative anatomy of marmoset and mouse cortex from genomic expression. J Neurosci 32:5039–5053. PMID 22496550

McKenna TM, Whitsel BL, Dreyer DA (1982) Anterior parietal cortical topographic organization in macaque monkey: a reevaluation. J Neurophysiol 48:289–317

Merlin S, Horng S, Marotte LR et al (2013) Deletion of Ten-m3 induces the formation of eye dominance domains in mouse visual cortex. Cereb Cortex 23:764–774. PMID 22499796

Meyer HS, Egger R, Guest JM et al (2013) Cellular organization of cortical barrel columns is whisker-specific. Proc Natl Acad Sci 110:19113–19118. PMID 24101458

Middlebrooks JC, Dykes RW, Merzenich MM (1980) Binaural response-specific bands in primary auditory cortex (A1) of the cat: topographical organization orthogonal to isofrequency contours. Brain Res 181:31–48. PMID 7350963

Moeller S, Freiwald WA, Tsao DY (2008) Patches with links: a unified system for processing faces in the macaque temporal lobe. Science 320:1355–1359. PMID 18535247

Molnár Z (2013) Cortical columns. In: Comprehensive developmental neuroscience: neural circuit development and function in the brain, vol 3. Elsevier, London, pp 109–129

Mountcastle VB (1957) Modality and topographic properties of single neurons of cat's somatic sensory cortex. J Neurophysiol 20:408–434. PMID 13439410

Mountcastle VB (1997) The columnar organization of the neocortex. Brain 120:701–722. PMID 9153131

Murphy KM, Jones DG, Van Sluyters RC (1995) Cytochrome oxidase blobs in cat primary visual cortex. J Neurosci 15:4196–4208. PMID 7790905

Mussa-Ivaldi FA, Bizzi E (2000) Motor learning through the combination of primitives. Philos Trans R Soc Lond B Biol Sci 355:1755–1769. PMID 11205339

Nelissen K, Luppino G, Vanduffel W et al (2005) Observing others: multiple action representation in the frontal lobe. Science 310:332–336. PMID 16224029

Olavarria JF (2001) Callosal connections correlate preferentially with ipsilateral cortical domains in cat areas 17 and 18, and with contralateral domains in the 17/18 transition zone. J Comp Neurol 433:441–457

Olavarria JF, Van Essen DC (1997) The global pattern of cytochrome oxidase stripes in visual area V2 of the macaque monkey. Cereb Cortex 7:395–404. PMID 9261570

Pinsk MA, DeSimone K, Moore T et al (2005) Representation of faces and body parts in macaque temporal cortex: a functional fMRI study. Proc Natl Acad Sci 102:6996–7001

Pons TP, Garraghty PE, Cusick CG, Kaas JH (1985) The somatotopic organization of area 2 in macaque monkeys. J Comp Neurol 241:445–466

Pons TP, Wall JT, Garraghty PE et al (1987) Consistent features of the representation of the hand in area 3b of macaque monkeys. Somatosens Res 4:309–331. PMID 3589287

Powell TP, Mountcastle VB (1959) Some aspects of the functional organization of the cortex of the postcentral gyrus of the monkey: a correlation of findings obtained in a single unit analysis with cytoarchitecture. Bull Johns Hopkins Hosp 105:133–162. PMID 14434571

Preuss TM, Kaas JH (1996) Cytochrome oxidase 'blobs' and other characteristics of primary visual cortex in a lemuroid primate, Cheirogaleus medius. Brain Behav Evol 47:103–112. PMID 8866707

Purves D, Riddle DR, LaMantia AS (1992) Iterated patterns of brain circuitry (or how the cortex gets its spots). Trends Neurosci 15:362–368. PMID 1279855

Qi HX, Kaas JH (2004) Myelin stains reveal an anatomica framework for the representation of the digits in somatosensory are 3b of macaque monkeys. J Comp Neurol 477:172–187. PMID 15300788

Qi HX, Gharbawie OA, Wong P, Kaas JH (2011) Cell-poor septa separate representations of digits in the ventroposterior nucleus of the thalamus in monkeys and prosimian galagos. J Comp Neurol 519:738–758. PMID 21246552

Rakic P (2007) The radial edifice of cortical architecture: from neuronal silhouettes to genetic engineering. Brain Res Rev 55:204–219. PMID 17467805

Rao SC, Toth LJ, Sur M (1997) Optically imaged maps of orientation preference in primary visual cortex of cats and ferrets. J Comp Neurol 387:358–370. PMID 9335420

Rockland KS (2010) Five points on columns. Front Neuroanat. eCollection 2010. PMID 20589097

Rockoff EC, Balaram P, Kaas JH (2014) Patchy distributions of myelin and VGLUT2 align with cytochrome oxidase blobs and interblobs in the superficial layers of primary visual cortex. Eye Brain 6:19–27

Roe AW, Tso DY (1995) Visual topography in primate V2: multiple representation across functional stripes. J Neurosci 15:3689–3715. PMID 7751939

Roe AW, Fritsches K, Pettigrew JD (2005) Optical imaging of functional organization of V1 and V2 in marmoset visual cortex. Anat Rec A: Discov Mol Cell Evol Biol 287:1213–1225. PMID 16235264

Schieber MH (2001) Constraints on somatotopic organization in the primary motor cortex. J Neurophysiol 86:2102–2108. PMID 11600665

Shatz CJ (1996) Emergence of order in visual system development. Proc Natl Acad Sci 93:602–608. PMID 8570602

Shipp S, Zeki S (1985) Segregation of pathways leading from area V2 to areas V4 and V5 of monkey visual cortex. Nature 315:322–325. PMID 2987702

Shostak Y, Ding Y, Mavity-Hudson J, Casagrande VA (2002) J Neurosci 22:2885–2893. PMID 11923453

Sincich LC, Horton JC (2002) Divided by cytochrome oxidase: a map of the projections from V1 to V2 in macaques. Science 295:1734–1737. PMID 11872845

Stepniewska I, Collins CE, Kaas JH (2005a) Reappraisal of DL/V4 boundaries based on connectivity patterns of dorsolateral visual cortex in macaques. Cereb Cortex 15:809–822. PMID 15459077

Stepniewska I, Fang PC, Kaas JH (2005b) Microstimulation reveals specialized subregions for different complex movements in posterior parietal cortex of prosimian galagos. Proc Natl Acad Sci 102:4878–4883. PMID 15772167
Stepniewska I, Gharbawie OA, Burish MJ, Kaas JH (2014) Effects of muscimol inactivations of functional domains in motor, premotor, and posterior parietal cortex on complex movements evoked by electrical stimulation. J Neurophysiol 111:1100–1119. PMID 24353298
Sur M, Wall JT, Kaas JH (1981) Modular segregation of functional cell classes within the postcentral somatosensory cortex of monkeys. Science 212:1059–1061. PMID 7233199
Sur M, Wall JT, Kaas JH (1984) Modular distribution of neurons with slowly adapting and rapidly adapting responses in area 3b of somatosensory cortex in monkeys. J Neurophysiol 51:724–744. PMID 6716121
Szentágothai J (1975) The 'module-concept' in cerebral cortex architecture. Brain Res 95:475–496. PMID 808252
Takahata T, Higo N, Kaas JH, Yamamori T (2009) Expression of immediate-early genes reveals functional compartments within ocular dominance columns after brief monocular inactivation. Proc Natl Acad Sci 106:12151–12155. PMID 19581597
Takahata T, Miyashita M, Tanaka S, Kaas JH (2014) Identification of ocular dominance domains in New World owl monkeys by immediate-early gene expression. Proc Natl Acad Sci USA 111:4297–4302
Tanaka S (1991) Theory of ocular dominance column formation. Mathematical basis and computer simulation. Biol Cybern 64:263–272. PMID 2025659
Tanaka K (2003) Columns for complex visual object features in the inferotemporal cortex: clusters of cells with similar but slightly different stimulus selectivities. Cereb Cortex 13:90–99. PMID 12466220
Tanigawa H, Lu HD, Roe AW (2010) Functional organization for color and orientation in macaque V4. Nat Neurosci 13:1542–1548. PMID 21076422
Towe AL (1975) Notes on the hypothesis of columnar organization in somatosensory cerebral cortex. Brain Behav Evol 11:16–47. PMID 1174930
Tsao DY, Freiwald WA, Knutsen TA et al (2003) Faces and objects in macaque cerebral cortex. Nat Neurosci 6:989–995. PMID 12925854
Tsao DY, Moeller S, Freiwald WA (2008) Comparing face patch systems in macaques and humans. Proc Natl Acad Sci 105:19514–19519. PMID 19033466
Van der Loos H, Dörfl J (1978) Does the skin tell the somatosensory cortex how to construct a map of the periphery? Neurosci Lett 7:23–30. PMID 19605083
Van Essen DC, Glasser MF, Dierker DL et al (2012) Parcellations and hemispheric asymmetries of human cerebral cortex analyzed on surface-based atlases. Cereb Cortex 22:2241–2262. PMID 22047963
Van Hooser SD (2007) Similarity and diversity in visual cortex: is there a unifying theory of cortical computation? Neuroscientist 13:639–656. PMID 17911223
Van Hooser SD, Heimel JA, Chung S et al (2005) Orientation selectivity without orientation maps in visual cortex of a highly visual mammal. J Neurosci 25:19–28. PMID 15634763
Waite PM, Marotte LR, Leamey CA, Mark RF (1998) Development of whisker-related patterns in marsupials: factors controlling timing. Trends Neurosci 21:265–269. PMID 9641540
Welker WI, Seidenstein S (1959) Somatic sensory representation in the cerebral cortex of the raccoon (Procyon lotor). J Comp Neurol 111:469–501. PMID 13843838
White LE, Bosking WH, Williams SM, Fitzpatrick D (1999) Maps of central visual space in ferret V1 and V2 lack matching inputs from the two eyes. J Neurosci 19:7089–7099. PMID 10436063
White LE, Bosking WH, Fitzpatrick D (2001) Consistent mapping of orientation preference across irregular functional domains in ferret visual cortex. Vis Neurosci 18:65–76. PMID 11347817
Wong P, Kaas JH (2009) An architectonic study of the neocortex of the short-tailed opossum (Monodelphis domestica). Brain Behav Evol 73:206–228. PMID 19546531

Wong P, Kaas JH (2010) Architectonic subdivisions of the neocortex in the Galago (Otolemur garnetti). Anat Rec 293:1033–1069. PMID 20201060

Wong-Riley M (1979) Changes in the visual system of monocularly sutured or enucleated cats demonstrable with cytochrome oxidase histochemistry. Brain Res 171:11–29. PMID 104776

Woolsey TA, Van der Loos H (1970) The structural organization of layer IV in the somatosensory region (SI) of mouse cerebral cortex. The description of a cortical field composed of discrete cytoarchitectonic units. Brain Res 17:205–242. PMID 4904874

Xu X, Bosking W, Sáry G et al (2004) Functional organization of visual cortex in the owl monkey. J Neurosci 24:9213–9214. PMID 15254078

Xu X, Collins CE, Khaytin I et al (2006) Unequal representation of cardinal vs. oblique orientations in the middle temporal visual area. Proc Natl Acad Sci 103:17490–17495. PMID 17088527

Yu YC, Bultje RS, Wang X, Shi SH (2009) Specific synapses develop preferentially among sister excitatory neurons in the neocortex. Nature 458:501–504. PMID 19204731

Zeki SM (1973) Colour coding in rhesus monkey prestriate cortex. Brain Res 53:422–427. PMID 4196224

Zilles K, Palomero-Gallagher N, Schleicher A (2005) Transmitter receptors and functional anatomy of the cerebral cortex. J Anat 205:417–432

Chapter 5
The Minicolumn in Comparative Context

Muhammad A. Spocter, Mary Ann Raghanti, Camilla Butti, Patrick R. Hof, and Chet C. Sherwood

Abstract The minicolumn has been defined as the smallest functional unit of the cortex with the widely-held view that there is a conservation of structure for this cortical processing unit. However, comparative data reveal significant differences among species in both the structure and composition of minicolumns. Here we review the available data on interspecific variation in minicolumn widths and the evidence in favor of phylogenetic variation in GABAergic interneurons, known to be a key component of the cortical microcircuit. Using data collated from the literature, we highlight the importance of variation in cortical column structure and build a framework towards further evolutionary explanations of minicolumn diversity. Although our preliminary analysis indicates that minicolumn width increases with increasing brain mass among anthropoid primates, this relationship is not constant when applied to other taxonomic orders. These findings highlight the need for further comparative analyses of minicolumn structure and their ecological, behavioral, and cognitive correlates.

Keywords Calcium-binding proteins • Evolution • Minicolumn • Variation • Diversity • Neuropathology • GABA

M.A. Spocter (✉)
Department of Anatomy, Des Moines University, Des Moines, IA 50312, USA
e-mail: muhammad.spocter@dmu.edu

M.A. Raghanti
Department of Anthropology and School of Biomedical Sciences, Kent State University, Kent, OH 44242, USA
e-mail: mraghant@kent.edu

C. Butti • P.R. Hof
Fishberg Department of Neuroscience and Friedman Brain Institute, Icahn School of Medicine at Mount Sinai, New York, NY 10029, USA
e-mail: butticamilla@gmail.com; patrick.hof@mssm.edu

C.C. Sherwood
Department of Anthropology, The George Washington University, Washington, DC 20052, USA
e-mail: sherwood@gwu.edu

M.F. Casanova, I. Opris (eds.), *Recent Advances on the Modular Organization of the Cortex*, DOI 10.1007/978-94-017-9900-3_5

5.1 Introduction: The Paradigm of Cortical Uniformity

Thomas Kuhn's 1962 *Structure of Scientific Revolutions* highlighted the observation that science is not merely an accumulation of facts, but rather that these facts and their interpretation provide a theoretical framework, a paradigm, which becomes deeply ingrained in the thinking of the scientific community. Paradigms contribute to how scientists interpret their findings, but may also constrain the acceptance of views that run counter to them.

In the neurosciences, the concept of 'cortical uniformity' as it pertains to the study of the cortical column, is a prime example of a widely accepted paradigm. The neocortex is composed of cell columns that unite the cells of each minicolumn (the minicolumn being the smallest unit of cortical function) into a coordinated functional unit (Mountcastle et al. 1955; Mountcastle 1997). The cortical uniformity hypothesis proposes that minicolumns are highly invariable in their architecture and function across species, a conclusion that was bolstered by a much disclaimed but still highly cited paper by Rockel et al (1980). In this highly influential paper, the estimate for the number of neurons within a minicolumn was based on cell counts in different cytoarchitectonic areas, without correcting for cell size, and using an arbitrary designation of the minicolumn (Rakic 2008; DeFelipe et al. 2002). The authors subsequently concluded that this was indicative of a basic uniformity in the structure of the neocortex and that functional and cytoarchitectonic differences were principally a result of wiring among cortical columns (Rockel et al. 1980). Although several studies have directly refuted the validity of this claim (e.g., Herculano-Houzel et al. 2008; DeFelipe et al. 2002; Beaulieu 1993; Skoglund et al. 1996; Preuss and Goldman-Rakic 1991), the concept of uniformity in cortical column structure has remained popular and is still used broadly in computational models of cortical operation.

The continued acceptance of this paradigm has been fueled by a more general belief in histological and connectional conservation rooted in the history of neuroscience (Preuss 2001). Early proponents of this view included Huxley and Darwin, who strongly defended the idea that the human brain and mind are simply an extension of those of other primates and that differences among species are largely a matter of degree rather than type (Huxley 1863; Darwin 1871). The concept of cortical uniformity provided a simple means by which to understand the cerebral cortex that aligned well with a view of the phyletic continuity of species. However, does an evolutionary perspective of the cortical column hinge on the acceptance of a conserved structure to the minicolumn or could embracing diversity in minicolumn structure widen the comparative framework for understanding variation and evolution in this system? Certainly several recent studies have shown that aspects of minicolumn structure vary among species and individuals (e.g., Raghanti et al. 2010) but how do we make sense of this variation and is it possible to articulate a comparative framework for generating and testing competing hypotheses on the evolution of the cortical column?

5.2 A Framework for Embracing Variation and Diversity in Minicolumn Structure

While variation in cortical columns has not been well studied, the exploration of inter-specific variation has the potential to explain individual-level and species-level differences in cortical function. Our current understanding of the cortical column is based on a small range of species, making it difficult to assess diversity in minicolumn structure (Buxhoeveden and Casanova 2002a; Raghanti et al. 2010). This is further complicated by the fact that different methodologies and measurement techniques have been used to define minicolumns. A comparative approach enables us to separate out minicolumn characteristics that were established early in mammalian evolution and are homologous across several species (Fig. 5.1) from those components that underwent modifications due to selective pressures or evolved independently in various lineages. For example, Fig. 5.1 shows a dendrogram depicting phylogenetic relationships for nine species from the mammalian clade Boreoeutheria in which minicolumn widths were derived from the frontal lobe using the pyramidal cell array as a defining feature. These data demonstrate the type of comparative framework facilitated by incorporating diversity into our understanding of the minicolumn. For example, *Pan troglodytes* and *Pan paniscus* are sister taxa that diverged from one another approximately 2 Ma ago and there has been relatively little alteration in their minicolumn width. Despite their evolutionary closeness, these two species are known to exhibit marked sociocognitive differences with *Pan paniscus* showing greater social tolerance (Hare et al. 2007), more adult play (Palagi 2006), less severe aggression (Goodall 1986; Wrangham 1999; Parish and de Waal 2000), and greater frequency of sexual behaviors (Wobber et al. 2010) in comparison to *Pan troglodytes*. While we can only speculate at this point, if other components of minicolumn structure do not differentiate between the two species, the dissimilarities in sociality and cognition between bonobos and chimpanzees are likely a result of rearrangements in other levels of organization (e.g., cortical area) rather than at the level of the cortical column. Furthermore, if we continue exploring variation in minicolumn width amongst the great apes and expand our comparison to include humans, it is apparent that this character state has undergone further modification in the human lineage and has likely been subject to selective forces, possibly in association with modifications to cognitive processing.

The comparative approach also reveals homoplasies, suggesting the independent emergence of a character or character state that is unrelated to ancestry but rather a result of similar environmental or ecological pressures. For example, in the comparison shown in Fig. 5.1, humans, domestic cats, and rats share a convergent pattern of enlarged minicolumn width that likely arose independently in these species. It is also important to point out in this broad comparative framework, that overall neocortical size does not appear to be typically associated with minicolumn widths across the range of mammalian phylogeny. The comparative approach thus challenges us to explore the various ways in which the cortical column differs among species and how these differences are manifest functionally.

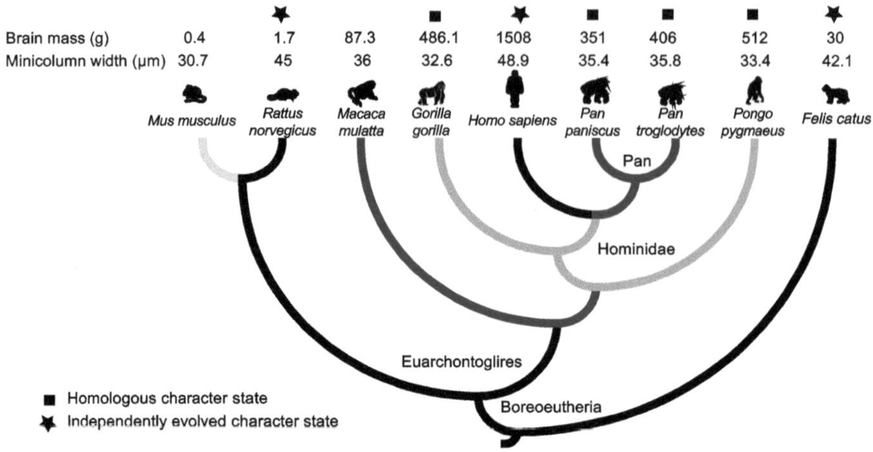

Fig. 5.1 A dendrogram depicting phylogenetic relationships for nine species from the mammalian clade Boreoeutheria. Minicolumn widths were derived from the frontal lobe using the pyramidal cell array as a defining feature (The following sources were used in collating data for Figs. 5.1, 5.2, and 5.3: Escobar et al. 1986; White and Peters 1993; Peters and Yilmaz 1993; Feldman and Peters 1974; Favorov and Diamond 1990; Tommerdahl et al. 1993; Gabbott and Bacon 1996; Peters and Walsh 1972; Peters and Kara 1987; Kohn et al. 1997; Fleischauer et al. 1972; DeFelipe et al. 1990; Peters and Sethares 1991, 1996, 1997; Buxhoeveden et al. 2001a; Von Bonin and Mehler 1971; Kaas et al. 1981; Buxhoeveden and Casanova 2002a; Schlaug et al. 1995; Seldon 1981; del Rio and DeFelipe 1997a, b; Buldyrev et al. 2000; and Casanova and Tilquist 2008. Note that in certain cases an average minicolumn width had been computed by calculating the mean based on the minimum and maximum values reported in each study)

A framework that embraces diversity in minicolumn structure would also enable the recognition of certain general organizational properties and identify constraints that limit evolutionary change in the cortical column. When combined with ecological, life history or brain: body size data, this can also be a powerful tool for exploring relationships between minicolumn structure and other aspects of species' biology. For instance, minicolumns have not been investigated across a wide range of brain sizes. Large comparative studies could establish the limits for cortical column dimensions and reveal if there exist patterns related to gyrencephaly or brain size (Buxhoeveden and Casanova 2002a, b). Figure 5.2 shows the scaling of minicolumn width (using the pyramidal cell array) to brain mass as reported in the literature for a range of anthropoid primate species encompassing ~30 Ma of evolutionary diversification. As indicated by this preliminary analysis, minicolumn width in anthropoid primates has a positive and significant allometry with brain size, i.e., with every gram increase in brain size, minicolumn width increases by 0.012 µm, suggesting that while slight, brain mass is a significant determinant of overall minicolumn size within this phylogenetic lineage and, as a result, a determinant of variation in cortical output within primates. These data seemingly align well with a recent analysis of neuronal complexity across primate species, indicating that larger brains (at least within the primates) tend to have increased neuronal

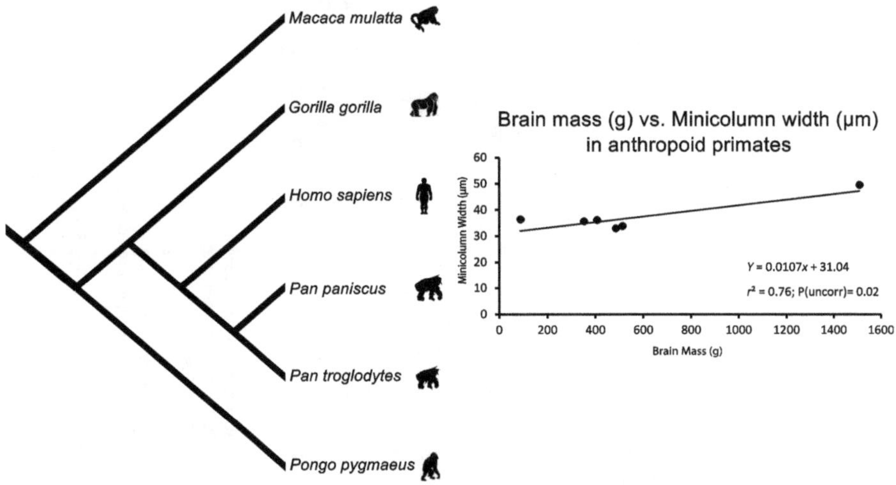

Fig. 5.2 The scaling of minicolumn width (using the pyramidal cell array) to brain mass as in a range of anthropoid primate species encompassing ~30 Ma of evolutionary diversification. Sources are the same as in Fig. 5.1 (Note that in certain cases an average minicolumn width had been computed by calculating the mean based on the minimum and maximum values reported in each study)

complexity (Manger et al. 2013). These findings support the claim that the human brain is not only larger but also more complexly organized from both a neuronal and minicolumn perspective, in a manner that is expected due to allometric scaling at the microstructural level.

A survey of the literature reveals that minicolumn width appears to vary substantially in mammals among cortical regions and among species, and that brain size alone cannot account for the full range of variation. Figure 5.3 summarizes the current knowledge of minicolumns among species. Studies that utilized the apical dendrite bundle in investigations of minicolumn width within the occipital lobe show an interesting pattern, which coincides with what we already know about the visual cortex for some of these species. For instance, column width in the primary visual cortex of the rhesus monkey (*Macaca mulatta*) is on average smaller (23–30 μm) than that reported for rats (30–40 μm, average = 35 μm), rabbits (40–50 μm, average = 45 μm) and cats (55–60 μm, average = 57.5 μm; see Fig. 5.3). This is remarkable given the difference in brain size observed among these species for example the rhesus monkey has a brain size of 106.4 mm^3 and the cat has a brain size of only 25.3 mm^3 (Peters and Yilmaz 1993). This suggests that selection acting to increase column complexity, through for example, modifications in receptor and neurotransmitter types, may also serve as one route through which cortical output could be increased whilst maintaining or reducing overall column width (Buxhoeveden and Casanova 2002a, b). Gustafsson (1997) has proposed that while wide columns are likely to have more neurons and a greater synaptic weight, narrow columns may also offer greater discriminatory ability useful for feature

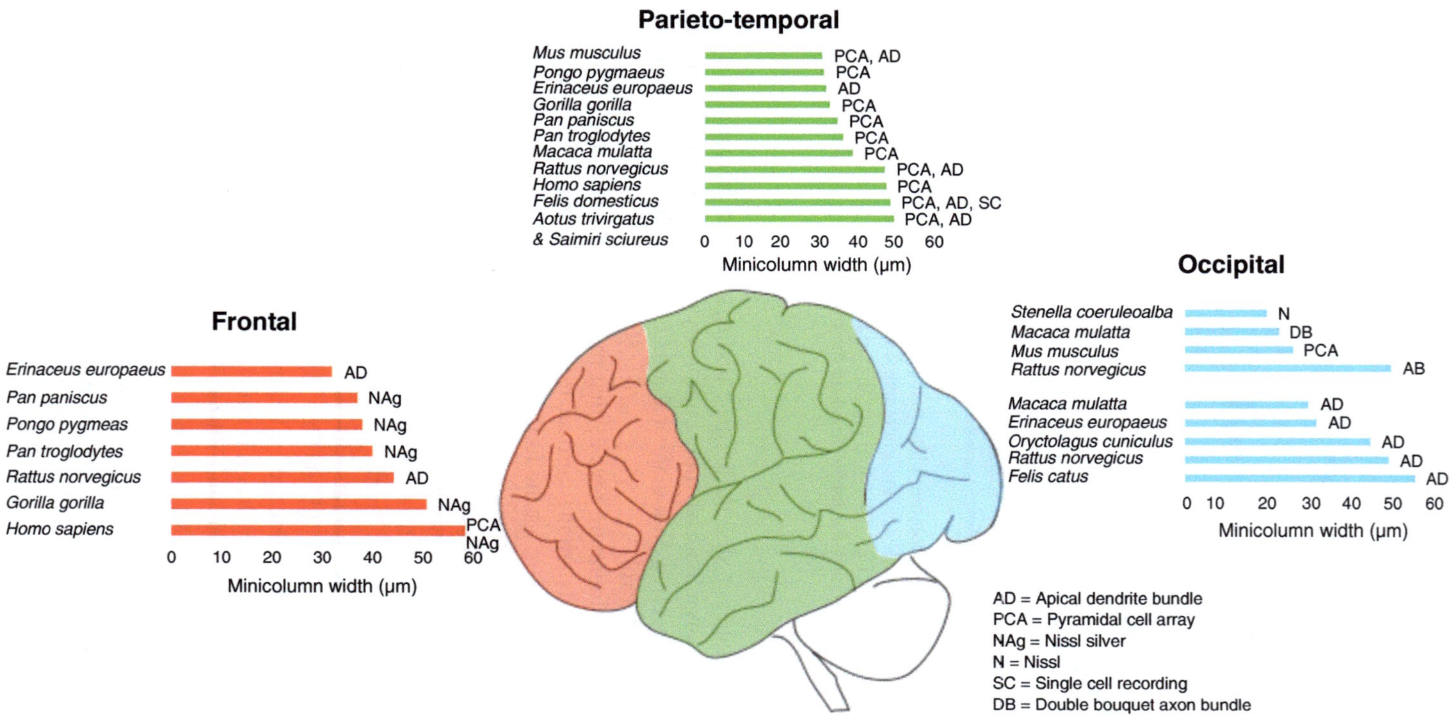

Fig. 5.3 A summary of average minicolumn width by brain region in all species studied to date (Sources are the same as in Fig. 5.1)

detection. According to Peters and Yilmaz (1993) this apparent contradiction in the size of the macaque minicolumn was facilitated through the evolution of more complex, narrower cell columns that allow a more detailed specification of information. This pattern of selection for greater complexity within the macaque minicolumn is likely to be a synapomorphy, a derived feature shared with several other extant primate species. Livingstone and Hubel (1984) have shown that primates have indeed evolved specific modules in the primary visual cortex that are utilized for discriminating and processing form, color, and motion and that these differ markedly from that observed in rodents and felids. Other collaborative evidence has also indicated that the primate visual system in fact surpasses that of cats in the display of color and visual acuity (Orban 1984), lending support to the idea that primates are indeed endowed with more complex minicolumns. Without a detailed survey of the cortical column across primates, we can only postulate that the emergence of this evolutionary change in the visual system likely occurred about 35 Ma ago, when higher level color vision in the form of trichromacy evolved at the origin of the catarrhine clade and that this was selected to enhance the tracking and foraging of highly variable and seasonal food sources.

That evolutionary reductions in minicolumn width may be offset by an increase in minicolumn complexity, is highlighted by the available data within another group of large brained social species, the cetaceans. Morgane et al. (1988) have reported relatively small minicolumn sizes within the visual cortex of cetaceans (*Stenella coeruleoalba* and *Tursiops truncatus*), paired with descriptions of a discontinuous minicolumn structure across cortical layers (i.e., interruptions in the continuity of cellular columns through all cortical layers) (Morgane et al. 1988; Manger 2006). Glezer and Morgane (1990) have suggested that this unique adaptation in the cetacean cortical column is likely facilitated by the integration of column activity occurring through layer 1 which in cetaceans is specialized and contains around 70 % of the total cortical synapses. Interestingly, the minicolumns within the visual cortex of humans and dolphins also contain roughly the same number of synapses (Morgane et al. 1988), potentially explaining some of the similarities in cetacean and primate behavioral complexity (Marino et al. 2008) despite two very different structural arrangements. This illustrates the need for caution in oversimplifying measurements of cell column width as a direct correlate with functional output and highlights why a comparative perspective is important in helping to understand the range of variation in column structure within and across multiple species.

Investigations of homologous areas in other taxa are likely to yield new insight into species-specific differences in cortical column structure. For example, analyses of minicolumn morphology in the temporoparietal cortex (area Tpt) of humans, macaques and chimpanzees (Buxhoeveden and Casanova 2000) suggests that this cortical region within humans underwent substantial rearrangement, resulting in cortical columns in the left hemisphere reportedly 30 % wider than those observed in the left hemisphere of chimpanzees and macaques and also reportedly having more neuropil space than that observed in nonhuman primates examined thus far (Buxhoeveden and Casanova 2000; Buxhoeveden et al. 2001a, b). As summarized in

Fig. 5.3, cortical column width measured in the temporoparietal cortex (area Tpt) is relatively uniform in the great apes and macaques, with humans possessing larger minicolumn sizes (50.4 μm). Such a restructuring in this extension of the human auditory cortex has been postulated to underlie our unique language abilities (Rilling et al. 2008). In addition, humans are also known to display significantly larger minicolumns in Broca's area (Brodmann areas 44 and 45) than that observed in other great apes (chimpanzee, bonobo, gorilla and orangutan), adding further support to the claim for restructuring in these language circuits (Schenker et al. 2008). It is notable, however, that minicolumn widths in the temporal and parietal cortex are matched or exceeded in humans by squirrel monkeys, owl monkey, cats, and rats (Fig. 5.3). Such observations place claims of human uniqueness in a broad phylogenetic context and encourage re-evaluation of assumptions that modifications of minicolumn morphology is necessarily linked to the evolution of cognitive function.

Adaptive accounts of minicolumn variability among species often overlook the possibility that certain aspects of column structure could simultaneously remained unchanged or be subject to non-adaptive forces. A prime example of this is that reported in a reanalysis of morphometric variability in minicolumns of the primary visual cortex in humans, chimpanzees, and macaques, which confirmed the existence of species differences in minicolumn widths, but also indicated that the core columns space was relatively invariable across these taxa (Casanova et al. 2009). Without a comprehensive analysis of column core variability across several species using a common methodology it is difficult to determine whether conservation in the column core is specific to this clade (catarrhine primates) or whether it is indicative of broader primate or mammalian bounds for this structure. While the above examples and patterns are tentative given the paucity of data, they highlight the utility of a broad comparative approach for revealing evolutionary patterns of cortical column variability and how an analysis sensitive to variation and diversity could provide adaptive explanations that take into consideration the environments within which species evolved.

Evolution by natural selection operates by differential fitness among variable phenotypes within a population. There is evidence indicating that minicolumn widths do show continuous variation within a normal population and may be under the influence of multiple independent factors (Casanova 2006). Casanova et al (2007), have reported significant differences in minicolumn width and mean cell spacing of a control human group ($n = 6$) and that of three distinguished scientists suggesting that genetic, environmental and developmental factors (i.e., the duration of cell division cycle, number of founder cells, selective cell death cycle) are likely to have an impact on the column phenotype. Cumulatively, both intra- and inter-specific variation in cortical column morphology points in favor of an interpretation of the cortical phenotype, which is more strongly rooted in an evolutionary view of diversity.

Aside from size, minicolumns in different species may also show variability in their structural subcomponents (i.e., fibers and neurons). The following section is aimed at reviewing the evidence for phylogenetic differences indicative of diversity in the microcircuitry of mammalian cortical structure, with a focus on different subtypes of γ-aminobutyric acid (GABA)-containing interneurons, and implications for mental illness in humans.

5.3 Phylogenetic Diversity in GABAergic Interneurons and Their Relation to Minicolumns

Inhibitory GABAergic interneurons are a critical component of cortical microcircuitry and play a fundamental role in intra- and intercolumnar processing (Hendry 1987; DeFelipe 2002; Casanova et al. 2003; Buzsaki et al. 2004; Ascoli et al. 2008; DeFelipe et al. 2013). Interneuron subpopulations are also known to show significant phylogenetic variation both in the diversity and density of interneuron subtypes and in their distribution and patterns of development (e.g., Hof et al. 1999; Preuss and Coleman 2002; Hof and Sherwood 2005; Sherwood et al. 2007). Qualitative comparisons of inhibitory interneurons across taxa, have also revealed Order specific patterns, with primates displaying a greater proportion of inhibitory interneurons (i.e., greater than 20 %) in comparison to that observed in rodents (i.e., less than 15 %).(Hendry et al. 1987; Beaulieu et al. 1992; Gabbott and Bacon 1996; Gabbott et al. 1997; DeFelipe et al. 1999, 2002). Furthermore, studies of the development and migration of GABAergic cells have also observed an expansion of neurogenesis into the lateral ventricular neurepithelium of primates (Petanjek et al. 2009). These observations are indicative of the effects of phylogeny in the evolution of this system and its role in supporting the elaborate behavioral and cognitive attributes definitive of primates.

Inhibitory interneurons can be classified into subpopulations based on their immunoreactivity for the three calcium-binding proteins, calbindin (CB), calretinin (CR), and parvalbumin (PV) (Hendry et al. 1989; Glezer et al. 1993; DeFelipe 1997; Zaitsev et al. 2005) with each of these subpopulations playing in turn interacting with pyramidal cells to modulate processing within the cortical circuit. The subpopulations of CB- and CR-immunoreactive(ir) neurons are known to be involved in intracolumnar communication while interneurons immunoreactive for PV are involved in transcolumnar signaling.

An example of a phylogenetic pattern in GABAergic interneurons, is that observed for the primate cortex, where CB-ir double bouquet cells are known to contribute significantly to the morphology and distribution of minicolumns (Buxhoeveden and Casanova 2002b; Casanova et al. 2009). As a potential source for minicolumn diversity, we outline below the range of variation known for GABAergic neurons in mammals. Calcium-binding protein- expressing interneurons are known to vary significantly in their distribution between species, (Glezer et al. 1992, 1993; Hof et al. 1996, 1999 Glezer et al. 1998; Hof and Sherwood 2005; Zaitsev et al. 2005; Sherwood et al. 2007; Sherwood et al. 2009), while variation in GABAergic cell phenotype and density would be a consequence of evolutionary alterations in local microcircuit processing. A number of studies have indicated that the proportion of cortical interneurons also varies among species, with members of the order primates having a higher overall percentage relative to rodents, afrotheria, and xenarthrans; while the cetaceans markedly have the greatest proportion of cortical interneurons of all species studied to date (Hof et al. 2000; DeFelipe et al. 2002; Sherwood et al. 2009). Furthermore, studies have also indicated that variation in electrophysiological response properties may contribute to species level differences in GABAergic interneurons. While a study of PV-ir fast-spiking basket cells reported no significant morphological differences between macaque monkeys and rats (Povysheva et al. 2008), a significant difference in excitability of PV-ir basket cells was observed with the neurons of the macaque monkey having a higher input resistance and lower firing threshold than those of rats. This finding coincides with the lower frequency firing rates reported for rats in prefrontal cortical neurons (Povysheva et al. 2008) and is intriguing given that the actions of PV-ir interneurons appear to be essential for the accomplishment of working memory tasks (Rao et al. 1999) and other cognitive functions (Constantinidis et al. 2002).

In addition, phenotypic variation within interneuron subpopulations have also been reported between species. An example of this species level variation, is that reported between primates and canids, with chandelier cells in the primate primary motor and somatosensory cortex expressing PV-ir (DeFelipe et al. 1990), while this was not observed in canids (Hof et al. 1996). Furthermore, CB-ir double bouquet cells were not observed in the following groups: rodents, lagomorphs, artiodactyls xenarthrans, and afrotherians (Sherwood et al. 2009; Ballesteros-Yañez et al. 2005) but are reportedly present in primates (human and macaque monkey), and to a lesser extent carnivores (Ballesteros-Yañez et al. 2005). The connectivity of these double bouquet neurons, characterized by long descending bundles of axon collaterals that form columns targeting pyramidal cells within a very narrow space, is argued to represent a specialization of minicolumn inhibition within primates (del Rio and DeFelipe 1997a; DeFelipe et al. 2002, 2006; Ballesteros-Yañez et al. 2005). Its also worth noting that minicolumns within the human cortex are known to be comprised of double bouquet axon bundles associated with myelinated axons although not all minicolumns were associated with CB-ir double bouquet cells (Ballesteros-Yañez et al. 2005).

Several neuropathological abnormalities in humans have been noted that involve minicolumns and interneurons. For example, decreased numbers of CB-ir interneurons have been reported consistently in the prefrontal cortex of patients with schizophrenia (Beasley and Reynolds 1997; Reynolds et al. 2001; Beasley et al. 2002; Cotter et al. 2002; Eyles et al. 2002; Chance et al. 2005; Sakai et al. 2008) while CR-ir interneuron density is preserved (Woo et al. 1998; Reynolds and Beasley 2001; Zhang and Reynolds 2002). The relationship between PV-ir neurons distributions and schizophrenia is less clear, with reports of a decrease in density (Beasley and Reynolds 1997) or no change in density (Woo et al. 1997; Cotter et al. 2002). Alterations of minicolumn width (i.e., neuropil space) are also well-documented in schizophrenia (Reynolds et al. 2004; Casanova et al. 2005,

2008; Chance et al. 2005; Di Rosa et al. 2009). These changes in minicolumn morphology appear to be consistent with a developmental abnormality rather than a progressive pathological process (Casanova et al. 2005, 2008).

Alzheimer's disease is also associated with a decrease of cortical CB-ir pyramidal neurons in some regions of human cortex (Ferrer et al. 1993; Nishiyama et al. 1993; Beasley et al. 2002) and has also been reported in the canine expression of dementia of the Alzheimer's type (Pugliese et al. 2004) while both PV- and CR-ir neuronal subpopulations are spared (Ferrer et al. 1993; Hof et al. 1991; 1993; Pugliese et al. 2004). However, it should be noted that not all forms of dementia are associated with a reduction in cortical calcium-binding protein-containing interneurons (Hof et al. 1994; Gomez-Tortosa et al. 2001). The structure of minicolumns is selectively disrupted in Alzheimer's disease, and the loss of columnar organization was related to the number of neurofibrillary tangles (Buldyrev et al. 2000). Tangles cluster into columns and their numbers are positively correlated with degree of cognitive loss in Alzheimer's disease (Nagy et al. 1996). Minicolumn thinning was also noted in normal human aging and is a process that may be continuous with the increased risk of Alzheimer's disease over the age of 65 (Nagy et al. 1996).

It has been postulated that cortical minicolumn GABAergic inhibitory control is also compromised in autism (Casanova et al. 2003). However, while dysregulation of the calcium-binding protein-ir interneuron populations was recently demonstrated within the hippocampus of patients with autism (Lawrence et al. 2010), comparable data are not available for neocortical regions. Nonetheless, autism and Asperger's syndrome are associated with a narrowing of the minicolumns, specifically, the peripheral neuropil space (Casanova et al. 2002a, b, c, 2003). Because the peripheral neuropil space is dependent upon inhibitory interneuron populations, a deficit in GABAergic control is suspected. In particular, the modulation of minicolumnar activity would be altered for both local and long-range connectivity, resulting in collateral over-excitation among minicolumns (Casanova 2008; Casanova and Trippe 2009). Such a deficit is also suspected to contribute to the incidence of seizures in autistic individuals (Casanova et al. 2003). This relationship finds support in recent reports of deficits in both PV- and CR-ir interneurons with focal cortical dysplasias associated with epilepsy (Zamecnik et al. 2006; Barinka et al. 2009).

5.4 Conclusions

Variability not only in the connectivity of the minicolumn but also in the subtle subcomponents of the columnar organization such as composition of interneuron subtypes, are a primary source of inter and intraspecific differences. The comparative approach highlights the need to go beyond the restraints enforced by a narrow view of the cortical minicolumn and challengers us to embrace diversity in our interpretations of minicolumn phenotype. It is only through the lens of variation that we are able to bring together discussions of the minicolumn with the evolutionary framework that unites the biological sciences.

Acknowledgements We thank Cheryl D. Stimpson for technical assistance. This work was supported by the National Science Foundation (BCS-0515484, BCS-0549117, BCS-0827531, BCS-0550209, BCS-0827546, DGE-0801634), the National Institutes of Health (NS042867), the Wenner-Gren Foundation for Anthropological Research, and the James S. McDonnell Foundation (22002078).

References

Ascoli GA, Alonso-Nanclares L, Anderson SA, Barrionuevo G, Benavides-Piccione R, Burkhalter A, Buzsáki G, Cauli B, DeFelipe J, Fairén A, Feldmeyer D, Fishell G, Fregnac Y, Freund TF, Gardner D, Gardner EP, Goldberg JH, Helmstaedter M, Hestrin S, Karube F, Kisvárday ZF, Lambolez B, Lewis DA, Marin O, Markram H, Muñoz A, Packer A, Petersen CCH, Rockland KS, Rossier J, Rudy B, Somogyi P, Staiger JF, Tamas G, Thomson AM, Toledo-Rodriguez M, Wang Y, West DC, Yuste R (2008) Petilla terminology: nomenclature of features of GABAergic interneurons of the cerebral cortex. Nat Rev Neurosci 9:557–568

Ballesteros-Yañez I, Munoz A, Contreras J, Gonzalez J, Rodriguez-Veiga E, DeFelipe J (2005) Double bouquet cell in the human cerebral cortex and a comparison with other mammals. J Comp Neurol 486:344–360

Barinka F, Druga R, Marusic P, Krsek P, Zamecnik J (2009) Calretinin immunoreactivity in focal cortical dysplasias and in non-malformed epileptic cortex. Epilepsy Res. doi:10.1016/j.eplepsyres.2009.09.021

Beasley CL, Reynolds GP (1997) Parvalbumin-immunoreactive neurons are reduced in the prefrontal cortex of schizophrenics. Schizophr Res 24:349–355

Beasley C, Zhang Z, Patten I, Reynolds G (2002) Selective deficits in prefrontal cortical GABAergic neurons in schizophrenia defined by the presence of calcium-binding proteins. Biol Psychiatry 52:708–715

Beaulieu C (1993) Numerical data on neocortical neurons in adult rat, with special reference to the GABA population. Brain Res 609:284–292

Beaulieu C, Kisvaraday Z, Somogyi P, Cuynader M, Cowey A (1992) Quantitative distribution of GABA-immunopositive and -immunonegative neurons and synapses in the monkey striate cortex. Cereb Cortex 2:295–309

Buldyrev SV, Cruz L, Gomez-Isla T, Gomez-Tortosa E, Havlin S, Le R, Stanley HE, Urbanc B, Hyman BT (2000) Description of microcolumnar ensembles in association cortex and their disruption in Alzheimer and Lewy body dementias. Proc Natl Acad Sci U S A 97:5039–5043

Buxhoeveden D, Casanova MF (2000) Comparative lateralization patterns in the language area of human, chimpanzee, and rhesus monkey brains. Laterality 5:315–330

Buxhoeveden DP, Casanova MF (2002a) The minicolumn and evolution of the brain. Brain Behav Evol 60:125–151

Buxhoeveden DP, Casanova MF (2002b) The minicolumn hypothesis in neuroscience. Brain 125:935–951

Buxhoeveden DP, Switala AE, Litaker M, Roy E, Casanova MF (2001a) Lateralization of minicolumns in human planum temporale is absent in nonhuman primate cortex. Brain Behav Evol 57:349–358

Buxhoeveden DP, Switala AE, Roy E, Litaker M, Casanova MF (2001b) Morphological differences between minicolumns in human and nonhuman primate cortex. Am J Phys Anthropol 115:361–371

Buzsaki G, Geisler C, Henze DA, Wang XJ (2004) Interneuron diversity series: circuit complexity and axon wiring economy of cortical interneurons. Trends Neurosci 27:186–193

Casanova MF (2006) Neuropathological and genetic findings in autism: the significance of a putative minicolumnopathy. Neuroscientist 12:435–441

Casanova MF (2008) The significance of minicolumnar size variability in autism: a perspective from comparative anatomy. In: Zimmerman A (ed) Autism current theories and evidence, Current clinical neurology. The Human Press, Inc., Totowa, pp 349–360

Casanova MF, Tilquist CR (2008) Encephalisation, emergent properties, and psychiatry: a minicolumnar perspective. Neuroscientist 14:101–118

Casanova MF, Trippe J (2009) Radial cytoarchitecture and patterns of cortical connectivity in autism. Phil Trans R Soc B 364:1433–1436

Casanova MF, Buxhoeveden D, Brown C (2002a) Clinical and macroscopic correlates of minicolumnar pathology in autism. J Child Neurol 17:692–695

Casanova MF, Buxhoeveden D, Switala AE, Roy E (2002b) Asperger's syndrome and cortical neuropatholgy. J Child Neurol 17:142–145

Casanova MF, Buxhoeveden D, Switala AE, Roy E (2002c) Minicolumnar pathology in autism. Neurology 58:428–432

Casanova MF, Buxhoeveden D, Gomez J (2003) Disruption in the inhibitory architecture of the cell minicolumn: implications for autism. Neuroscientist 9:496–507

Casanova MF, de Zeeuw L, Switala A, Kreczmanski P, Korr H, Ulfig N, Heinsen H, Steinbusch HWM, Schmitz C (2005) Mean cell spacing abnormalities in the neocortex of patients with schizophrenia. Psychiatry Res 133:1–12

Casanova MF, Switala A, Trippe J, Fitzgerald M (2007) Comparative minicolumnar morphometry of three distinguished scientists. Autism 11:557–569

Casanova MF, Kreczmanski P, Trippe J, Switala A, Heinsen H, Steinbusch HWM, Schmitz C (2008) Neuronal distribution in the neocortex of schizophrenic patients. Psychiatry Res 158:267–277

Casanova MF, Trippe J, Tillquist C, Switala AE (2009) Morphometric variability of minicolumns in the striate cortex of *Homo sapiens*, *Macaca mulatta*, and *Pan troglodytes*. J Anat 214:226–234

Chance SA, Walker M, Crow TJ (2005) Reduced density of calbindin-immunoreactive interneurons in the planum temporale in schizophrenia. Brain Res 1046:32–37

Constantinidis C, Williams GV, Goldman-Rakic PS (2002) A role for inhibition in shaping the temporal flow of information in prefrontal cortex. Nat Neurosci 5:175–180

Cotter D, Landau S, Beasley C, Stevenson R, Chana G, MacMillan L, Everall I (2002) The density and spatial distribution of GABAergic neurons, labelled using calcium binding proteins, in the anterior cingulate cortex in major depressive disorder, bipolar disorder, and schizophrenia. Biol Psychiatry 51:377–386

Darwin C (1871) The descent of man, and selection in relation to sex. John Murray, London (Facsimile edition: Princeton, NJ: Princeton University Press, 1981)

DeFelipe J (1997) Types of neurons, synaptic connections and chemical characteristics of cells immunoreactive for calbindin-D28k, parvalbumin and calretinin in the neocortex. J Chem Neuroanat 14:1–19

DeFelipe J (2002) Cortical interneurons: from Cajal to 2001. Prog Brain Res 136:215–238

DeFelipe J, Hendry SH, Hashikawa T, Molinari M, Jones EG (1990) A microcolumnar structure of monkey cerebral cortex revealed by immunocytochemical studies of double bouquet cell axons. Neuroscience 37:655–673

DeFelipe J, Gonzalez-Albo MC, Del Rio MR, Elston GN (1999) Distribution and patterns of connectivity of interneurons containing calbindin, calretinin, and parvalbumin in visual areas of the occipital and temporal lobes of the macaque monkey. J Comp Neurol 412:515–526

DeFelipe J, Alonso-Nanclares L, Arellano JI (2002) Microstructure of the neocortex: comparative aspects. J Neurocytol 31:299–316

DeFelipe J, Ballesteros-Yañez I, Inda MC, Munoz A (2006) Double-bouquet cells in the monkey and human cerebral cortex with special reference to areas 17 and 18. Prog Brain Res 154:15–32

DeFelipe J, López-Cruz PL, Benavides-Piccione R, Bielza C, Larrañaga P, Anderson S, Burkhalter A, Cauli B, Fairén A, Feldmeyer D, Fishell G, Fitzpatrick D, Freund TF, González-Burgos G, Hestrin S, Hill S, Hof PR, Huang J, Jones EG, Kawaguchi Y, Kisvárday Z, Kubota Y, Lewis DA, Marín O, Markram H, McBain CJ, Meyer HS, Monyer H, Nelson SB, Rockland K, Rossier J, Rubenstein JL, Rudy B, Scanziani M, Shepherd GM, Sherwood CC, Staiger JF, Tamás G, Thomson A, Wang Y, Yuste R, Ascoli GA (2013) New insights into the classification and nomenclature of cortical GABAergic interneurons. Nat Rev Neurosci 14(3):202–216

del Rio MR, DeFelipe J (1997a) Double bouquet cell axons in the human temporal neocortex: relationship to bundles of myelinated axons and colocalization of calretinin and calbindin D-28k immunoreactivities. J Chem Neuroanat 13:243–251

del Rio MR, DeFelipe J (1997b) Synaptic connections of calretinin-immunoreactive neurons in the human neocortex. J Neurosci 17:5143–5154

Di Rosa E, Crow TJ, Walker MA, Black G, Chance SA (2009) Reduced neuron density, enlarged minicolumn spacing and altered ageing effects in fusiform cortex in schizophrenia. Psychiatry Res 166:102–115

Escobar MI, Pimenta H, Caviness VS Jr, Jacobson M, Crandall JE, Kosik KS (1986) Architecture of apical dendrites in the murine neocortex: dual apical dendritic systems. Neuroscience 17:975–989

Eyles DW, McGrath JJ, Reynolds GP (2002) Neuronal calcium-binding proteins and schizophrenia. Schizophr Res 57:27–34

Favorov OV, Diamond ME (1990) Demonstration of discrete place-defined columns-segregates-in the cat SI. J Comp Neurol 298:97–112

Feldman ML, Peters A (1974) A study of barrels and pyramidal dendritic clusters in the cerebral cortex. Brain Res 77:55–76

Ferrer I, Tuñon T, Serrano MT, Casas R, Alcantara S, Zujar MJ, Rivera RM (1993) Calbindin D-28k and parvalbumin immunoreactivity in the frontal cortex in patients with frontal lobe dementia of non-Alzheimer type associated with amyotrophic lateral sclerosis. J Neurol Neurosurg Psychiatry 56:257–261

Fleischauer K, Petsche H, Wittkowski W (1972) Vertical bundles of dendrites in the neocortex. Z Anat Entwickl Gesch 136:213–223

Gabbott PL, Bacon SJ (1996) Local circuit neurons in the medial prefrontal cortex (areas 24a,b,c, 25 and 32) in the monkey: II. Quantitative areal and laminar distributions. J Comp Neurol 364:609–636

Gabbott PL, Dickie BG, Vaid RR, Headlam AJ, Bacon SJ (1997) Local-circuit neurones in the medial prefrontal cortex (areas 25, 32 and 24b) in the rat: morphology and quantitative distribution. J Comp Neurol 377:465–499

Glezer II, Morgane PJ (1990) Ultrastructure of synapses and golgi analysis of neurons in neocortex of the lateral gyrus (visual cortex) of the dolphin and pilot whale [published erratum appears in Brain Res. Bull. (1990) 24:639–640]. Brain Res Bull 24: 401–427

Glezer II, Hof PR, Morgane PJ (1992) Calretinin-immunoreactive neurons in the primary visual cortex of dolphin and human brains. Brain Res 595:181–188

Glezer II, Hof PR, Leranth C, Morgane PJ (1993) Calcium-binding protein-containing neuronal populations in mammalian visual cortex: a comparative study in whales, insectivores, bats, rodents, and primates. Cereb Cortex 3:249–272

Glezer II, Hof PR, Morgane PJ (1998) Comparative analysis of calcium-binding protein-immunoreactive neuronal populations in the auditory and visual systems of the bottlenose dolphin (*Tursiops truncatus*) and the macaque monkey (*Macaca fascicularis*). J Chem Neuroanat 15:203–237

Gomez-Tortosa E, Sanders JL, Newell K, Hyman BT (2001) Cortical neurons expressing calcium binding proteins are spared in dementia with Lewy bodies. Acta Neuropathol (Berl) 101:36–42

Goodall J (1986) The chimpanzees of gombe: patterns of behavior. Harvard University Press, Cambridge, MA

Gustafsson L (1997) Inadequate cortical feature maps: a neural circuit theory of autism. Biol Psychiatry 42:1138–1147

Hare B, Melis AP, Woods V, Hastings S, Wrangham R (2007) Tolerance allows bonobos to outperform chimpanzees on a cooperative task. Curr Biol 17:619–623

Hendry SHC (1987) Recent advances in understanding the intrinsic circuitry of the cerebral cortex. In: Wise SP (ed) Higher brain functions: recent explorations of the brain's emergent properties. Wiley, New York, pp 241–283

Hendry SHC, Schwark HD, Jones EG, Yan J (1987) Number and proportions of GABA-immunoreactive neurons in different areas of monkey cerebral cortex. J Neurosci 7:1503–1519

Hendry SHC, Jones EG, Emson PC, Lowson DEM, Heizmann CW, Streit P (1989) Two classes of cortical GABA neurons defined by differential calcium binding protein immunoreactivities. Exp Brain Res 76:467–472

Herculano-Houzel S, Collins CE, Wong P, Kaas JH, Lent R (2008) The basic nonuniformity of the cerebral cortex. Proc Natl Acad Sci U S A 105:12593–12598

Hof PR, Sherwood CC (2005) Morphomolecular neuronal phenotypes in the neocortex reflect phylogenetic relationships among certain mammalian orders. Anat Rec 287A:1153–1163

Hof PR, Cox K, Young WG, Celio MR, Rogers J, Morrison JH (1991) Parvalbumin-immunoreactive neurons in the neocortex are resistant to degeneration in Alzheimer's disease. J Neuropathol Exp Neurol 50:451–462

Hof PR, Morrison JH (1991) Neocortical neuronal subpopulations labeled by a monoclonal antibody to calbindin exhibit differential vulnerability in Alzheimer's disease. Exp Neurol 111:293–301

Hof PR, Nimchinsky EA, Celio MR, Bouras C, Morrison JH (1993) Calretinin-immunoreactive neocortical interneurons are unaffected in Alzheimer's disease. Neurosci Lett 152:145–148

Hof PR, Nimchinsky EA, Buée-Scherrer V, Buée L, Nasrallah J, Hottinger AF, Purohit DP, Loerzel AJ, Steele JC, Delacourte A, Bouras C, Morrison JH, Perl DP (1994) Amyotrophic lateral sclerosis/parkinsonism-dementia complex of Guam: quantitative neuropathology, immunohistochemical analysis of neuronal vulnerability, and comparison with related neurodegenerative disorders. Acta Neuropathol 88:397–404

Hof PR, Bogaert YE, Rosenthal RE, Fiskum G (1996) Distribution of neuronal populations containing neurofilament protein and calcium-binding proteins in the canine neocortex: regional analysis and cell typology. J Chem Neuroanat 11:81–98

Hof PR, Glezer II, Condé F, Flagg RA, Rubin MB, Nimchinsky EA, Vogt Weisenhorn DM (1999) Cellular distribution of the calcium-binding proteins parvalbumin, calbindin, and calretinin in the neocortex of mammals: phylogenetic and developmental patterns. J Chem Neuroanat 16:77–116

Hof PR, Glezer II, Nimchinsky EA, Erwin JM (2000) Neurochemical and cellular specializations in the mammalian neocortex reflect phylogenetic relationships: evidence from primates, cetaceans, and artiodactyls. Brain Behav Evol 55:300–310

Huxley TH (1863) Evidence as to man's place in nature. Williams and Norgate, London [1959, Ann Arbor, University of Michigan]

Kaas JH, Nelson RJ, Sur M, Merzenich MM (1981) Organisation of somatosensory cortex in primates. In: Schmitt O, Worden FG, Adelman G, Dennis SG (eds) The organization of the cerebral cortex. MIT Press, Cambridge MA, pp 237–261

Kohn A, Pinheiro A, Tommerdahl MA, Whitsel BL (1997) Optical imaging in vitro provides evidence for the minicolumnar nature of cortical response. Neuroreport 8:3513–3518

Lawrence YA, Kemper TL, Bauman ML, Blatt GJ (2010) Parvalbumin-, calbindin-, and calretinin-immunoreactive hippocampal interneuron density in autism. Acta Neurol Scand. 121(2):99–108

Livingstone MS, Hubel DH (1984) Specificity of intrinsic connections in primate connections of primary visual cortex. J Neurosci 4:2830–2835

Manger PR (2006) An examination of cetacean brain structure with a novel hypothesis correlating thermogenesis to the evolution of a big brain. Biol Rev 81:293–338

Manger PR, Spocter MA, Patzke N (2013) The evolutions of large brain size in mammals: the over 700 gram club quartet. Brain Behav Evol 82(1):68–78

Marino L, Butti C, Connor RC, Fordyce RE, Herman LM, Hof PR, Lefebvre L, Lusseau D, McCowan B, Nimchinsky EA, Pack AA, Reidenberg JS, Reiss D, Rendell L, Uhen MD, Van der Gucht E, Whitehead H (2008) A claim in search of evidence: reply to Manger's thermogenesis hypothesis of cetacean brain structure. Biol Rev 83:417–440

Morgane PJ, Glezer II, Jacobs MS (1988) Visual cortex of the dolphin: an image analysis study. J Comp Neurol 273:3–25

Mountcastle VB (1997) The columnar organization of the neocortex. Brain 120:701–722

Mountcastle VB, Berman AL, Davies P (1955) Topographic organization and modality representation in the first somatic area of cat's cerebral cortex by method of single unit analysis. Am J Physiol 183:464

Nagy Z, Esiri MM, Jobst KA, Morris JH, King EM, McDonald B, Litchfield S, Barneston L (1996) Clustering of pathological features in Alzheimer's disease: clinical and neuroanatomical aspects. Dementia 7:121–127

Nishiyama E, Ohwada J, Iwamoto N, Arai H (1993) Selective loss of calbindin D28K-immunoreactive neurons in the cortical layer II in brains of Alzheimer's disease: a morphometric study. Neurosci Lett 163:223–226

Orban GA (1984) Studies in brain function, vol II, Neuronal operation in the visual cortex. Springer, Berlin

Palagi E (2006) Social play in bonobos (Pan paniscus) and chimpanzees (Pan troglodytes): implications for natural social systems and interindividual relationships. Am J Phys Anthropol 129:418–426

Parish AR, de Waal FB (2000) The other "closest living relative". How bonobos (Pan paniscus) challenge traditional assumptions about females, dominance, intra- and intersexual interactions, and hominid evolution. Ann NY Acad Sci 907:97–113

Petanjek Z, Kostovic I, Esclapez M (2009) Primate-specific origins and migration of cortical GABAergic neurons. Front Neuroanat. 27:3–26

Peters A, Kara D (1987) The neuronal composition of area 17 in rat visual cortex. The organization of pyramidal cells. IV. J Comp Neurol 260:573–590

Peters A, Sethares C (1991) Organization of pyramidal neurons in area 17 of monkey visual cortex. J Comp Neurol 306:1–23

Peters A, Sethares C (1996) Myelinated axons and the pyramidal cell modules in monkey primary visual cortex. J Comp Neurol 365:232–255

Peters A, Sethares C (1997) The organization of double bouquet cells in monkey striate cortex. J Neurocytol 26:779–797

Peters A, Walsh TM (1972) A study of the organization of apical dendrites in the somatic sensory cortex of the rat. J Comp Neurol 144:253–268

Peters A, Yilmaz E (1993) Neuronal organization in area 17 of cat visual cortex. Cereb Cortex 3:49–68

Povysheva NV, Zaitsev AV, Rotaru DC, Gonzalez-Burgos G, Lewis DA, Krimer LS (2008) Parvalbumin-positive basket interneurons in monkey and rat prefrontal cortex. J Neurophysiol 100:2348–2360

Preuss TM (2001) The discovery of cerebral diversity: an unwelcomed scientific revolution. In: Falk D, Gibson KR (eds) Evolutionary anatomy of primate cerebral cortex. Cambridge University Press, Cambridge, pp 138–164

Preuss TM, Coleman GQ (2002) Human-specific organization of primary visual cortex: alternating compartments of dense Cat-301 and calbindin immunoreactivity in layer 4A. Cereb Cortex 12:671–691

Preuss TM, Goldman-Rakic PS (1991) Architectonics of the parietal and temporal association cortex in the strepsirhine primate Galago compared to the anthropoid primate Macaca. J Comp Neurol 310:475–506
Pugliese M, Carrasco JL, Geloso MC, Mascort J, Michetti F, Mahy N (2004) γ-aminobutyric acidergic interneuron vulnerability to aging in canine prefrontal cortex. J Neurosci Res 77:913–920
Raghanti MA, Spocter MA, Butti C, Hof PR, Sherwood CC (2010) A comparative perspective on minicolumns and inhibitory GABAergic interneurons in the neocortex. Front Neuroanat 4:3
Rakic P (2008) Confusing cortical columns. Proc Natl Acad Sci U S A 105:12099–12100
Rao SG, Williams GV, Goldman-Rakic PS (1999) Isodirectional tuning of adjacent interneurons and pyramidal cells during working memory: evidence for microcolumnar organization in PFC. J Neurophysiol 81:1903–1916
Reynolds GP, Beasley CL (2001) GABAergic neuronal subtypes in the human frontal cortex–development and deficits in schizophrenia. J Chem Neuroanat 22:95–100
Reynolds GP, Zhang ZJ, Beasley CL (2001) Neurochemical correlates of cortical GABAergic deficits in schizophrenia: selective losses of calcium binding protein immunoreactivity. Brain Res Bull 55:579–584
Reynolds G, Abdul-Monim Z, Neill J, Zhang Z (2004) Calcium binding protein markers of GABA deficits in schizophrenia- postmortem studies and animal models. Neurotox Res 6:57–61
Rilling JK, Glasser MF, Preuss TM, Ma X, Zhao T, Hu X, Behrenns TEJ (2008) The evolution of the arcuate fasciculus revealed with comparative DTI. Nat Neurosci 11:426–428
Rockel AJ, Hiorns RW, Powell TP (1980) The basic uniformity in the structure of the neocortex. Brain 103:221–244
Sakai T, Oshima A, Nozaki Y, Ida I, Haga C, Akiyama H, Nazazato Y, Mikuni M (2008) Changes in density of calcium-binding-protein-immunoreactive GABAergic neurons in prefrontal cortex in schizophrenia and bipolar disorder. Neuropathology 28:143–150
Schenker NM, Buxhoeveden D, Blackmon WL, Amunts K, Zilles K, Semendeferi K (2008) A comparative quantitative analysis of cytoarchitecture and minicolumnar organization in Broca's area in humans and great apes. J Comp Neurol 510:117–128
Schlaug G, Schleicher A, Zilles K (1995) Quantitative analysis of the columnar arrangement of neurons in the human cingulate cortex. J Comp Neurol 351:441–452
Seldon HL (1981) Structure of human auditory cortex I: cytoarchitectonics and dendritic distributions. Brain Res 229:277–294
Sherwood CC, Raghanti MA, Stimpson CD, Bonar CJ, de Sousa AJ, Preuss TM, Hof PR (2007) Scaling of inhibitory interneurons in areas V1 and V2 of anthropoid primates as revealed by calcium-binding protein immunohistochemistry. Brain Behav Evol 69:176–195
Sherwood CC, Stimpson CD, Butti C, Bonar CJ, Newton AL, Allman JM, Hof PR (2009) Neocortical neuron types in Xenarthra and Afrotheria: implications for brain evolution in mammals. Brain Struct Funct 213:301–328
Skoglund TS, Pascher R, Berthold CH (1996) Heterogeneity in the columnar number of neurons in different neocortical areas in the rat. Neurosci Lett 208:97–100
Tommerdahl M, Favorov OV, Whitsel BL, Nakhle B, Gonchar YA (1993) Minicolumnar activation patterns in cat and monkey S1 cortex. Cereb Cortex 3:399–411
Von Bonin G, Mehler WR (1971) On columnar arrangement of nerve cells in cerebral cortex. Brain Res 27:1–9
White EL, Peters A (1993) Cortical modules in the posteromedial barrel subfield (Sml) of the mouse. J Comp Neurol 334:86–96
Wobber V, Wrangham R, Hare B (2010) Bonobos exhibit delayed development of social behavior and cognition relative to chimpanzees. Curr Biol 20:226–230
Woo TU, Miller JL, Lewis DA (1997) Schizophrenia and the parvalbumin-containing class of cortical local circuit neurons. Am J Psychiatry 154:1013–1015

Woo TU, Whitehead RE, Melchitzky DS, Lewis DA (1998) A subclass of prefrontal Gamma-aminobutyric acid axon terminals are selectively altered in schizophrenia. Proc Natl Acad Sci U S A 95:5341–5346

Wrangham RW (1999) The evolution of coalitionary killing. Yearb Phys Anthropol 42:1–30

Zaitsev AV, Gonzalez-Burgos G, Povysheva NV, Kroner S, Lewis DA, Krimer LS (2005) Localization of calcium-binding proteins in physiologically and morphologically characterized interneurons of monkey dorsolateral prefrontal cortex. Cereb Cortex 15:1178–1186

Zamecnik J, Krsek P, Druga R, Marusic P, Benes V, Tichy M, Komarek V (2006) Densities of parvalbumin-immunoreactive neurons in non-malformed hippocampal sclerosis-temporal neocortex and in cortical dysplasias. Brain Res Bull 68:474–481

Zhang ZJ, Reynolds GP (2002) A selective decrease in the relative density of parvalbumin-immunoreactive neurons in the hippocampus in schizophrenia. Schizophr Res 55:1–10

Chapter 6
Unusual Cortical Lamination Patterns in the Sengis (Elephant Shrews) Do Not Appear to Influence the Presence of Cortical Minicolumns

Paul Manger, Nina Patzke, Nadine Gravett, Katarina Medger, Consolate Kaswera, Emmanuel Gilissen, and Nigel Bennett

Abstract The current study used a range of standard and immunohistochemical neuroanatomical stains to investigate the architectural organization of the cerebral cortex of the sengis (elephant shrews). We were able to identify four distinct cortical morphotypes – a medial neocortical, a lateral neocortical, a cingulate and a piriform. While the architectural organization of the medial neocortical, lateral neocortical and piriform morphotypes were similar to that seen in many other mammals, the cingulate cortical morphotype displayed a lamination pattern unique to the order Macroscelidae order. The cingulate cortex of the sengis displayed a very wide layer 4 and a reduced and granular-looking layer 3. Despite this variation in the laminar architecture of the cingulate cortex, cortical minicolumns were present in all regions of the cerebral cortex. Interestingly, all regions of cortex displayed apical dendrites immunoreactive to uncoupling protein 2, a marker of thermogenesis. Thus, the sengis display a unique mixture of morphologies that are standard across mammals and unique to their order. This mixture indicates that the development of horizontal lamination patterns and vertical columnar organization

P. Manger (✉) • N. Patzke • N. Gravett
School of Anatomical Sciences, University of the Witwatersrand, 7 York Road Parktown, 2193 Johannesburg, South Africa
e-mail: Paul.Manger@wits.ac.za; Nina.Patzke@wits.ac.za; Nadine.Gravett@wits.ac.za

K. Medger • N. Bennett
Department of Zoology and Entomology, University of Pretoria, Pretoria, South Africa
e-mail: kmedger@zoology.up.ac.za; ncbennett@zoology.up.ac.za

C. Kaswera
Faculté des Sciences, University of Kisangani, Kisangani, Democratic Republic of the Congo
e-mail: consolatekyams@googlemail.com

E. Gilissen
Department of African Zoology, Royal Museum for Central Africa; Laboratory of Histology and Neuropathology, Université libre de Bruxelles, Brussels, Belgium

Department of Anthropology, University of Arkansas, Fayetteville, AR, USA
e-mail: Emmanuel.gilissen@africamuseum.be

M.F. Casanova, I. Opris (eds.), *Recent Advances on the Modular Organization of the Cortex*, DOI 10.1007/978-94-017-9900-3_6

may be distinctly controlled and not interdependent aspects of cortical development. The sengis may provide a unique animal model of cortical development that could unlock clues to the development of cortical architectural variation across mammalian species.

Keywords Sengi • Elephant shrew • Cerebral cortex • Architecture • Evolution • Cortical minicolumn

6.1 Introduction

The sengis, or elephant shrews, belong to the order Macroscelidea, which is part of the afrotherian superorder (Stanhope et al. 1998; Asher et al. 2009). These small, insectivorous mammals possess a range of behavioural, morphological and physiological specializations that allow them to inhabit a range of environments throughout sub-Saharan Africa from deserts through to rainforests, with one species found in the semi-arid north-west of the continent (Nowack 1999; Skinner and Chimimba 2005; Rathbun 2009). Despite their interesting behavioural, physiological and morphological characteristics, studies of the brains of these species using modern techniques are not common (Stephan et al. 1981; Dengler-Crish et al. 2006; Sherwood et al. 2009; Pieters et al. 2010; Calvey et al. 2013; Kaufmann et al. 2013; Slomianka et al. 2013). Kaufmann et al. (2013) demonstrated that in comparison to species with similar body sizes and diets, the sengis have relatively large brains. The encephalization quotients (EQ) of the sengis ranges between 1.11 to 1.55 (*Elephantulus fuscipes* 1.30; *Elephantulus myurus* 1.48; *Rhynchocyon stuhlmanni* 1.24; *Rhynchocyon petersi* 1.13; *Rhynchocyon udzugwensis* 1.11; *Petrodromus tetradactylus* 1.55; based on data from the current study for *P. tetradactylus* and that provided in Kaufmann et al. 2013, using the regression derived in Manger 2006 to calculate EQ) indicating that they do indeed have somewhat larger brains than would be expected for their body masses. Slomianka et al. (2013) describe the sengis as having larger than normal hippocampi, and while not providing data for their conclusion, the recent study by Patzke et al. (2015) indicates that the relative size of the sengi hippocampus does indeed fall slightly above the upper 95 % confidence intervals of mammalian hippocampal size. Thus, the sengis appear to have somewhat larger relative brain masses than expected and larger relative hippocampal volumes than expected when compared to other mammals. In addition, cholinergic neurons are found in the colliculi and the cochlear nuclei (Pieters et al. 2010; Calvey et al. 2013), and these are not normally observed in other mammals (Dell et al. 2010).

Two recent studies, those of Dengler-Crish et al. (2006) and Sherwood et al. (2009), are the only ones that provide any data using modern techniques on the structure and organization of the sengi cerebral cortex. The study of Dengler-Crish et al. (2006), examining the Cape elephant shrew (*E. edwardii*), revealed that the general areal organization of the cerebral cortex of this species of elephant shrew was, for the most part, not dramatically dissimilar to that seen in many other small mammals. Dengler-Crish et al. (2006) noted that a significant amount of the cortical mantle was occupied by the visual cortex, with distinct primary and secondary somatosensory areas observed; however, unlike in many other mammalian species, the flattened cortex technique combined with cytochrome oxidase staining did not reveal distinct representations of specific body parts. In their study of a range of immunohistochemically identifiable cortical neurons, Sherwood et al. (2009) detailed that the presence and organization of a range of neuronal types, including those distinguished by non-phosphorylated neurofilament H (SMI-32), parvalbumin, calbindin, calretinin and neuropeptide Y immunohistochemistry, were not dramatically different in the individual giant elephant shrew (*R. petersi*) they studied when compared to the related rock hyrax (*Procavia capensis*) and xenarthrans. Interestingly, Sherwood et al. (2009) indicate the presence of cortical minicolumns in the afrotherian and xenarthran species studied, but this appears to be associated with a greater diversity of cortical projection neurons in these Atlantogenata species when compared with the Boreoeutheria mammalian lineage. This contrasts with the less distinct minicolumnar, or vertical, organization of the monotreme and marsupial mammalian radiations. Thus, it would appear that within the different mammalian lineages, distinct groupings of minicolumnar, or vertical, cortical organization are present.

In our earlier studies on the sengi brain (Pieters et al. 2010; Calvey et al. 2013) we noticed unusual cortical lamination patterns, especially in the cingulate cortex. The current report details our observations of the patterns of cortical lamination in the sengis (*E. myurus* and *P. tetradactylus*) and provides observations on the minicolumnar organization of the cerebral cortex in relation to these varied lamination patterns.

6.2 Materials and Methods

Three male eastern rock sengis, *Elephantulus myurus* (average body mass 45.08 g, average brain mass 1.27 g) and three male four-toed sengis, *Petrodromus tetradactylus* (average body mass 137.37 g, average brain mass 3.01 g), were used in this study. All animals were treated and used according to the guidelines of the University of the Witwatersrand Animal Ethics Committee, which parallel those of the NIH for the care and use of animals in scientific experimentation (clearance number 2008/36/1). The animals were caught under appropriate governmental permission in South Africa and the Democratic Republic of the Congo. Following capture the animals were euthanized (Euthanaze, 200 mg sodium pentobarbital/kg, i.p.). Upon cessation of breathing the animals were perfused intracardially, initially with a rinse solution of 0.9 % saline at 4 °C, followed by 4 % paraformaldehyde in 0.1 M phosphate buffer (PB) at 4 °C. The brain was then removed from the skull and post-fixed in 4 % paraformaldehyde in 0.1 M PB overnight at 4 °C. The brains were then allowed to equilibrate in 30 % sucrose in

0.1 M PB at 4 °C, and subsequently stored in an antifreeze solution at – 20 °C until use (Manger et al. 2009). The brains were re-equilibrated in 30 % sucrose, frozen in crushed dry ice and sectioned in the coronal plane (50 μm section thickness) using a sliding microtome.

Serial sections were stained for Nissl substance, myelin, parvalbumin (PV, PV28, Swant, raised in rabbit, 1:10,000 dilution), calbindin (CB, CB38a, Swant, raised in rabbit, 1:10,000 dilution), calretinin (CR, 7699/3H, Swant, raised in rabbit, 1:10,000 dilution), uncoupling protein 2 (UCP-2, Santa Cruz, raised in goat, 1:100 dilution) and neurofilament H non-phosphorylated (SMI-32, COVANCE, raised in mouse, 1:1,000 dilution). Sections to be stained for Nissl substance were mounted on 0.5 % gelatine coated glass slides, cleared in a 1:1 solution of chloroform and absolute alcohol (24 h), after which they were stained using a 1 % cresyl violet solution. Sections used for the myelin series were stored for 2 weeks in a 5 % formalin solution at 4 °C, mounted on 1.5 % gelatine coated glass slides, and stained using a modified silver stain (Gallyas 1979).

The sections used for immunohistochemical staining were first treated with an endogenous peroxidase inhibitor (49.2 % methanol, 49.2 % of 0.1 M PB: 1.6 % of 30 % H_2O_2) for 30 min, followed by a series of three 10 min rinses in 0.1 M PB. This was followed by a 2 h pre-incubation, at room temperature, in a blocking buffer solution containing 3 % normal goat serum (NGS, Chemicon)/3 % normal rabbit serum (NRS, Chemicon), 2 % bovine serum albumin (BSA, Sigma), and 0.25 % Triton X-100 (Merck) in 0.1 M PB. The sections were then placed in a primary antibody solution containing the appropriately diluted antibody in blocking buffer (see above), for 48 h at a temperature of 4 °C under gentle shaking. Following this the sections underwent three 10 min rinses in 0.1 M PB, followed by incubation in a secondary antibody at room temperature for 2 h. The secondary antibody solution contained a 1:1,000 dilution of biotinylated anti-rabbit IgG/anti-goat IgG (BA-1000/BA-5000, Vector Labs) in 3 % NGS/NRS, and 2 % BSA in 0.1 M PB. Following three 10 min rinses in 0.1 M PB, the sections underwent a one-hour incubation in AB solution (Vector Labs), and were again rinsed three times in 0.1 M PB. Sections were then placed in a 0.05 % diaminobenzidine (DAB) in 0.1 M PB solution for 5 min, followed by the addition of 3 μl of 3 % hydrogen peroxide to each 1 ml of solution in which each section was immersed. Chromatic precipitation was visually monitored and verified under a low power stereomicroscope. Staining was allowed to continue until such time that the background stain was at a level that would assist analysis without obscuring the immunopositive structures. Precipitation was arrested by placing sections in 0.1 M PB, followed by two more rinses in this solution. Sections were then mounted on 0.5 % gelatine coated glass slides, dried overnight, dehydrated in a graded series of alcohols, cleared in xylene and coverslipped with Depex. Omission of the primary or secondary antibody in selected sections was employed as controls for the immunohistochemical protocol, for which no staining was evident.

6.3 Results

The range of basic and immunohistochemical stains used in the current study allowed the identification of four distinct cerebral cortical morphotypes based on laminar variations including: (1) the medial neocortical morphotype; (2) the lateral neocortical morphotype; (3) the cingulate cortical morphotype; and (4) the piriform cortical morphotype. While the lamination pattern varied between these four cortical morphotypes, in all regions of the cerebral cortex of the sengis, minicolumns were evident although they were more distinct in some regions than others. An unexpected immunopositivity of the apical dendrites of pyramidal cells to uncoupling protein 2 (UCP 2) points to a unique aspect in the vertical processing of information in the sengi cortical minicolumns. In most cases the staining and lamination patterns were similar between the two species studied, so unless otherwise noted, the following description applies to both species.

6.3.1 The Medial Neocortical Morphotype

The medial neocortical morphotype occupied the majority of the dorsal medial surface of the cerebral hemisphere from a millimetre or two from the mid-sagittal fissure, spreading laterally to a similar distance from the rhinal sulcus. Apart from the more anterior portions of the cortex, which lacked a layer 4 and corresponds to the motor, pre-motor and prefrontal cortex, the medial neocortical morphotype presented with 6 distinct cortical layers (Fig. 6.1). A cell sparse layer 1 exhibited several tangentially oriented myelinated fibres, and interestingly, a number of calretinin immunopositive neurons and fibres were clearly evident in this layer. A distinct layer 2 formed of granular cells, numbering approximately 4–6 in depth was observed. Within layer 2, neurons immunopositive to parvalbumin, calbindin and calretinin were observed with a slightly higher neuropil staining of both parvalbumin and calbindin being evident. Layer 3 formed the thickest of the cortical layers and was made up of a mixture of pyramidal and other neurons, with an increasing size of pyramidal neuron somata with increasing depth in the layer. Occasional parvalbumin and calretinin immunopositive neurons were observed and a significantly higher number of calbinin immunopositive neurons were present. The deeper half of layer 3 revealed a denser neuropil immunoreactivity for calbindin when compared with adjacent layers and suggests a sublamination of layer 3. This suggested sublamination is supported by similar more intense SMI-32 cellular staining in deep layer 3 compared with superficial layer 3. In the posterior two thirds, or sensory regions of the medial neocortical morphotype, a distinct granular layer 4 was observed. A number of tangentially oriented myelinated fibres and parvalbumin immunopositive neurons were observed in layer 4, but only a few calbindin or calrentinin immunopositive neurons were found. A distinct strong neuropil staining for parvalbumin delineated the

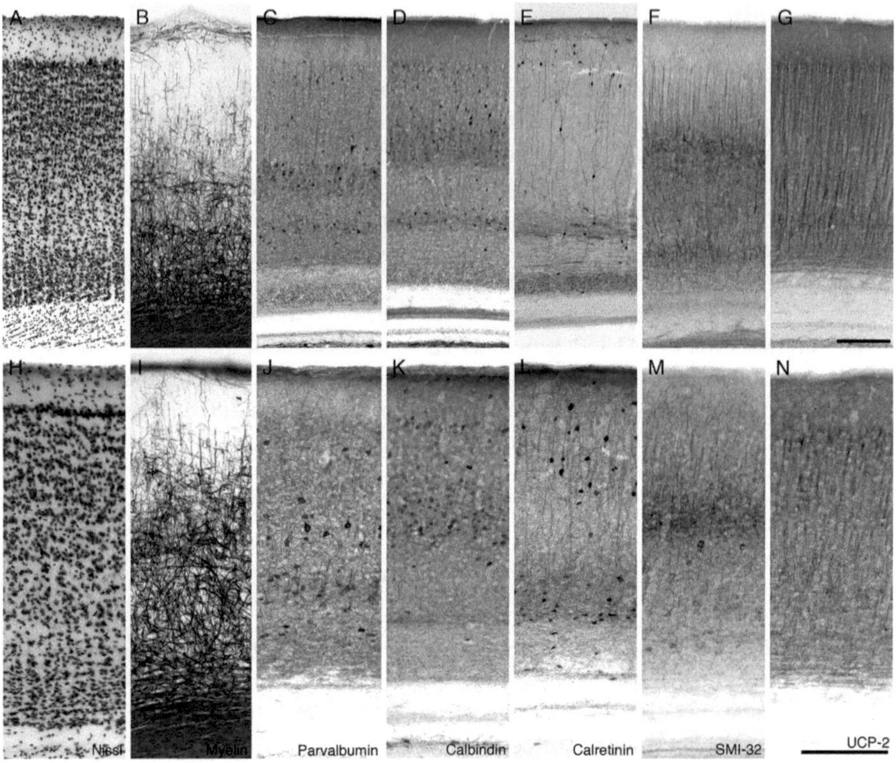

Fig. 6.1 Photomicrographs of the medial cerebral neocortical morphotype of *Petrodromus tetradactylus* (*A-G*) and *Elephantulus myurus* (*H-N*) stained with Nissl (*A, H*), myelin (*B, I*), parvalbumin (*C, J*), calbindin (*D, K*), calretinin (*E, J*), SMI-32 (*F, M*) and uncoupling protein 2 (*G, N*). Note the six distinct cortical layers typical of mammals, with layer 4 being distinguished by intense parvalbumin neuropil immunoreactivity. Also note the clear presence of cortical minicolumns, the vertical calretinin dendritic immunoreactivity and the intense staining of the apical dendrites of the pyramidal cells for uncoupling protein 2. Scale bar in G = 250 μm and applies to A-G, scale bar in N = 200 μm and applies to H-N

upper and lower border of layer 4 and this layer was also associated with a strong neuropil staining for SMI-32. In comparison to layers 2–4, a thin layer 5 was relatively cell sparse and evinced the occasional large pyramidal neuron as well as extensive tangentially oriented myelinated fibres. Within layer 5 parvalbumin, calbindin and calretinin immunopositive neurons were present and distinct neuropil staining for these three markers was also evident. Interestingly, with the calretinin stain, large tangentially oriented fascicles were observed in *P. tetradactylus*, but this was not evident in *E. myurus*. In both species a moderate staining of cells and fibres with SMI-32 was observed. Palisades of 2–3 cell widths were clearly evident in the heavily myelinated layer 6. Staining for parvalbumin, calbindin and calretinin was sparse in this layer, but a moderate staining of neuronal soma and dendrites was observed with SMI-32. Of the four cortical morphotypes observed in the current study, the medial neocortical morphotype showed the clearest minicolumnar organization, which spanned layers 2–6. While all the stains used in the current study reflected this vertical aspect of cortical organization, perhaps this was most clearly observed in the calretinin staining (Fig. 6.1f, i) where vertical dendritic processes, often spanning several layers, were readily observed.

6.3.2 *The Lateral Neocortical Morphotype*

The lateral neocortical morphotype occupied the lateral most few millimetres of the neocortex from the rhinal sulcus. While 6 cortical layers were evident, the borders between these cortical layers were not as clearly defined as those observed in the medial neocortical morphotype (Fig. 6.2). Layer 1 was again cell sparse, had a few tangentially oriented myelinated fibres, and lacked any structures immunopositive for parvalbumin, calbindin and calretinin; however, an overall dense neuropil staining for calbindin and calretinin was observed. Layer 2 was formed by a distinct band of 2–3 granular cells with the occasional parvalbumin and calretinin immunopositive neuron being observed in *P. tetradactylus*, but these were absent in *E. myurus*. Layer 3 was the thickest layer, and again, potential sublamination, evinced by denser myelin and SMI-32 staining, was observed. Neurons immunopositive for parvalbumin, calbindin and calretinin were scattered throughout layer 3 in *P. tetradactylus*, but were not present in *E. myurus*. The upper and lower borders of layer 4 were indistinct, but a denser neuropil staining for parvalbumin appeared to correspond with layer 4. In both species a number of parvalbumin immunopositive neurons were observed and in *P. tetradactylus* a high density of moderately strongly stained calretinin immunopositive neurons was present, but this was lacking in *E. myurus*. Layer 5 was relatively cell sparse compared to adjacent layers, was myelin dense with tangential fibres and contained the occasional parvalbumin and calbindin immunopositive neuron. In *P. tetradactylus* a high density of moderately strongly stained calretinin immunopositive neurons was present in layer 5, but this was absent in *E. myurus*. Layer 6 was again formed by a series of vertically oriented palisades, but compared to the medial neocortical morphotype, these palisades were wider, being 4–6 cells in width, and more loosely associated than in the medial neocortical morphotype. Layer 6 was heavily myelinated, had the occasional parvalbumin, calbindin and calretinin immunopositive neuron, and in *P. tetradactylus* had a higher density of neuronal structures immunopositive for SMI-32, although this was much weaker in *E. myurus*. As with the medial neocortical morphotype, the minicolumnar organization was apparent in the Nissl stained sections (Fig. 6.2a, h), however, the compactness and observable tight organization of these minicolumns was not as apparent, especially in *E. myurus*. In addition, the calretinin staining observed in the lateral neocortical morphotype in *P. tetradactylus* evinces clear vertical processing, but this staining was absent in *E. myurus*. Interestingly, it appears that the overall organization of the lateral neocortical morphotype, while similar between species,

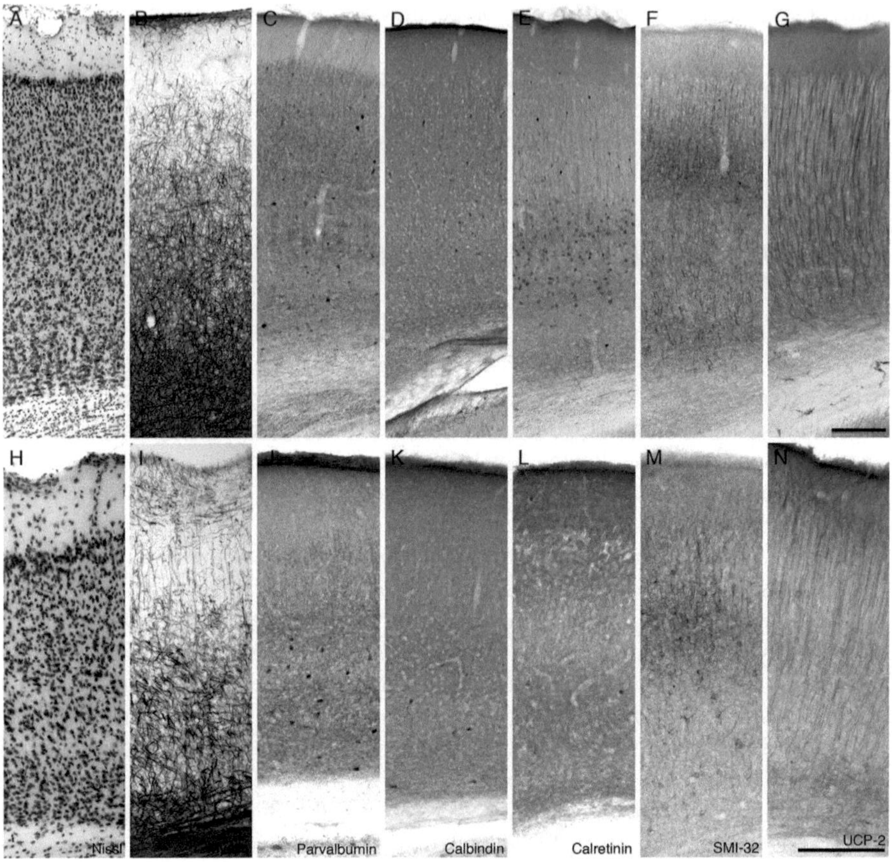

Fig. 6.2 Photomicrographs of the lateral cerebral neocortical morphotype of *Petrodromus tetradactylus* (*A-G*) and *Elephantulus myurus* (H-N) stained with Nissl (*A, H*), myelin (*B, I*), parvalbumin (*C, J*), calbindin (*D, K*), calretinin (*E, J*), SMI-32 (*F, M*) and uncoupling protein 2 (*G, N*). Note the six cortical layers typical of mammals, but layer 4 is less distinct than in the medial neocortical morphotype. Also note the presence of cortical minicolumns, the vertical calretinin dendritic immunoreactivity in *P. tetradactylus*, and the intense staining of the apical dendrites of the pyramidal cells for uncoupling protein 2. Scale bar in G = 250 μm and applies to A-G, scale bar in N = 200 μm and applies to *H-N*

appears to be more distinctly organized in *P. tetradactylus* than *E. myurus*, especially in terms of the calcium binding proteins.

6.3.3 The Cingulate Cortical Morphotype

The cingulate cortical morphotype observed in the sengis was very distinctive and had an unusual lamination pattern (Fig. 6.3), something we have not observed in

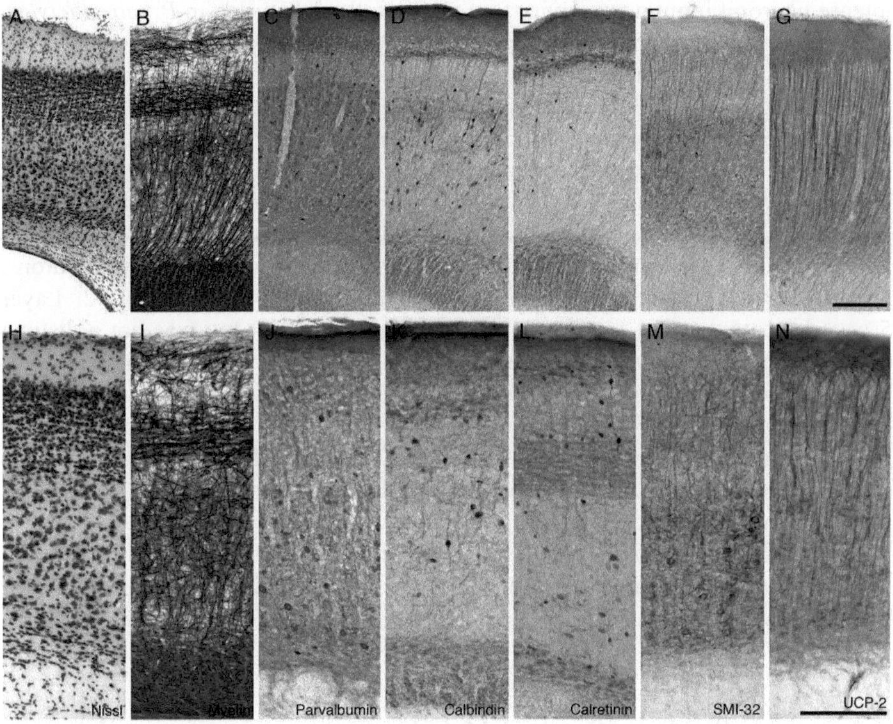

Fig. 6.3 Photomicrographs of the cingulate cerebral cortical morphotype of *Petrodromus tetradactylus* (*A-G*) and *Elephantulus myurus* (H-N) stained with Nissl (*A, H*), myelin (*B, I*), parvalbumin (*C, J*), calbindin (*D, K*), calretinin (*E, J*), SMI-32 (*F, M*) and uncoupling protein 2 (*G, N*). Note the unusual lamination pattern, especially of layer 3, with the greatly thickened layer 4. Also note the presence of cortical minicolumns although not as distinct as in the neocortical regions, the vertical calbindin dendritic immunoreactivity, and the intense staining of the apical dendrites for uncoupling protein 2. Scale bar in G = 250 μm and applies to A-G, scale bar in N = 200 μm and applies to H-N

other afrotherian species such as elephants, hyraxes, golden moles, and otter shrews. This cortical morphotype was found on the medial wall of the hemisphere, extending onto the dorsal surface for approximately 2 mm. Layer 1 was typically cell sparse, but was quite dense in tangentially oriented myelinated fibres. A high density neuropil staining was observed for parvalbumin, calbindin and calretinin, with the occasional calretinin immunopositive neuron being observed. Layer 2 was made up of a dense packing of granular cells with a low myelin density, but interesting in *P. tetradactylus* there was dense calbindin and calretinin neuropil immunoreactivity, but in *E. myurus* there was only dense calbindin immunoreactivity. Unlike the medial and lateral neocortical morphotypes described above, layer 3 was relatively thin and composed only of a high density of granular cells that formed a distinct band of 200–250 μm in depth. In both species this band was dense with tangentially oriented myelinated fibres and had a sparse neuropil staining for parvalbumin and calbindin. In *E. myurus*, layer 3 exhibited a more intense neuropil immunoreactivity for calretinin than that seen in *P. tetradactylus*. In both species of sengi the occasional parvalbumin, calbindin and calretinin immunopositive neuron was observed. Layer 4 was by far the thickest of the layers in the cingulate cortex, and in the transition from the medial neocortical morphotype to the cingulate cortical morphotype, the increase in thickness of layer 4 was evident from the dorsoventral spread of intense parvalbumin neuropil immunoreactivity. As a whole, this thicker layer 4 was myelin dense with both tangential and radial fibres, a higher number of parvalbumin and calbindin immunopositive neurons and a dearth of calretinin immunopositive neurons. Layer 4 also showed intense staining of neural structures for SMI-32. Layer 5 was present as a thin cell sparse region deep to layer 4, but was not a distinct lamina in this region of the cortex. Layer 6 was again made up of palisades, however, these were more loosely organized than in the other cortical morphotypes described above and were made up of 5–7 cells in width. Weak immunostaining for parvalbumin, calbindin, calretinin and SMI-32 distinguished this layer from the more superficial layers. Despite this unusual lamination pattern, the presence of minicolumns and vertical cortical processing were clearly evident with the Nissl stained sections, the radial fascicles in the myelin stained sections and the distinct vertical oriented dendrites of the calbindin immunopositive neurons in layer 4 (Fig. 6.3).

6.3.4 The Piriform Cortical Morphotype

The piriform cortical morphotype was found lateral to the rhinal sulcus and evinced three distinct layers (Fig. 6.4). Layer 1, as in other cortical regions, was cell sparse but possessed a number of tangentially oriented myelinated fibres. For the most part, immunostaining in layer 1 was weak, however, in *P. tetradactylus* a dense network of SMI-32 immunopositive dendrites was observed. Layer 2 was made up of a dense lamina of tightly packed granular cells, approximately 15 cells in depth. In both species this layer was myelin sparse, an intense parvalbumin neuropil was evident and the very occasional calbindin immunopositive neuron was observed. Interestingly, extraverted calretinin immunpositive neurons were present with dendrites that extended into and ramified within layer 1 (Fig. 6.4e, l). The third layer was the thickest layer and appeared to be composed of a range of neuronal morphologies. In *P. tetradactylus* this layer was quite dense with myelinated fibres, but in *E. myurus* the myelin density was lower. Large multipolar parvalbumin immunopositive neurons were present (Fig. 6.4c, j), as were a greater number of smaller multipolar calbindin immunopositive neurons (Fig. 6.4d, k). The occasional small calretinin immunopositive neuron was observed, as well as a moderate general staining of neuronal structures for SMI-32. The staining revealed the possibility of a superficial and deep sublamination of layer 3 based on the distribution of neurons and structures immunopositive for parvalbumin, calbindin, calretinin and SMI-32. Despite only being composed of three layers, evidence for

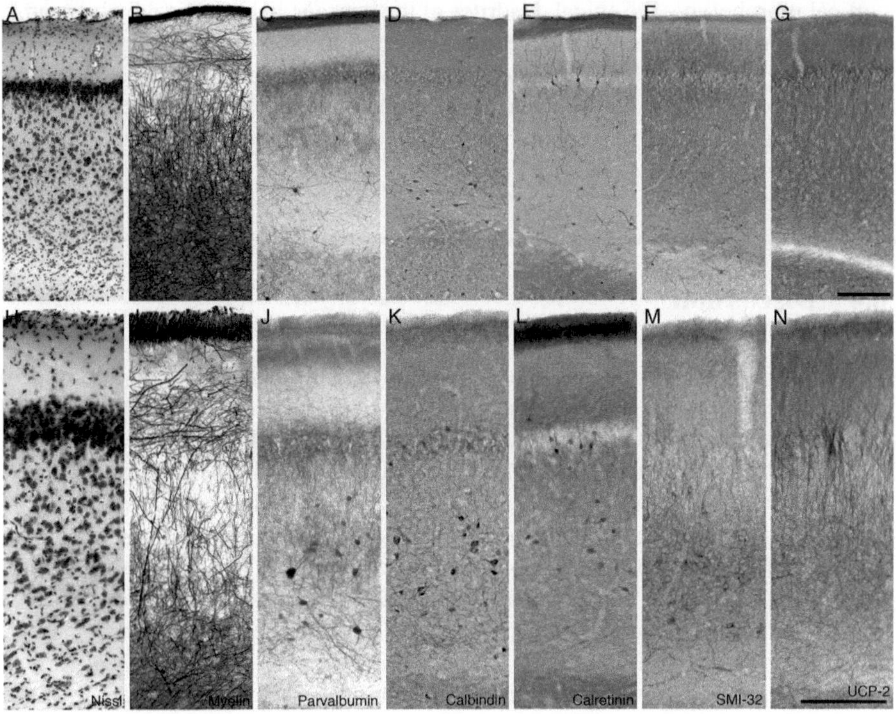

Fig. 6.4 Photomicrographs of the piriform cerebral cortical morphotype of *Petrodromus tetradactylus* (*A-G*) and *Elephantulus myurus* (*H-N*) stained with Nissl (*A, H*), myelin (*B, I*), parvalbumin (*C, J*), calbindin (*D, K*), calretinin (*E, J*), SMI-32 (*F, M*) and uncoupling protein 2 (*G, N*). Note that only three layers are present, that the cortical minicolumns are present but quite indistinct, the extraverted calretinin immunoreactive layer 2 neurons, and the intense staining of the apical dendrites arising from layer 3 for uncoupling protein 2. Scale bar in G = 250 μm and applies to *A-G*, scale bar in N = 200 μm and applies to *H-N*

the presence of minicolumnar, or vertical, cortical organization was present, especially with the myelin and calretinin stains.

6.3.5 Reactivity of Apical Dendrites to Uncoupling Protein 2 (UCP 2)

In all four cortical morphotypes, distinct staining of the apical, or vertically oriented, dendrites of pyramidal and other neurons was observed. In the medial and lateral neocortical morphotypes these were observed to mainly arise from the pyramidal neurons found in layers 3 and 5, and extended as tightly packed bundles (dendrons) through the cortical layers to layer 1 where the tips of these apical dendrites could be observed to ramify (Fig. 6.1g, 1 N, 2G, 2 N). In the cingulate cortical morphotype, the apical dendrites of the neurons in layer 4 were observed to form similar tightly packed bundles that ramified in layer 1 (Fig. 6.3g, 3 N). The packing of UCP-2 vertically oriented primary dendrites was not as compact in the piriform cortex as in the other cortical regions, but was still present (Fig. 6.4g, 4 N). These apical dendrites were again observed to traverse the cortical layers to ramify within the deeper levels of layer 1.

6.4 Discussion

The current study describes our observations of the cortical architecture of two previously unstudied species of sengis, *Petrodromus tetradactylus* and *Elephantulus myurus*. Our observations allow us to describe the existence of four distinct cortical morphotypes, and while, especially in the neocortex, these morphotypes may house several distinct cortical areas, the overall similarities in the patterns of staining allow this distinction to be readily made. The presence of medial and lateral neocortical morphotypes coincides with functional (Dengler-Crish et al. 2006) and connectional (Butler 1994, 1995) aspects of these cortical regions, while the cingulate and piriform cortical morphotypes are common features of mammalian cerebral cortical architecture. The presence of unusual cortical lamination patterns in the cingulate cortex and the staining of apical dendrites for uncoupling protein 2, appear to be unique features of the sengi cortex as we could not find similar patterns of staining in other afrotherians we have examined, including African elephant (*Loxodonta africana*), rock hyrax (*Procavia capensis*), golden mole (*Amblysomus hottentotus*), giant otter shrew (*Potomogale velox*) and manatee (*Trichechus manatus*) (unpublished observations).

6.4.1 Comparison to Previous Studies of Sengi Cerebral Cortex

As noted earlier, Dengler-Crish et al. (2006) provided evidence for a range of cortical areas in the cerebral cortex of the sengi, whose overall organization does not differ dramatically from that seen in many other mammalian species. In the current study the regions defined as medial and lateral neocortical morphotypes appear to correlate directly with the findings of Dengler-Crish and colleagues. The primary visual and primary somatosensory cortical areas described by Dengler-Crish et al. (2006) appear to be coincident with the medial neocortical morphotype, while the secondary somatosensory and auditory cortical areas appear to be coincident with the lateral neocortical morphotype. This distinction is reminiscent of the lemnothalamic and collothalamic cortical projection zones as defined by Butler (1994, 1995). Thus, the sengis may provide an interesting model species with which to investigate the distinction of these lemno- and collo- neocortical projection zones, as the cortical architecture is quite distinct between these regions and relates directly to the hypothesized functional differentiation of these cortical projection pathways. Interestingly, Dengler-Crish et al. (2006) also indicate, in comparison to a related afrotherian species (the tenrec, *Echinops telfairi*, Figure 6 of Dengler-Crish et al.), that the primary somatosensory and visual neocortical areas do not extend into the medial wall of the cerebral hemisphere, but are located entirely on the lateral surface of the cerebral hemisphere. The medial border of the primary somatosensory and visual cortex is coincident with the lateral border of the cingulate cortical morphotype as defined in the current study, which was found to extend onto the lateral surface of the cerebral hemisphere. Moreover, the lateral borders of the secondary somatosensory and auditory cortical regions are coincident with the piriform cortical morphotype defined here. Thus, the cortical morphotypes defined in the current study have distinct correlations with the functional subdivisions of the cerebral cortex of the sengi.

In their study of neocortical architecture of Xenarthrans and Afrotheria, Sherwood et al. (2009) report the results of a range of immunohistochemical stains in one species of sengi, namely *Rhynchocyon petersi*. Several of the stains used by Sherwood et al. (2009) are similar to those used in the current study. In the medial and lateral neocortical morphotypes, the immunostaining for SMI-32 appears similar across all three species, as does the staining for pavalbumin, calbindin and calretinin. This indicates that the cortical morphotypes described herein are likely to apply to the order Macrosceledidae as a whole (Manger 2005). It also indicates that the phylogenetic analysis undertaken by Sherwood et al. (2009) would produce the same results if the extra species of sengis studied herein were added. Thus, there appears to be consistency across the more recent studies of cortical architecture in the sengi, and that this organization shows an order specific suite of characters.

6.4.2 Unusual Cingulate Cortical Lamination

In this study we observed an unusual lamination pattern in the cingulate cerebral cortex of the sengi that has not been observed in previous studies. We have not observed this unusual lamination pattern in the anterior cingulate cortex of the African elephant, rock hyrax, golden mole, giant otter shrew and manatee (unpublished results). Thus, the lamination of the cingulate cortex in the sengis, with the thickened layer 4 and granular appearance of the layer 3 neurons, appears to be a feature distinct to the sengis. Without functional studies it is difficult to hypothesize as to why this region of cortex displays such an unusual lamination pattern, however, the presence of this unusual lamination pattern and the persistence of cortical minicolumns does indicate that while some aspects of the cortical architecture may change dramatically (lamination patterns), other aspect (minicolumns) appear to be retained. This indicates that these two features of cortical architecture, while clearly forming essential features of cortical information processing, are not interdependent in their evolutionary history, i.e. you can alter one without necessarily having to alter the other. While this may be a seemingly minor point, it does indicate that evolutionary change in the appearance of the cortical architecture is both consistent and malleable, and independent and interdependent, across mammalian species. Such changes in adult morphology, tracked through development, may help to elucidate mechanisms of cortical change across mammalian species and lead to a richer understanding of the diversity of cortical architectures observed across adult mammals.

6.4.3 Uncoupling Protein 2 Staining in the Apical Dendrites of Cortical Neurons

The observation of distinctly immunoreactive apical dendrites of cortical neurons for uncoupling protein 2 (UCP2) in the cerebral cortex of both species of sengi investigated is somewhat surprising. Our unpublished studies of immunostaining of UCP2 in other afrotherian mammals (African elephant, manatee and rock hyrax) do not reveal a similar pattern of staining, where the occasional neuron is stained as seen in the rodent cortex (e.g. Kim-Han and Dugan 2005). The uncoupling proteins belong to the mitochondrial anion carrier proteins that separate oxidative phosphorylation from the ATP cycle with the energy being dissipated as heat, and in the brain may create small temperature gradients (Bouillaud et al. 1985; Palou et al. 1998; Horvath et al. 2003). Thus, it appears that in the sengis there are two possibilities. The first is that the UCP2 staining seen here is an artefact and has no functional relationship to vertical cortical processing; however, we deem this unlikely, since there was clear staining in both species, across all cortical regions and not in other regions of the brain, although it is not possible to rule this explanation out at present. Despite this, if we accept that the staining is real staining, then this may have important functional consequences for vertical cortical processing in the sengis. The apical dendrites of the pyramidal and other neurons are in receipt of a great deal of specific synaptic input. If the cytoplasm of the apical dendrite was warmer than the surrounding neuronal and other structures, it is possible that the rate at which the apical dendritic potential reaches the neuronal soma may be increased. This would increase the rate at which vertical processing would occur in the cerebral cortex of the sengis. This may provide some adaptive advantage to these highly active and very fast moving animals (Rathbun 2009); however, this is all highly speculative and needs to be examined in more detail.

6.4.4 Summary

The present study of the cortical architecture in the sengi has revealed features that are distinct to this mammalian order, as well as features that are consistent across mammals. The three recent studies of cortical organization and architecture of the sengi cerebral cortex (Dengler-Crish et al. 2006; Sherwood et al. 2009; current study) appear to show good concordance using different approaches. Moreover, these studies indicate distinct lemno- and collo- cortical projections zones that would be interesting to study in more detail. The typical appearance of cortical lamination patterns (in neocortex and piriform cortex) is contrasted with an unusual lamination pattern in cingulate cortex, but there is a consistent presence of minicolumnar organization across all four cortical morphotypes. The sengis would appear to present an interesting model animal to unravel cortical architectural development as certain specific aspects of cortical architecture are distinct to this mammalian order. Such a study, especially applied to the unusual morphology of the cingulate cortex, may provide clues to the development of the more typical cortical morphologies observed in many other mammalian species. Moreover, it may allow a more precise distinction between the mechanisms involved in the formation of horizontal and vertical aspects of cortical architectural organization.

References

Asher RJ, Bennett NC, Lehmann T (2009) The new framework for understanding placental mammal evolution. BioEssays 31:853–864

Bouillaud F, Ricquier D, Thibault J, Wissenbach J (1985) Molecular approach to thermogenesis in brown adipose tissue: cDNA cloning of the mitochondrial uncoupling protein. Proc Natl Acad Sci U S A 82:445–448

Butler AB (1994) The evolution of the dorsal pallium in the telencephalon of amniotes – cladistic-analysis and a new hypothesis. Brain Res Rev 19:66–101

Butler AB (1995) The dorsal thalamus of jawed vertebrates – a comparative viewpoint. Brain Behav Evol 46:209–223

Calvey T, Patzke N, Kaswera C, Gilissen E, Bennett NC, Manger PR (2013) Nuclear organization of some immunohistochemically identifiable neural systems in three Afrotherian species – *Potomogale velox*, *Amblysomus hottentotus* and *Petrodromus tetradactylus*. J Chem Neuroanat 50–51:48–65

Dell LA, Kruger JL, Bhagwandin A, Jillani NE, Pettigrew JD, Manger PR (2010) Nuclear organization of cholinergic, putative catecholaminergic and serotonergic systems in the brains of two megachiropteran species. J Chem Neuroanat 40:177–195

Dengler-Crish CM, Crish SD, O'Riain MJ, Catania KC (2006) Organization of the somatosensory cortex in elephant shrews (*E. edwardii*). Anat Rec 288A:859–866

Gallyas F (1979) Silver staining of myelin by means of physical development. Neurol Res 1:203–209

Horvath TL, Diano S, Barnstable C (2003) Mitochondrial uncoupling protein 2 in the central nervous system: neuromodulator and neuroprotector. Biochem Pharmacol 65:1917–1921

Kaufmann JA, Turner GH, Holroyd PA, Rovero F, Grossman A (2013) Brain volume of the newly-discovered species *Rynchocyon udzungwensis* (Mammalia: Afrotheria: Macroscelidea): implications for encephalization in sengis. PLoS One 8:e58667

Kim-Han JS, Dugan LL (2005) Mitochondrial uncoupling proteins in the central nervous system. Antioxid Redox Signal 7:1171–1181

Manger PR (2005) Establishing order at the systems level in mammalian brain evolution. Brain Res Bull 66:282–289

Manger PR (2006) An examination of cetacean brain structure with a novel hypothesis correlating thermogenesis to the evolution of a big brain. Biol Rev 81:293–338

Manger PR, Pillay P, Maseko BC, Bhagwandin A, Gravett N, Moon DJ, Jillani N, Hemingway J (2009) Acquisition of brains from the African elephant (*Loxodonta africana*): perfusion-fixation and dissection. J Neurosci Methods 179:16–21

Nowack RM (1999) Walker's mammals of the world, 6th edn. The Johns Hopkins University Press, Baltimore

Palou A, Pico C, Bonet ML, Oliver P (1998) The uncoupling protein, thermogenin. Int J Biochem Cell Biol 30:7–11

Patzke N, Spocter MA, Karlsson K, Bertelsen MF, Haagensen M, Chawana R, Streicher S, Kaswera C, Gilissen E, Alagaili A, Mohammed OB, Reep RL, Bennett NC, Siegel JM, Ihunwo AO, Manger PR (2015) In contrast to many other mammals, cetaceans have relatively small hippocampi that appear to lack adult neurogenesis. Brain Struct Funct 220:361–383

Pieters RP, Gravett N, Fuxe K, Manger PR (2010) Nuclear organization of cholinergic, putative catecholaminergic and serotonergic nuclei in the brain of the eastern rock elephant shrew, *Elephantulus myurus*. J Chem Neuroanat 39:175–188

Rathbun GB (2009) Why is there discordant diversity in sengi (Mammalia: Afrotheria: Macroscelidea) taxonomy and ecology? Afr J Ecol 47:1–13

Sherwood CC, Stimpson CD, Butti C, Bonar CJ, Newton AL, Allman JM, Hof PR (2009) Neocortical neuron types in Xenarthra and Afrotheria: implications for brain evolution in mammals. Brain Struct Funct 213:301–328

Skinner JD, Chimimba CT (2005) The mammals of the Southern African subregion, 3rd edn. Cambridge University Press, Cape Town

Slomianka L, Drenth T, Cavegn N, Menges D, Lazic SE, Phalanndwa M, Chimimba CT, Amrein I (2013) The hippocampus of the eastern rock sengi: cytoarchitecture, markers of neuronal function, principal cell numbers, and adult neurogenesis. Front Neuroanat 7:34

Stanhope MJ, Waddell VG, Madsen O, de Jong W, Hedge SB, Cleven GC, Kao D, Springer MS (1998) Molecular evidence for multiple origins of Insectivora and for a new order of endemic African insectivore mammals. Proc Natl Acad Sci U S A 95:9967–9972

Stephan H, Frahm H, Baron G (1981) New and revised data on volumes of brain structures in insectivores and primates. Folia Primatol 35:1–29

Chapter 7
Active Inference, Predictive Coding and Cortical Architecture

Rick A. Adams, Karl J. Friston, and Andre M. Bastos

Abstract This chapter discusses how many features of cortical anatomy and physiology can be understood in the light of a predictive coding theory of brain function. In Sect. 7.1, we briefly discuss the theoretical reasons to suppose that the brain is likely to use predictive coding. One key theoretical underpinning of predictive coding is the free energy principle, which argues that brains must maximize the evidence for their (generative) model of sensory inputs: a process of 'active inference'. In Sect. 7.2, we discuss how active inference predicts commonalities in the *extrinsic* connections of sensory and motor systems. Such commonalities are found in their hierarchical structure (shown by laminar characteristics), their topography, their pharmacology and physiology. In Sect. 7.3, we show how the equations describing hierarchical message passing within a predictive coding scheme can be mapped on to key features of *intrinsic* connections, namely the canonical cortical microcircuit, and their implications for the oscillatory dynamics of different cell populations. In Sect. 7.4, we briefly review some empirical evidence for predictive coding in the brain.

Keywords Active inference • Predictive coding • Free energy • Hierarchy • Microcircuit

R.A. Adams (✉) • K.J. Friston
The Wellcome Trust Centre for Neuroimaging, University College London, 17 Queen Square, London WC1N 3BG, UK
e-mail: rick.adams@ucl.ac.uk; k.friston@ucl.ac.uk

A.M. Bastos
Ernst Strüngmann Institute (ESI) for Neuroscience in Cooperation with Max Planck Society; Picower Institute for Learning and Memory, MIT, Cambridge, MA, USA
e-mail: abastos@mit.edu

M.F. Casanova, I. Opris (eds.), *Recent Advances on the Modular Organization of the Cortex*, DOI 10.1007/978-94-017-9900-3_7

7.1 The Free Energy Principle and Predictive Coding

It might be thought impossible to specify the computations performed by the brain. However, there are some fairly fundamental constraints on the basic form of neuronal dynamics. The argument goes as follows – and can be regarded as a brief summary of the free energy principle (see Friston (2010) for details):

- Biological systems are homoeostatic (or allostatic), which means that they minimise the dispersion (entropy) of their interoceptive and exteroceptive states.
- Entropy is the average of surprise over time, which means biological systems minimise the surprise associated with their sensory states at each point in time.
- In statistics, surprise is the negative logarithm of Bayesian model evidence, which means biological systems – like the brain – must continually maximise the Bayesian evidence for their (generative) model of sensory inputs.
- Maximising Bayesian model evidence corresponds to Bayesian filtering of sensory inputs. This is also known as predictive coding.

These arguments mean that by minimising surprise, through selecting appropriate sensations, the brain is implicitly maximising the evidence for its own existence – this is known as active inference. In other words, to maintain a homeostasis the brain must predict its sensory states on the basis of a model. Fulfilling those predictions corresponds to accumulating evidence for that model – and the brain that embodies it.

The most popular scheme – for Bayesian filtering in neuronal circuits – is predictive coding (Rao and Ballard 1999). In this context, surprise corresponds (roughly) to prediction error. In predictive coding, top-down predictions are compared with bottom-up sensory information to form a prediction error. This prediction error is used to update higher-level representations – upon which top-down predictions are based. These optimised predictions then reduce prediction error at lower levels.

To predict sensations, the brain must be equipped with a generative model of how its sensations are caused (Helmholtz 1866). A generative model describes how variables or causes in the environment conspire to produce sensory input. Generative models map from (hidden) causes to (sensory) consequences. Perception then corresponds to the inverse mapping from sensations to their causes, while action can be thought of as the selective sampling of sensations.

A special case of these models are hierarchical dynamic models (see Fig. 7.1), which grandfather most parametric models in statistics and machine learning (see Friston 2008). These models explain sensory data in terms of hidden causes and states, which mediate structural and dynamic dependencies respectively.

Conditional dependencies among hidden states and causes are responsible for generating sensory input. These dependencies mean that we can interpret neuronal activity as message passing among the nodes of a generative model, wherein each node contains a canonical microcircuit that represents expectations about hidden states and causes (see Sect. 7.3). We now look at how perception or model inversion – recovering the hidden states and causes of this model given sensory data – might be implemented at the level of a microcircuit:

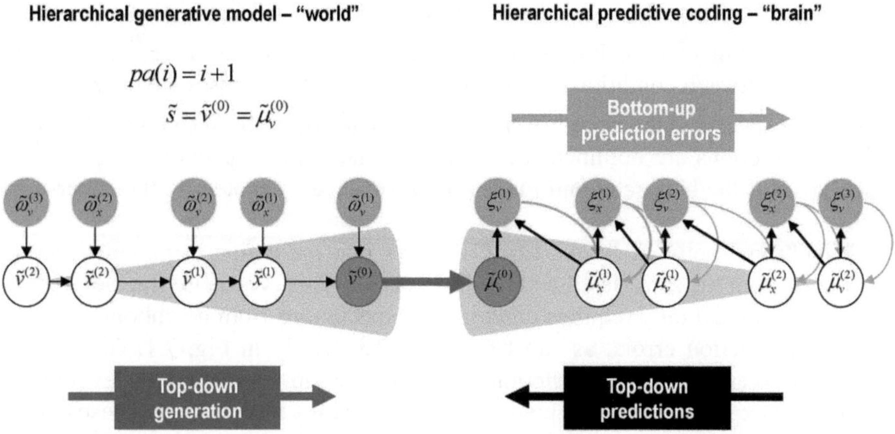

Fig. 7.1 Hierarchical structure in the world is represented in a cortical hierarchy
This figure describes the predictive coding scheme associated with a simple hierarchical model shown on the *left*. In this model each node has a single parent. The ensuing inversion or generalised predictive coding scheme is shown on the *right*. The key quantities in this scheme are (conditional) expectations of the hidden states and causes and their associated prediction errors. The basic architecture – implied by the inversion of the graphical (hierarchical) model – suggests that prediction errors (caused by unpredicted fluctuations in hidden variables) are passed up the hierarchy to update conditional expectations. These conditional expectations now provide predictions that are passed down the hierarchy to form prediction errors. We presume that the forward and backward message passing between hierarchical levels is mediated by *extrinsic* (feedforward and feedback) connections. Neuronal populations encoding conditional expectations and prediction errors now have to be deployed in a canonical microcircuit to understand the computational logic of *intrinsic* connections – within each level of the hierarchy – as shown in Fig. 7.4 (From Bastos et al. 2012)

In predictive coding, representations (or conditional expectations) generate top-down predictions to produce prediction errors. These prediction errors are then passed up the hierarchy in the reverse direction, to update conditional expectations. This ensures an accurate prediction of sensory input and all its intermediate representations. This hierarchical message passing can be expressed mathematically as a gradient descent on the (sum of squared) prediction errors $\xi^{(i)} = \Pi^{(i)}\tilde{\varepsilon}^{(i)}$, where the prediction errors are weighted by their precision (inverse variance).

$$\begin{aligned}
\dot{\tilde{\mu}}_v^{(i)} &= \mathcal{D}\tilde{\mu}_v^{(i)} - \partial_{\tilde{v}}\tilde{\varepsilon}(i) \cdot \xi^{(i)} - \xi_v^{(i+1)} \\
\dot{\tilde{\mu}}_x^{(i)} &= \mathcal{D}\tilde{\mu}_x^{(i)} - \partial_{\tilde{x}}\tilde{\varepsilon}(i) \cdot \xi^{(i)} \\
\xi_v^{(i)} &= \Pi_v^{(i)}\tilde{\varepsilon}_v^{(i)} = \Pi_v^{(i)}\left(\tilde{\mu}_v^{(i-1)} - g^{(i)}\left(\tilde{\mu}_x^{(i)}, \tilde{\mu}_v^{(i)}\right)\right) \\
\xi_x^{(i)} &= \Pi_x^{(i)}\tilde{\varepsilon}_x^{(i)} = \Pi_x^{(i)}\left(\mathcal{D}\tilde{\mu}_x^{(i)} - f^{(i)}\left(\tilde{\mu}_x^{(i)}, \tilde{\mu}_v^{(i)}\right)\right)
\end{aligned} \tag{7.1}$$

The first pair of equalities just says that conditional expectations about hidden causes and states $\left(\tilde{\mu}_v^{(i)}, \tilde{\mu}_x^{(i)}\right)$ are updated based upon the way we would predict them to change – the first term – and subsequent terms that minimise prediction error. The second pair of equations simply expresses prediction error ($\xi_v^{(i)}, \xi_x^{(i)}$) as the difference between conditional expectations about hidden causes and (the changes in) hidden states and their predicted values, weighed by their precisions ($\Pi_v^{(i)}, \Pi_x^{(i)}$). These predictions are nonlinear functions of conditional expectations ($g^{(i)}, f^{(i)}$) at each level of the hierarchy and the level above (levels denoted by the superscript 'i').

In neuronal network terms, Eq. (7.1) says that prediction error units receive messages from the same level and the level above. This is because the hierarchical form of the model only requires conditional expectations from neighbouring levels to form prediction errors, as can be seen schematically in Fig. 7.1. Conversely, expectations are driven by prediction error from the same level and the level below – updating expectations about hidden states and causes respectively. These constitute the bottom-up and lateral messages that drive conditional expectations to provide better predictions – or representations – that suppress prediction error. This updating corresponds to an accumulation of prediction errors – in that the rate of change of conditional expectations is proportional to prediction error. This is the essence of recurrent message passing between hierarchical levels to suppress prediction error (see Friston (2008) for a more detailed discussion).

7.2 Predictive Coding and Extrinsic Connection Properties in Sensory and Motor Areas

The brain can minimise prediction error in one of two ways. It can either change its predictions to better cohere with sensory input, or change the sampling of the environment such that sensory samples conform to predictions. The former process corresponds to perceptual inference – discussed in the previous section as predictive coding – the latter to action: together, they constitute 'active inference' (Friston et al. 2010). The free energy principle thus dictates that the perceptual and motor systems should not be regarded as separate but instead as a single active inference machine that tries to predict its sensory input in all domains: visual, auditory, somatosensory, interoceptive and, in the case of the motor system, proprioceptive.

Descending messages in the somatomotor system might therefore be best characterized as predictions of proprioceptive input and not motor commands. These proprioceptive predictions should not be corrected but fulfilled, by the automatic peripheral transformation of proprioceptive prediction errors into movement, by classical reflex arcs (see also Fig. 7.2a). If both systems are minimising prediction error, descending hierarchical projections in the motor cortex should share the laminar, topographic and physiological characteristics of backward connections in exteroceptive (sensory) systems.

A brain implementing predictive coding ought to have particular attributes (Friston 2005). These include: (i) a hierarchical organization with reciprocal connections between areas (conveying predictions and prediction errors) that are (ii) topographically asymmetrical (predictive connections being more divergent because a high level cause has multiple sensory consequences) and (iii) functionally asymmetrical. The functional asymmetry is important because descending predictions have to embody nonlinearities in the generative model (e.g., to model visual occlusion) that require them to interact or modulate each other, whereas ascending connections that drive higher representations do not.

We now review these attributes of the architecture of extrinsic connections in sensory and motor systems in the brain.

7.2.1 Hierarchical (Laminar) Characteristics

The cerebral neocortex consists of six layers of neurons, defined by differences in neuronal composition (pyramidal or stellate excitatory neurons, and numerous inhibitory classes) and packing density (Shipp 2007). Layer 4 is known as the 'granular layer' (due to its appearance), and the layers above and below it are known as 'supragranular' and 'infragranular' respectively.

Felleman and Van Essen (1991) surveyed 156 corticocortical pathways and specified criteria by which projections could be classified as forward, backward or lateral. They defined forward projections as originating predominantly (i.e. >70 % cells of origin) in supragranular layers, or occasionally with a bilaminar pattern (meaning <70 % either supra- or infragranular, but excluding layer 4 itself). Forward projections terminate preferentially in layer 4. Backward projections are predominantly infragranular or bilaminar in origin with terminations in layers 1 and 6 (especially the former), and always evading layer 4.

Felleman and Van Essen's (1991) analysis included some somatomotor areas but concentrated on sensory – especially visual – hierarchies. Continuing work by Shipp (2005), we have recently performed a more fine grained analysis of projections in somatomotor areas (Adams et al. 2013), of which the major findings are as follows:

(a) The terminations of projections ascending the somatomotor hierarchy are intermediate in character (terminate in all layers) apart from those originating in the sensory areas of parietal cortex, which have the characteristics of forward projections.
(b) The terminations of projections descending the somatomotor hierarchy have an overall backward character. The pattern is notably more distinct for terminations within postcentral granular areas, but the available evidence leans toward a backward pattern in the precentral agranular areas (i.e. from secondary to primary motor cortices) as well. Projection termination patterns are illustrated in Fig. 7.2b.
(c) The origins of projections ascending or descending the somatomotor hierarchy are qualitatively similar to each other; the projecting neurons are typically

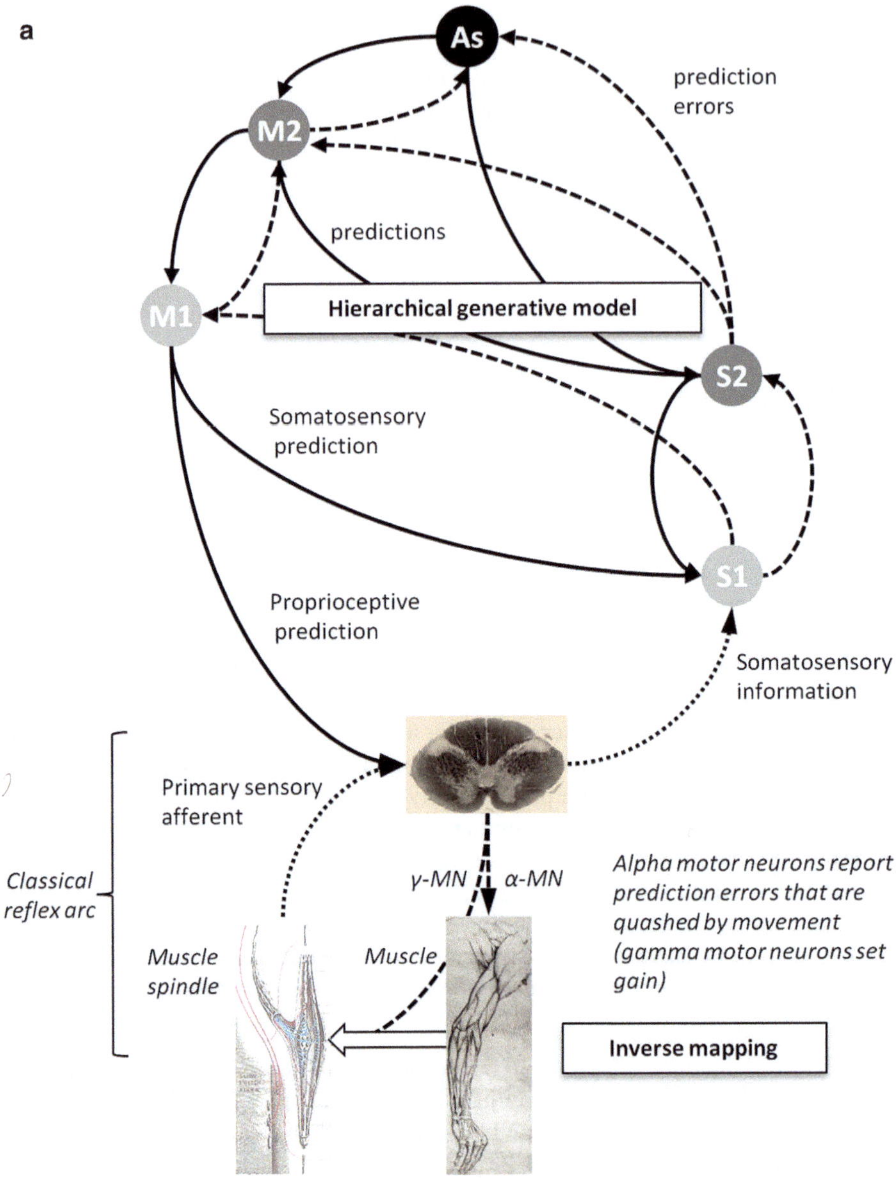

Fig. 7.2 Connections in the somatomotor hierarchy: functional and laminar characteristics (**a**) This simplified schematic ignores the contributions of spinal circuits and subcortical structures; and omits many hierarchical levels (especially on the sensory side). *M1*, *S1*, *M2* and *S2* signify primary and secondary motor and sensory cortex, while As signifies prefrontal association cortex. *Dashed arrows* denote driving '*forward*' projections, and *unbroken arrows* modulatory '*backward*' projections. Afferent somatosensory projections are *dotted*. *α-MN* and *γ-MN* signify alpha- and gamma motor neuron output (From Adams et al. 2013)

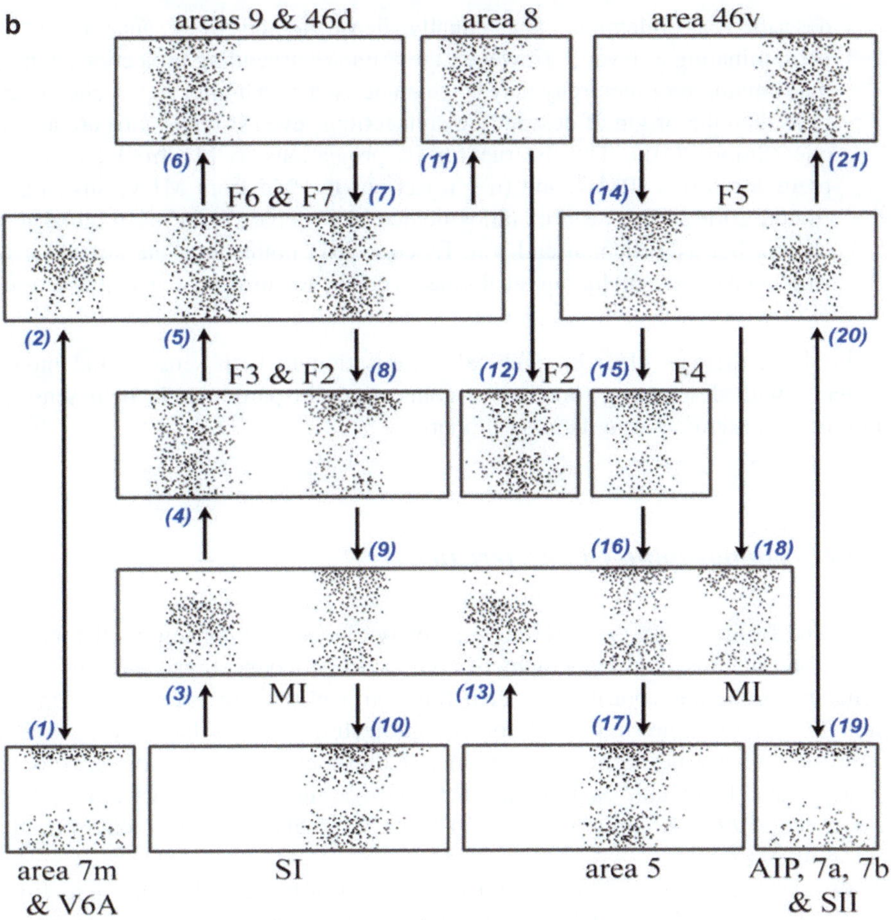

Fig. 7.2 (continued) (**b**) The diagram shows patterns of extrinsic projection terminations in selected areas comprising the somatomotor hierarchy. Not all connections are shown, only those for which an adequate indication of laminar characteristics is obtainable. The relevant citations (corresponding to the *bracketed numbers*) and a diagram of the patterns of projection origins have been omitted due to space constraints: they can be found in Adams et al. (2013). In order to compile data across studies with variable terminology and placement of injected tracers, or with similar outcomes, some areas are combined into single blocks; the '&' symbol should be interpreted as 'and/or'. The diagrams are intended to give an indication of forward or backward relationships, but not the precise number of pathways or levels involved. The sensory tiers, for instance, are compressed into a single level: *S1*, shown as a single block, comprises four separate areas (*3a*, *3b*, *1* and *2*) that precede higher order parietal areas in a sensory hierarchy. The figure depicts schematic illustrations of terminal patterns – forward (*2, 3, 13 & 20*); intermediate (*4, 5, 6, 11 & 21*); and backward (*1, 7–10, 12 & 14–19*). Forward patterns have a concentration in layer 3. Intermediate patterns are described as columnar, with little or no laminar differentiation. Backward patterns are concentrated in layers 1 (and 6) and/or tend to avoid the lower part of layer 3. Feedback from *M1* to *S1* tends to avoid layer 4 (This figure is updated from Shipp (2005) and taken from Adams et al. (2013))

described as bilaminar and equally dense in layers 3 and 5, or as predominating in layer 3. However, the origin of ascending projections within the somatomotor hierarchy may be characterised by a higher superficial : deep ratio than the origin of descending connections, even if both ratios are above one (Shipp 2005). This is true for (i) projections to M1 from S1 versus premotor cortex (PMd), and (ii) projections to PMd from M1 versus rostral frontal cortex. The ratio of ratios device may depart from the original test criteria but as Felleman and Van Essen (1991) point out: "the key issue is whether a consistent hierarchical scheme can be identified using a modified set of criteria".

In short, there is clear hierarchical organisation in both sensory and motor systems, with descending (predictive) connections in particular in both sensory and motor systems sharing laminar properties.

7.2.2 Topographic Characteristics

Salin and Bullier (1995) reviewed a large body of evidence concerning the microscopic and macroscopic topography of corticocortical connections, and how these structural properties contribute to their function; e.g. their receptive fields. In cat area 17, for example, <3 % of forward projecting neurons have axons which bifurcate to separate cortical destinations. Conversely, backward projections to areas 17 and 18 include as much as 30 % bifurcating axons (Bullier et al. 1984; Ferrer et al. 1992). A similar relationship exists in visual areas in the monkey (Salin and Bullier 1995).

Rockland and Drash (1996) contrasted a subset of backward connections from late visual areas (TE and TF) to primary visual cortex with typical forward connections in the macaque. The forward connections concentrated their synaptic terminals in 1–3 arbours of around 0.25 mm in diameter, whilst backward connections were distributed over a "wand-like array" of neurons, with numerous terminal fields stretching over 4–10 mm, and in one case, 21 mm (Fig. 7.3a). This very diffuse pattern was only found in around 10 % of backward projections, but it was not found in any forward projection.

These microscopic properties of backward connections reflect their greater macroscopic divergence. Zeki and Shipp (1988) reviewed forward and backward connections between areas V1, V2 and V5 in macaques, and concluded that backward connections showed much greater convergence and divergence than their forward counterparts (Fig. 7.3b). This means that cells in higher visual areas project back to a wider area than that which projects to them, and cells in lower visual areas receive projections from a wider area than they project to. Whereas forward connections are typically patchy in nature, backward connections are more diffuse (Salin and Bullier 1995; Shipp and Zeki 1989a, b). These attributes mean that visuotopy preserved in the forward direction is eroded in the backward

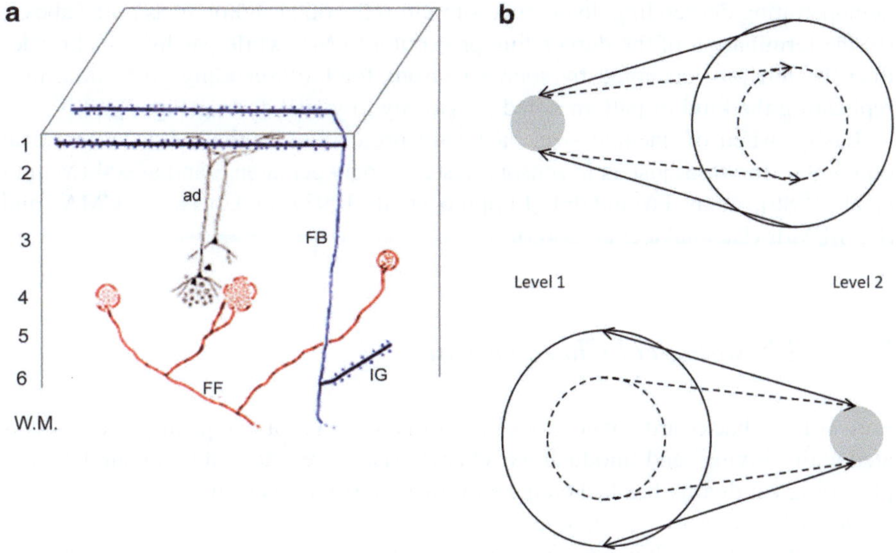

Fig. 7.3 Topographic characteristics of forward and backward projections
(**a**) This schematic is adapted from Rockland and Drash (1996), and illustrates the terminal fields of 'typical' forward (axon FF) and backward (axon FB) connections in the visual system. *IG* represents infragranular collaterals of a backward connection, and '*ad*' an apical dendrite; cortical layers are labelled on the *left*. Note the few delimited arbours of terminals on the forward connection, and the widely distributed "wand-like array" of backward connection terminals
(**b**) This schematic illustrates projections to and from a lower and higher level in the visual hierarchy (Adapted from Zeki and Shipp 1988). *Dashed arrows* signify forward connections and *continuous arrows* backward connections. Note that there is a much greater convergence (from the point of view of neurons receiving projections) and divergence (from the point of view of neurons sending projections) in backward relative to forward connections (From Adams et al. 2013)

direction, allowing backward projections to contribute significantly to the extra classical receptive field of a cell (Angelucci and Bullier 2003).

Salin and Bullier (1995) also noted that in the macaque ventral occipitotemporal pathway (devoted to object recognition), backward connections outnumber forward connections. Forward projections from the lateral geniculate nucleus (LGN) to V1 are outnumbered 20 to 1 by those returning in the opposite direction; and backward projections outweigh forward projections linking central V1 to V4, TEO to TE, and TEO and TE to parahippocampal and hippocampal areas.

Now let us examine corticocortical connections within the sensorimotor system. Studies using dual retrograde tracers have examined the sources of input to the parts of M1 innervating the distal and proximal parts of the monkey forelimb (Tokuno and Tanji 1993) and hindlimb (Hatanaka et al. 2001), and showed that divergence in the descending (motor) input to M1 exceeds divergence in the ascending (sensory) input to M1.

In a later study, Dancause et al. (2006) found that: (a) the distal forelimb part of premotor (PMv) connects with both distal and proximal forelimb parts of M1, demonstrating descending divergence similar to the other motor areas noted above; (b) the termination of the descending projection to M1, while patchy, was broader than the territory occupied by source neurons for the ascending projection, thus replicating the kind of pattern noted previously in visual cortex (Fig. 7.3b).

Lastly, when of unequal size, backward projections in the motor system also outnumber forward, just as in sensory systems: between area 6 and area 4 (Matelli et al. 1986), areas F6 and F3 (Luppino et al. 1993), and between CMAr and SMA/PMdr (Hatanaka et al. 2003).

7.2.3 *Physiological Characteristics*

Forward and backward connections in sensory systems have typically been associated with 'driving' and 'modulatory' characteristics, respectively, though the latter physiological duality has lacked the empirical clarity of its anatomical counterpart, particularly for cortical interactions.

Recent evidence, however, suggests that feedback connections do more than modulate lower level responses: Sherman and colleagues recorded cells in mouse area V1/V2 and A1/A2, while stimulating feedforward or feedback afferents. In both cases, driving-like responses as well as modulatory-like responses were observed (Covic and Sherman 2011; De Pasquale and Sherman 2011). This indicates that – for these hierarchically proximate areas – feedback connections can drive their targets just as strongly as feedforward connections. Imaging studies of top-down influences acting on area V1 imply that backward connections can sustain or even initiate activity, in the absence of a retinal signal (e.g. Harrison and Tong 2009; Muckli et al. 2005). In summary, backward connections can mediate modulatory and driving effects. This is important from the point of view of predictive coding because, as noted above, top-down predictions have to drive cells that explain away prediction error. From a computational perspective, the key role of modulatory effects is to model the context-sensitive and nonlinear way in which causes interact to produce sensory consequences.

We now consider evidence linking nonlinear (modulatory) effects to backward connections in both sensory and motor systems, much of which depends on a closer consideration of the roles played by the different types of postsynaptic glutamate receptors:

Glutamate is the principal excitatory neurotransmitter in the cortex and activates both ionotropic and metabotropic receptors. Metabotropic receptor binding is clearly modulatory in action (Pin and Duvoisin 1995). Ionotropic glutamate receptors are classified according to their AMPA, Kainate and NMDA agonists (Traynelis et al. 2010). AMPA activation is fast and stereotyped, with onset times <1 msec, and deactivation within 3 msec; recombinant Kainate receptors have AMPA receptor-like kinetics, although they can be slower in vivo. NMDA currents, by contrast, are smaller but more prolonged: the onset and deactivation are one and two orders of magnitude slower, respectively.

Unlike non-NMDA receptors, NMDA receptors (NMDA-Rs) are both ligand-gated and voltage-dependent – to open their channel they require both glutamate binding and membrane depolarisation to displace the blocking Mg^{2+} ion. The voltage dependence makes NMDA transmission non-linear and the receptors function, in effect, as postsynaptic coincidence detectors. These properties may be particularly important in governing the temporal patterning of network activity (Durstewitz 2009). Once activated, NMDA-Rs play a critical role in changing long term synaptic plasticity (via Ca^{2+} influx) and increase the short-term gain of AMPA/Kainate receptors (Larkum et al. 2004). In summary, NMDA-Rs are nonlinear and modulatory in character, whereas non- NMDA-Rs have more phasic, driving properties.

NMDA-Rs are ubiquitous in distribution, and clearly participate in forward, intrinsic and backward signal processing. Their subunit composition differs, however, and there is evidence in the macaque sensory cortex that the most non-linear variants are densest in the layers receiving backward connections (Muñoz et al. 1999). Predictive coding requires descending non-linear predictions to negate ascending prediction errors, and interestingly it seems that the inhibitory effects of backward projections to macaque V1 are mediated by NR2B-containing NMDA-R's (Self et al. 2012). By contrast, the least non-linear NMDA-R variants are found in layer 4 of macaque area 3B (Muñoz et al. 1999) and of rodent S1 (Binshtok et al. 2006).

Studies with pharmacological manipulation of NMDA-R in vivo are rare. However application of an NMDA-R agonist to cat V1 raised the gain of response to stimulus contrast (Fox et al. 1990). The effect was observed in all layers, except layer 4. Application of an NMDA-R antagonist had the reverse effect, however the gain-reduction effect was only observed in layers 2 and 3. To interpret these results, the NMDA-R agonist may have simulated a recurrent enhancement of responses in the layers exposed to backward connections (i.e. all layers save layer 4). The restriction of the antagonist effect to layer 2/3 could indicate that NMDA-R plays a more significant role in nonlinear intrinsic processing in these layers.

NMDA-Rs in motor cortices have similar anatomical and functional properties. Zilles et al. (1995) demonstrated that human motor cortex has the same distribution of NMDA and non- NMDA-Rs as is found elsewhere in the brain: the former are concentrated in supragranular layers, whereas the latter have a uniform (AMPA-R) or infragranular (KA-R) distribution.

Ghosh et al. (1987) counted the relative numbers of neurons projecting to monkey forelimb M1. In the 3 animals they examined, 11–31 % of neurons projecting to M1 came from premotor cortex (lateral area 6), whereas 1–17 % of neurons originated in area 5 (higher sensory cortex). Ghosh and Porter (1988) then stimulated these two cortical areas using surface electrodes, and recorded EPSP's and IPSP's in M1. They found that despite the bias in numbers towards descending projections, stimulation of area 5 neurons elicited responses in 90 % of recorded M1 neurons, whereas the same stimulation of premotor cortex caused only 30 % of recorded M1 neurons to respond. One can infer from this that ascending projections from sensory cortex are the more driving in character, despite their lesser number.

Likewise, it is known that inactivation of M1 has a more significant effect on the activity in PMv and SMA than vice versa (Schmidlin et al. 2008), which one would expect if descending connections to M1 were more modulatory, and ascending connections from M1 more driving in character.

What of the receptor types mediating these connections? By administering receptor antagonists, Shima and Tanji (1998) showed that the influence of SMA over M1 depends to a much greater extent upon NMDA-R transmission, whilst the ascending connections from S1 to M1 rely more heavily on AMPA or Kainate receptors.

Shima and Tanji (1998) speculated that SMA – via NMDA-R's – might modulate the gain of driving S1 inputs to M1. Evidence for higher motor areas modulating the gain of M1 neurons has actually been provided by Shimazu et al. (2004), who recorded corticospinal outputs following stimulation of the ventral premotor area (F5) and/or M1. M1 stimulation alone evoked several corticospinal volleys, whereas F5 stimulation alone evoked minimal output. If F5 stimulation directly preceded that of M1, however, the later corticospinal volleys were powerfully facilitated, as were the resulting EPSE's in 92 % of intrinsic hand motor neurons. Similarly, Zagha et al. (2013) have shown in the mouse that descending connection from primary motor cortex can modulate responses in primary sensory cortex according to context.

In summary, predictive coding requires forward (prediction error) connections to be driving and backward (predictive) connections to have both driving and non-linear modulatory properties. This pattern is seen in sensory and motor cortices in both the anatomical distribution of non-linear NMDA-Rs (which are particularly present in lamina to which backward connections project), and in the responses of neurons to electrophysiological stimulation.

7.3 Section 3: Predictive Coding and Intrinsic Connections of the 'Canonical Microcircuit'

In this section, we show how the equations describing hierarchical message passing within a predictive coding scheme in Sect. 7.1 can be mapped on to key features of intrinsic connections, namely the canonical cortical microcircuit.

7.3.1 Microcircuits in the Visual Cortex

The seminal work of Douglas and Martin (1991), in the cat visual system, produced a model of how information flows through the cortical column. Their model contained superficial and deep pyramidal cells with a common pool of inhibitory cells, and was able to explain how recurrent connections between the recorded cell populations gave rise to their observed data, in terms of intracellular membrane potentials. Their circuit, although based on recordings from cat visual cortex, was also proposed as a basic theme that might be present and replicated, with minor variations, throughout the cortical sheet (Douglas et al. 1989).

Subsequent studies have used intracellular recordings and histology to measure spikes (and intracellular membrane potentials) in pre and post-synaptic cells, whose cellular morphology can be determined. This approach quantifies both the connection probability – defined as the number of observed connections divided by total number of pairs recorded – and connection strength – defined in terms of post-synaptic responses. Thomson et al. (2002) used these techniques to study layers 2 to 5 (L2 to L5) of the cat and rat visual systems. The most frequently connected cells were located in the same cortical layer, where the largest interlaminar projections were the 'feedforward' connections from L4 to L3 and from L3 to L5. Excitatory reciprocal 'feedback' connections were not observed (L3 to L4) or less commonly observed (L5 to L3), suggesting that excitation spreads within the column in a feedforward fashion. Feedback connections were typically seen when pyramidal cells in one layer targeted inhibitory cells in another (see Thomson and Bannister 2003 for a review).

Recent advances in optogenetics have also made it possible to more easily study inhibitory cells: Kätzel et al. (2011) combined optogenetics and whole-cell recording to investigate the intrinsic connectivity of inhibitory cells in mouse cortical areas M1, S1, and V1. They transgenically expressed channelrhodopsin in inhibitory neurons and activated them, while recording from pyramidal cells. This allowed them to assess the effect of inhibition as a function of laminar position relative to the recorded neuron.

Several conclusions can be drawn from this approach (Kätzel et al. 2011): first, L4 inhibitory connections are more restricted in their lateral extent, relative to other layers. This supports the notion that L4 responses are dominated by thalamic inputs, while the remaining laminae integrate afferents from a wider cortical patch. Second, the primary source of inhibition originates from cells in the same layer, reflecting the prevalence of inhibitory intralaminar connections. Third, several interlaminar motifs appeared to be general – at least in granular cortex: principally, a strong inhibitory connection from L4 onto supragranular L2/3 and from infragranular layers onto L4. Figure 7.4 provides a summary of key excitatory and inhibitory intralaminar connections.

7.3.2 Microcircuits in the Sensorimotor Cortex

Do the features of visual microcircuits generalize to other cortical areas? Recently, two studies have mapped the intrinsic connectivity of mouse sensory and motor cortices: Lefort et al. (2009) used multiple whole-cell recordings in mouse barrel cortex to determine the probability of monosynaptic connections – and the corresponding connection strength. As in visual cortex, the strongest connections

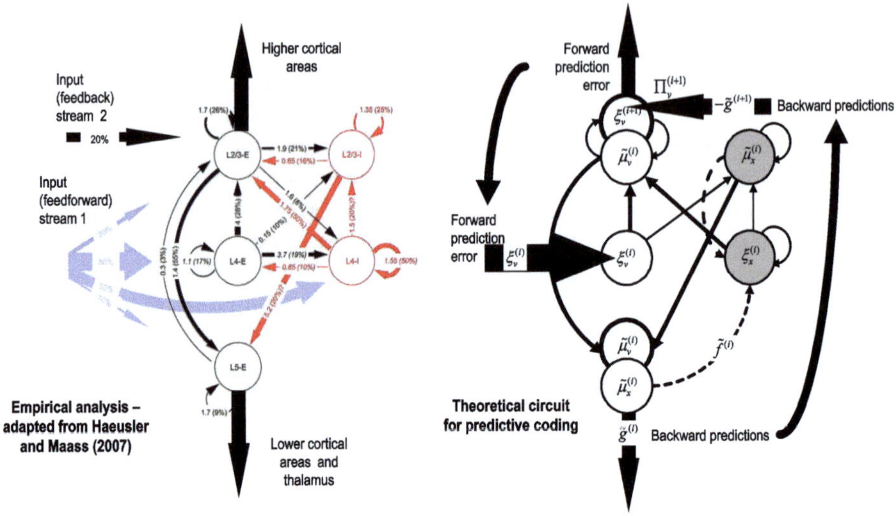

Fig. 7.4 A canonical microcircuit for predictive coding
The upper panel is the canonical microcircuit based on Haeusler and Maass (2007), where we have removed inhibitory cells from the deep layers – because they have very little interlaminar connectivity. The numbers denote connection strengths (mean amplitude of PSPs measured at soma in mV) and connection probabilities (in parentheses) according to Thomson et al. (2002). Excitatory and inhibitory populations are indicated as E and I, respectively. The lower panel shows the proposed cortical microcircuit for predictive coding, where the quantities from Fig. 7.1 have been associated with various cell types. Here, extrinsic connections are shown in *thick* lines. Intrinsic connections are shown in *thinner* lines. Prediction error populations are the populations encoded by $\xi_v^{(i)}, \xi_x^{(i)}$, and populations encoding conditional expectations are encoded by $\tilde{\mu}_v^{(i)}, \tilde{\mu}_x^{(i)}$. Intrinsic inhibitory populations are shown in light gray, while excitatory populations are shown in white. The *dotted* lines refer to connections that are not present in the microcircuit on the *left*. In this scheme, expectations (about causes and states) are assigned to (excitatory and inhibitory) interneurons in the supragranular layers, which are passed to infragranular layers. The corresponding prediction errors occupy granular layers, while superficial pyramidal cells encode prediction errors that are sent forward to the next hierarchical level. Conditional expectations and prediction errors on hidden causes are associated with excitatory cell types, while the corresponding quantities for hidden states are assigned to inhibitory cells. *Circles with dark outlines* indicate pyramidal cells. Finally, we have placed the precision of the feedforward prediction errors against the superficial pyramidal cells. This quantity controls the postsynaptic sensitivity or gain to (intrinsic and top-down) pre-synaptic inputs. We have previously discussed this in terms of attentional modulation, which may be intimately linked to the synchronisation of pre-synaptic inputs and ensuing postsynaptic responses (Fries et al. 2001; Feldman and Friston 2010) (From Bastos et al. 2012)

were intralaminar and the strongest interlaminar connections were the ascending L4 to L2, and descending L3 to L5.

One puzzle about canonical microcircuits is whether motor cortex has a local circuitry that is qualitatively similar to sensory cortex. This question is important because motor cortex lacks a clearly defined granular L4. Weiler et al. (2008) combined whole-cell recordings in mouse motor cortex with photo-stimulation to uncage Glutamate, and mapped the excitatory influence that each layer exerts over the others. They found that the L2/3 to L5A/B was the strongest connection – accounting for one-third of the total synaptic current in the circuit. The second strongest interlaminar connection was the reciprocal L5A to L2/3 connection. This pathway may be homologous to the prominent L4/5A to L2/3 pathway in sensory cortex. Also – as in sensory cortex – recurrent (intralaminar) connections were prominent, particularly in L2, L5A/B and L6. The largest fraction of synaptic input arrived in L5A/B, consistent with its key role in accumulating information from a wide range of afferents – before sending its output to the corticospinal tract. In summary, strong input-layer to superficial, and superficial to deep connectivity, together with strong intralaminar connectivity, suggests that the intrinsic circuitry of motor cortex is similar to other cortical areas.

7.3.3 Predictive Coding in the Canonical Microcircuit

Recently, we showed that the variables shown in Eq. (7.1) (Sect. 7.1) can be associated with specific populations in the laminar circuit (Bastos et al. 2012). The proposed mapping in Fig. 7.4 illustrates a remarkable correspondence between the form of Eq. (7.1) and the connectivity of the canonical microcircuit. Furthermore, the resulting scheme corresponds almost exactly to the computational architecture proposed by Mumford (1992).

The architecture bears a striking correspondence to the microcircuit in Haeusler and Maass (2007) in the left panel of Fig. 7.4 – in the sense that nearly every connection required by the predictive coding scheme appears to be present in terms of quantitative measures of intrinsic connectivity. However, there are two exceptions that both involve connections to the inhibitory cells in the granular layer (shown as dotted lines in Fig. 7.4). Predictive coding requires that these cells (that encode prediction errors on hidden states) compare the expected changes in hidden states with the actual changes. This suggests that there should be interlaminar projections from supragranular (inhibitory) and infragranular (excitatory) cells. In terms of their synaptic characteristics, one would predict that these intrinsic connections would be of a feedback sort – in the sense that they convey predictions. Although the infragranular to granular connection was not included in the Haeusler and Maass circuit, this is likely due to the omission of layer 6 from their analysis, as feedback connections from layer 6 to layer 4 are an established component of the cortical microcircuit (e.g., Usrey and Fitzpatrick 1996; Olsen et al. 2012).

7.3.4 Functional Asymmetries in the Microcircuit and Oscillatory Dynamics

The circuitry in Fig. 7.4 appears consistent with the broad scheme of ascending (feedforward) and descending (feedback) intrinsic connections: feedforward prediction errors from a lower cortical level arrive at granular layers and are passed forward to excitatory and inhibitory interneurons in supragranular layers, encoding expectations. Strong and reciprocal intralaminar connections couple superficial excitatory interneurons and pyramidal cells. Excitatory and inhibitory interneurons in supragranular layers then send strong feedforward connections to the infragranular layer. These connections enable deep pyramidal cells and excitatory interneurons to produce (feedback) predictions, which ascend back to L4 or descend to a lower hierarchical level. This arrangement recapitulates the functional asymmetries between extrinsic feedforward and feedback connections and is consistent with the empirical characteristics of intrinsic connections. Note, however, that deviations from this circuitry (e.g., activation of the cortical column independent of layer 4: Constantinople and Bruno 2013) do exist and may play important roles in predictive coding and canonical cortical computations, but it is beyond the scope of this chapter to review all such exceptions. What we wish to highlight here is the remarkable correspondence between the theoretically predicted microcircuit and the currently established models of intrinsic cortical processing.

Furthermore, if we focus on the superficial and deep pyramidal cells, the form of the recognition dynamics in Eq. (7.1) tells us something quite fundamental: We would anticipate higher frequencies in the superficial pyramidal cells, relative to the deep pyramidal cells. One can see this easily by taking the Fourier transform of the first equality in Eq. (7.1)

$$(j\omega)\widetilde{\mu}_v^{(i)}(\omega) = \mathcal{D}\widetilde{\mu}_v^{(i)}(\omega) - \partial_{\widetilde{v}}\widetilde{\varepsilon}^{(i)} \cdot \xi^{(i)}(\omega) - \xi_v^{(i+1)}(\omega) \tag{7.2}$$

This equation says that the contribution of any (angular) frequency ω in the prediction errors (encoded by superficial pyramidal cells) to the expectations (encoded by the deep pyramidal cells) is suppressed in proportion to that frequency (Friston 2008). In other words, high frequencies should be attenuated when passing from superficial to deep pyramidal cells. There is nothing mysterious about this attenuation – it is a simple consequence of the fact that conditional expectations accumulate prediction errors, thereby suppressing high-frequency fluctuations to produce smooth estimates of hidden causes. This smoothing – inherent in Bayesian filtering – leads to an asymmetry in frequency content of superficial and deep cells: for example, superficial cells should express more gamma relative to beta, and deep cells should express more beta relative to gamma (Maier et al. 2010; Roopun et al. 2008; Roopun et al. 2006).

Figure 7.5 provides a schematic illustration of the spectral asymmetry predicted by Eq. (7.2). Note that predictions about the relative amplitudes of high and low frequencies in superficial and deep layers pertain to all frequencies – there is nothing in predictive coding per se to suggest characteristic frequencies in the gamma and beta ranges. However, one might speculate the characteristic frequencies of canonical microcircuits have evolved to model and – through active inference – create the sensorium (Berkes et al. 2011; Engbert et al. 2011; Friston 2010).

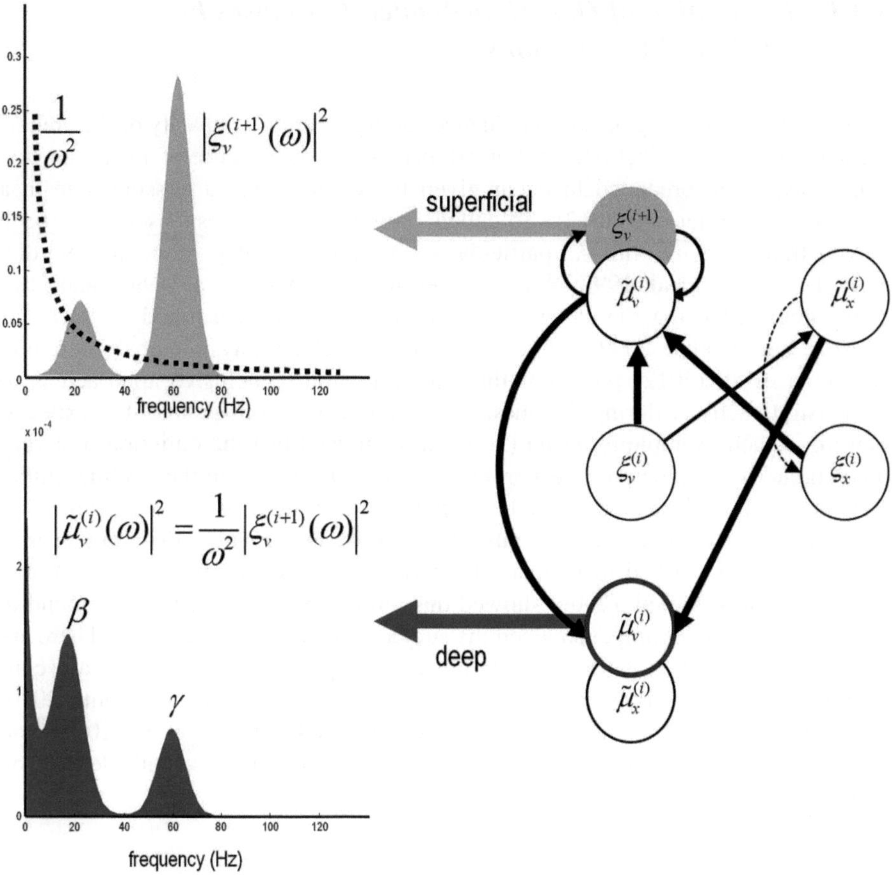

Fig. 7.5 Spectral asymmetries in superficial and deep cells
This schematic illustrates the functional asymmetry between the spectral activity of superficial and deep cells predicted theoretically. In this illustrative example, we have ignored the effects of influences on the expectations of hidden causes (encoded by *deep* pyramidal cells), other than the prediction error on causes (encoded by *superficial* pyramidal cells). The *lower* panel shows the spectral density of deep pyramidal cell activity, given the spectral density of superficial pyramidal cell activity in the *upper* panel. The equation expresses the spectral density of the deep cells as a function of the spectral density of the superficial cells; using Eq. (7.2). This schematic is meant to illustrate how the relative amounts of *low* (beta) and *high* (gamma) frequency activity in superficial and deep cells can be explained by the evidence accumulation implicit in predictive coding (From Bastos et al. 2012)

7.4 Evidence for Predictive Coding

In this section, we briefly review some empirical evidence consistent with predictive coding itself, rather than just its architecture (as reviewed in the previous Sections).

7.4.1 Generation of Omission-Related Responses by Backward Connections

One of the most striking pieces of evidence for the predictive capacity of the brain is the generation of an electrophysiological response to the absence of a stimulus that could have been predicted to occur given the context, e.g. a missing note in a musical scale (Nemoto 2012). So-called omission-related responses have been demonstrated many times, particularly in the auditory domain (Nordby et al. 1994; Raij et al. 1997; Wacongne et al. 2011; Yabe et al. 1997) and have been simulated using the same active inference scheme described in Sect. 7.1 (Friston and Kiebel 2009). In a review of the electrophysiological literature, Bendixen et al. (2012) point out that an even more conclusive approach is to demonstrate activity during the missing stimulus which conforms to the 'expectation': one such example is Janata (2001) who showed that the omission of a very predictable tone generated responses which were very similar to the initial (N1) processing of the actual tone being presented. Brain activity during a missing fragments of a familiar song is greater than that during an unfamiliar song even if the subject is not asked to recreate the missing part (Kraemer et al. 2005), and likewise Bendixen et al. (2009) showed that similar predictive activity is found in early auditory processing even when the subject's attention is elsewhere. There are clear connections here with the literature exploring the ability of backward connections to activate early visual areas in visual imagery (Esterman and Yantis 2010; Kosslyn and Thompson 2003; Mechelli et al. 2004; Muckli et al. 2005) and although imagery and omission responses are distinct entities, both depend on predictions.

7.4.2 Suppression of Predictable (and Enhancement of Unpredictable) Activity by Backward Connections

Neural responses to deviant stimuli – that violate sensory predictions established by a regular stimulus sequence – are enhanced relative to predicted stimuli (Garrido et al. 2009). Similarly, violating expectations of auditory repetition causes enhanced gamma-band responses in early auditory cortex (Todorovic et al. 2011). These enhanced responses are thought to reflect an inability of higher cortical areas to predict, and thereby suppress, the activity of populations encoding prediction error (Garrido et al. 2007; Wacongne et al. 2011). The suppression of predictable responses can also be regarded as repetition suppression, observed in single unit recordings (Desimone 1996). Furthermore, neurons in monkey inferotemporal cortex respond significantly less to a predicted sequence of natural images, compared to an unpredicted sequence (Meyer and Olson 2011).

Suppression of expected responses is thought to be an effect of feedback connections: supported by neuroimaging studies (Alink et al. 2010; Harrison et al. 2007; Murray et al. 2002; Summerfield et al. 2008). These studies show that predictable stimuli evoke smaller responses in early cortical areas. Crucially, this suppression cannot be explained in terms of local adaptation, because the attributes of the stimuli that can be predicted are not represented in early sensory cortex (e.g., Harrison et al. 2007). It should be noted that the suppression of responses to predictable stimuli can coexist with (top-down) attentional enhancement of signal processing (Wyart et al. 2012): mediated by increasing the gain of populations encoding prediction error. The resulting attentional modulation (e.g., Hopfinger et al. 2000) can interact with top-down predictions to override their suppressive influence – as demonstrated empirically (Wyart et al. 2012).

Further evidence for the suppressive effect of feedback connections on predictable stimuli comes from neuropsychology: Patients with damage to the prefrontal cortex (PFC) show disinhibition of event related potential responses (ERP) to repeating stimuli (Knight et al. 1989; Yamaguchi and Knight 1990; but see Barceló et al. 2000). In contrast, they show reduced-amplitude P300 ERPs in response to novel stimuli – as if there were a failure to communicate top-down predictions to sensory cortex (Knight 1984). Furthermore, normal subjects show a rapid adaptation to deviant stimuli as they become predictable – an effect not seen in prefrontal patients.

Several invasive studies complement these human studies in suggesting an overall inhibitory role for feedback connections. In a seminal study, Olsen et al. (2012) studied corticothalamic feedback between L6 of V1 and the LGN. By driving L6 cells optogenetically – while recording in V1 and the LGN – the authors showed that deep L6 principal cells inhibited their extrinsic targets in the LGN and their intrinsic targets in cortical layers 2 to 5. This suppression was powerful – in the LGN, visual responses were suppressed by 76 %, and in V1, around 80–84 %.

Backward connections can be either excitatory or inhibitory depending on the content of classical and extra-classical receptive fields. Hupé and colleagues cooled area V5/MT while recording from areas V1, V2, and V3 in the monkey. When visual stimuli were presented in the classical receptive field (CRF), cooling of area V5/MT decreased unit activity in earlier areas, suggesting an excitatory effect of extrinsic feedback (Hupé et al. 1998). However, when the authors used a stimulus that spanned the extra-classical RF the responses of V1 neurons were – on average – enhanced after cooling area V5, consistent with the suppressive role of feedback connections. Similar effects were observed when area V2 was cooled and neurons were measured in V1 (Bullier et al. 1996). These results are easily explained by predictive coding.

7.4.3 *Generation of Typical Receptive Field Properties by Predictive Coding Networks*

Rao and Ballard (1999) trained a hierarchical predictive coding network to recognise natural images. They showed that higher levels in the hierarchy learn to predict visual features that extend across many CRFs in the lower levels (e.g. tree trunks or horizons). Hence, higher visual areas come to predict that visual stimuli will span the receptive fields of cells in lower visual areas. In this setting, a stimulus that is confined to a single CRF would elicit a strong prediction error signal (because it cannot be predicted). This provides a simple explanation for the findings of Hupé et al. (1998) and Bullier et al. (1996): when feedback connections are deactivated, there are no top-down predictions to explain responses in lower areas – leading to a disinhibition of responses in earlier areas, when – and only when – stimuli can be predicted over multiple CRFs.

The inhibitory effects of feedback connections could be mediated via their termination in L1 (Anderson and Martin 2006; Shipp 2007): L1 cells are almost all inhibitory and interconnect strongly with each other, via electrical connections and chemical synapses (Chu et al. 2003). Simultaneous whole cell patch clamp recordings show that they provide strong monosynaptic inhibition to L2/3 pyramidal cells, whose apical dendrites project into L1 (Chu et al. 2003; Wozny and Williams 2011). This means L1 inhibitory cells are in a prime position to mediate inhibitory effects of extrinsic feedback (Meyer et al. 2011). Indeed, a study of rat barrel cortex – that stimulated (and inactivated) L1 – showed that it exerts a powerful inhibitory effect on whisker-evoked responses (Shlosberg et al. 2006).

Leading on from Rao and Ballard's work, Michael Spratling has demonstrated in a series of papers that a predictive coding network trained on lifelike images can recapitulate many properties of the receptive fields of V1 neurons, such as orientation selectivity, centre-surround effects, spatial and temporal frequency, cross-orientation and surround suppression (Spratling 2010), and also psychophysical data relating to the saliency of contours in textured regions (Spratling 2012, see also Friston et al. 2012).

7.5 Conclusion

In this chapter we hope to have shown why it makes sense to think of the brain as a predictive coding network which performs Bayesian filtering on the causes of its sensory data, for both theoretical and empirical reasons. We have also highlighted the important implications of this framework for understanding commonalities of extrinsic connections in both sensory and motor hierarchies, and for interpreting the function intrinsic connections. Together, these architectures define the basic properties of the cortical microcircuit. This analysis converges on the notion that many aspects of this circuit repeat themselves in multiple cortical areas, and therefore can be thought to in some sense be "canonical." Predictive coding offers a compelling hypothesis of what the computational role of this canonical microcircuit may be. Future work will seek to describe the underlying computations of these circuits in terms of the specific predictions made by the predictive coding model.

References

Adams RA, Shipp S, Friston KJ (2013) Predictions not commands: active inference in the motor system. Brain Struct Funct 218:611–643. doi:10.1007/s00429-012-0475-5

Alink A, Schwiedrzik CM, Kohler A et al (2010) Stimulus predictability reduces responses in primary visual cortex. J Neurosci 30:2960–2966. doi:10.1523/JNEUROSCI. 3730-10.2010

Anderson JC, Martin KAC (2006) Synaptic connection from cortical area V4 to V2 in macaque monkey. J Comp Neurol 495:709–721. doi:10.1002/cne.20914

Angelucci A, Bullier J (2003) Reaching beyond the classical receptive field of V1 neurons: horizontal or feedback axons? J Physiol Paris 97:141–154. doi:10.1016/j.jphysparis.2003.09.001

Barceló F, Suwazono S, Knight RT (2000) Prefrontal modulation of visual processing in humans. Nat Neurosci 3:399–403. doi:10.1038/73975

Bastos AM, Usrey WM, Adams RA et al (2012) Canonical microcircuits for predictive coding. Neuron 76:695–711. doi:10.1016/j.neuron.2012.10.038

Bendixen A, Schröger E, Winkler I (2009) I heard that coming: event-related potential evidence for stimulus-driven prediction in the auditory system. J Neurosci 29:8447–8451. doi:10.1523/JNEUROSCI. 1493-09.2009

Bendixen A, SanMiguel I, Schröger E (2012) Early electrophysiological indicators for predictive processing in audition: a review. Int J Psychophysiol 83:120–131. doi:10.1016/j.ijpsycho.2011.08.003

Berkes P, Orbán G, Lengyel M, Fiser J (2011) Spontaneous cortical activity reveals hallmarks of an optimal internal model of the environment. Science 331:83–87. doi:10.1126/science.1195870

Binshtok AM, Fleidervish IA, Sprengel R, Gutnick MJ (2006) NMDA receptors in layer 4 spiny stellate cells of the mouse barrel cortex contain the NR2C subunit. J Neurosci 26:708–715. doi:10.1523/JNEUROSCI. 4409-05.2006

Bullier J, Kennedy H, Salinger W (1984) Branching and laminar origin of projections between visual cortical areas in the cat. J Comp Neurol 228:329–341. doi:10.1002/cne.902280304

Bullier J, Hupé JM, James A, Girard P (1996) Functional interactions between areas V1 and V2 in the monkey. J Physiol Paris 90:217–220

Chu Z, Galarreta M, Hestrin S (2003) Synaptic interactions of late-spiking neocortical neurons in layer 1. J Neurosci 23:96–102

Constantinople CM, Bruno RM (2013) Deep cortical layers are activated directly by thalamus. Science 340:1591–1594. doi:10.1126/science.1236425

Covic EN, Sherman SM (2011) Synaptic properties of connections between the primary and secondary auditory cortices in mice. Cereb Cortex 21:2425–2441. doi:10.1093/cercor/bhr029

Dancause N, Barbay S, Frost SB et al (2006) Ipsilateral connections of the ventral premotor cortex in a new world primate. J Comp Neurol 495:374–390. doi:10.1002/cne.20875

De Pasquale R, Sherman SM (2011) Synaptic properties of corticocortical connections between the primary and secondary visual cortical areas in the mouse. J Neurosci 31:16494–16506. doi:10.1523/JNEUROSCI. 3664-11.2011

Desimone R (1996) Neural mechanisms for visual memory and their role in attention. Proc Natl Acad Sci U S A 93:13494–13499

Douglas RJ, Martin KA (1991) A functional microcircuit for cat visual cortex. J Physiol 440:735–769

Douglas RJ, Martin KAC, Whitteridge D (1989) A canonical microcircuit for neocortex. Neural Comput 1:480–488. doi:10.1162/neco.1989.1.4.480

Durstewitz D (2009) Implications of synaptic biophysics for recurrent network dynamics and active memory. Neural Netw 22:1189–1200. doi:10.1016/j.neunet.2009.07.016

Engbert R, Mergenthaler K, Sinn P, Pikovsky A (2011) An integrated model of fixational eye movements and microsaccades. Proc Natl Acad Sci U S A 108:E765–E770. doi:10.1073/pnas.1102730108

Esterman M, Yantis S (2010) Perceptual expectation evokes category-selective cortical activity. Cereb Cortex 20:1245–1253. doi:10.1093/cercor/bhp188

Feldman H, Friston KJ (2010) Attention, uncertainty, and free-energy. Front Hum Neurosci 4:215. doi:10.3389/fnhum.2010.00215

Felleman DJ, Van Essen DC (1991) Distributed hierarchical processing in the primate cerebral cortex. Cereb Cortex 1:1–47

Ferrer JM, Kato N, Price DJ (1992) Organization of association projections from area 17 to areas 18 and 19 and to suprasylvian areas in the cat's visual cortex. J Comp Neurol 316:261–278. doi:10.1002/cne.903160302

Fox K, Sato H, Daw N (1990) The effect of varying stimulus intensity on NMDA-receptor activity in cat visual cortex. J Neurophysiol 64:1413–1428

Fries P, Reynolds JH, Rorie AE, Desimone R (2001) Modulation of oscillatory neuronal synchronization by selective visual attention. Science 291:1560–1563. doi:10.1126/science.291.5508.1560

Friston K (2005) A theory of cortical responses. Philos Trans R Soc Lond B Biol Sci 360:815–836. doi:10.1098/rstb.2005.1622

Friston K (2008) Hierarchical models in the brain. PLoS Comput Biol 4:e1000211. doi:10.1371/journal.pcbi.1000211

Friston K (2010) The free-energy principle: a unified brain theory? Nat Rev Neurosci 11:127–138. doi:10.1038/nrn2787

Friston K, Kiebel S (2009) Cortical circuits for perceptual inference. Neural Netw 22:1093–1104. doi:10.1016/j.neunet.2009.07.023

Friston KJ, Daunizeau J, Kilner J, Kiebel SJ (2010) Action and behavior: a free-energy formulation. Biol Cybern 102:227–260. doi:10.1007/s00422-010-0364-z

Friston K, Adams RA, Perrinet L, Breakspear M (2012) Perceptions as hypotheses: saccades as experiments. Front Psychol 3:151. doi:10.3389/fpsyg.2012.00151

Garrido MI, Kilner JM, Kiebel SJ, Friston KJ (2007) Evoked brain responses are generated by feedback loops. Proc Natl Acad Sci U S A 104:20961–20966. doi:10.1073/pnas.0706274105

Garrido MI, Kilner JM, Stephan KE, Friston KJ (2009) The mismatch negativity: a review of underlying mechanisms. Clin Neurophysiol 120:453–463. doi:10.1016/j.clinph.2008.11.029

Ghosh S, Porter R (1988) Corticocortical synaptic influences on morphologically identified pyramidal neurones in the motor cortex of the monkey. J Physiol 400:617–629

Ghosh S, Brinkman C, Porter R (1987) A quantitative study of the distribution of neurons projecting to the precentral motor cortex in the monkey (M. fascicularis). J Comp Neurol 259:424–444. doi:10.1002/cne.902590309

Haeusler S, Maass W (2007) A statistical analysis of information-processing properties of lamina-specific cortical microcircuit models. Cereb Cortex 17:149–162. doi:10.1093/cercor/bhj132

Harrison SA, Tong F (2009) Decoding reveals the contents of visual working memory in early visual areas. Nature 458:632–635. doi:10.1038/nature07832

Harrison LM, Stephan KE, Rees G, Friston KJ (2007) Extra-classical receptive field effects measured in striate cortex with fMRI. NeuroImage 34:1199–1208. doi:10.1016/j.neuroimage.2006.10.017

Hatanaka N, Nambu A, Yamashita A et al (2001) Somatotopic arrangement and corticocortical inputs of the hindlimb region of the primary motor cortex in the macaque monkey. Neurosci Res 40:9–22

Hatanaka N, Tokuno H, Hamada I et al (2003) Thalamocortical and intracortical connections of monkey cingulate motor areas. J Comp Neurol 462:121–138. doi:10.1002/cne.10720

Hopfinger JB, Buonocore MH, Mangun GR (2000) The neural mechanisms of top-down attentional control. Nat Neurosci 3:284–291. doi:10.1038/72999

Hupé JM, James AC, Payne BR et al (1998) Cortical feedback improves discrimination between figure and background by V1, V2 and V3 neurons. Nature 394:784–787. doi:10.1038/29537

Janata P (2001) Brain electrical activity evoked by mental formation of auditory expectations and images. Brain Topogr 13:169–193

Kätzel D, Zemelman BV, Buetfering C et al (2011) The columnar and laminar organization of inhibitory connections to neocortical excitatory cells. Nat Neurosci 14:100–107. doi:10.1038/nn.2687

Knight RT (1984) Decreased response to novel stimuli after prefrontal lesions in man. Electroencephalogr Clin Neurophysiol 59:9–20

Knight RT, Scabini D, Woods DL (1989) Prefrontal cortex gating of auditory transmission in humans. Brain Res 504:338–342

Kosslyn SM, Thompson WL (2003) When is early visual cortex activated during visual mental imagery? Psychol Bull 129:723–746. doi:10.1037/0033-2909.129.5.723

Kraemer DJM, Macrae CN, Green AE, Kelley WM (2005) Musical imagery: sound of silence activates auditory cortex. Nature 434:158. doi:10.1038/434158a

Larkum ME, Senn W, Lüscher H-R (2004) Top-down dendritic input increases the gain of layer 5 pyramidal neurons. Cereb Cortex 14:1059–1070. doi:10.1093/cercor/bhh065

Lefort S, Tomm C, Floyd Sarria J-C, Petersen CCH (2009) The excitatory neuronal network of the C2 barrel column in mouse primary somatosensory cortex. Neuron 61:301–316. doi:10.1016/j.neuron.2008.12.020

Luppino G, Matelli M, Camarda R, Rizzolatti G (1993) Corticocortical connections of area F3 (SMA-proper) and area F6 (pre-SMA) in the macaque monkey. J Comp Neurol 338:114–140. doi:10.1002/cne.903380109

Maier A, Adams GK, Aura C, Leopold DA (2010) Distinct superficial and deep laminar domains of activity in the visual cortex during rest and stimulation. Front Syst Neurosci. doi:10.3389/fnsys.2010.00031

Matelli M, Camarda R, Glickstein M, Rizzolatti G (1986) Afferent and efferent projections of the inferior area 6 in the macaque monkey. J Comp Neurol 251:281–298. doi:10.1002/cne.902510302

Mechelli A, Price CJ, Friston KJ, Ishai A (2004) Where bottom-up meets top-down: neuronal interactions during perception and imagery. Cereb Cortex 14:1256–1265. doi:10.1093/cercor/bhh087

Meyer T, Olson CR (2011) Statistical learning of visual transitions in monkey inferotemporal cortex. Proc Natl Acad Sci U S A 108:19401–19406. doi:10.1073/pnas.1112895108

Meyer HS, Schwarz D, Wimmer VC et al (2011) Inhibitory interneurons in a cortical column form hot zones of inhibition in layers 2 and 5A. Proc Natl Acad Sci 108:16807–16812. doi:10.1073/pnas.1113648108

Muckli L, Kohler A, Kriegeskorte N, Singer W (2005) Primary visual cortex activity along the apparent-motion trace reflects illusory perception. PLoS Biol 3:e265. doi:10.1371/journal.pbio.0030265

Mumford D (1992) On the computational architecture of the neocortex. II. The role of cortico-cortical loops. Biol Cybern 66:241–251

Muñoz A, Woods TM, Jones EG (1999) Laminar and cellular distribution of AMPA, kainate, and NMDA receptor subunits in monkey sensory-motor cortex. J Comp Neurol 407:472–490

Murray SO, Kersten D, Olshausen BA et al (2002) Shape perception reduces activity in human primary visual cortex. Proc Natl Acad Sci U S A 99:15164–15169. doi:10.1073/pnas.192579399

Nemoto I (2012) Evoked magnetoencephalographic responses to omission of a tone in a musical scale. J Acoust Soc Am 131:4770–4784. doi:10.1121/1.4714916

Nordby H, Hammerborg D, Roth WT, Hugdahl K (1994) ERPs for infrequent omissions and inclusions of stimulus elements. Psychophysiology 31:544–552

Olsen SR, Bortone DS, Adesnik H, Scanziani M (2012) Gain control by layer six in cortical circuits of vision. Nature 483:47–52. doi:10.1038/nature10835

Pin JP, Duvoisin R (1995) The metabotropic glutamate receptors: structure and functions. Neuropharmacology 34:1–26

Raij T, McEvoy L, Mäkelä JP, Hari R (1997) Human auditory cortex is activated by omissions of auditory stimuli. Brain Res 745:134–143

Rao RP, Ballard DH (1999) Predictive coding in the visual cortex: a functional interpretation of some extra-classical receptive-field effects. Nat Neurosci 2:79–87. doi:10.1038/4580

Rockland KS, Drash GW (1996) Collateralized divergent feedback connections that target multiple cortical areas. J Comp Neurol 373:529–548. doi:10.1002/(SICI)1096-9861(19960930)373:4<529::AID-CNE5>3.0.CO;2–3

Roopun AK, Middleton SJ, Cunningham MO et al (2006) A beta2-frequency (20–30 Hz) oscillation in nonsynaptic networks of somatosensory cortex. Proc Natl Acad Sci U S A 103:15646–15650. doi:10.1073/pnas.0607443103

Roopun AK, Kramer MA, Carracedo LM et al (2008) Period concatenation underlies interactions between gamma and beta rhythms in neocortex. Front Cell Neurosci 2:1. doi:10.3389/neuro.03.001.2008

Salin PA, Bullier J (1995) Corticocortical connections in the visual system: structure and function. Physiol Rev 75:107–154

Schmidlin E, Brochier T, Maier MA et al (2008) Pronounced reduction of digit motor responses evoked from macaque ventral premotor cortex after reversible inactivation of the primary motor cortex hand area. J Neurosci 28:5772–5783. doi:10.1523/JNEUROSCI. 0944-08.2008

Self MW, Kooijmans RN, Supèr H et al (2012) Different glutamate receptors convey feedforward and recurrent processing in macaque V1. Proc Natl Acad Sci U S A 109:11031–11036. doi:10.1073/pnas.1119527109

Shima K, Tanji J (1998) Involvement of NMDA and non-NMDA receptors in the neuronal responses of the primary motor cortex to input from the supplementary motor area and somatosensory cortex: studies of task-performing monkeys. Jpn J Physiol 48:275–290

Shimazu H, Maier MA, Cerri G et al (2004) Macaque ventral premotor cortex exerts powerful facilitation of motor cortex outputs to upper limb motoneurons. J Neurosci 24:1200–1211. doi:10.1523/JNEUROSCI. 4731-03.2004

Shipp S (2005) The importance of being agranular: a comparative account of visual and motor cortex. Philos Trans R Soc Lond B Biol Sci 360:797–814. doi:10.1098/rstb.2005.1630

Shipp S (2007) Structure and function of the cerebral cortex. Curr Biol 17:R443–R449. doi:10.1016/j.cub.2007.03.044

Shipp S, Zeki S (1989a) The organization of connections between areas V5 and V1 in macaque monkey visual cortex. Eur J Neurosci 1:309–332

Shipp S, Zeki S (1989b) The organization of connections between areas V5 and V2 in macaque monkey visual cortex. Eur J Neurosci 1:333–354

Shlosberg D, Amitai Y, Azouz R (2006) Time-dependent, layer-specific modulation of sensory responses mediated by neocortical layer 1. J Neurophysiol 96:3170–3182. doi:10.1152/jn.00520.2006

Spratling MW (2010) Predictive coding as a model of response properties in cortical area V1. J Neurosci 30:3531–3543. doi:10.1523/JNEUROSCI. 4911-09.2010

Spratling MW (2012) Predictive coding as a model of the V1 saliency map hypothesis. Neural Netw 26:7–28. doi:10.1016/j.neunet.2011.10.002

Summerfield C, Trittschuh EH, Monti JM et al (2008) Neural repetition suppression reflects fulfilled perceptual expectations. Nat Neurosci 11:1004–1006. doi:10.1038/nn.2163

Thomson AM, Bannister AP (2003) Interlaminar connections in the neocortex. Cereb Cortex 13:5–14. doi:10.1093/cercor/13.1.5

Thomson AM, West DC, Wang Y, AP B (2002) Synaptic connections and small circuits involving excitatory and inhibitory neurons in layers 2–5 of adult rat and cat neocortex: triple intracellular recordings and biocytin labelling in vitro. Cereb Cortex 12:936–953

Todorovic A, van Ede F, Maris E, de Lange FP (2011) Prior expectation mediates neural adaptation to repeated sounds in the auditory cortex: an MEG study. J Neurosci 31:9118–9123. doi:10.1523/JNEUROSCI. 1425-11.2011

Tokuno H, Tanji J (1993) Input organization of distal and proximal forelimb areas in the monkey primary motor cortex: a retrograde double labeling study. J Comp Neurol 333:199–209. doi:10.1002/cne.903330206

Traynelis SF, Wollmuth LP, McBain CJ et al (2010) Glutamate receptor ion channels: structure, regulation, and function. Pharmacol Rev 62:405–496. doi:10.1124/pr.109.002451

Usrey WM, Fitzpatrick D (1996) Specificity in the axonal connections of layer VI neurons in tree shrew striate cortex: evidence for distinct granular and supragranular systems. J Neurosci 16:1203

von Helmholtz H (1866) Handbuch der physiologischen Optik: mit 213 in den Text eingedruckten Holzschnitten und 11 Tafeln. Voss

Wacongne C, Labyt E, van Wassenhove V et al (2011) Evidence for a hierarchy of predictions and prediction errors in human cortex. Proc Natl Acad Sci U S A 108:20754–20759. doi:10.1073/pnas.1117807108

Weiler N, Wood L, Yu J et al (2008) Top-down laminar organization of the excitatory network in motor cortex. Nat Neurosci 11:360–366. doi:10.1038/nn2049

Wozny C, Williams SR (2011) Specificity of synaptic connectivity between layer 1 inhibitory interneurons and layer 2/3 pyramidal neurons in the rat neocortex. Cereb Cortex 21:1818–1826. doi:10.1093/cercor/bhq257

Wyart V, Nobre AC, Summerfield C (2012) Dissociable prior influences of signal probability and relevance on visual contrast sensitivity. Proc Natl Acad Sci U S A. doi:10.1073/pnas.1120118109

Yabe H, Tervaniemi M, Reinikainen K, Näätänen R (1997) Temporal window of integration revealed by MMN to sound omission. Neuroreport 8:1971–1974

Yamaguchi S, Knight RT (1990) Gating of somatosensory input by human prefrontal cortex. Brain Res 521:281–288

Zagha E, Casale AE, Sachdev RNS et al (2013) Motor cortex feedback influences sensory processing by modulating network state. Neuron 79:567–578. doi:10.1016/j.neuron.2013.06.008

Zeki S, Shipp S (1988) The functional logic of cortical connections. Nature 335:311–317. doi:10.1038/335311a0

Zilles K, Schlaug G, Matelli M et al (1995) Mapping of human and macaque sensorimotor areas by integrating architectonic, transmitter receptor, MRI and PET data. J Anat 187(Pt 3):515–537

Chapter 8
Columnar Organization of the Motor Cortex: Direction of Movement

Apostolos P. Georgopoulos

Abstract The discovery by Vernon B. Mountcastle of the columnar organization of the cerebral cortex (Mountcastle VB, J Neurophysiol 20:408–434, 1957, Brain 120:701–722, 1997) was the single most important discovery of the twentieth century in cortical physiology. Not only did it serve as the framework for the orderly arrangement of knowledge concerning cortical organization and function (Edelman and Mountcastle, The mindful brain. MIT Press, Cambridge, MA, 1978) but also as a framework for exploring and investigating new ideas and for revisiting old ones about the organization of particular cortical areas. Here I review the history of facts and ideas about the organization of the motor cortex and discuss the evidence that the direction of movement is the principle governing motor cortical columnar organization.

Keywords Columnar organization • Motor activity • Motor cortex • Directional tuning

8.1 Introduction

Motor cortex controls movement. Hughlings Jackson put it elegantly in his paper titled “On some implications of dissolution of the nervous system” in 1882, as follows: “Nervous centres represent movement, not muscles. From negative lesions of motor centres there is not paralysis of muscles, but loss of movements.” (Jackson 1882, p. 411). Unfortunately, Jackson’s insight into the motor cortical representation of *integrated* movements gradually faded and was replaced by an “atomistic” perspective, namely that *components* of movements (about a joint) or just “musculature” are represented, rather than whole movements. In the famous “motor figurine chart” of Woolsey et al. (1952; Fig. 122, p. 239) “each of the figurines indicates the peripheral location of the musculature activated by stimulating electrically the cortical point to which the figurine corresponds” (Woolsey et al. 1952,

A.P. Georgopoulos (✉)
Brain Sciences Center, Minneapolis Veterans Affairs Health Care System; and Department of Neuroscience, University of Minnesota Medical School, One Veterans Drive, Minneapolis, MN 55417, USA
e-mail: omega@umn.edu

M.F. Casanova, I. Opris (eds.), *Recent Advances on the Modular Organization of the Cortex*, DOI 10.1007/978-94-017-9900-3_8

p. 239). I think there were two main conceptual reasons for this regressing to simple components. The first reason had to do with the general emphasis on the neuron as the structural and functional element of organization of the central nervous system: a component-based structural framework extended conceptually to encompass a component-based cortical mapping of function. The work and views of Sherrington (1906, 1940), Lashley (1933), Hebb (1949), and Pitts and McCulloch (1947) were notable exceptions, which failed, however, to shift the mainstream thinking from an atomistic to an integrative perspective. The second reason was a carry-over from the somatosensory system where sensation began by stimulation of discreet receptors in the skin and was ultimately integrated in higher cortical areas, such as the posterior parietal cortex. However, at the level of the first somatosensory cortex (S1), the representation of the sensory stimulus was discreet and atomistic, in the sense that cells there responded to stimulation of specific-modality receptors located in discreet receptive fields. It was natural, then, to assume implicitly that representations in the primary motor cortex (M1), separated from S1 only by the central sulcus, would similarly correspond to some atomistic motor attribute, such as motion about a joint. It was no wonder then that Woolsey's maps of S1 and M1 (Woolsey et al. 1952) strived to illustrate the correspondence between the sensory and motor atomistic representations in the two areas, respectively. Although somatotopically correct in the general sense of body part representation, the point as to *what* is represented in the two areas was missed, and M1 representations became as distinctly atomistic as those of S1, a viewpoint exemplified later by Fetz (1984). At the other extreme, protracted electrical stimulation of the motor cortex (Graziano et al. 2002) elicits a small number of stereotyped movement sequences, reflecting massive, simultaneous activations and apparently unrelated to specific cortical columns.

The work of Mountcastle on the columnar organization of S1 drew the attention of Asanuma and his colleagues who used their newly developed intracortical microstimulation (ICMS) technique (Asanuma and Sakata 1967) to investigate the matter in M1 of the cat (Asanuma and Sakata 1967) and monkey (Asanuma and Rosén 1972). It was found that ICMS elicited localized muscle contractions of the same muscles along anatomical columns. It was argued that this is the organizing principle of M1 columns and was attributed to the excitation of cells of origin of the pyramidal tract. However, it was shown later that "a weak intracortical stimulation is relatively ineffective in a direct excitation of pyramidal tract cells and that the effects of such a stimulation are mainly indirect, especially when repetitive stimuli are used." (Jankowska et al. 1975, p. 617). In fact, Jankowska (1975), after reviewing the then available literature on the effects of surface or ICMS electrical stimulation of the motor cortex, concluded that "that there is no satisfactory evidence for location of the pyramidal tract (PT) cells which project directly (monosynaptically) to motoneurones of one muscle in primates within small and separate cortical areas; such evidence is even weaker for PT cells which exert their effects via polysynaptic pathways in other species. *The available evidence shows in contrast a very pronounced overlap of cortical areas of projection to different motor nuclei*." (Jankowska 1975, p. 699; italics mine). This conclusion is very much in accord with a re-representation of muscles in various overlapping combinations underlying the representation of movement direction in motor cortex, as discussed below.

Other mapping studies in the motor cortex focused explicitly on joint movements. For example, in one study "The Movement about a joint was selected as the index of response." (Kwan et al. 1978, p. 1120). Unfortunately, this *a priori* focus restricted severely the possibility of exploring other hypotheses and ideas. The joint-focused research culminated in a paper by Murphy et al. (1982) titled "Precentral unit activity correlated with angular components of a compound arm movement" in which 3-D arm motions at various joints were recorded simultaneously with single activity in M1 of monkeys reaching to targets in 3-D space, an engineering feat at that time. The neuronal discharge rates were analyzed with respect to angular joint motions and muscle activity, with essentially negative results. The main conclusion was that "No simple relation between the EMG and single precentral neuron activities was observed in these experiments, even in those cases where the muscles recorded were activated by ICMS at the unit recording site. A similar lack of a simple relationship has been noted for presumed single joint movements ... Although the motions occurring at the two joints were produced quite differently, shoulder and elbow units tended to behave very similarly in this task." (Murphy et al. 1982, p. 144). Obviously, an analysis of M1 cell activity during multijoint movements using motion/muscle components as a reference did not lead to any meaningful insights. Ironically, it was that same year that our paper was published, reporting on the orderly relations between motor cortical activity and direction of arm movement in 2-D space (Georgopoulos et al. 1982).

8.2 Movement in Space

A different, integrative perspective was employed in studies of cell activity in the posterior parietal cortex (areas 5 and 7) (Mountcastle et al. 1975), recordings in which I was privileged to participate as a postdoctoral fellow in the early 1970s. Mountcastle's fame then was as a somatosensory neurophysiologist with a keen interest in cortical organization and the quantitative bridging of sensation and neural activity (Mountcastle 1967). His adoption of the behaving monkey experimental paradigm was the ultimate step in his attempt to explore, study and elucidate the neural mechanisms of sensation and perception. The natural step in this direction was to record cell activity in posterior parietal cortical areas 5 and 7, where integration of somatic perception was thought to occur, as suggested by the higher-order deficits seen in people with damage to the posterior parietal cortex (Critchley 1969) and by the convergence model of cortical organization put forward by Jones and Powell (1970) based on detailed neuroanatomical studies. The first aim in those recordings was to explore and identify adequate conditions for cell activation by examining the monkey and eliciting movements towards rewards (e.g. pieces of fruits). We then discovered a class of cells Vernon called "reach" neurons which became active when the monkey reached for a reward (Mountcastle et al. 1975), even in total darkness (Mountcastle et al. 1980). In classical Mountcastle style, we studied their activity in exquisite quantitative detail by designing a special apparatus that we called the "electric train" (Mountcastle et al. 1975, Fig. 2, p. 875). It had a semicircular rail on which a box ("train") with a light could move at different speeds; the box was also a press-switch. The monkey would initiate the train's motion by depressing a key at lap level. At some point along its travel, the light would dim, and this served as the "go" signal to the monkey to let go the key, reach towards the moving train, press its switch by hitting it, get a juice reward and return to the key to initiate a new trial. To our surprise, we found that the profile of cell activity was very similar when the train was hit at different locations along its travel, i.e. with reaching movements even when their motor profile was very different across those trials. We discussed those finding in the paper as indicating neural control of overall reaching movements, irrespective of their particular components (Mountcastle et al. 1975, Figs. 7–9, pp. 883–884). This was the first instance of relating neuronal cortical activity to a holistic, integrative aspect of arm movement *in extrapersonal space*. Now, since these were recordings from the parietal cortex, they did not influence concurrent studies by other groups in the motor cortex, which continued unabated on the atomistic platform.

8.3 The Importance of Movement Direction in Space

It was during that time that I made the basic observation about the direction of arm movement in space as the important variable for cell activity, as follows. Naturally, we used the electric train to study those cells that were active with reaching movements. Fortunately, we continued spike collection for a few hundred milliseconds after the delivery of reward, that is, while the arm was returning to the key in front of the monkey. Now, this return movement was in a direction opposite to that of the forward reach. I plotted the time course of cell activity during that return movement and saw that very frequently cell firing was very reduced: it seemed that cell firing was related to the direction of movement in space, and that in our forward reach paradigm we were using only a portion of the direction space. In fact, my colleagues and I confirmed that hypothesis when, several years later, we recorded cell activity in area 5 in the center → out task, where movement direction was varied systematically over the whole 2-D, 360 deg space (Kalaska et al. 1983, Fig. 3, p. 251). The broad directional tuning explained the finding of relative invariance of cell firing with reaching in the earlier study (Mountcastle et al. 1975): cell firing was consistently high around the cell's preferred direction, and the electric train design employed movements spread rather narrowly around the preferred direction, hence the apparent invariance.

8.4 The Center → Out Design

In 1978, I established my own laboratory next to Dr. Mountcastle's, in what used to be Clinton Woolsey's lab, where he had done his mapping experiments! I was keen to study systematically the relations of motor cortical cell activity to the direction of movement in space. Ideally, we needed to monitor free reaching movements in 3-D space but at that time I did not have the means to do that. So, I settled for the next best, namely drawing movements in 2-D space, on a tablet (Georgopoulos et al. 1981; Georgopoulos and Massey 1991). The essence of the design was for the monkey to make movements of the same amplitude whose direction would be equally (isotropically) distributed in 2-D space. Naturally, this suggested center → out movements, starting from the center of the tablet and ending at the circumference. It seemed that sampling 360 deg every 45 deg would be adequate, hence the 8-movement, starfish design (Georgopoulos et al. 1981, Fig. 1, p. 728; Georgopoulos et al. 1982, Fig. 1). Interestingly, a similar design had been employed by Fitts and Deininger (1954) in their study of stimulus-response compatibility; somehow, that paper reinforced my choice.

It should be realized that in 1978 planning an experiment on motor cortical control of movement outside the atomistic "box" of muscles and/or single-joint movements was unheard of, simply because the conceptual framework was so rigid and dominant. Nevertheless, I went ahead with the 2-D movement plan using a device constructed specially for this experiment at the Applied Physics Laboratory of the Johns Hopkins University (Georgopoulos and Massey 1991). The results were stunning: the activity of single cells in the motor cortex varied in an orderly fashion with the direction of movement *in space,* cell after cell (Georgopoulos et al. 1982). I remember how impressed Elwood Henneman (Dr. Mountcastle's old friend and collaborator) was when he visited us from Harvard (always wearing his bowtie), and sat in a recording session and couldn't believe his eyes at the regularity of this directional variation, as spike trains were plotted one after another at our large analog Tektronix screen and printed as hard copies. The regularity in the variation of cell discharge with direction in those plots (Figs. 8.1 and 8.2) were scientifically amazing and esthetically beautiful. Scatterplots of that orderly variation established the *directional tuning curve* (Figs. 8.2 and 8.3) and the findings were reported at the Annual meeting of the Society for Neuroscience in 1980 (Georgopoulos et al. 1980). Remarkably, that time a new wind started blowing in my direction(!). Two other, different labs were focusing on the planning and control of arm movements in 2-D space, claiming *space* as the proper framework: one lab was in Minnesota (Viviani and Terzuolo 1982) and the other at MIT (Abend et al. 1982). This was a conceptual meeting of minds. I presented our results at a conference in Henniker, NH in July 1981 (Engineering Foundation Conference on "Biomechanics and neural control of movement") where I was encouraged by Peter Strick to pursue this line of work. Another ideological supporter was Andras Pellionisz from New York University, who, with Rodolfo Llinás, had proposed an integrative role of the cerebellum in motor control (Pellionisz and Llinás 1979).

From within the neurophysiology field, Vernon Brooks was an enthusiastic supporter of the new idea and, surprisingly, motor physiologists of the spinal cord, including Anders Lundberg and Sten Grillner. The spinal cord is a hotbed of integrative motor functions, including central pattern generation, sensorimotor integration, online sensory control of locomotion, etc. So, spinal cord physiologists were more at home endorsing my integrative notion of movement direction as a basic motor parameter than were motor cortical neurophysiologists who were thinking mostly within the atomistic, movement component framework. The confluence of spinal and motor cortical integrative mechanisms resulted in a *Perspectives* paper in *Science* jointly authored by Sten Grillner and myself (Georgopoulos and Grillner 1989). Finally, it should be noted that, although in 1980 movement direction in space struck as a purely spatial measure, research during the subsequent 30 years showed that it captures many key motor attributes, including torques and EMG activity. We have reviewed these studies in detail in a recent "theory and hypothesis" paper (Mahan and Georgopoulos 2013) in which we also reviewed the evidence for the presence of directional tuning in all cortical and subcortical areas where it has been investigated ("directional motor resonance").

Since the 2-D device we used in those early experiments involved motion by the monkey of an articulated manipulandum, and took a few months to train the monkeys to use it, I was concerned as to what extent directional tuning would hold in 3-D space for free reaching movements. Indeed, it held amazingly well, as it was shown in several studies (Georgopoulos et al. 1986; Schwartz et al. 1988; Caminiti et al. 1990; Naselaris et al. 2006a, b). The 3-D tuning curve was also esthetically beautiful (Fig. 8.4).

8.5 Directional Columns

Directional tuning became firmly established during the 1980s. Two key issues were raised and began to be addressed during that period. The first had to do with the unique neural coding of movement direction: given that the directional tuning curve is broad and symmetric, it can provide unambiguous information only at its peak (i.e. at the cell's preferred direction), but it is unreasonable to suppose that only that point in the curve is used and all else discarded. This problem was solved by the neuronal population vector, an ensemble coding scheme (Georgopoulos et al. 1983, 1986, 1988) which became a nodal factor for the resurgence of computational neuroscience. The neuronal population vector has proved an effective way for decoding directional tuned motor cortical recordings currently used in neuroprosthetics (Collinger et al. 2013; Courtine et al. 2013) and for monitoring dynamic, time-varying cognitive operations (Georgopoulos et al. 1993; Pellizzer et al. 1995).

The second problem had to do with the cortical representation of the preferred direction (PD). Cells had different PDs which covered the whole directional range (Figs. 8.3 and 8.5) (Schwartz et al. 1988). Naturally, we asked the question of

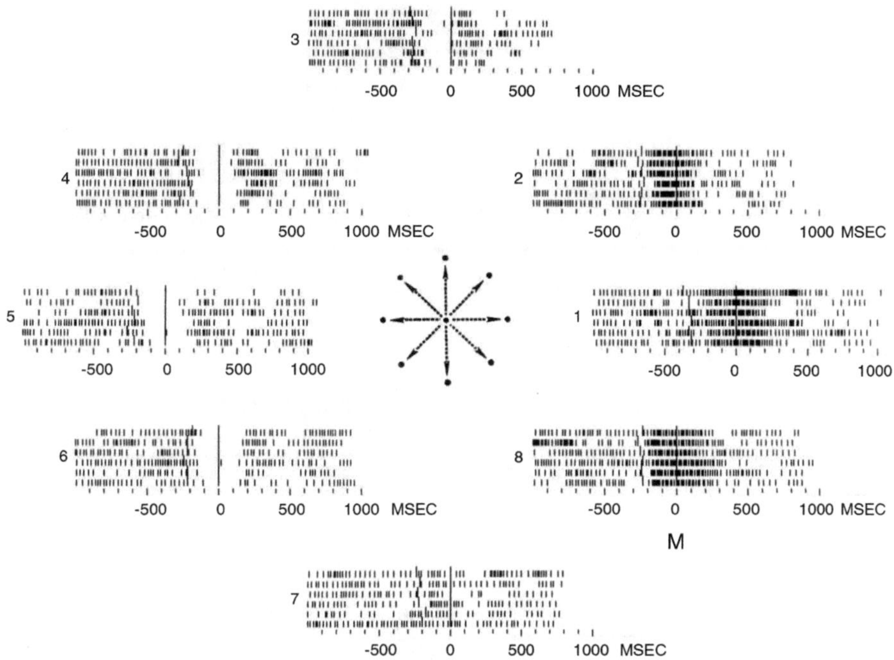

Fig. 8.1 Orderly variation of neuronal activity with movement direction in the center → out task (Georgopoulos et al. 1982), as indicated in the insert. Trials are aligned to the onset of movement (*M*); longer bars prior to movement onset denote the time of target onset. (Georgopoulos AP, Kalaska JF and Caminiti R, unpublished observations)

whether the preferred direction would be a feature (or, rather, *the* feature) of motor cortical columnar organization. Our first approach was similar to that used by Powell and Mountcastle (1959), namely to note the location of cells with specific PD along histologically identified penetrations and then observe possible *en block* changes in PD in penetrations at an angle with anatomical cortical columns. I presented our first results (Georgopoulos et al. 1984) at a meeting of the Neuroscience Research Program (NRP) held at the Salk Institute in 1983, after the NRP moved from the Rockefeller University to the Neurosciences Institute in La Jolla, CA. The results provided strong evidence for a columnar organization of the PD: in penetrations at the exposed cortex, PDs stayed very similar (Fig. 8.6), whereas they changed *en block* in penetrations at an angle with the anatomical columns (Amirikian and Georgopoulos 2003, Fig. 1). We went a step further and analyzed this relation quantitatively. For that purpose, we correlated two measures: one was this angle φ between the penetration and the anatomical columns that it crossed (Fig. 8.7) and the other was the spread of PDs along that penetration, measured as the circular standard deviation s_0 (Mardia 1972) of the PD distribution. If the columnar organization of PD holds, we argued that, at the one extreme, when $\varphi \approx 0$ (i.e. for penetrations along a column), then $s_0 \approx 0$, whereas, at the other extreme, when $\varphi \approx 90°$ (i.e. for penetrations perpendicular to columns, as in a

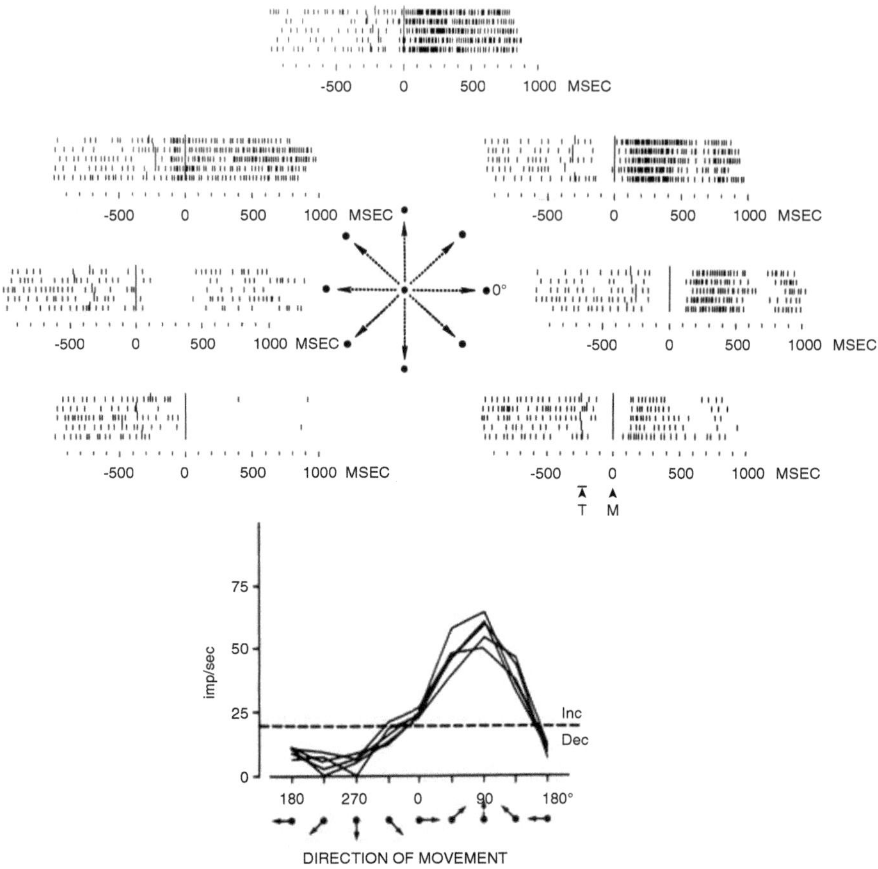

Fig. 8.2 Directional tuning of single cell activity and directional tuning curve. Five tuning curves, one for each trial (*raster row*) are plotted superimposed to illustrate their similarity. Conventions as in Fig. 8.1. The *dotted line* in the lower panel indicates the discharge rate preceding stimulus onset (Inc, Dec: increase, decrease of cell activity from that level) (Georgopoulos AP, Kalaska JF and Caminiti R, unpublished observations)

bank), then $s_0 \approx 180°$, with intermediate values in-between. Indeed, we found a statistically significant positive correlation between φ and s_0 ($r = 0.756$, $p < 0.01$) (Georgopoulos et al. 1984). This finding provided strong support for the columnar organization of the PD. Interestingly, at the same meeting, Bruce Dow presented the results of a similar analysis done for orientation selectivity in the visual cortex; their correlation coefficient was $r = 0.6$ (Dow et al. 1984). Therefore, our quantitative evidence for a columnar organization of the PD in the motor cortex was as good as, or better than, that for orientation selectivity in the visual cortex!

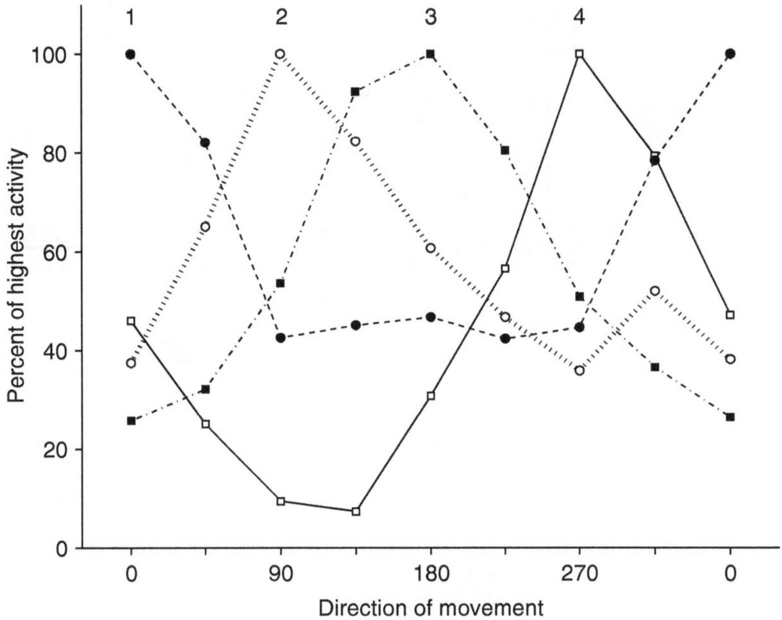

Fig. 8.3 Four directional tuning curves, normalized to their maximum, to illustrate the range of the preferred direction across the 360 deg direction space (Georgopoulos AP, Kalaska JF and Caminiti R, unpublished observations)

8.6 3-D Reaching Movements

The next major step was to extend the testing of the columnar hypothesis for PD to 3-D reaching movements. Our first attempt provided clear quantitative evidence in that direction (Amirikian and Georgopoulos 2003). However, it was clear that we needed an experimental arrangement designed specifically for this problem. Specifically, we needed to (a) have a 3-D reaching task, (b) insert microelectrodes in a regular grid on motor cortical surface, (c) identify (or approximate) the trajectory of microelectrode penetrations, and (d) record neural activity during 3-D reaching at regular spatial intervals (depths) along a penetration. We successfully implemented those objectives as follows (see Naselaris et al. 2005, 2006a for details). We (a) employed the original 3-D reaching task (Schwartz et al. 1988), (b) constructed precise location-aligned templates for inserting and advancing 16 microelectrodes simultaneously using the Eckhorn Multielectrode matrix, (c) used dyes to identify the edges of the penetration matrix, (d) recorded neural activity simultaneously from 16 electrodes every 150 μm during task performance, (e) approximated the location of recording sites along the penetrations, (f) flattened the cortex, and (g) projected the PD in the recording sites onto the flattened cortical surface (Fig. 8.8). Thus, a 2-D cortical map was constructed with the PD color coded, after they were binned to octants (Fig. 8.9). Figure 8.9 shows that PDs were

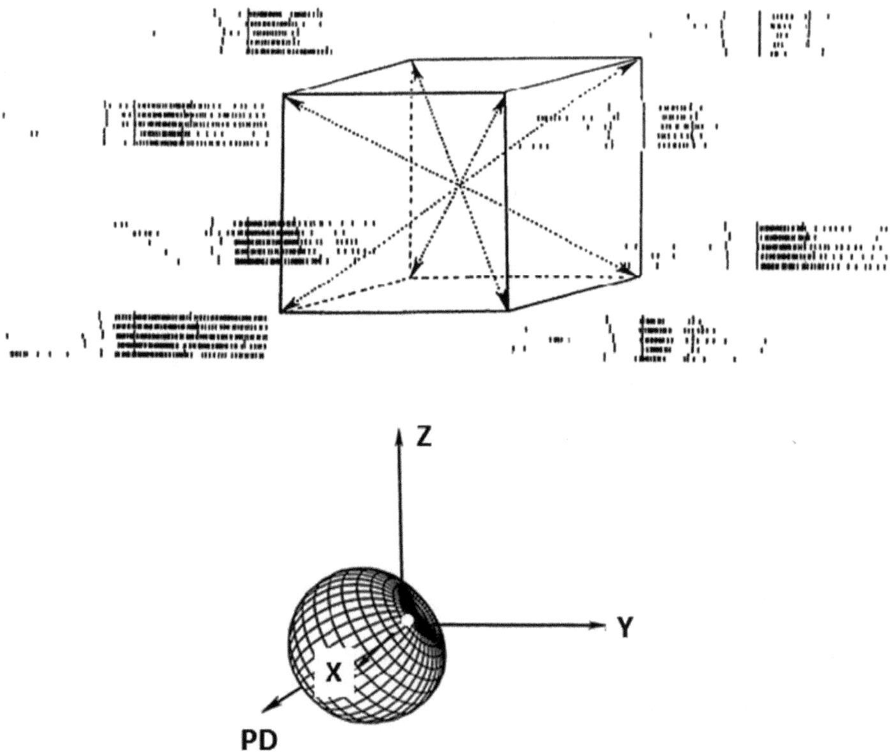

Fig. 8.4 Directional tuning in the 3-D reaching task and associated 3-D fitted tuning curve (Adapted from Schwartz et al. 1988)

repeatedly represented on the motor cortical surface such that, within a given locale, practically the full range of the PD continuum was represented. This meant that an accurate neuronal population estimate of the movement direction could be derived from within any one of these locales (Naselaris et al. 2006b).

The next challenge, of course, was to find out whether PDs are organized in a columnar fashion. For that purpose, we used the full precision of the PD determination (i.e. without binning) and carried out a spectral analysis of the distribution of the PDs on the cortical surface (Georgopoulos et al. 2007). We identified 3 major peaks in the periodogram, namely one at a period of ~240 μm, another at a period of ~90 μm, and a smaller peak at ~30 μm (see Figs. 8.4 and 8.5 in Georgopoulos et al. 2007). These findings suggested a columnar organization of the PD with an estimated unit-column width of ~ 30 μm and a repetition of the full PD range every ~240 μm (Fig. 8.10). In fact, a regression analysis revealed an orderly increasing angular difference of PDs away from a given locus, up to 120 μm, suggesting an orderly representation of a series of PDs on the cortical surface (Georgopoulos et al. 2007). Altogether, these findings suggested a lattice representation of PDs, as illustrated in Fig. 8.10.

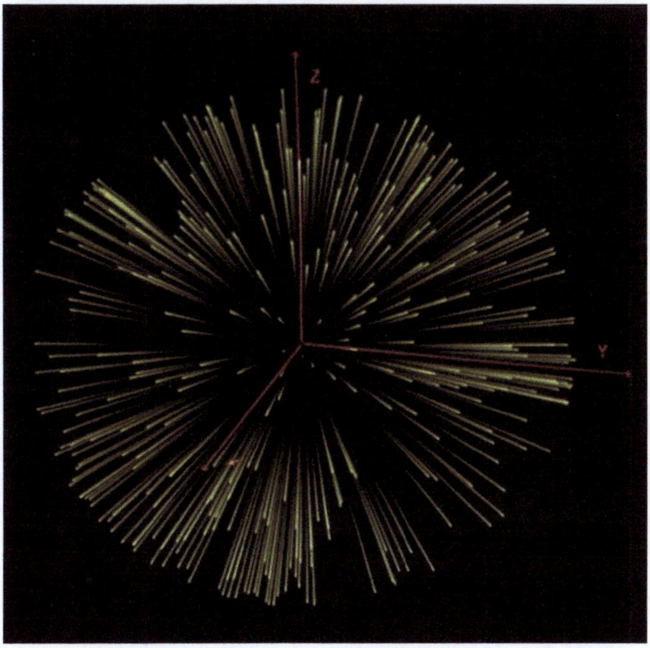

Fig. 8.5 Distribution of 475 3-D preferred directions (From Schwartz et al. 1988)

8.7 Concluding Remarks

There is no doubt that the column is *the* organizing unit of the cerebral cortex. Although the computer metaphor of the brain has been discredited in many attempts, prominently fought against relentlessly by the late giant of neuroscience and dear friend Gerry Edelman, there is something to be said for the operational usefulness of the analogy.[1] Assuming the metaphor for the sake of argument, one can consider the cortical column as a microprocessor, and the brain as a set of massively interacting such microprocessors, i.e. a high-performance computer cluster. Accordingly, intra-columnar processing (Anderson et al. 2010; Opris et al. 2011; Apicella et al. 2012; Chadderdon et al. 2014) would correspond to operations within the microprocessor, whereas inter-columnar operations would correspond to serial and parallel operations in the cluster. The most immediate case concerns local inter-columnar operations, i.e. within an area of a radius of ~500 μm around a column (Gatter and Powell 1978; Georgopoulos and Stefanis 2010). Such local interactions would serve, e.g., to shape the directional tuning curve (Lee

[1] I argued in favor of the resurgence of this idea, playing the Devil's advocate, at a meeting of the Neuroscience Research Program in La Jolla, CA in 2013. Dr. Edelman was incredulous and hardly believed his ears. Nevertheless, in his usual grand style, humor and compassion, he counter argued, and we had a lot of intellectual fun.

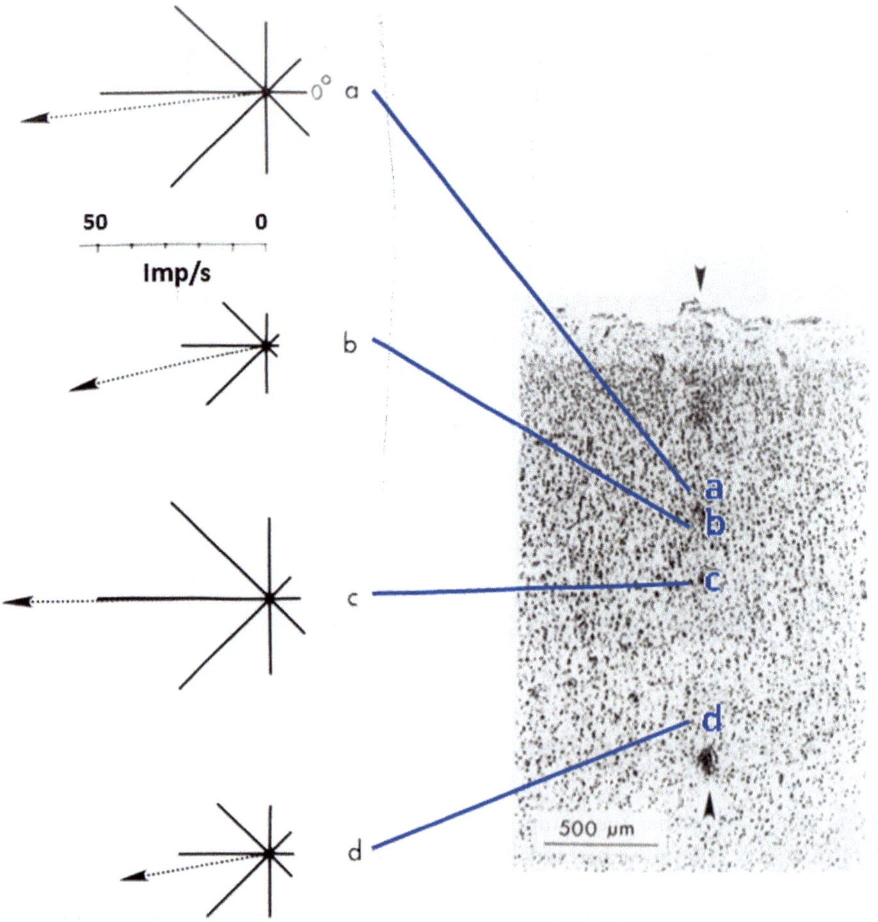

Fig. 8.6 Directional tuning of 4 neurons recorded along the histologically identified penetration shown. Preferred directions are very similar for this penetration, parallel to the cortical columns (Adapted from Georgopoulos et al. 1984)

et al. 2012; Mahan and Georgopoulos 2013; Georgopoulos 2014). Long-range interactions among columns (Caminiti et al. 1985, 1988), and with spinal systems (Georgopoulos 1996), would correspond to large-scale, parallel computer-cluster type of operations. Figure 8.11 gives an overall picture of columnar and intercolumnar/hemispheric organization, based on known anatomical and physiological facts, as follows. (a) Cortical layers (laminae) are color-coded and labeled with Roman numerals (layer IV is omitted for this agranular cortex). (b) Neurons in different layers of the same column interact and are also synchronized (Opris et al. 2011). (c) Neurons from different layers project predominantly to different targets: layer II → ipsilateral cortex; layer III → contralateral cortex; layer V → subcortical structures (basal ganglia, brainstem, spinal cord); layer VI → thalamus.

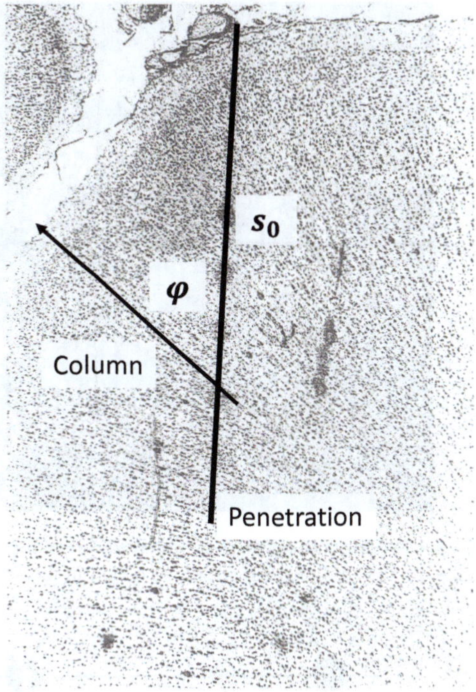

Fig. 8.7 Diagram to illustrate angle φ and a histologically identified penetration traversing the anterior bank of the central sulcus; s_0, circular standard deviation of the preferred directions of cells recorded at different depths along the penetration (See text for details)

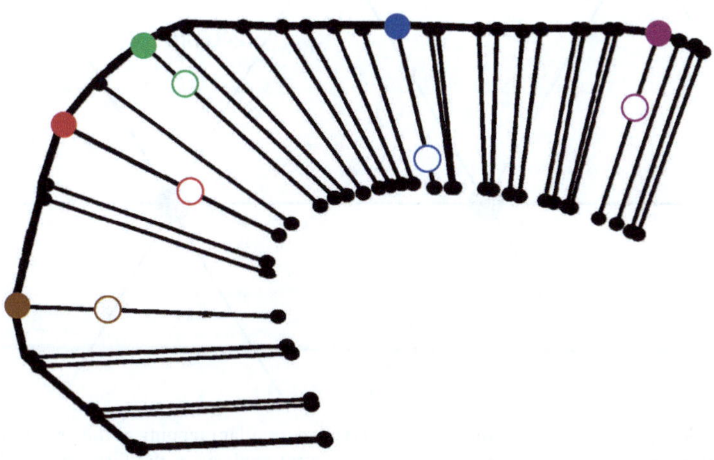

Fig. 8.8 Schematic diagram to illustrate the projection of the preferred direction from a recording site on the cortical surface, along the anatomical column of the recording site (From Georgopoulos et al. 2007)

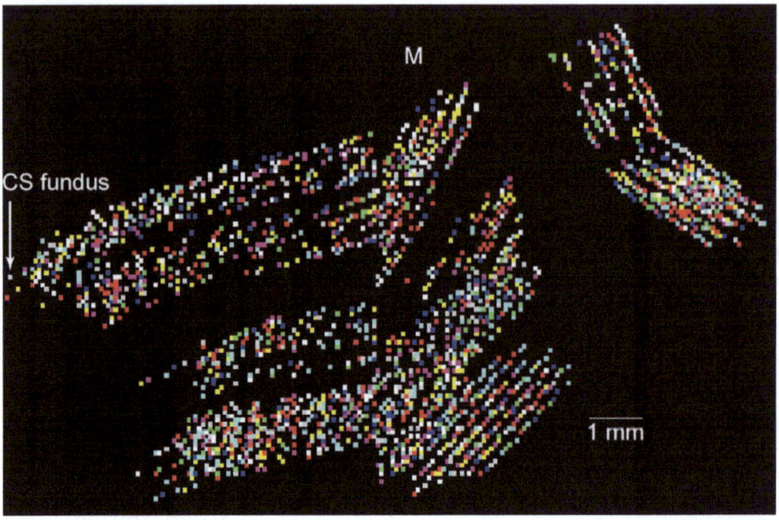

Fig. 8.9 Motor cortical map of preferred directions, constructed as indicated in Fig. 8.8. Colors denote preferred directions within an octant in the unit sphere (From Georgopoulos et al. 2007)

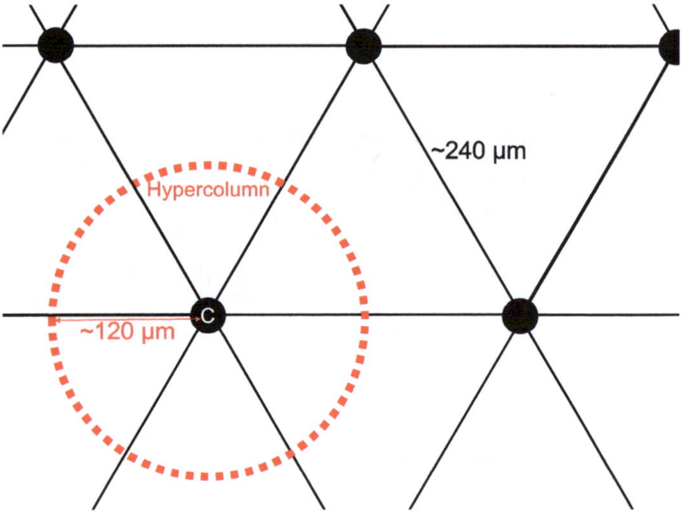

Fig. 8.10 A hypothesized lattice model of the repeated, regular mapping of the preferred direction in the motor cortex, based on the results of the spectral analysis of the distribution of preferred directions on the motor cortical surface (See text for details; from Georgopoulos et al. 2007)

The extensive cortical synchronization observed in studies using various technologies, including fMRI (Christova et al. 2011), magnetoencephalography (Leuthold et al. 2005; Langheim et al. 2006) and local field potentials (Merchant et al. 2014), is probably due to multiple factors, namely (i) local mechanisms (Stefanis and

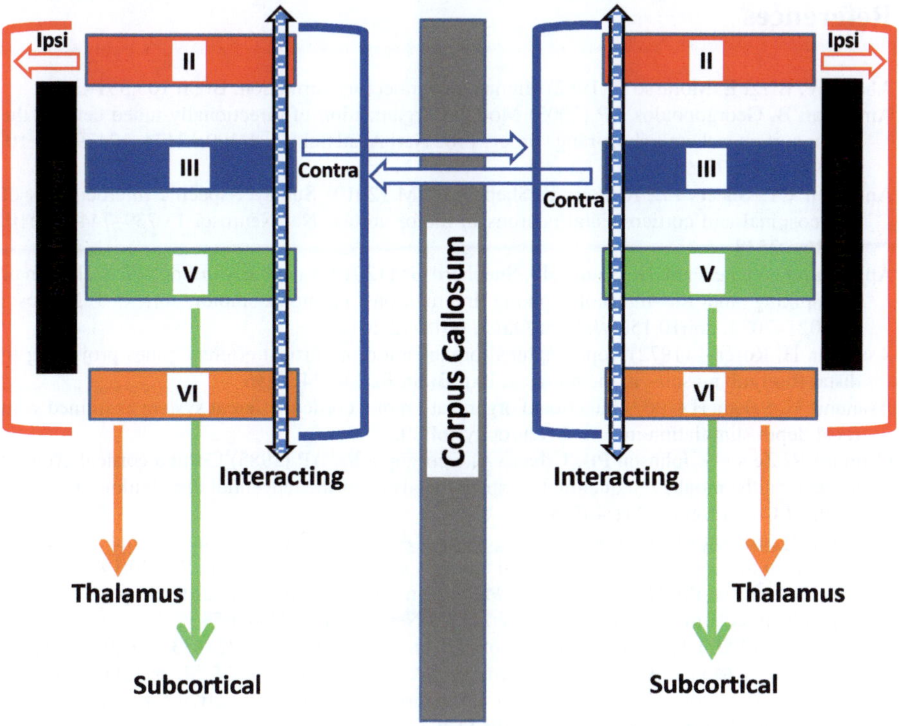

Fig. 8.11 Schematic diagram of intralaminar and inter-columnar interactions, based on known anatomical and physiological findings (See text for details)

Jasper 1964a, b), (ii) specific and non-specific thalamic afferents (Jones 2001), and (iii) synchronization among cortical layers, carrying over to their projections (Opris et al. 2011).

Finally, although the aspect of large-scale intercolumnar interactions has been extensively investigated, mainly by analyzing the effects of brain lesions or electrophysiological interventions (e.g. electrical stimulation, transcranial magnetic stimulation, etc.), possible behavioral consequences arising from disturbances in intra-columnar processing are gradually being recognized and investigated (Opris and Casanova 2014). The term "mini-columnopathy" has been coined to columnar abnormalities observed in autism (Casanova 2007), the beginning of a new era in connecting cortical columns to disease.

Acknowledgments This work was supported by the American Legion Brain Sciences Chair and the US Department of Veterans Affairs.

References

Abend W, Bizzi E, Morasso P (1982) Human arm trajectory formation. Brain 105:331–348
Amirikian B, Georgopoulos AP (2003) Modular organization of directionally tuned cells in the motor cortex: is there a short-range order? Proc Natl Acad Sci U S A 100:12474–12479. doi:10.1073/pnas.2037719100
Anderson CT, Sheets PL, Kiritani T, Shepherd GM (2010) Sublayer-specific microcircuits of corticospinal and corticostriatal neurons in motor cortex. Nat Neurosci 13:739–744. doi:10.1038/nn.2538
Apicella AJ, Wickersham IR, Seung HS, Shepherd GM (2012) Laminarly orthogonal excitation of fast-spiking and low-threshold-spiking interneurons in mouse motor cortex. J Neurosci 32:7021–7033. doi:10.1523/JNEUROSCI. 0011-12.2012
Asanuma H, Rosén I (1972) Topographical organization of cortical efferent zones projecting to distal forelimb muscles in the monkey. Exp Brain Res 14:243–256
Asanuma H, Sakata H (1967) Functional organization of a cortical efferent system examined with focal depth stimulation in cats. J Neurophysiol 30:35–54
Caminiti R, Zeger S, Johnson PB, Urbano A, Georgopoulos AP (1985) Cortico-cortical efferent systems in the monkey: a quantitative spatial analysis of the tangential distribution of cells of origin. J Comp Neurol 241:405–419
Caminiti R, Johnson PB, Urbano A, Georgopoulos AP, Zeger S (1988) Callosal and association neurons in the cortical space: a spectral analysis approach. Behav Brain Res 30:193–201
Caminiti R, Johnson PB, Urbano A (1990) Making arm movements within different parts of space: dynamic aspects in the primate motor cortex. J Neurosci 10:2039–2058
Casanova MF (2007) The neuropathology of autism. Brain Pathol 17:422–433. doi:BPA100 [pii]
Chadderdon GL, Mohan A, Suter BA, Neymotin SA, Kerr CC, Francis JT, Shepherd GM, Lytton WW (2014) Motor cortex microcircuit simulation based on brain activity mapping. Neural Comput 26:1239–1262. doi:10.1162/NECO_a_00602
Christova P, Lewis SM, Jerde TA, Lynch JK, Georgopoulos AP (2011) True associations between resting fMRI time series based on innovations. J Neural Eng 8:046025. doi:10.1088/1741-2560/8/4/046025
Collinger JL, Wodlinger B, Downey JE, Wang W, Tyler-Kabara EC, Weber DJ, McMorland AJ, Velliste M, Boninger ML, Schwartz AB (2013) High-performance neuroprosthetic control by an individual with tetraplegia. Lancet 381:557–564. doi:10.1016/S0140-6736(12)61816-9
Courtine G, Micera S, DiGiovanna J, Millan Jdel R (2013) Brain-machine interface: closer to therapeutic reality? Lancet 381:515–517. doi:10.1016/S0140-6736(12)62164-3
Critchley J (1969) The parietal lobes. Hafner, New York
Dow BM, Bauer R, Snyder AZ, Vautin RG (1984) Receptive fields and orientation shifts in foveal striate cortex of the awake macaque monkey. In: Edelman GM, Cowan WM, Gall WE (eds) Dynamic aspects of neocortical function. Wiley, New York, pp 41–65
Edelman GM, Mountcastle VB (1978) The mindful brain. MIT Press, Cambridge, MA
Fetz E (1984) The representation of movement direction in the motor cortex: single cell and population studies. In: Edelman GM, Cowan WM, Gall WE (eds) Dynamic aspects of neocortical function. Wiley, New York, pp 453–473
Fitts PM, Deininger RL (1954) S-R compatibility: correspondence among paired elements within stimulus and response codes. J Exp Psychol 48:483–492
Gatter KC, Powell TPS (1978) The intrinsic connections of the cortex of area 4 of the monkey. Brain 101:513–541
Georgopoulos AP (1996) On the translation of directional motor cortical commands to activation of muscles via spinal interneuronal systems. Cogn Brain Res 3:151–155
Georgopoulos AP (2014) Cell directional spread determines accuracy, precision, and length of the neuronal population vector. Exp Brain Res 232:2391–2405. doi:10.1007/s00221-014-396-7
Georgopoulos AP, Grillner S (1989) Visuomotor coordination in reaching and locomotion. Science 245:1209–1210

Georgopoulos AP, Massey JT (1991) Biomedical program research: the primate motor system. Johns Hopkins APL Tech Dig 12:105–114

Georgopoulos AP, Stefanis C (2010) The motor cortical circuit. In: Shepherd G, Grillner S (eds) Brain microcircuits. Oxford University Press, New York, pp 39–45

Georgopoulos AP, Kalaska JF, Massey JT (1980) Cortical mechanisms of two-dimensional aiming arm movements. I. Aiming at different target locations. Soc Neurosci Abstr 6:156

Georgopoulos AP, Kalaska JF, Massey JT (1981) Spatial trajectories and reaction times of aimed movements: effects of practice, uncertainty, and change in target location. J Neurophysiol 46:725–743

Georgopoulos AP, Kalaska JF, Caminiti R, Massey JT (1982) On the relations between the direction of two-dimensional arm movements and cell discharge in primate motor cortex. J Neurosci 2:1527–1537

Georgopoulos AP, Caminiti R, Kalaska JF, Massey JT (1983) Spatial coding of movement: a hypothesis concerning the coding of movement direction by motor cortical populations. Exp Brain Res Suppl 7:327–336

Georgopoulos AP, Kalaska JF, Crutcher MD, Caminiti R, Massey JT (1984) The representation of movement direction in the motor cortex: single cell and population studies. In: Edelman GM, Cowan WM, Gall WE (eds) Dynamic aspects of neocortical function. Wiley, New York, pp 501–524

Georgopoulos AP, Merchant H, Naselaris N, Amirikian B (2007) Mapping of the preferred direction in the motor cortex. Proc Natl Acad Sci U S A 104:11068–11072

Georgopoulos AP, Schwartz AB, Kettner RE (1986) Neuronal population coding of movement direction. Science 233:1416–1419

Georgopoulos AP, Kettner RE, Schwartz AB (1988) Primate motor cortex and free arm movements to visual targets in three-dimensional space. II. Coding of the direction of movement by a neuronal population. J Neurosci 8:2928–2937

Georgopoulos AP, Taira M, Lukashin A (1993) Cognitive neurophysiology of the motor cortex. Science 260:47–52

Graziano MS, Taylor CS, Moore T (2002) Complex movements evoked by microstimulation of precentral cortex. Neuron 34:841–851. doi:S0896627302006980

Hebb DO (1949) The organization of behavior: a neuropsychological theory. Wiley, New York

Jackson JH (1882) On some implications of dissolution of the nervous system. Medical Press and Circular, vol ii, p 411 (Reprinted in: Taylor J, Selected writings of John Hughlings Jackson, Stapes, London, UK, 1958, p. 29)

Jankowska E (1975) Cortical motor representation in view of recent experiments on cortico-spinal relations. Acta Neurobiol Exp (Wars) 35:699–706

Jankowska E, Padel Y, Tanaka R (1975) The mode of activation of pyramidal tract cells by intracortical stimuli. J Physiol Lond 249:617–636

Jones EG (2001) The thalamic matrix and thalamocortical synchrony. Trends Neurosci 24:595–601

Jones EG, Powell TP (1970) An anatomical study of converging sensory pathways within the cerebral cortex of the monkey. Brain 93:793–820

Kalaska JF, Caminiti R, Georgopoulos AP (1983) Cortical mechanisms related to the direction of two-dimensional arm movements: relations in parietal area 5 and comparison with motor cortex. Exp Brain Res 51:247–260

Kwan HC, MacKay WA, Murphy JT, Wong YC (1978) Spatial organization of precentral cortex in awake primates. II. Motor outputs. J Neurophysiol 41:1120–1131

Langheim FJ, Leuthold AC, Georgopoulos AP (2006) Synchronous dynamic brain networks revealed by magnetoencephalography. Proc Natl Acad Sci U S A 103:455–459. doi:10.1073/pnas.0509623102

Lashley KS (1933) Integrative functions of the cerebral cortex. Physiol Rev 13:1–42

Lee SH, Kwan AC, Zhang S, Phoumthipphavong V, Flannery JG, Masmanidis SC, Taniguchi H, Huang ZJ, Zhang F, Boyden ES, Deisseroth K, Dan Y (2012) Activation of specific interneurons improves V1 feature selectivity and visual perception. Nature 488:379–383. doi:10.1038/nature11312

Leuthold AC, Langheim FJ, Lewis SM, Georgopoulos AP (2005) Time series analysis of magnetoencephalographic data during copying. Exp Brain Res 164:411–422. doi:10.1007/s00221-005-2259-0
Mahan MY, Georgopoulos AP (2013) Motor directional tuning across brain areas: directional resonance and the role of inhibition for directional accuracy. Front Neural Circ 7:92. doi:10.3389/fncir.2013.00092
Mardia KV (1972) Statistics of directional data. Academic, New York
Merchant H, Crowe DA, Fortes AF, Georgopoulos AP (2014) Cognitive modulation of local and callosal neural interactions in decision making. Front Neurosci 8:245. doi:10.3389/fnins.2014.00245
Mountcastle VB (1957) Modality and topographic properties of single neurons of cat's somatic sensory cortex. J Neurophysiol 20:408–434
Mountcastle VB (1967) The problem of sensing and the neural coding of sensory events. In: Schmitt FO, Quarton G, Melnuchuk T (eds) The neurosciences: an intensive study program. Rockefeller University Press, New York, pp 393–407
Mountcastle VB (1997) The columnar organization of the neocortex. Brain 120:701–722
Mountcastle VB, Lynch JC, Georgopoulos A, Sakata H, Acuna C (1975) Posterior parietal association cortex of the monkey: command functions for operations within extrapersonal space. J Neurophysiol 38:871–908
Mountcastle VB, Motter BC, Andersen RA (1980) Some further observations of the functional properties of neurons in the parietal lobe of the waking monkey. Behav Brain Sci 3:485–534
Murphy JT, Kwan HC, MacKay WA, Wong YC (1982) Precentral unit activity correlated with angular components of a compound arm movement. Brain Res 246:141–145. doi:10.1016/0006-8993(82)90152-4
Naselaris T, Merchant H, Amirikian B, Georgopoulos AP (2005) Spatial reconstruction of trajectories of an array of recording microelectrodes. J Neurophysiol 93:2318–2330. doi:10.1152/jn.00581.2004
Naselaris T, Merchant H, Amirikian B, Georgopoulos AP (2006a) Large-scale organization of preferred directions in the motor cortex. I. Motor cortical hyperacuity for forward reaching. J Neurophysiol 96:3231–3236. doi:10.1152/jn.00487.2006
Naselaris T, Merchant H, Amirikian B, Georgopoulos AP (2006b) Large-scale organization of preferred directions in the motor cortex. II. Analysis of local distributions. J Neurophysiol 96:3237–3247. doi:10.1152/jn.00488.2006
Opris I, Casanova MF (2014) Prefrontal cortical minicolumn: from executive control to disrupted cognitive processing. Brain 137:1863–1875. doi:10.1093/brain/awt359
Opris I, Hampson RE, Stanford TR, Gerhardt GA, Deadwyler SA (2011) Neural activity in frontal cortical cell layers: evidence for columnar sensorimotor processing. J Cogn Neurosci 23:1507–1521. doi:10.1162/jocn.2010.21534
Pellionisz A, Llinás R (1979) Brain modeling by tensor network theory and computer simulation. The cerebellum: distributed processor for predictive coordination. Neurosci 4:323–348. doi:10.1016/0306-4522(79)90097-6
Pellizzer G, Sargent P, Georgopoulos AP (1995) Motor cortical activity in a context-recall task. Science 269:702–705
Pitts W, McCulloch WS (1947) How we know universals; the perception of auditory and visual forms. Bull Math Biophys 9:127–147
Powell TP, Mountcastle VB (1959) Some aspects of the functional organization of the cortex of the postcentral gyrus of the monkey: a correlation of findings obtained in a single unit analysis with cytoarchitecture. Bull Johns Hopkins Hosp 105:133–162
Schwartz AB, Kettner RE, Georgopoulos AP (1988) Primate motor cortex and free arm movements to visual targets in three-dimensional space. I. Relations between single cell discharge and direction of movement. J Neurosci 8:2913–2927
Sherrington CS (1906) The integrative action of the nervous system. Yale University Press, New Haven

Sherrington CS (1940) Man on his nature. Cambridge University Press, London
Stefanis C, Jasper H (1964a) Intracellular microelectrode studies of antidromic responses in cortical pyramidal tract neurons. J Neurophysiol 27:828–854
Stefanis C, Jasper H (1964b) Recurrent collateral inhibition in pyramidal tract neurons. J Neurophysiol 27:855–877
Viviani P, Terzuolo C (1982) Trajectory determines movement dynamics. Neurosci 7:431–437. doi:10.1016/0306-4522(82)90277-9
Woolsey CN, Settlage PH, Meyer DR, Sencer W, Pinto Hamuy T, Travis AM (1952) Patterns of localization in precentral and "supplementary" motor areas and their relation to the concept of a premotor area. Res Publ Assoc Res Nerv Ment Dis 30:238–264

Chapter 9
Discrete, Place-Defined Macrocolumns in Somatosensory Cortex: Lessons for Modular Organization of the Cerebral Cortex

Oleg V. Favorov, Barry L. Whitsel, and Mark Tommerdahl

Keywords Minicolumn • Macrocolumn • Receptive field • Segregate

9.1 Introduction

In 1957 Mountcastle introduced the concept of the *cortical column* as the vertical processing unit of the cerebral cortex. This idea, the "columnar hypothesis," was based on the then prevailing view that the cortex is most richly interconnected in its vertical dimension (Lorente de No 1949) and on Mountcastle's demonstration in single-unit recording experiments in cat (and later monkey; Powell and Mountcastle 1959) primary somatosensory cortex (SI) that neurons in ~0.5 mm wide vertical columns are activated by peripheral stimuli of the same submodality and have similar receptive fields (RFs). Mountcastle (1957) and Powell and Mountcastle (1959) also showed that their cortical columns – later named "macrocolumns" to distinguish them from single-cell-wide "minicolumns" (Mountcastle 1978) – can be separated from each other by abrupt boundaries, on the opposite sides of which neurons respond to stimuli of different submodalities and/or have prominently different RFs.

This paper reviews the evidence that SI cortex of cats and monkeys is partitioned into a honeycomb-like mosaic of discrete macrocolumns. The review then draws

O.V. Favorov (✉)
Department of Biomedical Engineering, University of North Carolina, 070 MacNider Hall, Chapel Hill, NC 27599, USA
e-mail: favorov@bme.unc.edu

B.L. Whitsel
Department of Cell Biology & Physiology, University of North Carolina, Chapel Hill, NC, USA
e-mail: barry_whitsel@med.unc.edu

M. Tommerdahl
Department of Biomedical Engineering, University of North Carolina, Chapel Hill, NC, USA
e-mail: tommerda@med.unc.edu

M.F. Casanova, I. Opris (eds.), *Recent Advances on the Modular Organization of the Cortex*, DOI 10.1007/978-94-017-9900-3_9

upon the lessons derived from studies of these columnar entities to update and clarify the original Mountcastle's definition of the macrocolumn and advance the possibility that macrocolumns constitute fundamental functional units of the cerebral cortex, as envisioned by Mountcastle (1978).

9.2 Columnar Organization of SI Cortex

9.2.1 SI Cortex is Partitioned into a Mosaic of Discrete Macrocolumns

The simplest way to demonstrate the sharp boundaries separating discrete macrocolumns is to use the "minimal RF" mapping method (Favorov and Diamond 1990). This method is designed to identify the skin locus that provides the strongest input to a local cluster of neurons (for practical details of this RF mapping method, which are critical for detecting macrocolumnar borders, see Favorov and Diamond 1990).

Figure 9.1 shows typical sequences of minimal RFs encountered in near-radial microelectrode penetrations of cat or monkey SI cortex. The penetrations shown in Fig. 9.1 were inserted into the forelimb region of cat SI and minimal RFs were mapped every 100–150 μm along the electrode track. In the penetration in Fig. 9.1a, all the minimal RFs mapped from the pial surface to the white matter occupy the same position on the skin and do not show any somatotopic drift. In the penetration in Fig. 9.1b, the first 4 minimal RFs occupy one position on the skin, whereas the last 3 minimal RFs occupy a new, prominently displaced skin position. The 5th minimal RF was mapped right at the obviously very sharp topographic transition, signaling a boundary separating two cortical columns. Similarly abrupt jumps of minimal RFs are also reliably observed in penetrations that travel tangentially through the SI cortex (Favorov et al. 1987; Favorov and Diamond 1990).

Minimal RFs mapped in arrays of closely spaced penetrations reveal the shape and size of macrocolumns delineated by such sharp minimal RF discontinuities (Favorov and Diamond 1990). An exemplary array of such penetrations is illustrated in Fig. 9.2. Twenty five penetrations were inserted in this experiment approximately perpendicular to the cortical surface and in each penetration two or three minimal RFs were mapped at or near layer 4 (Fig. 9.2a). Figure 9.2b shows the outlines of all 62 mapped minimal RFs on a single drawing of the skin of the forearm, revealing that they form 10 nonoverlapping clusters, labeled from *a* to *j*. Figure 9.2c shows a two-dimensional surface-view reconstruction of the cortical region sampled by the 25 penetrations. Positions of all the recording sites, projected radially onto layer 4, are indicated by filled circles. Recording sites mapped in the same penetration are connected by thin lines, and each recording site is assigned the letter of the RF cluster to which its minimal RF belongs. The reconstruction in Fig. 9.2c clearly shows that recording sites with the same minimal RF form

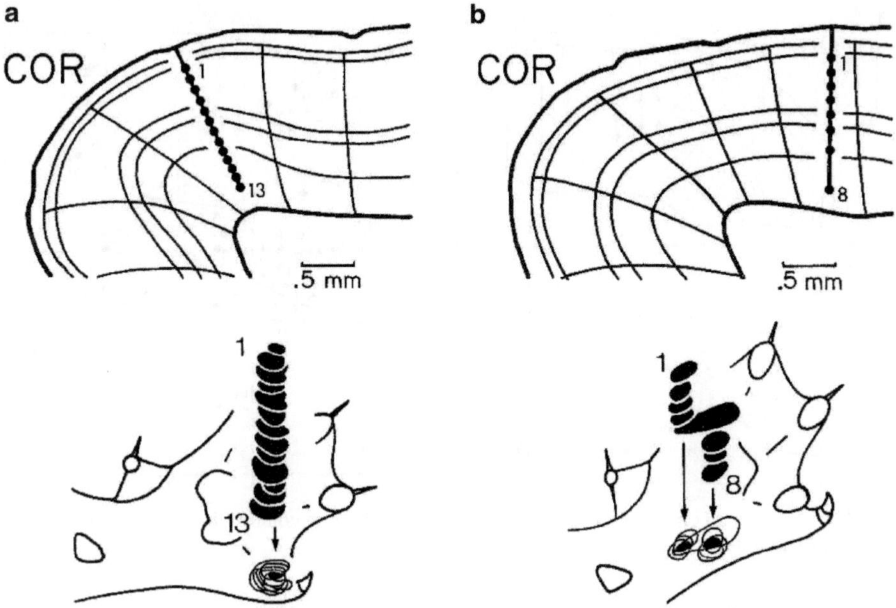

Fig. 9.1 Typical near-radial penetrations of cat SI. In this and the following figures illustrating histological sections, orientations of radial cords of cells are indicated by thin lines and recording sites are indicated by circles. Minimal RFs are drawn stacked in the order in which they were mapped in the penetration. *COR* coronal sulcus (From Favorov and Diamond 1990)

nonoverlapping groups up to 350 μm in width, separated from other such groups by just a few tens of micrometers in the plane of cortical surface. The most probable arrangement of the columnar boundaries is indicated in Fig. 9.2c by the set of solid and dotted lines, revealing a mosaic of sharply delineated topographic units, within each of which the minimal RF does not change, but jumps to a new skin location when crossing into another such unit.

To summarize, the minimal RF mapping method reveals sharp somatotopic discontinuities that subdivide SI cortex into a mosaic of 300–600 μm-diameter irregular hexagonally-shaped columns. The same minimal RF is mapped at any site within a column. In adjacent columns, minimal RFs occupy prominently displaced, nonoverlapping positions on the skin. The columns with more distally located minimal RFs tend to be larger in size. To emphasize their discrete nature, Favorov and colleagues called such place-defined macrocolumns "segregates."

9.2.2 RF Diversity Within SI Macrocolumns

A minimal RF is determined jointly by multiple neurons in a vicinity of the electrode tip and is the skin site that provides the most effective afferent drive to

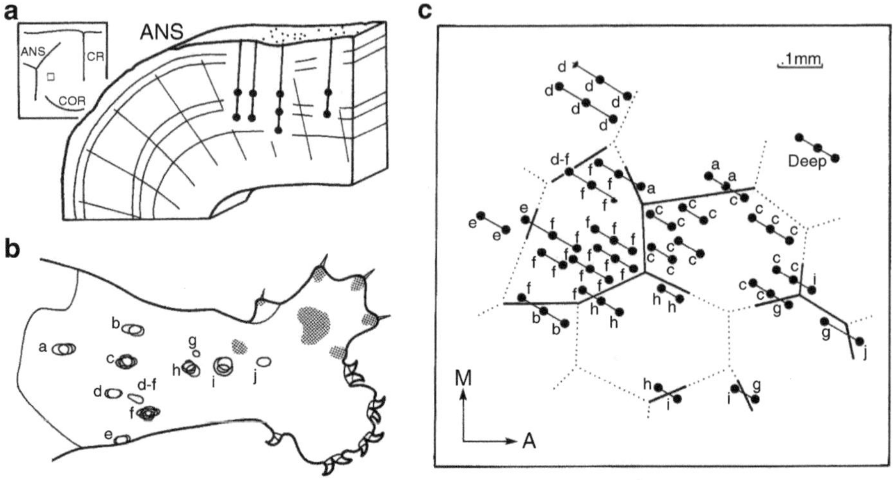

Fig. 9.2 Array of 25 closely spaced penetrations reveals discrete columns in cat SI (From Favorov and Diamond 1990)

those neurons as a group. Individual neurons in the recorded group obviously are driven from skin areas larger than their common minimal RF. For a single neuron, its *maximal* RF can be defined as the full extent of the skin area that provides suprathreshold input to that neuron (Favorov et al. 1987). Neurons located within a single macrocolumn/segregate all share in their maximal RFs that segregate's minimal RF, but also each neuron receives afferent input from some additional – and frequently very extensive – surrounding territories, which are different for different neurons (Favorov et al. 1987; Favorov and Whitsel 1988ab).

Figure 9.3 offers an example of the diversity of maximal RFs of neurons found in a single segregate. Shown in Fig. 9.3a, 21 single neurons were isolated in a near-radial penetration of a single segregate in area 3b in cat and their maximal RFs were mapped. All 21 maximal RFs are drawn in Fig. 9.3b, revealing prominent diversity of their sizes, shapes, and skin positions across the length of the penetration and even among close neighbors. However, there is one very small skin locus that is common to all 21 RFs (Fig. 9.3c). This skin locus coincides with the minimal RF mapped in this penetration.

To summarize, maximal RFs of all neurons in a segregate include a common skin locus (the minimal RF) and, in addition, extend for variable distances outward in all directions from that common locus. Neurons in a segregate differ from one another in how much their maximal RFs extend in each and every direction from the common skin locus (Favorov and Whitsel 1988b). Because of its central importance to segregate organization, the common skin locus of a segregate – its minimal RF – was given a special name, "segregate RF center" (Favorov and Whitsel 1988a).

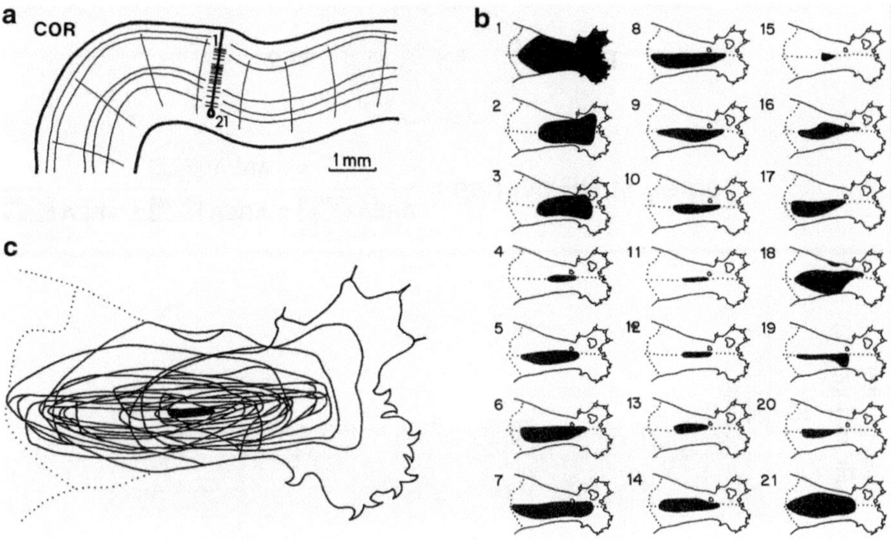

Fig. 9.3 Maximal RFs collected in a typical radial penetration of cat SI. Cortical locations of studied single units are indicated in panel A by tics (From Favorov and Diamond 1990)

9.2.3 Minicolumnar Organization of Maximal RFs

Mountcastle (1978) hypothesized that a radial cord of cells about 30–50 μm in diameter – a "minicolumn" – might be the smallest functional unit of neocortical organization. Structurally, minicolumns are attributable to the radially-oriented cords of neuronal cell bodies evident in Nissl-stained sections of cerebral cortex. Population analysis of maximal RFs of neurons isolated within the same SI segregate supports Mountcastle's minicolumnar hypothesis (Favorov and Whitsel 1988a; Favorov and Diamond 1990). According to this analysis, the maximal RFs of neurons within minicolumns are most similar in size, shape, and position on the skin. In contrast, neurons located even in adjacent minicolumns typically have RFs that differ significantly in size and shape, and frequently overlap only minimally on the skin. In other words, local RF diversity within segregates is mostly attributable to diversity *among* minicolumns and much less to within minicolumns.

For example, Fig. 9.4 plots the average degree of overlap of maximal RFs as a function of the tangential distance separating neurons within a segregate. The plot shows that neurons that are the closest neighbors in the tangential plane of the cortex have the most similar RFs, and that similarity declines with increased distance. At separations larger than 50 μm, RF overlap within segregates is independent of the distance between neurons. Instead, moving from one minicolumn to the next within a segregate, maximal RFs shift back and forth on the skin without yielding a net RF shift across the entire segregate. Only at a border separating

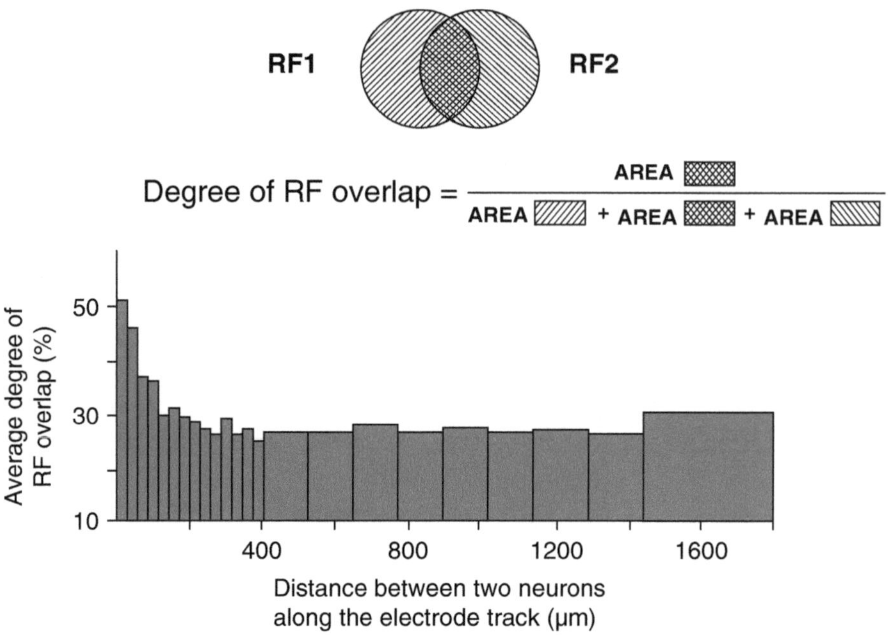

Fig. 9.4 Similarity of maximal RFs as a function of distance separating two neurons in the plane of the cortical surface in monkey SI (From Favorov and Whitsel 1988a)

adjacent segregates do RFs shift *en masse* to a new skin territory (Favorov and Whitsel 1988a; Favorov and Diamond 1990).

Because of the prominent differences in RF properties among neighboring minicolumns, even a point-like tactile stimulus ought to evoke a spatial pattern of activity in the responding SI region consisting of a mix of active and inactive minicolumns. High-resolution 2-deoxyglucose (2-DG) metabolic studies of monkey SI (Tommerdahl et al. 1993) indeed showed that column-shaped patches of 2-DG label evoked in SI cortex by natural skin stimuli comprise groupings of highly active minicolumns interdigitated with less active minicolumns.

9.3 Functional Significance of Discrete Place-Defined Macrocolumns

Mountcastle originally defined columns as functional entities comprising groups of minicolumns bound together by common input and short-range lateral connections. Since then, however, the term "column" has been frequently used more broadly to refer to any vertical cluster of cells that share the same tuning for any given RF attribute, not necessarily of any functional significance (Horton and Adams 2005). At the same time the functional significance of Mountcastle's cortical columns has been questioned (Swindale 1990; Purves et al. 1992; Horton and Adams 2005; Da Costa and Martin 2010). For example, Horton and Adams (2005) conclude their comprehensive critique of Mountcastle's columnar hypothesis by stating that "one must abandon the idea that columns are the basic functional entity of the cortex. It now seems doubtful that any single, transcendent principle endows the cerebral cortex with a modular structure. Each individual area is constructed differently..."

The minimal and maximal RF organization of SI segregates – Mountcastle's original macrocolumns – however offers a number of insights that counteract the criticism and clarify the nature of macrocolumns as functional entities:

1. **The structural evidence of discreteness associated with macrocolumns should be sought at the level of cell bodies, rather than at the level of dendrites and axon terminals**

The basal dendrites of pyramidal cells have a large horizontal spread – up to 400–500 μm in diameter (Feldman 1984). This means that a portion of the basal dendritic fields of the majority of pyramidal cells in an SI segregate invades neighboring segregates. If there were no restrictions on the inputs the dendrites of pyramidal and spiny stellate cells can receive from elements outside their own segregate, then the majority of the cells in a segregate would reflect the activity from the neighboring segregates invaded by their dendrites. Neurons in different parts of a given segregate would be influenced by activity of different surrounding segregates, and the RFs of a linear array of neurons across a segregate would show an orderly, gradual and continuous shift in skin position. In reality, segregates show no such somatotopic gradients within their confines (see above), thus indicating that although some dendritic branches of cells in one segregate invade the territories of adjacent segregates, they do not receive opportunistic synaptic contacts there from neurons residing there or from afferents innervating those segregates. On the other hand, the afferents innervating a given segregate and the local axon collaterals of neurons residing in that segregate will follow the dendritic branches originated in that segregate into the territories of the adjacent segregates. As a result, the systems of afferent and intrinsic connections wiring adjacent segregates can remain functionally separate while physically intermingled in each other's neuropil.

2. **Definition of the macrocolumn by common input and short-range connections should not be taken to imply that neurons making up the macrocolumn all have uniform RFs**

Instead of uniformity, neurons making up an SI segregate possess prominently diverse maximal RFs (see above), with each maximal RF differ in how much it extends in different directions away from the segregate RF center (i.e., the skin locus common to RFs of all the cells in the segregate). Thus, the segregate is recognized on statistical grounds by its possession of a certain assortment, or *distribution*, of maximal RFs centered on a particular skin point. Neurons in different sectors of the same segregate apparently have the same distribution of maximal RFs. The reason why the border between adjacent segregates can be detected is because RF distributions immediately across the border have clearly different central tendencies and occupy different, only partially overlapping skin territories (the RF center of one segregate is included in only 50 % of the maximal RFs found in an adjacent segregate). The reason why minimal RF mapping method is highly effective in detecting segregate borders is that it is designed to estimate the central tendency of the distribution of maximal RFs of a local group of neurons picked up by the tip of the recording electrode, and thus it directly reveals where in the course of an SI penetration the central tendency of local maximal RFs stays the same and where it jumps to a new skin location.

These statistical properties of segregate maximal RFs are depicted schematically in Fig. 9.5. It shows hypothetical distribution of maximal RFs across two adjacent segregates. For graphic clarity, the skin and RFs are treated as unidimensional. Hypothetical maximal RFs are plotted in Fig. 9.5b for 200 neurons sampled along a continuous path that spans the two segregates (Fig. 9.5a). The random variations in the size (length) of the RFs reflect the diversity of sizes of maximal RFs sampled in cat and monkey experiments (Favorov and Whitsel 1988a; Favorov and Diamond 1990). The first 100 RFs belong to neurons located in the first segregate. They all share in common skin point A, but otherwise they vary randomly how far they extend to the left and to the right from this central point of their distribution. The next 100 RFs belong to neurons located in the second segregate. Their distribution is similar to that of the first 100 RFs, except that it is centered on skin point B. The change in the central tendency occurs at the transition from the first segregate to the second, between the 100th and 101th RFs. The distance between the two central skin points, A and B, is such that the central point of one segregate is included in 50 % of RFs of the other segregate. This Fig. 9.5b plot captures the essential characteristics of the segregate organization of maximal RFs in cat and monkey SI cortex.

3. **The afferent input to the macrocolumn is provided by neurons in lower-level cortical areas and/or thalamic nuclei via axons all terminating extensively over the same territory**

Individual thalamocortical axons projecting to Layer 4 in SI have terminal arbors varying in size from 350 to 800 μm (~600 μm average) in macaque (Garraghty and Sur 1990; Raussel and Jones 1995) or from 300 to 600 μm (~500 μm average) in cat (Landry and Deschenes 1981; Landry et al. 1987). Such arbor sizes match or slightly exceed the range of widths of SI segregates in these species. The most parsimonious interpretation is that each thalamocortical axon targets a particular segregate and arborizes throughout its territory and also follows dendrites of the spiny stellate cells near the segregate borders into the adjacent segregates. Any given segregate is innervated in this manner by a group of thalamocortical neurons, which are expected to possess similar RFs. Because of their non-identical, partially shifted RFs, such a group of thalamocortical neurons together will cover a skin area greater than the size of an individual RF; we can call this total skin area the segregate's "*afferent RF*." By their axons all terminating over more or less the full extent of the segregate, while avoiding synaptic contacts with cells in the adjacent segregates, the group of thalamocortical neurons innervating a

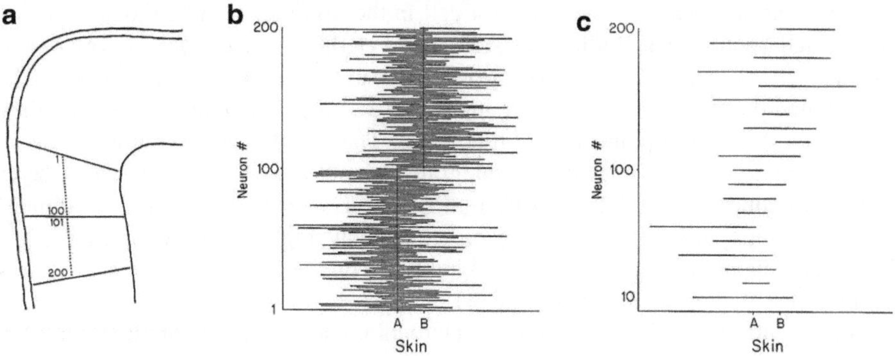

Fig. 9.5 The segregate plan of topographic organization of SI cortex. (**a**) A hypothetical track of a microelectrode penetration down the posterior bank of the central sulcus. The *solid lines* indicate the boundaries of two adjacent segregates. The *line of dots* indicates the locations of 200 single units mapped in the two segregates. (**b**) Hypothetical maximal RFs of the 200 mapped single units. Each maximal RF is drawn as a horizontal line showing its full extent and position on the skin. RFs of successive single units are drawn in a vertical order according to the single unit's position along the electrode track. Maximal RFs from 1 to 100 belong to the top segregate and they all share in common skin point A (indicated by a vertical line segment). Maximal RFs from 101 to 200 belong to the next segregate and they all share in common skin point B. (**c**) 1 out of 10 subsample of the maximal RFs plotted in panel B. While the *en masse* shift of RFs at the segregate border is clearly noticeable in panel B, it is unrecognizable in the smaller RF sample in panel C (From Favorov 1991)

segregate in fact creates that segregate: it gives that segregate its sharp boundaries while avoiding somatotopic gradients across its territory. Adjacent segregates are supplied by different groups of thalamocortical neurons.

4. **Having discrete macrocolumns might be a strategy used by cortical areas to devote larger numbers of cells to recognizing significant spatiotemporal patterns of activity in a reduced number of chosen subsets of afferent neurons innervating an area**

According to this strategy, a group of afferent neurons terminating over a particular macrocolumn constitutes one of the "chosen" subsets. In SI cortex, each segregate has as its chosen subset a group of thalamic neurons terminating over it, through which it receives information about patterns of tactile events taking place in that segregate's afferent RF (defined above). The spatiotemporal patterns of activity in the segregate's set of afferent neurons are first processed by the segregate in its layer 4 by roughly 1,000–1,500 spiny stellate cells residing there, and then by 2,500–3,500 pyramidal cells residing in the upper layers (estimated based on Beaulieu and Colonnier 1989, and Favorov and Diamond 1990). Layer 4 cells tune to different afferent input patterns as a step to generalization, which takes place in the upper layers (Poggio and Bizzi 2004). If we think of the segregate's set of N afferent neurons as defining an N-dimensional "*afferent pattern*" space of the segregate, each dimension of which corresponds to one of the afferent neurons, then each layer 4 cell in the segregate can be thought of as a functional analog of a Radial Basis Function (RBF) in that space (Favorov and Kursun 2011). The 1,000–1,500 layer 4 cells in the segregate have their RBFs distributed throughout the afferent pattern space of the segregate and together they create a 1,000–1,500-point map of that space. The 2,500–3,500 pyramidal cells in the segregate's upper layers next integrate the outputs of the segregate's layer 4 cells to compute 2,500–3,500 different functions over the segregate's *N*-dimensional afferent pattern space, thereby creating a new, higher-level representation of significant patterns of tactile events taking place on the territory of the segregate's afferent RF (Favorov and Kursun 2011).

In the absence of such discrete macrocolumnar compartmentalization of a cortical area, we can expect individual minicolumns or small groups of minicolumns to create their own afferent pattern spaces, but these spaces will be mapped by many fewer layer 4 cells/RBFs and will be much more redundant than afferent pattern spaces of macrocolumns. Whether having fewer, more densely mapped afferent pattern spaces is functionally advantageous and offers the functional rationale for existence of discrete macrocolumns will have to be determined in future studies.

5. **Segregate-like discrete macrocolumns might be present in the visual cortex, V1 and V2**

Visuotopic organization of the visual cortex has received much less experimental attention than its other properties, such as ocular dominance and orientation tuning. Nevertheless, Roe and Ts'o (1995) found prominent topographic discontinuities at the borders separating the thin, thick and pale stripes in macaque V2. Furthermore, they found topographic discontinuities not only between but also within stripes, where RF jumps were coincident with other signs of functional partitioning of individual stripes into discrete macrocolumn-scale domains. This is very suggestive that V2 – similar to SI – is partitioned into a mosaic of discrete place-defined macrocolumns.

V1 of cats and monkeys has a well-developed map of orientation tuning. This map has local regions in which the preferred orientation changes relatively slowly and monotonically, but these regions are separated by fractures and singularities where the orientation preference changes rapidly. The fracture lines tend to enclose ~400 μm-wide columnar regions, which Blasdel and Salama (1986) – who were the first to generate orientation maps using the voltage-sensitive dye technique – described as modular units of V1 organization. Studying the relationship between orientation and retinotopic maps in cat V1, Das and Gilbert (1997) found that orientation fractures are associated with retinotopic discontinuities, suggesting that Blasdel and Salama's (1986) fracture-defined modular units can also be place-defined as discrete macrocolumns comparable to SI segregates.

In disagreement with Das and Gilbert (1997), the map of visual space in V1 has been consistently described as smooth, although at a local level neighboring cells do exhibit prominent, but seemingly random, scatter in the positions of their RFs (Hubel and Wiesel 1974; Albus 1975; Bosking et al. 2002; Buzas et al. 2003; Yu et al. 2005). The magnitude of such local RF scatter is comparable to that observed in cat and monkey SI. However, as we see in SI, such RF scatter effectively hides topographic discontinuities at segregate boundaries in SI cortex. As is illustrated in Fig. 9.5b, an abrupt shift in the central tendency of local groups of maximal RFs at a segregate border is easy to see in plots involving large numbers of RFs. But such a high density of RF sampling is not practically achievable and a limited sample of RFs obtainable in an experiment is not likely to reveal any jumps in RF central tendency upon visual inspection. For example, in Fig. 9.5c only one out of every ten RFs shown in Fig. 9.5b is plotted and now, because of large RF variability, it is no longer apparent that RFs actually were sampled from two different, displaced distributions. A common practice in V1 studies of using the geometric centers of RFs to map the progression of RFs across a cortical territory would only further obscure topographic discontinuities. A much more effective approach to detecting topographic discontinuities is to use the minimal RF mapping method (see above), but it has not been popular in visual cortical studies.

To conclude, the studies of visuotopic organization of the visual cortex carried out up to date have not been suited for detection of topographic discontinuities and discrete place-defined macrocolumns. However, the limited evidence of Roe and Ts'o (1995) and Das and Gilbert (1997) suggests that such discontinuities and macrocolumns responsible for them are present in V1 and V2.

9.4 Conclusion

Mountcastle's concept of the discrete macrocolumn as a group of minicolumns bound together by common input and short-range lateral connections has been best exemplified so far by segregates, which are the discrete place-defined columns found in cat and monkey SI (Favorov et al. 1987; Favorov and Whitsel 1988a, b; Favorov and Diamond 1990). The available knowledge of segregate dimensions, morphological properties of thalamocortical axon arborizations in SI, dendritic and axonal fields of SI neurons, and RF composition of segregates allow us to refine the definition of Mountcastle's macrocolumn in terms of its common input and local connections. That is, the macrocolumn is as a group of 40–80 minicolumns that form a structural and functional union by imposing the following constraints on their connectivity:

1. The ascending input to this group of minicolumns is provided by a set of afferent axons each of which terminates on layer 4 cells across all the minicolumns (although not uniformly, but varying in the strength of its individual connections);
2. While dendrites and axons of the cells belonging to this group of minicolumns penetrate into the neuropil space of the neighboring minicolumns outside the group, they stay functionally isolated by avoiding indiscriminate connections with local neurons and afferent axons there;

3. Horizontal connections between macrocolumns are organized on the principle of functional relatedness and not simple physical proximity.

We propose that the reason for exposing all the layer 4 cells in a macrocolumn to the same set of afferents is to enable them, as a large group, to map at high resolution the state space, or the "afferent pattern" space, of this set of afferents (Favorov and Kursun 2011). Whether the functional benefits of such mapping strategy are limited to somatosensory and possibly visual cortex (Roe and Ts'o 1995; Das and Gilbert 1997; Favorov and Kursun 2011) or whether such strategy is used universally throughout the cerebral cortex remains to be determined. Unfortunately, because of interpenetration of dendritic and axonal fields of adjacent macrocolumns, direct visualization of the macrocolumnar segregation of afferent projections and of local connections is problematic. Instead, functional discontinuities, which are a telltale sign of macrocolumnar borders, offer a more practical means of detecting discrete macrocolumns. However, in a note of caution, not all discontinuities might be associated with macrocolumnar borders. In principle, feature maps, such as orientation or ocular dominance maps in V1, do not have to have a simple relationship to macrocolumns. Any feature map will comprise local maps developed by individual macrocolumns in a possible coordination across macrocolumnar borders. The maps of particular features, as well as relationships between maps of different features, might vary across cortical areas, mammalian species and even in different regions of the same area (Horton and Adams 2005), but this does not necessarily have to mean that these areas do not contain discrete macrocolumns.

References

Albus K (1975) A quantitative study of the projection area of the central and the paracentral visual field in area 17 of the cat: I. The precision of the topography. Exp Brain Res 24:159–179

Beaulieu C, Colonnier M (1989) Number of neurons in individual laminae of areas 3b, 4γ, and 6aα of the cat cerebral cortex: a comparison with major visual areas. J Comp Neurol 279:228–234

Blasdel GG, Salama G (1986) Voltage-sensitive dyes reveal a modular organization in monkey striate cortex. Nature 321:579–585

Bosking WH, Crowley JC, Fitzpatrick D (2002) Spatial coding of position and orientation in primary visual cortex. Nat Neurosci 5:874–882

Buzas P, Volgushev M, Eysel UT, Kisvarday ZF (2003) Independence of visuotopic representation and orientation map in the visual cortex of the cat. Eur J Neurosci 18:957–968

Da Costa NM, Martin KAC (2010) Whose cortical column would that be? Front Neuroanat 4:16

Das A, Gilbert CD (1997) Distortions of visuotopic map match orientation singularities in primary visual cortex. Nature 387:594–598

Favorov OV (1991) Detection and characterization of the mosaic body representation in SI cortex. In: Franzen O, Westman J (eds) Information processing in the somatosensory system. Macmillan Press, London, pp 224–232

Favorov OV, Diamond M (1990) Demonstration of discrete place-defined columns, segregates, in cat SI. J Comp Neurol 298:97–112

Favorov OV, Kursun O (2011) Neocortical layer 4 as a pluripotent function linearizer. J Neurophysiol 105:1342–1360

Favorov O, Whitsel BL (1988a) Spatial organization of the peripheral input to area 1 cell columns: I. The detection of "segregates". Brain Res Rev 13:25–42

Favorov O, Whitsel BL (1988b) Spatial organization of the peripheral input to area 1 cell columns: II. The forelimb representation achieved by a mosaic of segregates. Brain Res Rev 13:43–56

Favorov OV, Diamond ME, Whitsel BL (1987) Evidence for a mosaic representation of the body surface in area 3b of the somatic cortex of cat. Proc Natl Acad Sci U S A 84:6606–6610

Feldman ML (1984) Morphology of the neocortical pyramidal neuron. In: Peters A, Jones EG (eds) Cerebral Cortex. Vol. 1. Plenum, New York, pp 123–200

Garraghty PE, Sur M (1990) Morphology of single intracellularly stained axons terminating in area 3b of macaque monkeys. J Comp Neurol 294:583–593

Horton CH, Adams DL (2005) The cortical column: a structure without a function. Philos Trans R Soc Lond B Biol Sci 360:837–862

Hubel DH, Wiesel TN (1974) Uniformity of monkey striate cortex: a parallel relationship between field size, scatter, and magnification factor. J Comp Neurol 158:295–305

Landry P, Deschenes M (1981) Intracortical arborizations and receptive fields of identified ventrobasal thalamocortical afferents to the primary somatic sensory cortex in the cat. J Comp Neurol 199:345–371

Landry P, Diadori P, Leclerc S, Dykes RW (1987) Morphological and electrophysiological characteristics of somatosensory thalamocortical axons studied with intra-axonal staining and recording in the cat. Exp Brain Res 65:317–330

Lorente de No R (1949) Cerebral cortex: architecture, intracortical connections, motor projections. In: Fulton JF (ed) Physiology of the nervous system, 3rd edn. Oxford University Press, New York, pp 288–312

Mountcastle VB (1957) Modality and topographic properties of single neurons of cat's somatic sensory cortex. J Neurophysiol 20:374–434

Mountcastle VB (1978) An organizing principle for cerebral function: the unit module of the distributed system. In: Mountcastle VB, Edelman GM (eds) The mindful brain. MIT Press, Cambridge, MA, pp 7–50

Poggio T, Bizzi E (2004) Generalization in vision and motor control. Nature 431:768–774

Powell TPS, Mountcastle VB (1959) Some aspects of the functional organization of the cortex of the postcentral gyrus of the monkey: a correlation of findings obtained in a single unit analysis with cytoarchitecture. Bull Johns Hopkins Hosp 105:133–162

Purves D, Riddle DR, La Mantia A-S (1992) Iterated patterns of brain circuitry (or how the cortex gets its spots). Trends Neurosci 15:362–368

Raussel E, Jones EG (1995) Extent of intracortical arborization of thalamocortical axons as a determinant of representational plasticity in monkey somatic sensory cortex. J Neurosci 15:4270–4288

Roe AW, Ts'o DY (1995) Visual topography in primate V2: multiple representation across functional stripes. J Neurosci 15:3689–3715

Swindale NV (1990) Is the cerebral cortex modular? Trends Neurosci 13:487–492

Tommerdahl M, Favorov OV, Whitsel BL, Nakhle B, Gonchar YA (1993) Minicolumnar activation patterns in cat and monkey SI cortex. Cereb Cortex 3:399–411

Yu H, Farley BJ, Jin DZ, Sur M (2005) The coordinated mapping of visual space and response features in visual cortex. Neuron 47:267–280

Chapter 10
Prefrontal Cortical Microcircuits for Executive Control of Behavior

Ioan Opris, Ioana L. Popa, and Manuel F. Casanova

Abstract During the perception-to-action cycle, our cerebral cortex mediates the interactions between the environment and the perceptual-executive systems of the brain. At the top of the executive hierarchy, prefrontal cortical microcircuits are assumed to bind perceptual and executive control information to guide goal-driven behavior. Here, we discuss new results that show the involvement of prefrontal cortical inter-laminar microcircuits in the executive control of behavior. Recent results show that during perception and executive selection phases, cell firing in the localized prefrontal layers and caudate-putamen region exhibited a similar location preference on spatial-trials, but less on object- trials. When the perceptual-executive microcircuit became facilitated by electrically micro-stimulating the prefrontal infra-granular-cell layers with signal patterns previously derived from neuron firing in the supra-granular-layers, it was shown to produce stimulation-induced spatial preference (similar to neural tuning) in the percent correct performance only during spatial trials. These results suggested that inter-laminar prefrontal microcircuits play causal roles to the executive control of behavior across the perception-to-action cycle.

Keywords Cortical microcircuits • Executive control • Minicolumn • Prefrontal cortex • Primates • MIMO model • Causal relationship • Perception-to-action

---

I. Opris, Ph.D. (✉)
Department of Physiology and Pharmacology, Wake Forest University Health Sciences, Winston-Salem, NC 27103, USA
e-mail: ioanopris.phd@gmail.com

I.L. Popa
Department of Cognitive Science, University of Vienna, Vienna, Austria
e-mail: a0748526@unet.univie.ac.at

M.F. Casanova, M.D.
Department of Psychiatry, University of Louisville, Louisville, KY, USA
e-mail: manuel.casanova@louisville.edu

M.F. Casanova, I. Opris (eds.), *Recent Advances on the Modular Organization of the Cortex*, DOI 10.1007/978-94-017-9900-3_10

10.1 Introduction

As it was initially proposed by Mountcastle, the primate neocortical circuitry has a modular architecture that sub-serves a multitude of sensory (visual, auditory, touch), motor, cognitive (attention, memory, decision) and emotional functions (Barbas et al. 2005; Buschman et al. 2012; Buxhoeveden et al. 2002; Favorov et al. 1987; Gilbert and Wiesel 1989; Mountcastle 1957, 1997; Shepherd and Grillner 2010; Opris and Bruce 2005). These modules are composed of elementary building blocks formed by vertical arrangements of cortical neurons, called minicolumns (Casanova et al. 2008; Mountcastle 1997; Szentágothai and Arbib 1975; Hubel and Wiesel 1969). Within minicolumns, cortical neurons are aggregated into six horizontal layers (or laminae): three supra-granular layers (L1-L3), a granular layer (L4) and two infra-granular layers (L5/L6) (Fig. 10.1a). The granular layer receives sensory input from thalamus (Constantinople and Bruno 2013), while the minicolumnar outputs in the infra-granular layers send top-down signals to the subcortical structures (including striatum and thalamus; Alexander et al. 1986; Mountcastle et al. 1955, 1957; Opris et al. 2013). The supra-granular layers consist of small pyramidal neurons that form a complex network of intra-cortical connections, particularly the connections to the infra-granular layers of larger pyramidal neurons that generate most of the output from cerebral cortex to other parts of the brain (Buxhoeveden and Casanova 2002). According to this three stratum functional module, infra-granular layers execute the associative computations elaborated in supra-granular layers (Buxhoeveden and Casanova 2002; Casanova et al. 2011).

10.2 Prefrontal Cortical Minicolumns

Vernon Mountcastle described for the first time the electrophysiological basis of the cortical minicolumn and suggested it as an elemental unit of information processing (Mountcastle 1998, 1957, 1997; DeFelipe et al. 2012). According to this model of cortical organization, neurons and their connections form part of a vertical system which unites the cells of each minicolumn (Fig. 10.1a, b) into a coordinated functional unit (Mountcastle 1978, 1997). In this context, the smallest unit of cortical organization is the minicolumn, usually defined in Nissl stained sections by a narrow radial array of pyramidal neurons traversing laminae II-VI (Rakic 1988; Mountcastle 1997). Minicolumns are composed of vertical chains of excitatory neurons surrounded by inhibition (Fig. 10.1c, d) in cylindrical arrangement that constitutes the smallest module capable of information processing (Mountcastle 1957, 1998; Rakic 2008). The human neocortex is composed of a large number of minicolumns in parallel vertical arrays (Casanova et al. 2007). Minicolumns are the first step in a nested ensemble of nodes or "echelons" of increasing complexity (Opris and Casanova 2014). Other levels of modular organization include multiple minicolumns, macrocolumns, and large-scale networks of macrocolumns that are interconnected with the entire brain (Buxhoeveden and

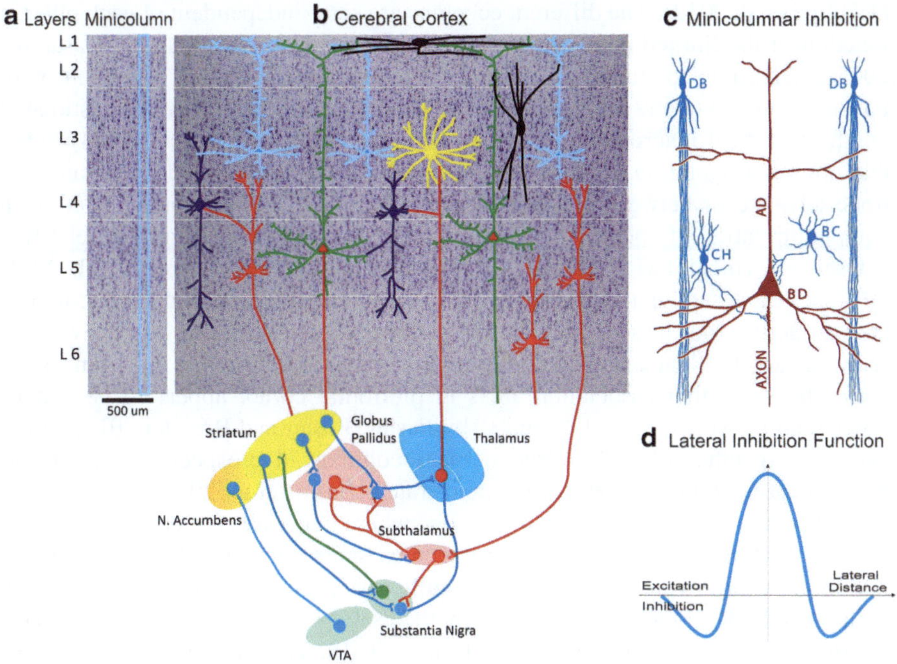

Fig. 10.1 Illustration of prefrontal cortical minicolumns, layers and loops. (**a**) Prefrontal cortical minicolumn highlighted in a human brain slice. (**b**) Laminar and columnar display of the major cell types in prefrontal cortex together with its connections to basal ganglia and thalamus. The six-layered cortex is showing on a Nissl background the pyramidal cells and the cortical interneurons. The cells are connected to the thalamus and basal ganglia, as shown on the lower panel. VTA is ventral tegmental area; N. accumbens is nucleus accumbens. (**c**) Pyramidal cell surrounded by interneurons illustrates the minicolumnar curtain o inhibition. AD is apical dendrite, BC is basket cell, BD is basal dendrite, CH is chandelier cell, DB is double bouquet cell. (**d**) Lateral inhibition function shows the neural activation (excitation and inhibition) with horizontal distance from a minicolumn's center (Adapted from Opris and Casanova, 2014)

Casanova 2002). The somas of pyramidal cells are not randomly distributed in space; rather, they are organized into layers and different-sized columns or modules. Similarly, some dendritic and axonal ramifications that begin or end in these somas are wired in parallel groups of fibers (Mountcastle 1997, 2003). Minicolumns are often considered highly repetitive, even clone-like, units; however, they display considerable heterogeneity between areas and species, perhaps even within a given macrocolumn.

10.2.1 Columns and Minicolumns

Minicolumns are arranged within larger columns or macrocolumns (e.g., barrel somatosensory cortex of the rodent) bound together by short-range horizontal connections (Jones 2000; Zhang and Alloway 2006; Jones and Rakic 2010;

DeFelipe et al. 2012). The different echelons are semi-independent of each other, a function of the limited number of information channels between them (Casanova 2005). According to Buxhoeveden and Casanova (2002), and in agreement with Favorov and coworkers (1987, 1990) and Mountcastle (1957, 1997), the estimated width of cortical macrocolumns is 350–600 μm. Hubel and Wiesel (1974) found that optimal orientation tuning changes systematically through 180° with an electrode advance of between 0.5 and 1.0 μm. The term "hypercolumn" refers to a complete rotation of columns (e.g. 0°, 10°, 20°,. . .,180°; Hubel and Wiesel 1974; Wiesel and Hubel 1974). Recently, Opris and collaborators (Opris et al. 2011, 2012b, c) suggested that interlaminar interaction in prefrontal cortical minicolumns take an active role in sensorimotor integration and the selection of behaviorally relevant targets. Sparse distributed representations of inputs flowing from visual cortex to the higher association areas in prefrontal cortex appear to be part of distributed networks, named "cognits" by Fuster (Fuster and Bressler 2012). However, the specific role of prefrontal macrocolumns in prospective coding and representation storage is yet to be demonstrated (Bastos et al. 2012).

A sparse distributed representation is one where items are encoded by activation of a small set of the available representing units. Sparse encoding does not reduce to a straight majority vote scheme. Anatomically, sparse encoding may be enforced by variability between minicolumns which themselves suggest differences in their internal architecture (Casanova 2008; Rinkus 2010). This variability among components of a minicolumn may contribute to the fault tolerance of larger networks such as macrocolumns. McCulloch (1959) has shown that when failure of individual components occur under a certain threshold (e.g., cell loss at the beginning stages of Alzheimer's disease) redundant networks of unstable nets could be designed for greater reliability than redundant systems of stable nets of the same size.

10.2.2 Cortical Modules and Maps

Early studies by (Mountcastle et al. 1955; Mountcastle 1957, 1997) and Hubel and Wiesel (Hubel and Wiesel 1974; Wiesel and Hubel 1974) showed that neurons with similar response properties are grouped in vertical columns, about 0.5–1 mm in diameter, with each column oriented perpendicular to the surface of the cortex and spanning its thickness. Correspondingly, anatomical studies (using histological staining for the enzyme cytochrome oxidase) showed a modular organization in the primate visual cortex with periodically spaced patches about 350 μm apart (Horton and Adams 2005). In the primate prefrontal cortex the diameter of columns vary between 300 and 500 μm, but it does not differ significantly in size between brains with over three orders of magnitude difference in volume (Bugbee and Goldman-Rakic 1983). This common periodicity means that any block of cortex, approximately the size of a single hypercolumn, contains cells tuned to all values of every receptive field variable (Swindale et al. 2000). Hubel and Wiesel

(Hubel 1982; Katz et al. 1989) applied the term "module" to this tissue block comprising multiple, overlapping hypercolumns. Mountcastle (1997) has used the term 'module' interchangeably with 'column'.

The most salient feature of cortical organization is the presence of an orderly topographic map of visual space that is remapped sequentially as information flows from visual cortex to prefrontal cortex (Salinas 2004). Neighboring neurons tend to have receptive fields in similar positions in visual space, and these positions "change predictably" as a function of position on the cortex (Swindale et al. 2000). In addition to orientation, visual neurons vary in their preference for the direction of motion of an oriented bar or edge, in their preference for stimuli delivered to one eye or the other (ocular dominance), and in their preference for low versus high spatial frequencies in the visual image (Swindale 1998). All of these properties have been found to vary in an orderly way with position on the cortical surface, so that, typically, a complete set of values occurs at least once every mm or so.

10.3 Prefrontal Cortical Microcircuits

As proposed by Mountcastle, the primate neocortical circuitry has a modular architecture that subserves a multitude of sensory (visual, auditory, touch), motor, cognitive (attention, memory, decision) and emotional functions (Mountcastle 1957, 1997; Shepherd and Grillner 2010; Opris and Bruce 2005). According to this "modular" view of the prefrontal cortex, Goldman-Rakic (1996) provided evidence from experimental studies in nonhuman primates that the central executive decomposes into segregated information processing modules each with its own sensory, mnemonic, and motor control features.

Inter-laminar microcircuits, consisting of interconnected pyramidal neurons between the supra- and infra-granular layers, form a three stratum functional module that connects both worlds via sensory and motor circuits, in which infra-granular layers execute the associative computations elaborated in supra-granular layers (Buxhoeveden and Casanova 2002; Casanova et al. 2011; Thomson and Bannister 2003; Opris et al. 2011, 2012b, 2013). These microcircuits receive input from neurons in layer L4, which project to L2-L3, or through direct thalamic projections to the supra-granular layers in the higher-order cortical areas. Neurons in L2-L3 then project top-down to L5, where they target specific types of pyramidal cells and inhibitory interneurons. Some L5 neurons project back to L2-L3 neurons, forming an inter-laminar loop (Weiler et al. 2008) back to L4, targeting mostly interneurons (Thomson and Bannister 2003). The outputs from cortical microcircuits, cortico-striatal projections arise mostly from L5, whereas cortico-thalamic projections arise from L6.

Cortical microcircuits are connected by cortico-cortical connections into a macro-network that links areas within the same hemisphere, as well as across hemispheres (Van Essen et al. 1982). This super network subserves the 'perception-to-action' cycle – a group of processes that handle environmental stimuli and convert them into actions (Fuster and Bressler 2012; Romo et al. 2002). Microcircuits within the same hemisphere are interconnected (from low level sensory to high level associative processes) through horizontal connections in lamina 2/3, spanning over many cortical areas (Das and Gilbert 1995; Kritzer and Goldman-Rakic 1995; Fuster and Bressler 2012).

Inter-area connectivity of cortical microcircuits preserves spatial topography suggesting a column-to-column match from one area to another (Goldman-Rakic 1996). Additionally, the topography is preserved within minicolumns owing to the inter-laminar projections (Opris et al. 2013). Interhemispheric connectivity is formed by neural interconnections of lamina 3b (Jones 2000; Van Essen et al. 1982). Cortical microcircuits are also interconnected across the two hemispheres through cortico-cortical connections from lamina 3b (Jones 2000; Van Essen et al. 1982). This implies that the firing of a single callosal neuron might influence several cortical columns within the opposite hemisphere. However, microcircuits in each hemisphere are symmetrically interconnected (from low level sensory to high level associative processes) through horizontal connections in lamina 2/3 across many cortical areas (Das and Gilbert 1995; Kritzer and Goldman-Rakic 1995; Fuster and Bressler 2012). This imply a bottom-up remapping of a spatial topography with a column-to-column match from one area to another on the dorsal visual stream from V1 to prefrontal cortical area 46 (Goldman-Rakic 1996). In the same time each topography is preserved across layers through top-down inter-laminar projections within minicolumns (Opris et al. 2013).

Recent research conducted in nonhuman primates indicates that a variety of sensory, motor and executive functions emerge from the interactions between frontal, parietal, temporal and occipital cortical microcircuits (Atencio and Schreiner 2010; Buffalo et al. 2011; Opris et al. 2012b, c, 2013; Mahan and Georgopoulos 2013; Hirabayashi et al. 2013a, b; Takeuchi et al. 2011; Hansen et al. 2012). This suggests that cortical microcircuits perform elementary computations while cognitive functions are sub-served by a broader network comprising multiple cortical areas (Fuster and Bressler 2012). This implies that cortical microcircuit integrate, represent or select the relevant signals from a multitude of incoming inputs.

10.4 Integration, Representation and Selection in Cortical Microcircuits

The functional role of the cortical minicolumn is a continuing source of research and debate more than half a century after it was identified as a component of brain organization (Mountcastle 1957, 1997; Mountcastle et al. 1955). Nevertheless, the cognitive ability of prefrontal cortical mechanism is hypothesized to emerge from the "laminar-columnar" architecture of the prefrontal cortical minicolumns, that are

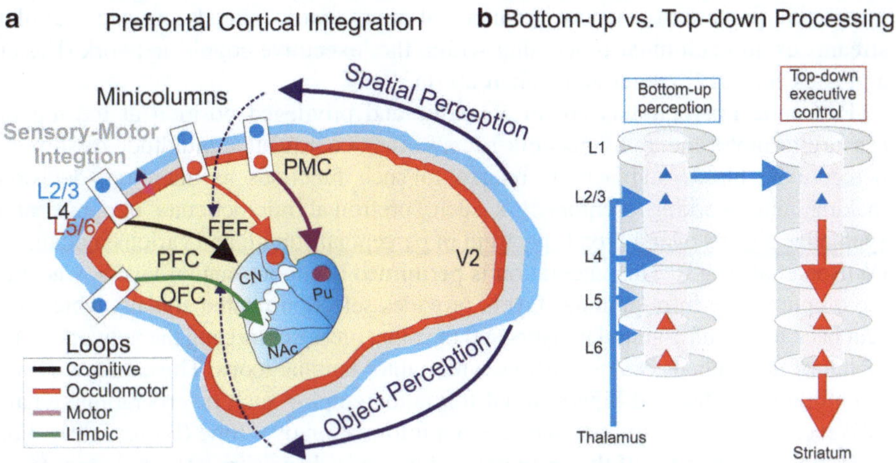

Fig. 10.2 Inter-laminar microcircuits for integration, representation and selection. (**a**) Prefrontal cortical integration of cognitive, oculomotor, motor and limbic structures in four loops. (**b**) Illustration of bottom-up and top-down processing within cortical microcircuits

interconnected (Fig. 10.2) with basal ganglia and thalamus in recurrent loops (Bugbee and Goldman-Rakic 1983; Goldman-Rakic 1996; Opris et al. 2011, 2012d, 2013). Such columnar recurrent 'microcircuit' may be regarded as the 'basic functional unit' of the cognitive function. Its cognitive relevance emerges from the 'computational' ability of the inter-laminar microcircuit to: (a) integrate incoming signals of the input layers, (b) store information through feedback connections in reverberatory loops, and (c) to compare input signals to a threshold criterion, triggering an output response i.e. the ability to make a decision.

10.4.1 Integration

The integrative role of cortical minicolumns as a module (Leise 1990) stems from connecting the horizontal and vertical components of the cortex within the same columnar space. The supragranular layers L2/3 which are the major source of corticocortical projections also receive sensory information, while the infragranular layer L5 is the output to subcortical structures involved in behavior (Miller and Cohen 2001). Thus, interlaminar connections form microcircuits that bind sensory-related signals with behavior/movement related outputs (Opris et al. 2011). This sensorimotor integration (Fig. 10.2a) was demonstrated by Opris et al. (2011, 2013) by means of interlaminar correlated firing between supragranular layers that carry perceptual/visual spatial information and the infragranular layers that carry action related information. Such transformations of neural signals may likely reduce the output degrees of freedom within the cortical minicolumn by selecting only the relevant signals for action/behavior. This integrative process that occurs in canonical microcircuits binds/segregates parallel streams of minicolumnar processing within the 'executive cognit' network (Fuster and Bressler 2012; Miller and Cohen 2001).

Prefrontal microcircuits are in a unique and privileged position at the top of sensory-to-motor hierarchy network because they coordinate a multitude of stimuli, perceptions, biases and actions related to such functions as attention, decision making, and working memory. As such, prefrontal microcircuits integrate and synthetize signals over a broad spectrum of perceptual stimuli and various modalities (Wilson et al. 1993). This integration is performed in supra-granular layers, whereas the output of the infra-granular layers provides selection-related signals, which are sent back to the infra-granular layers and the other areas comprising the network. As a matter of fact, signals can reverberate within inter-laminar loops. Thus, microcircuits in entorhinal cortex and hippocampal formation employ such reverberating signals (Takeuchi et al. 2011) to integrate relevant information over time (Fuster 2001). For example, microcircuits of the temporal cortex use such synchrony to maintain items in long term memory (Takeuchi et al. 2011; Hirabayashi et al. 2013a), while the microcircuits in the prefrontal cortex perform elementary computations for the executive control of behavior (Opris et al. 2012b, c);

10.4.2 Representation

The prefrontal cortex is crucial for the representation of stimulus properties (spatial, object features) in working memory and the hippocampus is essential for the long term memory (Funahashi et al. 1989; Miller and Cohen 2001; Takeuchi et al. 2011; Naya and Suzuki 2011). Prefrontal cortical "modules" (minicolumns) are composed of neurons with "memory fields" (Goldman-Rakic 1996), with isodirectional tuning for minicolumnar cells (Rao et al. 1999). Prefrontal neurons exhibit persistent "delay period" activity with enhanced firing rate in animals performing delay response tasks, being regarded as a neural "signature" of working memory (Funahashi et al. 1989). Such persistent firing of prefrontal cortical neurons is hypothesized to emerge from functional interactions between cells in different cortical layers, wired together in reverberatory loops (inter-laminar and/or thalamo-cortical) (Alexander et al. 1986). As a matter of fact, signals can reverberate within inter-laminar loops (Weiler et al. 2008). Thus, cortical microcircuits for long term memory in entorhinal cortex employ such reverberating signals (Takeuchi et al. 2011) to represent relevant information over time (Fuster 2001). Takeuchi et al. (2011) demonstrated a "reversal" of interlaminar signal between sensory and memory processing in monkey temporal cortex. Thus, during the sensory "cue" epoch, the canonical microcircuit signals flowed "feed-forward" from granular to supragranular layers and from supragranular to infragranular layers, while during the "memory" delay epoch, however, the signal flow reversed to the "feed-back" direction: from infragranular to supragranular layers. Such reversal of signal flow highlights a neat dissociation of the sensory and mnemonic processing in the temporal cortex that differentially recruits its laminar circuits.

10.4.3 Selection and Decision Making

The available evidence suggests that the prefrontal cortical minicolumn might be the first stage bottleneck in the cortical-striatal-palidal-thalamo-cortical loop (Alexander et al. 1986). It is obvious that layers 2/3 and 4 cells integrate a lot of inputs from virtually all of the brain, while the number of outputs from layer 5/6 pyramidal cells to subcortical structures participating in behavior is much less. Also, a key role in selection is played by the GABAergic interneurons of minicolumns (Raghanti et al. 2010) that shape the tuning for preferred direction/location by means of lateral inhibition. The minicolumnar role in plasticity is associated not only with the tuning (Fig. 10.3) of synaptic activity states but also with optimal selection among alternate subnetworks of microcircuits developing within a given context. Such parallel subnetworks may process complementary submodalities within a defined receptive/memory field; alternatively they may provide overlapping response characteristics to a common input. Competition among networks allows for circuit optimization (selection), in particular by means of learning.

We hypothesize that minicolumnar diversity provides the substrate for this competition and the basis for adapting learned behavior to context. During development, neurogenetic programs interact with epigenetic factors to regulate formation of cortical microcircuit templates (Rakic 1988; Jones 2000; Jones and Rakic 2010; Kaas 2012), which are then shaped and pruned by differential patterns of sensory activity. Thus, increased minicolumnar diversity may give rise to greater potential for combinatorial activity of microcircuits within overlapping networks, resulting in enhanced learning and behavioral flexibility (Casanova 2008). Cortical minicolumns may therefore play a crucial role in behavioral selection that is in fact the substrate of executive function (e.g., attention, decision making).

Is cortical minicolumn a decision module? A decision circuit is defined as a closed neural network that measures the probable value of a signal element and makes an output signal based on the value of the input signal and a predetermined criterion or threshold (Ratclif et al. 2003; Opris and Casanova 2014). Minicolumns in PFC are interconnected to each other through horizontal "long range" projections in layer 2/3 (Kritzer and Goldman-Rakic 1995; Rao et al. 1999) and interlaminar mini-loops (Weiler et al. Takeuchi et al. 2011). The loop is then closed ("reverberatory loops") through projections to the subcortical basal ganglia nuclei and thalamus (Alexander et al. 1986; Swadlow et al. 2002). Such "reverberatory loops" may be regarded as the "basic functional unit" of cognitive/executive mechanism because they: (i) combine incoming signals of the different input layers (Casanova et al. 2007); (ii) store mnemonic information through feedback connections in "persistent" spiking activity (Wang 2012); (iii) compare input signals to a threshold criterion triggering an output response (selection), which constitutes the

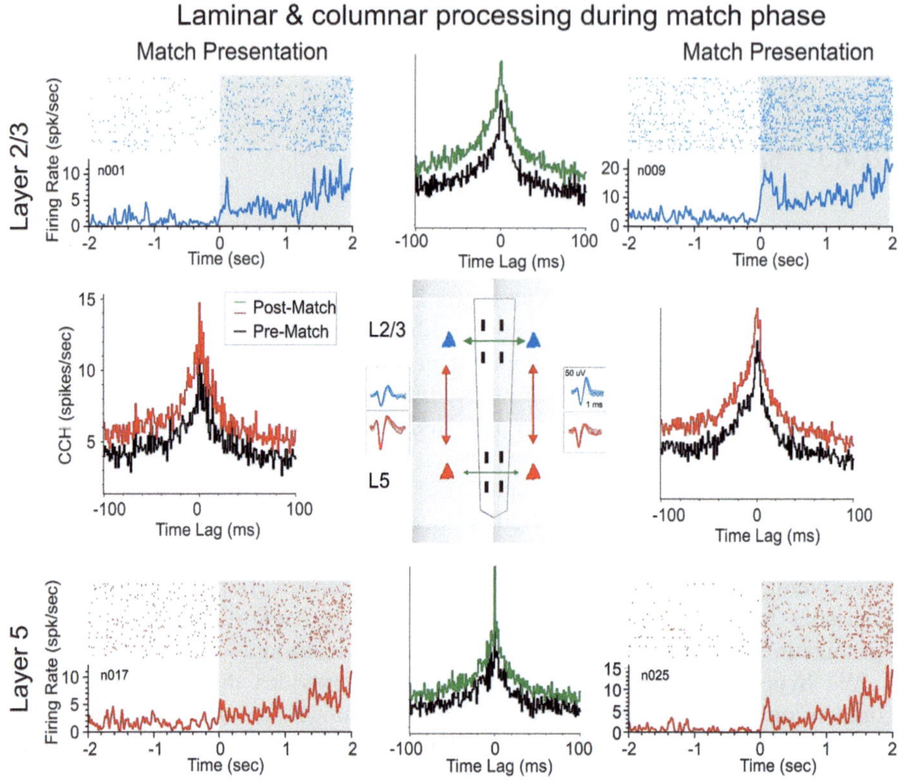

Fig. 10.3 Simultaneous recordings of cortical minicolumns. Minicolumnar interlaminar processing of target selection during the match phase of the task. Peri-event and cross-correlation (post- vs. pre-match) histograms depict the functional role in integration/selection of prefrontal cortical layers and minicolumns

ability to make a decision (Ratcliff et al. 2003). Thus, a cortical minicolumn with integrative, selective and threshold abilities can play the role of a decision module.

10.5 Correlated vs. Causal Relationships to Executive Control

Several approaches based on microcircuits have been implemented. These advances have been possible owing to of the development of new multi-electrode arrays (MEA) fitted for recordings from neural elements of cortical columns (Hampson et al. 2004; Moxon et al. 2004). Thus, MEAs with linear or bi-linear geometry have been successfully employed for simultaneous recordings from supra- and infragranular cortical laminae in adjacent minicolumns, resulting in unprecedented insights into the function of cortical microcircuits (Mo et al. 2011, Opris et al. 2011, 2012b, c, 2013).

10.5.1 Inter-laminar Interactions and Emergence of Executive Control

The relevance of minicolumnar activity to executive function has been investigated with different approaches under several conditions (Hirata and Sawaguchi 2008; Opris et al. 2011, 2012b, c, 2013; Hampson et al. 2012). Our recent results in nonhuman primates show for the first time interlaminar processing in PFC (Fig. 10.3) during target selection (Opris et al. 2011, 2012b, c) and sensorimotor integration (Opris et al. 2011). An example of this interlaminar interaction during target selection (Opris et al. 2012b, c) in delay match to sample (DMS) task is shown in Fig. 10.3 for two cell pairs with rasters and perievent histograms (PEHs) bracketing the temporal interval of image presentation (Match Phase onset) and completion of the target selection Match Response (0–2 s). The cell pairs were recorded on appropriate sets of adjacent pads (minicolumns 1&2) in the conformal multiple electrode array (MEA) shown in the illustration (Opris et al. 2011, 2012b, c) of both interlaminar cell pairs in L2/3 and L5 (Fig. 10.3 center). Neurons in both layers showed significant increases in mean firing in supra- and infragranular layers as a function of Match presentation (Post Match: 0 to +2 s) and during subsequent movements associated with target selection. Demonstration of precise functional connections between individual cells within each minicolumn was provided by cross correlation histograms (CCHs; Opris et al. 2011, 2012b, c); constructed for individual L2/3 and L5 cell pairs recorded on vertically positioned pads of the MEA. Normalized CCHs for both minicolumn cell pairs and between minicolumns are shown in Fig. 10.3 for cell firing in the displayed PEHs: (i) prior to (black) Match phase onset (2 s to 0, Pre) or (ii) after (green) Match phase onset (0 to +2 s, Post) for the same cell pairs. Both CCHs show significantly correlated firing (Opris et al. 2012b).

10.5.2 Causal Relations Involving Inter-laminar Microcircuits

The unique properties of conformal MEAs (Fig. 10.4a; Moxon et al. 2004) together with prior microstimulation work (Opris et al. 2001, 2005; Hampson et al. 2004) has provided a basis for showing functional relationships to executive function in prefrontal cortex of nonhuman primates. The conformal multi-electrode array (MEAs) also provide the basis for applying a system specific model to control firing of cells via application of electrical stimulation (Opris et al. 2012c; Hampson et al. 2012) to the same loci in which columnar firing has been detected and analyzed with respect to DMS task performance (Opris et al. 2012c; Hampson et al. 2012). This same model was implemented to test whether it could facilitate performance on trials that show a distinctive difference in correct performance (Hampson et al. 2012) as a function of the prior instructions as to type of response to make in the Match phase (i.e. Object *vs.* Spatial trials). Figure 10.4 shows the integration of a multi-input multi-output (MIMO) nonlinear math model to assess the patterns of firing in L2/3 and L5 cells recorded in the columnar manner with the MEA shown with adjacent vertical pads (Opris et al. 2011, 2012b; Hampson et al. 2012). Figure 10.4b reflects the type of input and output firing patterns recorded and analyzed by the MIMO model and also illustrates how the output pattern of L5 cell firing is duplicated via a multichannel stimulator that is capable of delivering predetermined patterns of pulses to the same L5 pads to mimic firing on correct trials. The advantage of the MIMO model is that the online recording provides the means to detect when the inappropriate L2/3 firing pattern occurs which triggers the delivery of the appropriate L5 stimulation pattern providing the means to override errors and enhance performance (Hampson et al. 2012). Stimulation consisted of 1.0 ms bipolar pulses (20–50 μA) delivered to L5 recording locations following presentation of the Match phase screen and prior to the completion of the Match Response.

The results of microstimulation delivery are shown in Fig. 10.4c, d, in which the effects on performance are compared to trials in which stimulation was not delivered, respective of trial type. Figure 10.4c compares the change in % correct performance as a function of processing time (reaction time + movement time) on stimulation (Stim) trials with respect to the no stimulation (No stim) case. Figure 10.4d showed for the first time the increase in correct performance on trials as a function of the number of distracter images in the Match phase. The results indicate that MIMO derived stimulation induces enhanced cognitive processing (Opris et al. 2012c, 2013; Hampson et al. 2012) required to retrieve the "rule for successful selection" of the appropriate item. The distribution of microstimulation induced spatial bias (Fig. 10.4e) will be shown in Fig. 10.5 of the next section.

10.6 Prefrontal Microcircuits Bind Perception to Executive Control

A broad range of brain functions, from perceptual to executive actions encode, represent, monitor and select information that is either spatial- and/or object-specific for effective behavioral performance (Quintana and Fuster 1999; Goldman-Rakic 1996; Posner and Snyder 1975; Shallice and Burgess 1996; Botvinick et al. 1999; Selemon and Goldman-Rakic 1988; Opris and Bruce 2005). Such constellations of brain abilities use large scale neural circuits consisting of thalamo-cortical loops and cortical microcircuits with functional roles in the integration, representation and selection of information (Fuster and

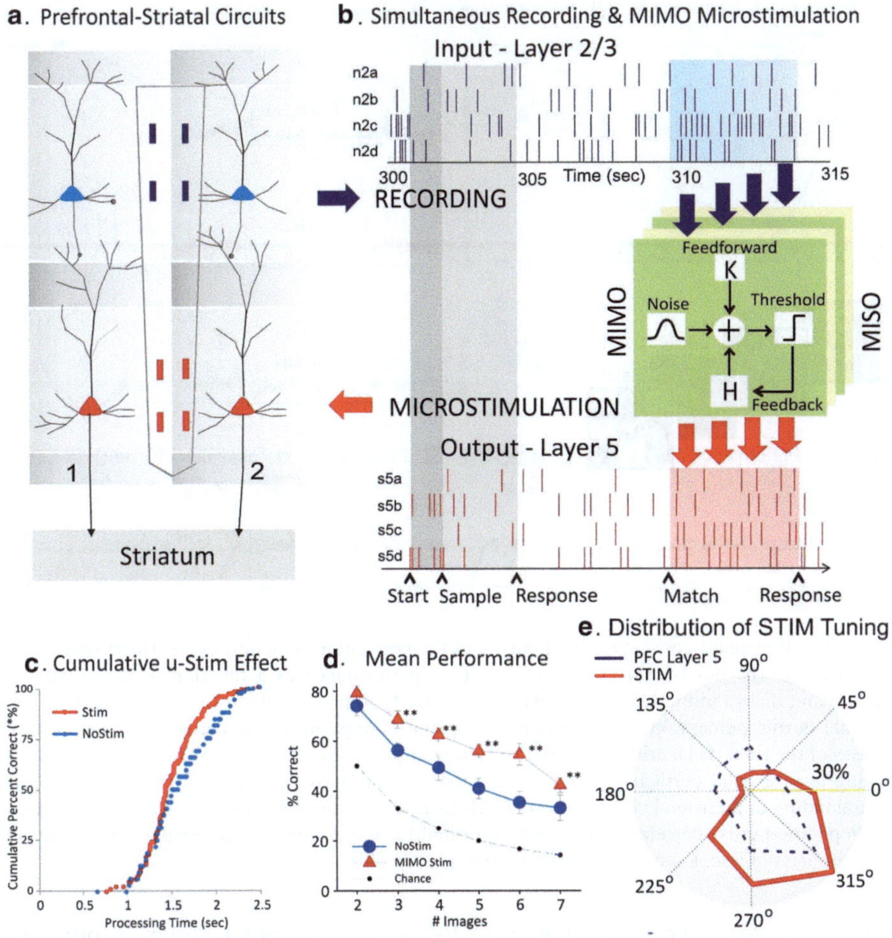

Fig. 10.4 Manipulation of executive control in PFC microcircuits. (**a**) Prefrontal striatal neuronal circuits. (**b**) Simultaneous recording and MIMO model microstimulation. (**c**) Cumulative microstimulation effect neuronal circuits. (**d**) Population mean performance as a function of the number of images in the task. (**e**) Distribution of stimulation induced tuning vectors

Bressler 2012; Alexander et al. 1986; Opris et al. 2011). Cortical microcircuits topographically connected by cortico-cortical and subcortical connections into a super network subserves the 'perception-to-action' cycle – an ensemble of processes that sense/perceive the environmental stimuli and convert them into actions (Fuster and Bressler 2012; Romo et al. 2002).

As previously shown the dorsal visual stream from the striate cortex to the posterior parietal region carries the spatial information (Fig. 10.5a) required for sensorimotor transformations in visually guided actions, while the ventral stream

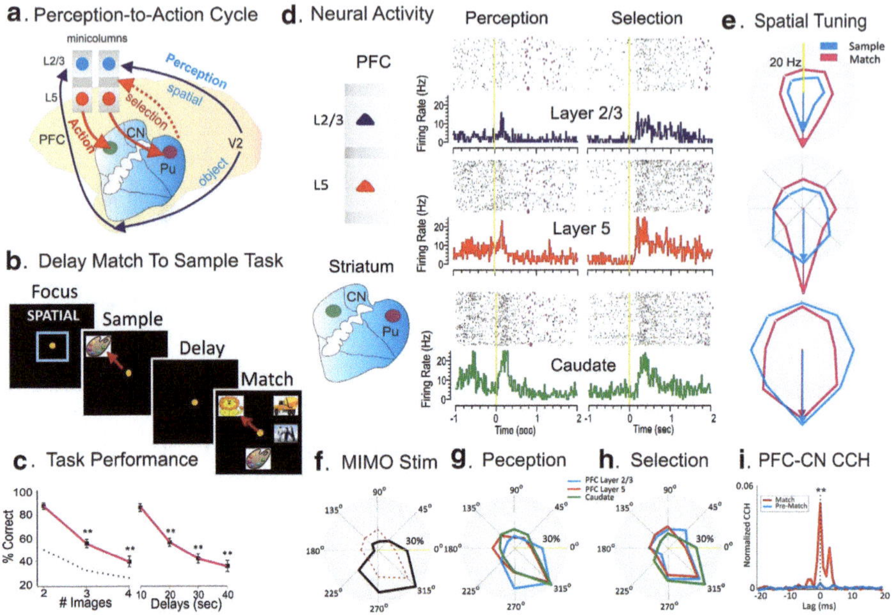

Fig. 10.5 Prefrontal cortical microcircuits and the perception to action cycle. (**a**) Perception to action cycle. (**b**) Behavioral DMS task. (**c**) Task performance as a function of the number of images and delays in the DMS task. (**d**) Neural activity in prefrontal cortical layers *L2/3*, *L5* and caudate during perception and selection. (**e**) Spatial tuning during the perception and selection phases of the task. (**f**) Distribution of MIMO stimulation induced tuning. (**g**) Distribution of neural tuning in prefrontal cortical layers *L2/3*, *L5* and caudate during perception. (**h**) Distribution of neural tuning in prefrontal cortical layers *L2/3*, *L5* and caudate during executive control/selection. (**i**) Population cross-correlation between prefrontal layers *L5* and caudate nucleus during match vs. pre-match phase. Compiled figure from Opris et al. 2013

projections from the striate cortex to the inferior temporal cortex is primarily responsible for perceptual identification of objects (Goodale and Milner 1992; Ungerleider and Mishkin 1982). Thus, a visual object's qualities and its spatial location depend on the processing of different types of visual information in the inferior temporal and posterior parietal cortex, respectively. However, object and spatial information carried in these two separate pathways has been shown to be integrated into a unified 'visual percept' in prefrontal cortex which receives connections from both circuits (Goodale and Milner 1992; Ungerleider and Mishkin 1982; Rao et al. 1997).

Basal ganglia participate in multiple parallel segregated circuits or 'thalamo-cortical loops' that make connections with motor, sensory and cognitive areas of the cerebral cortex (Alexander et al. 1986; Middleton and Strick 2002; Hoover and Strick 1993). Prefrontal cortical areas seem to be the target of extensive, topographically organized outputs from the basal ganglia (Middleton and Strick 2002). Such thalamo-cortical projections from basal ganglia to the superficial and deep prefrontal cortical layers can directly activate specific inputs to the re-entrant loop

(McFarland and Haber 2002; Swadlow et al. 2002). Thus, the outputs from the inter-laminar microcircuits of prefrontal cortex are in ideal position to support the decision to act via the synchronous excitation of the constellation of circuits in the executive hierarchy (Quintana and Fuster 1999; Fuster and Bressler 2012).

Executive control is a fundamental function of the brain that mediates the integration of perception and action during behaviorally relevant environmental events. It has been proposed that executive control involves a broad network of brain areas, including frontal and parietal/temporal cortex, as well as striatum and other subcortical structures (Fuster and Bressler 2012). These structures have been consistently associated with roles in sensorimotor integration and selection of task specific behavioral responses, commonly considered to be the regions necessary for 'executive decisions' (Opris et al. 2012b, d). However, recent evidence has shown that these brain structures are part of functional loops in which inter-laminar microcircuits or 'minicolumns' in dorsolateral prefrontal cortex 'bind' perception and executive selection of spatial targets to guide goal-specific behavior.

To demonstrate this Opris et al. (2013) have trained nonhuman primates (rhesus monkeys) to perform a delayed match to sample (DMS) task with the instruction to select the remembered spatial location of the image on the screen, each presented in the Sample phase of the task (Fig. 10.5b). Subjects made hand tracking movements to the appropriate visual targets for rewards in the Match phase of the task (Fig. 10.5b). The DMS task incorporated key features like the number of distracter images (2–4) which could appear in any of eight locations on the screen in the Match phase after variable durations of the intervening delay period (1–40 s). These factors were reflected in the animal's behavioral performance levels during encoding and selection of spatial or object stimuli as shown in Fig. 10.5c.

Neurons were recorded simultaneously in PFC (in layer L2/3 and in layer L5) and in the striatum (caudate and putamen) while the animals performed the DMS task. Consistent with previous reports (Opris et al. 2012b, d) firing of cells in prefrontal layers and minicolumns reflected differential encoding of spatial location in the DMS task. Figure 10.5d shows raster and peri-event histograms of cells recorded in prefrontal cortical layers 2/3 and 5, together with cells recorded simultaneously in caudate/putamen. For each of these cells (Fig. 10.5d) firing patterns were compared during (a) encoding of sample target's location of sample presentation on the screen (Perception), and during: (b) selection of the spatial location of the sample image presentation on the screen (spatial trials), in the Match phase of the task (Selection). The polar plots in Fig. 10.5e show that neurons in layer 2/3 and 5 fired similarly with caudate neurons and were synchronized and spatially tuned to the same screen locations (black arrows). When compared during match phase presentation (Match Tuning) neural tuning directions for the 3 regions were again similar.

To further test whether inter-laminar firing links spatial perception to executive selection we applied a novel type of closed loop patterned MIMO stimulation previously shown to facilitate performance of the same task (Opris et al. 2012b; Hampson et al. 2012). This is shown in Fig. 10.4 as a functional diagram in which neural firing in PFC layer 2/3 was recorded with a multielectrode array (Opris et al. 2012b, d) and fed into a nonlinear multi-input–multi-output (MIMO) math model (Fig. 10.3a), which processed and simultaneously delivered a pattern of electrical pulses from a multi-channel stimulator that mimicked the correlated firing of PFC layer 5 cells on successful trials (Hampson et al. 2012). MIMO stimulation methods and associated control procedures proving columnar activation have been previously published in detail (Opris et al. 2012d; Hampson et al. 2012). These controls included delivery of stimulation pulse patterns that were different than what the MIMO model derived for correct trials. In this case the intensity and the number of pulses, plus the area (L5) that was stimulated were identical.

The effectiveness of MIMO stimulation delivered to this particular region of PFC is shown in Fig. 10.5f where the preference effect on stimulated (Stim) vs. nonstim trials is compared for all Spatial ($n = 40$ sessions) trials within the same session. The difference in mean % correct performance for all stim vs. nonstim trials (ALL) is shown in comparison to stim vs. nonstim trials in which performance at locations was significantly above that at all other locations (Facilitated; see also Opris et al. 2013). The marked difference in the degree of increase in % correct trials produced by MIMO stimulation at preferred vs. non-preferred (ALL) locations indicates that in addition to facilitating performance at all response locations, the stimulation enhanced the innate directional preference (spatial tuning) which corresponded to the anatomic location of the PFC layer 2/3 minicolumn. This demonstrated that the MIMO stimulation delivered during the match/selection phase of the task was likely to have facilitated discharge of Layer 5 neurons in the same recorded minicolumns and that is what improved spatial target selection in this phase of the task.

The unique feature of these experiments is that they allow us to tap into the perception-to-action cycle (Quintana and Fuster 1999). As a final validation of microcircuit tuning in PFC and caudate we compared polar firing across the same three nodes in the perception and selection phases on spatial trials in which MIMO stimulation induced increases in performance. Figure 10.5g, h shows nearly complete overlap (between 81 % and 91 %) in spatially tuned firing indicating that the majority of neural tuning vectors for the preferred microcircuit target location (315°) facilitated task performance when subjected to MIMO stimulation during spatial trials. The anatomic link between prefrontal cortex and striatum is demonstrated physiologically through normalized cross-correlations pairs of cells in PFC layer 5 and Caudate displaying synchronized firing during Match target presentation epoch (0, 2 s; red) compared to the pre-Match epoch (−2 s,0; blue). Therefore, such synchronized firing of PFC and Caudate neurons during the match phase (dealing with target selection and executive control; Fig. 10.5i) is telling us that these key nodes in the prefrontal cortical striatal loop show the modulation of executive control signals in the cortical-striatal executive loop (Alexander et al. 1986).

These novel findings demonstrate a robust involvement of cortical layers and striatum in the perception-to-action cycle (Quintana and Fuster 1999; Fuster and Bressler 2012). This is supported by implementation of the MIMO model which extracts the percept from prefrontal layer 2/3 and imparts the appropriate signal to columnar related layer 5 cells, thereby strengthening activation via the executive loop through the caudate nucleus (shown in Fig. 10.4a) to manifest selection of a particular target location. Given these findings, the functional specificity of the perceptual circuit is likely determined via "tuned" inter-laminar microcircuits connected to executive prefrontal cortico-striatal, thalamo-cortical loops, that are translated into action via "cognits" that coordinate information in large scale networks (Fuster and Bressler 2012; Alexander et al. 1986; Fuster 2007).

The enhancement in cognitive performance by the MIMO stimulation may be explained by induced changes in the balance between excitation and inhibition in cortical-striatal loop (Hampson et al. 2012; Opris et al. 2005) and by the temporal specificity of the PFC layered L2/3-L5 firing pattern (Hampson et al. 2012), since stimulation in a "scrambled" (random) pattern with the same pulses impaired performance in prior studies (Opris et al. 2012d; Hampson et al. 2012). The microstimulation current activates the neighboring minicolumns around the micro-electrode pad/tip causing the preference of this group of minicolumns to win the competition for the behavioral output (Opris et al. 2005). Therefore, these results clearly indicate the need for inter-laminar microcircuits to bind perception and action.

10.7 Implications for a Modular Approach to Neuroscience

Recent developments in nanotechnological tools and in the design and synthesis of nano-materials have generated optical, electrical, and chemical methods that can be adapted for use in neuroscience. Nanotechnology was instrumental to nanofabricated planar electrode arrays for high-density neuronal voltage recording (Du et al. 2011; Suyatin et al. 2009). Nanofabrication technology raises the prospect for creating vastly greater numbers of electrodes and smaller, less invasive implantable devices. Among the promising tools for the study of brain microcircuits is the planar electrode array (Viventi et al. 2011; Alivisatos et al. 2013), which can be patterned on a crystalline, ceramic, or polymer support structure. The recording of neuronal activity with three-dimensional (3D) microelectrode arrays (Zorzos et al. 2012) represents a major advance in brain activity mapping techniques, by providing a tool to probe how intra and inter-laminar/regional neural circuits cooperate to process information. In the next decade or so, the modular approach to neuroscience research will gain an unprecedented momentum.

One reason would be the building of prosthetic minicolumns as basic modules to repair the damaged cortical tissue. Microcircuit-based approaches could be implemented in various cortical areas for building cognitive prosthetics (Berger et al. 2011) in order to reverse cognitive deficits in a broad spectrum of diseases like schizophrenia (Dobbs 2010; Casanova 2007), dementia (Chance et al. 2006, 2008, 2011; Di Rosa et al. 2009), autism (Casanova 2012; Casanova et al. 2002a, b, 2003a, 2006a, b, 2010, 2012, 2013; Sokhadze et al. 2010, 2012), ADHD (Brennan and Arnsten 2008), addiction (Tomasi et al. 2010), aging (Wang et al. 2011;

Opris et al. 2009) and executive dysfunction (Duncan et al. 1997; Shallice and Burgess 1991) in which inter-laminar processing is likely disrupted due to cortical tissue damage or malfunction (Opris et al. 2012a, d; Casanova et al. 2003a; Duncan et al. 1997; Shallice and Burgess 1991). Therefore, a better understanding of the function of inter-laminar microcircuits across the neocortex is needed to develop treatments for neurological disorders, as well as to develop methods for brain augmentation.

10.8 Conclusion

In summary, these concepts provide support for modular approaches to executive brain functions in key nodes of the prefrontal loop (including PFC layer 2/3, layer 5 and caudate nucleus), as well as causal relationships involving the inter-laminar microcircuits across neocortex. These findings suggest that prefrontal inter-laminar microcircuits play a causal role in linking perception to the executive selection of spatial targets to spatial locations to which such microcircuitry has been tuned via past experience. The fact that activation of an innate PFC minicolumnar bias (shown via MIMO model-controlled stimulation) resulted in improved performance provides the real support for extending the modular approach across many areas/regions of the brain. This modular approach provides an important basis for building cognitive prosthetics (Berger et al. 2011) in order to reverse cognitive deficits in a broad spectrum of diseases.

References

Alexander GE, DeLong ME, Strick PL (1986) Parallel organization of functionally segregated circuits linking basal ganglia and cortex. Annu Rev Neurosci 9:357–381

Alivisatos AP, Andrews AM, Boyden ES, Chun M, Church GM, Deisseroth K, Donoghue JP, Fraser SE, Lippincott-Schwartz J, Looger LL, Masmanidis S, McEuen PL, Nurmikko AV, Park H, Peterka DS, Reid C, Roukes ML, Scherer A, Schnitzer M, Sejnowski TJ, Shepard KL, Tsao D, Turrigiano G, Weiss PS, Xu C, Yuste R, Zhuang X (2013) Nanotools for neuroscience and brain activity mapping. ACS Nano 7:1850–1866

Atencio CA, Schreiner CE (2010) Columnar connectivity and laminar processing in cat primary auditory cortex. PLoS One 5(3), e9521

Jones EG (2000) Microcolumns in the cerebral cortex. A historical overview of the concept of micro- or minicolumns. Proc Natl Acad Sci U S A 97(10):5019–5021

Barbas H, Hilgetag CC, Saha S, Dermon CR, Suski JL (2005) Parallel organization of contralateral and ipsilateral prefrontal cortical projections in the rhesus monkey. BMC Neurosci 6:32

Bastos AM et al (2012) Canonical microcircuits for predictive coding. Neuron 76:695–711

Berger TW et al (2011) A cortical neural prosthesis for restoring and enhancing memory. J Neural Eng 8(4):046017

Botvinick M, Nystrom LE, Fissell K, Carter CS, Cohen JD (1999) Conflict monitoring versus selection-for-action in anterior cingulate cortex. Nature 402(6758):179–181

Brennan AR, Arnsten AF (2008) Neuronal mechanisms underlying attention deficit hyperactivity disorder: the influence of arousal on prefrontal cortical function. Ann N Y Acad Sci 1129:236–245

Buffalo EA, Fries P, Landmanc R, Buschman TJ, Desimone R (2011) Laminar differences in gamma and alpha coherence in the ventral stream. Proc Natl Acad Sci U S A 108:11262–11267

Bugbee NM, Goldman-Rakic PS (1983) Columnar organization of corticocortical projections in squirrel and rhesus monkeys: similarity of column width in species differing in cortical volume. J Comp Neurol 220:355–364

Buschman TJ, Denovellis EL, Diogo C, Bullock D, Miller EK (2012) Synchronous oscillatory neural ensembles for rules in the prefrontal cortex. Neuron 76(4):838–846

Buxhoeveden DP, Casanova MF (2002) The minicolumn hypothesis in neuroscience. Brain 125 (5):935–951

Buxhoeveden D, Fobbs A, Roy E, Casanova MF (2002) Quantitative comparison of radial cell columns in children with Down syndrome and controls. J Intellect Disabil Res 46(1):76–81

Casanova MF (2005) An apologia for a paradigm shift in neurosciences. In: Casanova MF (ed) Neocortical modularity and the cell minicolumn. Nova Biomedical Publishers, New York, pp 33–55

Casanova MF (2007) Schizophrenia seen as a deficit in the modulation of cortical minicolumns by monoaminergic systems. Int Rev Psychiatry 19(4):361–372

Casanova MF (2008) The significance of minicolumnar size variability in autism: a perspective from comparative anatomy (ch. 16). In: Zimmerman A (ed) Autism current theories and evidence. Current clinical neurology. The Humana Press, Totowa, pp 349–360

Casanova MF (2012) The minicolunopathy of autism. In: Buxbaum JD, Hof PR (eds) The neuroscience of autism spectrum disorder. Academic, Amsterdam, pp 327–334

Casanova MF, Buxhoeveden DP, Switala AE, Roy E (2002a) Minicolumnar pathology in autism. Neurology 58:428–432

Casanova MF, Buxhoeveden DP, Switala AE, Roy E (2002b) Neuronal density and architecture (gray level index) in the brains of autistic patients. J Child Neurol 17(7):515–521

Casanova MF, Buxhoeveden D, Gomez J (2003a) Disruption in the inhibitory architecture of the cell minicolumn: implications for autism. Neuroscientist 9(6):496–507

Casanova MF, Buxhoeveden D, Switala A, Roy E (2003b) Rett syndrome as a minicolumnopathy. Clin Neuropathol 22:163–168

Casanova MF, van Kooten I, Switala AE, van Engeland H, Heinsen H, Steinbusch HWM, Hof PR, Trippe J, Stone J, Schmitz C (2006a) Minicolumnar abnormalities in autism. Acta Neuropathol 112(3):287–303

Casanova MF, van Kooten I, Switala AE, van Engeland H, Heinsen H, Steinbusch HWM, Hof PR, Schmitz C (2006b) Abnormalities of cortical minicolumnar organization in the prefrontal lobes of autistic patients. Clin Neurosci Res 6(3):127–133

Casanova MF, Trippe JT II, Switala AE (2007) A temporal continuity to the vertical organization of the human neocortex. Cereb Cortex 17(1):130–137

Casanova MF et al (2008) Neuronal distribution in the neocortex of schizophrenic patients. Psychiatry Res 158(3):267–277

Casanova MF, El-Baz A, Vanbogaert E, Narahari P, Switala A (2010) A topographic study of minicolumnar core width by lamina comparison between autistic subjects and controls: possible minicolumnar disruption due to an anatomical element in-common to multiple laminae. Brain Pathol 20:451–458

Casanova MF, El-Baz A, Switala AE (2011) Laws of conservation as related to brain growth, aging, and evolution: symmetry of the minicolumn. Front Neuroanat 5:66

Casanova MF, Baruth JM, El-Baz A, Tasman A, Sears L, Sokhadze EM (2012) Repetitive TMS (rTMS) modulates ERP indices of attention in autism. Transl Neurosci 3(2):170–180

Casanova MF, El-Baz A, Kamat SS, Dombroski BA, Khalifa F, Elnakib A, Soliman A, Allison-McNutt A, Andrew AE (2013) Focal cortical dysplasias in autism spectrum disorders. Acta Neuropathol Commun 1:67. doi:10.1186/2051-5960-1-67

Chance SA, Casanova MF, Switala AE, Crow TJ, Esiri MM (2006) Minicolumn thinning in temporal lobe association cortex but not primary auditory cortex in normal human ageing. Acta Neuropathol 111(5):459–464

Chance SA, Casanova MF, Switala AE, Crow TJ (2008) Auditory cortex asymmetry, altered minicolumn spacing and absence of ageing effects in schizophrenia. Brain 131:3178–3192

Chance SA, Clover L, Cousijn H, Currah L, Pettingill R, Esiri MM (2011) Microanatomical correlates of cognitive ability and decline: normal ageing, MCI, and Alzheimer's disease. Cereb Cortex 21(8):1870–1878

Constantinople CM, Bruno RM (2013) Deep cortical layers are activated directly from thalamus. Science 340(6140):1591–1594

Das A, Gilbert CD (1995) Long-range horizontal connections and their role in cortical reorganization revealed by optical recording of cat primary visual cortex. Nature 375(6534):780–784

DeFelipe J, Markram H, Rockland KS (2012) The neocortical column. Front Neuroanat 6:22

Di Rosa E, Crow TJ, Walker MA, Black G, Chance SA (2009) Reduced neuron density, enlarged minicolumn spacing and altered ageing effects in fusiform cortex in schizophrenia. Psychiatry Res 166(2–3):102–115

Dobbs D (2010) Schizophrenia: the making of a troubled mind. Nature 468:154–156

Duncan J, Johnson R, Swales M, Freer C (1997) Frontal lobe deficits after head injury: unity and diversity of function. Cognit Neuropsychol 14:713–741

Du J, Blanche TJ, Harrison RR, Lester HA, Masmanidis S (2011) Multiplexed, high density electrophysiology with nanofabricated neural probes. PLoS One 6(10), e26204

Favorov OV, Diamond ME (1990) Demonstration of discrete place-defined columns-segregates-in the cat SI. J Comp Neurol 298:97–112

Favorov OV, Diamond ME, Whitsel BL (1987) Evidence for a mosaic representation of the body surface in area 3b of the somatic cortex of cat. Proc Natl Acad Sci U S A 84(18):6606–6610

Funahashi S, Bruce CJ, Goldman-Rakic PS (1989) Mnemonic coding of visual space in the monkey's dorsolateral prefrontal cortex. J Neurophysiol 61(2):331–349

Fuster JM (2001) The prefrontal cortex–an update: time is of the essence. Neuron 30:319–333

Fuster JM (2007) Jackson and the frontal executive hierarchy. Int J Psychophysiol 64:106–107

Fuster JM, Bressler SL (2012) Cognit activation: a mechanism enabling temporal integration in working memory. Trends Cogn Sci 16(4):207–218

Gilbert CD, Wiesel TN (1989) Columnar specificity of intrinsic horizontal and corticocortical connections in cat visual cortex. J Neurosci 9:2432–2442

Goldman-Rakic PS (1996) The prefrontal landscape: implications of functional architecture for understanding human mentation and the central executive. Philos Trans R Soc Lond B Biol Sci 351:1445–1453

Goodale MA, Milner AD (1992) Separate visual pathways for perception and action. Trends Neurosci 15(1):20–25

Hampson RE, Coates TD Jr, Gerhardt G,A, Deadwyler SA (2004) Ceramic-based micro-electrode neuronal recordings in the rat and monkey. Proc Annu Int Conf IEEE Eng Med Biol Soc (EMBS) 25:3700–3703

Hampson RE, Gerhardt GA, Marmarelis V, Song D, Opris I, Santos L, Berger TW, Deadwyler SA (2012) Facilitation and restoration of cognitive function in primate prefrontal cortex by a neuroprosthesis that utilizes minicolumn-specific neural firing. J Neural Eng 9(5):056012

Hansen BJ, Chelaru MI, Dragoi V (2012) Correlated variability in laminar cortical circuits. Neuron 76(3):590–602. doi:10.1016/j.neuron.2012.08.029

Hirabayashi T, Takeuchi D, Tamura K, Miyashita Y (2013a) Microcircuits for hierarchical elaboration of object coding across primate temporal areas. Science 341(6142):191–195

Hirabayashi T, Takeuchi D, Tamura K, Miyashita Y (2013b) Functional microcircuit recruited during retrieval of object association memory in monkey perirhinal cortex. Neuron 77(1):192–203

Hirata Y, Sawaguchi T (2008) Functional columns in the primate prefrontal cortex revealed by optical imaging in vitro. Neurosci Res 61(1):1–10

Hoover JE, Strick PL (1993) Multiple output channels in the basal ganglia. Science 259 (5096):819–821
Horton JC, Adams DL (2005) The cortical column: a structure without a function. Philos Trans R Soc Lond B Biol Sci 360(1456):837–862
Hubel DH (1982) Cortical neurobiology: a slanted historical perspective. Annu Rev Neurosci 5:363–370
Hubel DH, Wiesel TN (1969) Anatomical demonstration of columns in the monkey striate cortex. Nature 221:747–750
Hubel DH, Wiesel TN (1974) Sequence regularity and geometry of orientation columns in the monkey striate cortex. J Comp Neurol 158(3):267–293
Jones EG (2000) Microcolumns in the cerebral cortex. A historical overview of the concept of micro- or minicolumns. Proc Natl Acad Sci U S A 97(10):5019–5021
Jones EG, Rakic P (2010) Radial columns in cortical architecture: it is the composition that counts. Cereb Cortex 20(10):2261–2264
Kaas JH (2012) Evolution of columns, modules, and domains in the neocortex of primates. Proc Natl Acad Sci U S A 109(Suppl 1):10655–10660
Katz LC, Gilbert CD, Wiesel TN (1989) Local circuits and ocular dominance columns in monkey striate cortex. J Neurosci 9:1389–1399
Kritzer MF, Goldman-Rakic PS (1995) Intrinsic circuit organization of the major layers and sublayers of the dorsolateral prefrontal cortex in the rhesus monkey. J Comp Neurol 359:131–143
Leise EM (1990) Modular construction of nervous systems: a basic principle of design for invertebrates and vertebrates. Brain Res Brain Res Rev 15(1):1–23
Mahan MY, Georgopoulos AP (2013) Motor directional tuning across brain areas: directional resonance and the role of inhibition for directional accuracy. Front Neural Circ 7:92. doi:10.3389/fncir.2013.00092
McCulloch WS (1959) Agatha Tyche of nervous nets – the lucky reckoners. In: McCulloch WS (ed) Embodiments of mind. MIT Press, Cambridge, 1965, pp 203–215 (Reprint of: National Physical Laboratory. Mechanisation of thought processes. London: H.M. Stationery Office, pp 611–625)
McFarland NR, Haber SN (2002) Thalamic relay nuclei of the basal ganglia form both reciprocal and nonreciprocal cortical connections, linking multiple frontal cortical areas. J Neurosci 22 (18):8117–8132
Middleton FA, Strick PLB-g (2002) Projections to the prefrontal cortex of the primate. Cereb Cortex 12(9):926–935
Miller EK, Cohen JD (2001) An integrative theory of prefrontal cortex function. Annu Rev Neurosci 24:167–202
Mo J, Schroeder CE, Ding M (2011) Attentional modulation of alpha oscillations in macaque inferotemporal cortex. J Neurosci 31(3):878–882
Mountcastle VB (1957) Modality and topographic properties of single neurons of cat's somatic sensory cortex. J Neurophysiol 20:408–434
Mountcastle VB (1978) An organizing principle for cerebral function: the unit module and the distributed system. In: Edelman GM, Mountcastle VB (eds) The mindful brain. MIT Press, Cambridge, MA, pp 7–50
Mountcastle VB (1997) The columnar organization of the neocortex. A comprehensive review of the literature indicating the modular architecture of the cortex. Brain 120(4):701–722
Mountcastle VB (1998) Perceptual neuroscience. The cerebral cortex. Harvard University Press, Cambridge
Mountcastle VB (2003) Introduction. Cereb Cortex 13:2–4
Mountcastle VB, Berman A, Davies P (1955) Topographic organization and modality representation in the first somatic area of cat's cerebral cortex by method of single unit analysis. Am J Physiol 183:464

Mountcastle VB, Berman AL, Davies P (1957) Response properties of neurons of cats' somatic sensory cortex to peripheral stimuli. J Neurophysiol 20(4):374–407
Moxon KA, Leiser SC, Gerhardt GA, Barbee KA, Chapin JK (2004) Ceramic-based multisite electrode arrays for chronic single-neuron recording. IEEE Trans Biomed Eng 51:647–656
Naya Y, Suzuki WA (2011) Integrating what and when across the primate medial temporal lobe. Science 333:773–776
Opris I, Bruce CJ (2005) Neural circuitry of judgment and decision mechanisms. Brain Res Rev 48:509–526
Opris I, Barborica A, Ferrera VP (2001) A gap effect during microstimulation in the prefrontal cortex of monkeys. Exp Brain Res 138:1–7
Opris I, Barborica A, Ferrera VP (2005) Microstimulation of dorsolateral prefrontal cortex biases saccade target selection. J Cogn Neurosci 17(6):893–904
Opris I, Hampson RE, Deadwyler SA (2009) The encoding of cocaine vs. natural rewards in the striatum of nonhuman primates: categories with different activations. Neuroscience 163 (1):40–54
Opris I, Hampson RE, Stanford TR, Gerhardt GA, Deadwyler SA (2011) Neural activity in frontal cortical cell layers: evidence for columnar sensorimotor processing. J Cogn Neurosci 23 (6):1507–1521
Opris I, Santos LM, Song D, Gerhardt GA, Berger TW, Hampson RE, Deadwyler SA (2012a) Prefrontal cortical microcircuits bind perception to executive control. Sci Rep 2013 3:2285
Opris I, Hampson RE, Gerhardt GA, Berger TW, Deadwyler SA (2012b) Columnar processing in primate pFC: evidence for executive control microcircuits. J Cogn Neurosci 24(12):2334–2347
Opris I, Fuqua JL, Huettl PF, Gerhardt GA, Berger TW, Hampson RE, Deadwyler SA (2012c) Closing the loop in primate prefrontal cortex: inter-laminar processing. Front Neural Circuits 6:88
Opris I et al (2012d) Closing the loop in primate prefrontal cortex: inter-laminar processing. Front Neurosci 6
Opris I, Santos L, Gerhardt GA et al (2013) Prefrontal cortical microcircuits bind perception to executive control. Sci Rep 3:2285. doi:10.1038/srep02285
Posner M, Snyder C (1975) Attention and cognitive control. In: Solso R (ed) Information processing and cognition: the Loyola symposium. Lawrence Erlbaum, Hillsdale
Quintana J, Fuster JM (1999) From perception to action: temporal integrative functions of prefrontal and parietal neurons. Cereb Cortex 9:213–221
Raghanti MA, Spocter MA, Butti C, Hof PR, Sherwood CC (2010) A comparative perspective on minicolumns and inhibitory GABAergic interneurons in the neocortex. Front Neuroanat 4:3
Rakic P (1988) Specification of cerebral cortical areas. Science 241(4862):170–176
Rakic P (2008) Confusing cortical columns. Proc Natl Acad Sci U S A 105(34):12099–12100
Rao SC, Rainer G, Miller EK (1997) Integration of what and where in the primate prefrontal cortex. Science 276:821–824
Rao SG, Williams GV, Goldman-Rakic PS (1999) Isodirectional tuning of adjacent interneurons and pyramidal cells during working memory: evidence for microcolumnar organization in PFC. J Neurophysiol 81:1903–1916
Ratcliff R, Cherian A, Segraves M (2003) A comparison of macaque behavior and superior colliculus neuronal activity to predictions from models of two-choice decisions. J Neurophysiol 90(3):1392–1407
Rinkus GJ (2010) A cortical sparse distributed coding model linking mini- and macrocolumn-scale functionality. Front Neuroanat 4:17
Romo R, Hernández A, Zainos A, Lemus L, Brody CD (2002) Neuronal correlates of decision-making in secondary somatosensory cortex. Nat Neurosci 5(11):1217–1225
Salinas E (2004) Fast remapping of sensory stimuli onto motor actions on the basis of contextual modulation. J Neurosci 24(5):1113–1118

Selemon LD, Goldman-Rakic PS (1988) Common cortical and subcortical targets of the dorsolateral prefrontal and posterior parietal cortices in the rhesus monkey: evidence for a distributed neural network subserving spatially guided behavior. J Neurosci 8:4049–4068

Shallice T, Burgess PW (1991) Deficits in strategy application following frontal lobe damage in man. Brain 114(2):727–741

Shallice T, Burgess P (1996) The domain of supervisory processes and temporal organization of behaviour. Philos Trans R Soc Lond B Biol Sci 351:1405–1411

Shepherd G, Grillner S (2010) Handbook of brain microcircuits. Oxford University Press, New York

Sokhadze E et al (2010) Impaired error monitoring and correction function in autism. J Neurother 14:79–95

Sokhadze E et al (2012) Prefrontal neuromodulation using rTMS improves error monitoring and correction function in autism. Appl Psychophysiol Biofeedback 37:91–102

Suyatin DB, Hallstram W, Samuelson L, Montelius L, Prinz CN, Kanje M (2009) Gallium phosphide nanowire arrays and their possible application in cellular force investigations. J Vac Sci Technol B 27:3092–3094

Swadlow HA, Gusev AG, Bezdudnaya T (2002) Activation of a cortical column by a thalamocortical impulse. J Neurosci 22:7766–7773

Swindale NV (1998) Cortical organization: modules, polymaps and mosaics. Curr Biol 8:R270–R273

Swindale NV, Shoham D, Grinvald A, Bonhoeffer T, Hübener M (2000) Visual cortex maps are optimized for uniform coverage. Nat Neurosci 3:822–826

Szentágothai J, Arbib MA (1975) Conceptual models of neural organization. MIT Press, Cambridge, MA

Takeuchi D, Hirabayashi T, Tamura K, Miyashita Y (2011) Reversal of interlaminar signal between sensory and memory processing in monkey temporal cortex. Science 331:1443–1447

Thomson AM, Bannister AP (2003) Interlaminar connections in the neocortex. Cereb Cortex 13(1):5–14

Tomasi D, Volkow ND et al (2010) Disrupted functional connectivity with dopaminergic midbrain in cocaine abusers. PLoS One 5:e10815

Ungerleider LG, Mishkin M (1982) Two cortical visual systems. In: Ingle DJ, Goodale MA, Mansfield RJW (eds) Analysis of visual behavior. MIT, Cambridge, MA, pp 549–586

Van Essen DC, Newsome WT, Bixby JL (1982) The pattern of interhemispheric connections and its relationship to extrastriate visual areas in the macaque monkey. J Neurosci 2(3):265–283

Viventi J, Kim D-H, Vigeland L, Frechette ES, Blanco JA, Kim Y-S, Avrin AE, Tiruvadi VR, Hwang S-W, Vanleer AC, Wulsin DF, Davis K, Gelber CE, Palmer L, Van der Spiegel J, Wu J, Xiao J, Huang Y, Contreras D, Rogers JA, Litt B (2011) Flexible, foldable, actively multiplexed, high-density electrode array for mapping brain activity in vivo. Nat Neurosci 14:1599–1605

Wang XJ (2012) Neural dynamics and circuit mechanisms of decision-making. Curr Opin Neurobiol 22(6):1039–1046

Wang M et al (2011) Neuronal basis of age-related working memory decline. Nature 476:210–213

Weiler N, Wood L, Yu J, Solla SA, Shepherd GM (2008) Top-down laminar organization of the excitatory network in motor cortex. Nat Neurosci 11:360–366

Wiesel TN, Hubel DH (1974) Ordered arrangement of orientation columns in monkeys lacking visual experience. J Comp Neurol 158(3):307–318

Wilson FA, O'Scalaidhe SP, Goldman-Rakic PS (1993) Dissociation of object and spatial processing domains in primate prefrontal cortex. Science 260:1955–1958

Zhang M, Alloway KD (2006) Intercolumnar synchronization of neuronal activity in rat barrel cortex during patterned airjet stimulation: a laminar analysis. Exp Brain Res 169(3):311–325

Zorzos AN, Scholvin J, Boyden ES, Fonstad CG (2012) Three-dimensional multiwaveguide probe array for light delivery to distributed brain circuits. Opt Lett 37:4841–4843

Chapter 11
Cytoarchitectural Modules: Functional Specialisation and Disruption in Neuropsychiatric Disorders

Steven Chance

Abstract Central to the concept of a modular brain is the notion that 'units' of function can be identified. Although structural units corresponding to these functions are not a necessary consequence of the model, the presence of modular structures suggests a neural basis. Three levels of structural modular organisation–cortical region, macrocolumn, minicolumn–are considered here, with a focus on the smallest units, the minicolumns, which provide an index of the specialisation and functional integrity of the larger modules. The issue of processing specialisation is discussed with regard to functions that may themselves be considered to be modular: face processing and language. The issue of functional integrity is approached through the association between pathology, minicolumn disruption and functional abnormality in Alzheimer's disease and schizophrenia. In such disorders there is evidence of pervasive functional disruption across multiple domains which suggests commonalties across modules. In dementia and old age a process of 'dedifferentiation' has been observed whereby functions become simplified and less distinct. This appears to be the opposite of 'emergent modularisation' that is observed during development. In this context, the degree of modularity of the brain may be considered to change over time, initially increasing and then decreasing across the lifespan. The widespread structural motif of the (mini)column is a simple modular component, common to most brain regions but varying in relation to regional processing biases. Systematic regional variations may develop to support emergent modular function and acquisition of expertise, or may be eroded, reflecting vulnerability to disease and loss of specialisation.

Keywords Minicolumn • Macrocolumn • Alzheimer's disease • Schizophrenia

S. Chance, D.Phil. (✉)
Neuropathology Department, University of Oxford, West Wing, Level 1 John Radcliffe Hospital, Headington, Oxford OX3 9DU, United Kingdom
e-mail: steven.chance@ndcn.ox.ac.uk

M.F. Casanova, I. Opris (eds.), *Recent Advances on the Modular Organization of the Cortex*, DOI 10.1007/978-94-017-9900-3_11

11.1 Vision and Face Processing

The visual cortex is an archetypal example of modular brain organisation. Its hierarchical, well-characterised structure has been taken as representative of cortical organisation in general, although the clarity of its systematic organisation may, in fact, be rather exceptional. The feedforward-feedback arrangement of visual processing stages from primary to secondary to tertiary and onwards, spans several cytoarchitectonically distinct brain areas. The visual cortex is also famous for its columnar functional organisation constituting modular organisation at a sub-regional scale consisting of patches or 'macrocolumns' with a diameter of approximately 500–800 μm. Smaller still are the minicolumnar units with a diameter of approximately 50 μm, which have been measured structurally in both the primary visual cortex and at the far end of the cortical visual stream in the object-specific, face processing cortex. Of these three levels of modular organisation–cortical region, macrocolumn, minicolumn–this chapter considers the evidence that the smallest units, the minicolumns, are building blocks, providing an index of the specialisation and functional integrity of the larger modules. The issue of processing specialisation is considered with regard to functions that may themselves be considered to be modular: face processing and language, by focusing on aspects of these that are asymmetrical between the cerebral hemispheres. The issue of functional integrity is approached by considering the association between pathology, minicolumn disruption and functional abnormality in Alzheimer's disease and schizophrenia.

Central to the concept of a modular brain is the notion that 'units' of function can be identified. Although structural modules corresponding to these functions are not a necessary consequence of the model, the presence of modular structures suggests a potential link and an opportunity to understand the neural basis of these functions. Indeed, past failures to match measures of structure and function may be due to attempts to match incompatible levels of modular organisation (e.g. attempting to match the variation of a small structural unit with characteristics of a function that actually emerge from the interaction of multiple larger brain regions).

In visual cortex, the most commonly investigated modular components are the macro-columns: ocular dominance and orientation preference columns. This columnar organization is dependent on excitatory competition and cell plasticity. Glutamate (NMDAR1) receptor function is required for the development of orientation preference (Ramoa et al. 2001) and manipulation of GABAergic inhibition perturbs the development of columnar architecture (Hensch and Stryker 2004). Markers of dendritic arborization also reflect macrocolumn organization during development (Fenstemaker et al. 2001). For example, SMI32 neurofilament protein is absent at ocular dominance column borders and as a result of monocular stimulus deprivation its distribution is altered. In monocularly deprived dominance columns, MAP2 is reduced, concomitant with a reduction in size of the deprived columns. Myelin staining also varies in distribution according to macro-columnar organization during development. It is more intense in ocular dominance column centers compared to the column borders and becomes darker and wider in columns dominant due to monocular stimulus deprivation.

The macro-columns observed in primary visual cortex do not have clear homologues in many other cortical regions. However, the smaller units, the minicolumns, have been measured throughout cerebral cortex. The horizontal expansion of cortical surface during development (within individual brains), and across evolutionary time (between species), is largely due to the proliferation and spacing of these radial minicolumns of cells (Rakic 1995). These structures persist throughout most cortical areas in the mature brain, where they span the 3–4 mm depth of the cortex with a horizontal width of approximately 50 μm. Column-like radial organization is found for cell bodies and their axonal and dendritic connections. Recent single unit electrophysiological recordings have demonstrated that cells within the same minicolumn share greater similarity of stimulus sensitivity than with cells in neighbouring columns (Opris et al. 2012).

Further down the ventral visual stream, the face processing area in the mid-fusiform region of the temporal cortex (approximately Brodmann area 37 in human brain), is part of the brain network supporting social cognition in humans and other primates. This area specialisation arises from expertise in certain sub-categories of visual object. The high heritability of face processing (Zhu et al. 2010) makes it plausible that there is a detectable neuroanatomical correlate. Although humans and other apes perceive faces holistically (see Taubert and Parr 2010), this process is clearly lateralised in humans—individual facial features are detected in the left hemisphere whereas holistic analysis is biased to the right hemisphere (Rossion et al. 2000). The right hemisphere is dominant for making categorical (face vs non-face) distinctions (Meng et al. 2012). It is reasonable to consider that structural asymmetry in this region may be related to its known asymmetrical function. In humans, cells have become large and less densely packed in the evolution of mid-fusiform cortex compared to the chimpanzee and this is accentuated in the left hemisphere to generate an inter-hemispheric asymmetry. In a study of minicolumn structure narrower minicolumns (and smaller neurons) were, correspondingly, found in the right hemisphere, i.e. the hemisphere that is usually dominant for face perception (Chance et al. 2013).

In the visual domain of face processing it has been proposed that wider minicolumn spacing is associated with detailed feature processing whereas narrow minicolumns may facilitate holistic, configural processing of the type usually associated with face processing (Chance et al. 2013). A mechanistic model is required to explain this. It has been suggested that greater spacing of minicolumns in human association cortex results in less-overlapping dendritic trees and allows more independent minicolumn function (Seldon 1981a, b). Applying this principle, holistic, configural processing for face recognition potentially benefits from the computational overlap generated by narrow minicolumns in the fusiform gyrus. Such a mechanistic interpretation is consistent with the correspondence between the rightward lateralisation of holistic face processing and the narrow minicolumns found in the right hemisphere in humans and is replicated in the structure-function correspondence found in the auditory domain although the processing demands of acoustic processing lead to different hemispheric dominance (described below). The hemispheric processing bias for a given task may depend, therefore, on the degree to which task success emphasizes local or global processing and the hemispheric asymmetry of minicolumnar units in the brain region associated with that functional domain. This concept refines the simple notion that, in terms of cerebral asymmetry, a larger brain area is associated with dominance for a function and offers an alternative, mechanistic explanation associated with 'processing type' (Chance et al. 2013). As a result, minicolumn width appears to be dissociated from 'dominance', per se, and instead relates to the type of processing: featural or holistic.

11.2 Auditory Cortex and Language

Functional modules are often characterized as domain specific. This specificity depends on a combination of the type of processing that is uniquely associated with the given module and the particular connections restricting the input to the module. In neural structure, the restricted inputs may be seen in the form of specific white matter connections. With regards to processing type, the specificity may also be seen, as suggested in the mechanistic model (above), in the minicolumnar organisation. The minicolumns are structural modules in the cortex that are almost ubiquitous but their systematic variation in width and spacing between brain regions reflects processing specialisation. This variation, thought to be related to the computational independence of units in the local network, confers certain processing characteristics utilized by multiple modules across different domains of processing. An example is seen in the comparison of face-processing with auditory language processing for which similar patterns of structural components confer functional biases with different hemispheric dominance depending on the domain of processing.

Auditory cortex in the superior temporal gyrus (STG) develops a clear columnar cell distribution by the third trimester of fetal life, which is established in early childhood, although axonal maturation continues up to at least 12 years of age (Moore and Guan 2001) and probably later in more associative regions. The planum temporale (PT) in the posterior STG is the most consistently asymmetrical brain area since Geschwind and Levitsky's original observations (Geschwind and Levitsky 1968). It may be further subdivided into medial, lateral and caudal parts, each associated with different aspects of speech processing (Tremblay et al. 2013). Anterior STG is sensitive to syntactic word category violation in a sentence (Friederici et al. 1993) while the posterior STG supports a left-hemisphere bias for phonological processing (e.g. Robson et al. 2012). Meanwhile, the right hemisphere auditory areas are dominant for music perception in untrained listeners (Ono et al. 2011). Therefore the evidence suggests that lateralised functions (e.g. language) often depend on multiple modular components (e.g. phonology, prosodic intonation etc.) and structural asymmetry (e.g. Sylvian fissure length) depends on structural sub-region components (e.g. anterior, posterior STG and sub-regions of PT).

In the human planum temporale, minicolumn width asymmetry is associated with surface area asymmetry (Chance et al. 2006a). Notably, the asymmetry of minicolumn spacing is absent in the equivalent areas of the brains of other apes (Buxhoeveden et al. 2001). Although the human minicolumn asymmetry is not large (Buxhoeveden et al. 2001; Hutsler 2003), it is estimated to account for a surface area asymmetry of eight to nine percent of the region's size (Chance et al. 2006a). The asymmetries appear to be amplified in more recently evolved association cortex where the phase of expansion has greater influence on region size and asymmetry (Chance and Crow 2007). Pearlson et al. (1997) have suggested that measurement of surface area is more important than volume and Barta et al. (1997) detected asymmetries by surface area measurements that were not detected by volume measures. The microscopic asymmetry in humans is also detected at the slightly larger scale of inter-connected 'macrocolumn' patches in auditory cortex which are more widely spaced in the left than in the right auditory association cortex (Galuske et al. 2000). The cells of a macro-column are selectively interconnected with those of other macro-columns which share a similar stimulus sensitivity.

The wider minicolumn spacing in the left STG is thought to facilitate fine temporal discrimination because minicolumns function as more discrete computational elements, whereas narrow minicolumn spacing in the right STG supports broad spectral processing, due to the minicolumns' greater computational overlap. In such a scheme, music processing is therefore similar to face processing, based on overlapping, holistic processing. We have found that asymmetry of the minicolumnar organisation of cells in PT reflects axonal interhemispheric connectivity through its connecting region–the isthmus–of the corpus callosum (Chance et al. 2006a). Greater asymmetry is associated with fewer axons, presumably reflecting more independent hemisphere function.

There is not such a tight association between minicolumn spacing and cortical surface area in Heschl's gyrus (HG) as there is in PT (Chance et al. 2008). Variation in HG size appears to be more dependent on the early established proliferation in number rather than spacing of minicolumns. By contrast, the size and asymmetry of PT is linked more closely to the spacing of its minicolumns. It has been proposed, therefore, that cortical surface expansion in association cortex depends more on the later developmental expansion of minicolumns after their initial proliferation than in primary sensory brain regions (Chance et al. 2008). Regional differences between HG and PT may relate to the hierarchical relationship between them in which PT is the recipient of feed-forward projections from the primary auditory area of HG and plays a role in more integrative, associative processing than HG. The two regions differ in maturation (Guillery 2005; Chance 2006; Toga et al. 2006), dendritic arborization (Elston et al. 1999), asymmetry and neuroplasticity (Arendt 2004). There are also different relationships between minicolumn asymmetry and callosal axon number for HG and PT. An increased number of interhemispheric axons is associated with a more rightward bias for PT and a leftward bias for HG. This may relate to the observation that auditory processing which varies in the temporal domain is processed preferentially by the primary auditory region in the left hemisphere whereas variation in the spectral domain is preferentially processed by the auditory associative areas in the right hemisphere (Jamison et al. 2006). Therefore, it is possible that the domain-sensitive processing bias for each region also depends on callosal interaction.

The normally faster maturing female brain (Kretschmann et al. 1979) is associated with more narrow minicolumns relative to the male brain. It appears that the prolonged development in males contributes to wider minicolumns, larger region size and greater asymmetries in the mature brain compared with that of females (Chance et al. 2008). The corpus callosum goes on developing in size later in females than in males (Cowell et al. 1992; Pujol et al. 1993) continuing through the third and fourth decades of life. Consequently, the sexes differ in their rates of maturation of cortex compared to interhemispheric connectivity. There may be an association between the arrest of axodendritic plasticity seen in the minicolumn data and the peak of callosal maturation (i.e. myelination).

11.3 Schizophrenia: The Disruption of Lateralised Modules and Semantic Memory

The auditory region offers one of the clearest associations between the psychotic symptoms of schizophrenia and brain structure as it is activated during one of the most prominent symptoms of the disorder: auditory hallucinations (Shergill et al. 2000; Ropohl et al. 2004). Reduced cortical grey matter in this area, including the PT, is one of the most replicated structural changes in the disorder (Honea et al. 2005). Minicolumn asymmetry of this region is also altered, in particular in male patients (in whom illness is usually more severe), in such a way that both hemispheres are configured more like the typical right hemisphere (Chance et al. 2008). PT inter-hemispheric connections are implicated by the changed cerebral asymmetry, altered callosal white matter (Diwadkar et al. 2004) and abnormal relationship between callosal axon number and magnopyramidal neuron density in schizophrenia (Simper et al. 2011). We have also reported disturbed associations between cytoarchitectural minicolumn asymmetry and callosal axon distribution in schizophrenia (Chance et al. 2008).

Although neuron density is reported to be unchanged, altered clustering (Beasley et al. 2005) and reduced volume of layer III pyramidal neurons have been found in this area (Sweet et al. 2003). It has been suggested that these findings implicate impaired feed-forward connections in schizophrenia (Sweet et al. 2003) perhaps affecting the interaction between other language areas including prefrontal cortex in Broca's area. There is a reported asymmetry in the size of magnopyramidal neurons in motor language cortex (Hayes and Lewis 1993) and one may expect a similar asymmetry in Wernicke's language area (including the PT), which is the sensory equivalent. Indeed, magnopyramidal neurons tend to be larger and denser in the left PT (Simper et al. 2011)

Reduced inhibitory interneuron density has been found in the PT in schizophrenia (Chance et al. 2005). Double bouquet cells are a major class of calbindin containing interneurons. The majority of calbindin-containing cells in the mature brain are double-bouquet cells with vertically oriented dendrites and axon bundles. By exercising inhibitory modulation of pyramidal cells in a columnar arrangement they make possible cohesive vertical inhibition of minicolumns. A bilateral reduction (20 %) in calbindin cell density has been found in patients (Chance et al. 2005). Loss of columnar inhibition may result in reduced minicolumnar segregation and altered cell size appears to reflect altered minicolumn size - calbindin cells are typically larger in subjects with larger minicolumns. High immunoreactivity to calbindin is also found (in layers III and IV) at the borders of macro-columns where it is thought to be associated with synaptic competition on excitatory neurons (Fenstemaker et al. 2001).

In addition to having wider minicolumns, patients do not show the normal minicolumn thinning found in old age. It has been shown that minicolumn thinning occurs during normal ageing without cell loss (Chance et al. 2006b) indicating a loss of neuropil between minicolumns that is likely to be substantially due to dendrite remodelling. Notably, the normal ageing effect is found in association cortex, including the auditory association cortex of PT, but not in the neighbouring primary auditory region of Heschl's gyrus. The absence of age associated minicolumn thinning in association cortex in schizophrenia suggests a failure of the processes of neuroplastic remodelling. Evidence from neuroimaging, electrophysiology, cytoarchitectural and molecular studies all indicate aberrant neuroplasticity in schizophrenia (Broadbelt et al. 2002; Jones et al. 2002; Stephan et al. 2006). A plausible interpretation is that, as a result of arrested plasticity, the patients retain a larger amount of unmodified neuropil compared to the equivalent elderly controls.

It is worth noting that altered minicolumn asymmetry in the PT contributes to a body of evidence reporting asymmetric changes in schizophrenia. With reference to the face processing regions discussed above, asymmetric volume reduction of fusiform cortex (McDonald et al. 2000) and an asymmetric reduction of minicolumn density in the left fusiform cortex have been reported (Di Rosa et al. 2009). In that study of minicolumns, wider and more asymmetrical minicolumns were found in males compared to females. Findings in the PT have also indicated greater anatomical asymmetry in males (e.g., Chance et al. 2006a). The reduced density of minicolumns in the fusiform cortex in schizophrenia was coincident with a low pyramidal cell density–a logical corollary of a more sparse neuronal distribution–although this was not a statistically significant correlation. These effects are consistent with the notion of anomalous development and the absence of age-associated neuropil restructuring in schizophrenia.

Disrupted minicolumn organisation, reduced inhibition and altered neuroplasticity appear to contribute to the neuropsychological deficits seen in the disorder in both auditory stimulus processing (Rojas et al. 1997) and face processing (Johnston et al. 2005). However, the abnormalities extend beyond these particular domains into multi-modal functions. Indeed, some neuropsychological models of the disorder have struggled to account for the breadth of cognitive deficits identified. Yet, the widespread effect on minicolumn structure suggests a common underlying disruption. As described, minicolumns represent simple modular components with variation related to processing biases. These biases may be scaled up to support emergent modular function in multiple domains. This formulation is consistent with the neuroconstructivist approach (Karmiloff-Smith 2009) whereby domain relevant biases in early development are shaped by experience during subsequent development into emergent 'modules'. Therefore, minicolumns may not be synonymous with functional modules, but rather, as units in a network they provide a neural correlate for the biases from which functional modules emerge.

In schizophrenia, an example may be seen in the extension of anomalies of auditory processing into the more complex emergent deficits in interpreting linguistic meaning. The loss of left hemisphere ERP mismatch responses to anomalous words at the end of a sentence, based on incongruous word meaning (Spironelli et al. 2008), provides a link between the sensory, phonological abnormalities and semantic category understanding. Patients with schizophrenia have been shown to have altered organisation of the semantic network that encodes basic knowledge about the meanings of words (Paulsen et al. 1996; Rossell et al. 1999). This is supported by several studies indicating that semantic category boundaries are less clear in schizophrenia (Paulsen et al. 1996). The interpretation of this is informed by identifying how the mechanistic description of asymmetrical minicolumn function applies to the domain of semantic processing. In healthy control subjects, a reduced discrimination between primary and secondary word meanings is typically found in the right hemisphere compared to the left (Weisbrod et al. 1998) and this lack of discrimination is consistent with the holistic, overlapping activation in the right hemisphere, as described in the sections above. It also appears to be consistent with the observation that the right hemisphere usually uses more dimensions than the left hemisphere to represent the semantic map in healthy subjects (Taylor et al. (1999)). An alteration of semantic mapping in schizophrenia is identified with less effective mapping of semantic space in low dimensions in patients. Word generation (semantic fluency) tests the integrity of the semantic network and in schizophrenia there is a requirement for more dimensions (Rossell et al. 1999; Paulsen et al. 1996).

If increased number of dimensions is taken to be indicative of more diffuse activation in the right hemisphere network in normal healthy subjects (Taylor et al. 1999), then the hypothesis that schizophrenia patients have unusually diffuse semantic associations in the left hemisphere as well as the right hemisphere (Weisbrod et al. 1998) predicts that patients use more, poorly discriminative dimensions overall. Therefore, alterations in the dimensions of conceptual space, consistent with disruption of lateralised cognitive processing biases, accompany abnormal anatomical structure of the cortex, including altered asymmetrical minicolumn cytoarchitecture in schizophrenia.

Normally a developmental shift occurs in cognition: whereas older children and adults perceive dimensions such as high and tall, or big and bright, to be separable, young children tend to confuse these concepts (Carey 1978). Goldstone and Barsalou (1998) have described the development of reasoning about dimensions: "dimensions that are easily separated by adults, such as the brightness and size of a square, are treated as fused together for children... [they] have difficulty identifying whether two objects differ on their brightness or size even though they can easily see that they differ in some way. Both differentiation and dimensionalization occur throughout one's lifetime." Differences in conceptual organisation can be interpreted as differences in the metrics underlying the psychological space (Gardenfors 2000). Gardenfors (2000) has proposed that this is a developmental shift from a Euclidean cognitive metric to the more separable dimensions of the 'city-block' metric. The development of more separable dimensions, therefore, is associated with more sophisticated cognitive discriminative ability. Aspects of brain structural maturation and plasticity presumably relate to this process of cognitive maturation. The increase in discrimination associated with separable dimensions is similar to the acquisition of expertise, which is often associated with left hemisphere specialisation for fine-grained difference judgements, e.g. for faces, word meaning and music.

The developmental shift from the Euclidean cognitive metric to the more separable dimensions of the city-block metric proposed by Gardenfors (2000) may be relevant in the neurodevelopmental context of schizophrenia. Although there is a clear genetic component in the aetiology of schizophrenia, onset of illness is not identified until adolescence or early adulthood. It has been proposed that, structurally, this may be linked to the time-course of myelination and altered minicolumn organisation (Chance et al. 2008; Crow et al. 2007). Functionally, it may be linked to the shift in cognitive metric and as dimensionalisation matures, in comparison with typically developing individuals, the anomalies associated with psychosis are exposed, leading to the recognition of 'onset' and diagnosis.

11.4 Alzheimer's Disease: Modular Pathology

It is possible that the modular organization of the cortex from minicolumns, to macrocolumns, to surface regions, determines the pattern of pathological spread in Alzheimer's disease and, consequently, the pattern of function loss. The progression of pathological changes through anatomically connected regions in AD is consistent with the concept of the disease exploiting the brain's modular organization at the regional level.

The widely accepted amyloid hypothesis of Alzheimer's disease is based on the demonstration that mutations in the gene coding for amyloid-β protein precursor (AβPP) cause AD (Goate et al. 1991). Fibrillary amyloid-β, or a precursor form of it, is considered to be toxic to neurons and capable of inducing neurofibrillary tangle (NFT) formation, although the exact mechanism by which amyloid-β causes NFT formation remains unclear. One problem with the amyloid cascade hypothesis is that it does not readily provide an explanation for the regionally selective distribution of NFT observed in AD. NFT accumulation typically appears initially and is most severe in medial temporal cortex. It has been suggested that the vulnerability of the medial temporal lobe, including hippocampus and olfactory areas may be related to the high degree of local synaptic plasticity in these brain regions. There is regional variation in the potential for neuroplasticity in the adult brain. Age-related dendritic growth is most pronounced in limbic cortical areas, while the primary sensory and motor cortex show either dendritic stability or regression (Arendt et al. 1998; Arendt 2003; Arendt 2004). An intermediate degree of dendritic remodeling occurs in association cortex. This variation in dendrites is structurally related to the regional variation in minicolumn spacing as a result of the space-filling properties of the dendritic arbors (Chance 2006).

It has been found that NFT clustering is highly correlated with symptoms: "high numbers of NFTs restricted to small areas are more important in disturbing function than NFTs in a widespread distribution (with the same mean values)" (Nagy et al. 1996). The finding that NFTs are clustered coincidentally across supragranular and infragranular layers provides an important component of the argument that the pathology is distributed to some degree in a modular and columnar fashion. Although plaques also appear to show a degree of clustering (Armstrong 1995) between dendritic clusters (Kosik et al. 1987), the proximity of symptoms to pathology distribution is less striking than it is for tangles.

As an anatomical module the macro-column has a diameter reflecting the tangential spread of afferent projections to the cortex (Seldon 1981b). NFT clustering occurs at a similar scale to that of macrocolumn size. Hiorns et al. (1991) observed that the distribution of the cells of origin of ipsilateral cortico-cortical projections to a given cortical area were clustered in bands of a similar size to macro-columns and a distribution comparable to that of NFTs in AD. Casanova (2003) has anticipated that "the spread of neurofibrillary tangles (NFT) preferentially to corticopetal neurons causes clustering of the neurodegenerative changes which reflect modular structure". The development of NFT in neurons of layers III and V is consistent with the spatial coincidence of supragranular and infragranular NFT clusters due to shared connectivity of cells in the same macro-column. Consequently NFT clustering may reflect the selective grouping of connections which is fundamental to columnar organization.

Subsequently, the issue of columnar organization and clustering has been pursued with respect to other pathological structures. Disruption of the columnar organization of glial processes has been reported for AD (Colombo et al. 2002). Lewy bodies and Pick bodies (Armstrong et al. 1997; Armstrong et al. 1998) also exhibit clustered distributions. The potency of a link between pathology that clusters and minicolumn abnormality is also borne out by the finding that minicolumns were disrupted even in the absence of an overall loss of neurons, in a group of patients with Lewy body dementia (Buldyrev et al. 2000). In AD, the finding of NFT clustering offers some insight into the roles that plasticity and connectivity play in the spread of pathology.

One of the first groups of cells to be lost in Alzheimer's disease are cholinergic neurons that normally project to the hippocampus and medial temporal region (Geula 1998; Wynn and Cummings 2004). The cholinergic system stimulates both muscarinic and nicotinic receptors and the nicotinic $\alpha 7$ receptor is the receptor for amyloid. Amyloid binding to the receptor may stimulate pathology (such as tau phosphorylation) in AD (Wang et al. 2000). Acetylcholine projections are known to innervate the cerebral cortex in a modular distribution, spanning macro-columns, and are believed to contribute to relevant stimulus detection by excitation of the stimulus-sensitive column and inhibition of the surrounding columns. Auditory cortex is particularly densely innervated by cholinergic projections, and this may contribute to this region's comparative resistance to pathology. It may also confer sensitivity to the effect of anticholinesterase medication on this region since conditioning-related responses in auditory cortex can be manipulated in humans by altering levels of acetylcholine. Acetylcholine projections from the Basal nucleus innervate a proportion of inhibitory interneurons. The innervation of nicotinic receptors on inhibitory cells in layer V excites the vertical axons of inhibitory cells to upper layers, therefore, enhancing columnar inhibition. Many of the inhibitory neurons in receipt of cholinergic projections have the columnar dendritic morphology and electrophysiological properties of double bouquet cells (Xiang et al. 1998). Calbindin in neurons appears to be relatively protective in aging, but may be lost as AD advances (Greene et al. 2001; Iritani et al. 2001). Therefore an interaction between acetylcholine, calbindin cell inhibition, and columnar organization in AD is plausible.

Another group of cortical neurons, usually pyramidal cells, has also been found to be vulnerable in AD: neurons that stain positively for the SMI-32 neurofilament protein (Hof et al. 1990; Morrison et al. 1987). By contrast a different group of neurons tends to be relatively resistant to tangle formation: neurons that stain positively for the N200 neurofilament protein (Radenahmad et al. 2003; Law and Harrison 2003). Both sub-populations of cells tend to be found in cortical layers III and V. Humans and chimpanzees show similar cytoarchitectural patterns. Notably, among the resistant N200 cells, the majority of immunolabelled neurons are found in lower layer III, and usually immunoreactivity is restricted to one pyramidal neuron in a minicolumn. Whereas, the vulnerable SMI-32 immunoreactive neurons are spread more widely in the bottom and middle areas of cortical layer III and often include several cells within the same minicolumn. The reported vulnerability of minicolumns to disruption in old age and dementia (Chance et al. 2011) may be related to this observation that SMI-32+ neurons are often grouped vertically within the same minicolumn so that pathology will damage multiple neurons that would be expected to share similar stimulus response properties (Chance et al. 2013).

The relationship of discrete brain regions to cognitive scores in aging and dementia has been examined for several cortical areas in humans. Previously, studies in aging rhesus macaque monkeys demonstrated that a correlation between minicolumn organisation and prefrontal function is selective to the region of prefrontal cortex supporting the particular functional measure rather than with another region (Cruz et al. 2009). In humans, regional minicolumn thinning and plaque load have been associated more closely with decline in different aspects of function: prefrontal cortex (PFC) minicolumns and plaque load corresponded more closely with IQ, medial temporal lobe minicolumns (parahippocampal gyrus–PHG) and plaque load corresponded more closely with MMSE score (Van Veluw et al. 2012). The timing of changes also appeared to differ between brain regions. PHG minicolumns tended to thin with increasing impairment in a steady way when compared across healthy elderly controls, mild cognitively impaired (MCI) subjects and AD patients. Whereas dlPFC minicolumns did not reduce between controls and MCI, and were only thinner in more advanced AD (Van Veluw et al. 2012). The argument has been made that PHG minicolumn thinning occurs during early impairment, accompanying the overall cognitive decline observed in mildly impaired and AD patients. By contrast, the late change in dlPFC is consistent with a later spread of pathology to this region and with a delayed effect on IQ. There is also a coincident pattern of regional plaque load and minicolumn thinning. PHG plaque load was shown to be increased in MCI, and further in AD. Similarly, plaque load in the dlPFC was not increased in MCI and was only substantially greater in more severe AD. This relationship between minicolumn thinning and increasing plaque load is consistent with a relationship detected originally in normal aging (Chance et al. 2006b). The findings have confirmed that significant interactions may be found between cognition and minicolumn width that are independent of overall brain size. Such microanatomical effects related to regional function rather than general brain atrophy in humans are consistent with Cruz et al.'s (2009) assertion of the specific regional, functional significance of minicolumn integrity in other primates.

In normal aging a distributed synaptic loss may be absorbed as cortical shrinkage with only a subtle effect on cognitive performance. However, a threshold appears to be reached that marks the boundary between normal decline with aging and the onset of dementia. In AD, synapse loss exceeds that which can be accounted for by neuron loss through NFT formation (Hof and Morrison 2004). The minicolumn structure of the cortex is defined by the contacts between its constituent cells and it represents a larger unit that can compensate for the lost connections of a proportion of its individual cells. However, as more synapses and dendrites are lost the column structure will begin to break down. In monkeys minicolumns in the frontal cortex become more disorganized with increasing age and this disorganization is correlated with age-related cognitive decline (Cruz et al. 2004). In normal human aging, there is minicolumn narrowing in middle temporal and superior temporal auditory association cortex, while the normally more narrowly spaced mini-columns of primary auditory cortex show little change (Chance et al. 2006b). A threshold will be reached sooner in areas of cortex with minimal redundancy at which point the coordinated connectivity of mini-columnar units may confer a weakness as it provides a basis for the focal clustering of pathological features with the result that the clustering of tangles is positively correlated with the degree of minicolumn disruption (Buldyrev et al. 2000).

Thinning of minicolumns is therefore associated with reduced cognitive function. However, narrow minicolumns do not necessarily indicate functional deficit and in young, healthy adults regional variation in minicolumns may relate to normal processing biases in those regions as discussed above (e.g. Jung-Beeman 2005). The correspondence between regional specialization and microanatomy suggests a possible neural basis for the process of "dedifferentiation" in old age (Goh 2011). By this general process, brain regions that show specialized responses for specific cognitive processes in young adults tend to become less specialized, and respond more similarly. Goh (2011) has described reduced ability at face discrimination as an instance of "dedifferentiation" in old age and Di Rosa et al. (2009) have found marked minicolumn thinning in old age in the fusiform gyrus which is the structure that contains the fusiform face area. Meanwhile, as described, some regions of cortex which already have relatively narrow minicolumns do not show further thinning in normal old age (Chance et al. 2006b). Therefore, the extent of age associated shrinkage and its accompanying functional deficit may depend on the degree of developmental expansion and processing specialisation for that region, with implications for the effect of "dedifferentiation" for that region. The process of dedifferentiation has also been reported for the prefrontal cortex (Cabeza 2002).

The development, through education and training, of alternative problem solving strategies likely to be associated with PFC function, has been linked to enhancement of cognitive reserve (Steinerman 2010). In this context, the thinning of minicolumns in the PFC may reflect the loss of the initial neural reserve (e.g. loss of neuropil and/or neuronal connections) in the early stages of aging. The cognitive reserve hypothesis proposes that the cognitive deficit due to accretion of pathology in AD may be offset either by 'neural reserve' wherein brain networks can absorb a degree of damage without noticeable effects on cognition or by active 'neural compensation' whereby damaged networks are supplemented by recruitment of additional resources (Stern 2009). There is evidence that individuals with high IQ are better able to cope with pathological aging, enabling preservation of cognitive abilities (Lindenberger and Baltes 1994; Tucker and Stern 2011), probably due to plastic reorganization of neurocognitive networks. The relationship with plaques rather than tangles suggests the possibility that, in addition to 'neural reserve' and 'neural compensation', the reserve may be associated with 'neural resistance' to the spread of tangle pathology (the feature of pathology that is most closely linked to disease progression), while at the same time, the using-up of reserve is associated with a cost indicated by plaque accumulation.

11.5 Conclusion

Advances in behavioural measurement and neuropsychology have clarified that many, apparently unitary, cognitive functions are composed of multiple sub-components. Functional neuroimaging has demonstrated that frequently the same brain region may be recruited to sub-serve multiple different behaviours. Consequently, the definitions of modules and their boundaries have become increasingly uncertain. Similarly, the investigation of brain disorders such as schizophrenia and dementia (but also autism and others not discussed here) has found that there is evidence of pervasive functional disruption across multiple domains often not considered in the original diagnostic criteria. To address this, in the field of developmental disorders, the concept of neuro-constructivism has been adopted by some researchers as a useful description of the emergent modularity that develops over the early lifespan and which may explain how disruption across apparently diverse modules may emerge from an original disruption in the common, underlying cognitive architecture (Karmiloff-Smith 2009). In the field of neurodegenerative conditions, 'cognitive reserve' is referred to as a common, perhaps multi-modal, cognitive resource available for flexible deployment to bolster degenerating faculties (Esiri and Chance 2012). Both cognitive reserve and emergent modularity indicate an underlying flexibility in the use of cognitive resources even while constrained by some degree of specificity in functional modules. A similar combination of uniformity and difference is also found at the anatomical level as cytoarchitectural boundaries are identifiable even while almost all of the cerebral cortex shares the minicolumnar structural motif with few other differences (for example, there is little evidence of regional difference in the presence of distinct cell types). Yet there are patterns of variation in columnar structures which suggest biases that relate to functional specialisation. The relationship between structural units and processing biases may reflect certain underlying rules of information processing consistent with generalizations identified across different modalities in psychometric studies (Shepard 1987). The commonalities may be modified to diverge during development, or to converge by dedifferentiation in old age, or by disease. The widespread structural motif of the (mini) column represents a simple modular component with variation related to processing bias which appears to be scaled up to support emergent modular function, which may be eroded to generate vulnerability to disease and loss of specialisation, or may be flexibly recruited to contribute to cognitive reserve and expertise.

References

Arendt T (2003) Synaptic plasticity and cell cycle activation in neurons are alternative effector pathways: the Dr. Jekyll and Mr. Hyde concept of Alzheimer's disease or the yin and yang of neuroplasticity. Prog Neurobiol 71:83–248

Arendt T (2004) Neurodegeneration and plasticity. Int J Dev Neurosci 22:507–514

Arendt T, Bruckner MK, Gertz HJ, Marcova L (1998) Cortical distribution of neurofibrillary tangles in Alzheimer's disease matches the pattern of neurons that retain their capacity of plastic remodelling in the adult brain. Neuroscience 83:991–1002

Armstrong RA (1995) Is the clustering of beta-amyloid (A beta) deposits in the frontal cortex of Alzheimer patients determined by blood vessels? Neurosci Lett 195:121–124

Armstrong RA, Cairns NJ, Lantos PL (1997) Dementia with Lewy bodies: clustering of Lewy bodies in human patients. Neurosci Lett 224:41–44

Armstrong RA, Cairns NJ, Lantos PL (1998) Clustering of pick bodies in patients with Pick's disease. Neurosci Lett 242:81–84

Barta PE, Pearlson GD, Brill LB 2nd, Royall R, McGilchrist IK, Pulver AE et al (1997) Planum temporale asymmetry reversal in schizophrenia: replication and relationship to gray matter abnormalities. Am J Psychiatry 154:661–667
Beasley CL, Chana G, Honavar M, Landau S, Everall IP, Cotter D (2005) Evidence for altered neuronal organisation within the planum temporale in major psychiatric disorders. Schizophr Res 73:69–78
Broadbelt K, Byne W, Jones LB (2002) Evidence for a decrease in basilar dendrites of pyramidal cells in schizophrenic medial prefrontal cortex. Schizophr Res 58(1):75–81
Buldyrev SV, Cruz L, Gomez-Isla T, Gomez-Tortosa E, Havlin S, Le R, Stanley HE, Urbanc B, Hyman BT (2000) Description of microcolumnar ensembles in association cortex and their disruption in Alzheimer and Lewy body dementias. Proc Natl Acad Sci U S A 97:5039–5043
Buxhoeveden DP, Switala AE, Litaker M, Roy E, Casanova MF (2001) Lateralization of minicolumns in human planum temporale is absent in nonhuman primate cortex. Brain Behav Evol 57:349–358
Cabeza R (2002) Hemispheric asymmetry reduction in older adults: the HAROLD model. Psychol Aging 17(1):85–100
Carey S (1978) The child as a word learner. In: Halle M, Bresnan J, Miller G (eds) Linguistic theory and psychological reality. MIT Press, Cambridge, MA
Casanova MF (2003) Modular concepts of brain organization and the neuropathology of psychiatric conditions. Psychiatry Res 118:101–102
Chance SA (2006) Subtle changes in the ageing human brain. Nutr Health 18:217–224
Chance SA, Crow TJ (2007) Distinctively human: cerebral lateralisation and language in Homo sapiens. J Anthropol Sci 85:83–100
Chance SA, Walker M, Crow TJ (2005) Reduced density of calbindin immunoreactive interneurons in the planum temporale in schizophrenia. Brain Res 1046:32–37
Chance SA, Casanova MF, Switala AE, Crow TJ (2006a) Minicolumnar structure in Heschl's gyrus and planum temporale: asymmetries in relation to sex and callosal fiber number. Neuroscience 143(4):1041–1050
Chance SA, Casanova MF, Switala AE, Crow TJ, Esiri MM (2006b) Minicolumn thinning in temporal lobe association cortex but not primary auditory cortex in normal human ageing. Acta Neuropathol 111:459–464
Chance SA, Casanova MF, Switala A, Crow TJ (2008) Auditory cortex asymmetry, altered minicolumn spacing and absence of ageing effects in schizophrenia. Brain 131(12):3178–3192
Chance SA, Clover L, Cousijn H, Currah L, Pettingill R, Esiri MM (2011) Micro-anatomical correlates of cognitive ability and decline: normal ageing, MCI and Alzheimers disease. Cereb Cortex 21(8):1870–1878
Chance SA, Sawyer EK, Clover LM, Wicinski B, Hof PR, Crow TJ (2013) Hemispheric asymmetry in the fusiform gyrus distinguishes Homo sapiens from chimpanzees. Brain Struct Funct 218:1391–1405
Colombo JA, Quinn B, Puissant V (2002) Disruption of astroglial interlaminar processes in Alzheimer's disease. Brain Res Bull 58:235–242
Cowell PE, Allen LS, Zalatimo NS, Denenberg VH (1992) A developmental study of sex and age interactions in the human corpus callosum. Brain Res Dev Brain Res 66:187–192
Crow TJ, Paez P, Chance SA (2007) Callosal misconnectivity and the sex difference in psychosis. Int Rev Psychiatry 19(4):449–457
Cruz L, Roe DL, Urbanc B, Cabral H, Stanley HE, Rosene DL (2004) Age-related reduction in microcolumnar structure in area 46 of the rhesus monkey correlates with behavioral decline. Proc Natl Acad Sci U S A 101:15846–15851
Cruz L, Roe DL, Urbanc B, Inglis A, Stanley HE, Rosene DL (2009) Age-related reduction in microcolumnar structure correlates with cognitive decline in ventral but not dorsal area 46 of the rhesus monkey. Neuroscience 158(4):1509–1520

Di Rosa E, Crow TJ, Walker MA, Black G, Chance SA (2009) Reduced neuron density, enlarged minicolumn spacing and altered ageing effects in fusiform cortex in schizophrenia. Psychiatry Res 166:102–115

Diwadkar VA, DeBellis MD, Sweeney JA, Pettegrew JW, Keshavan MS (2004) Abnormalities in MRI-measured signal intensity in the corpus callosum in schizophrenia. Schizophr Res 67:277–282

Elston GN, Tweedale R, Rosa MG (1999) Cellular heterogeneity in cerebral cortex: a study of the morphology of pyramidal neurones in visual areas of the marmoset monkey. J Comp Neurol 415(1):33–51

Esiri MM, Chance SA (2012) Cognitive reserve, cortical plasticity and resistance to Alzheimer's disease. Alzheimers Res Ther 4(2):7. doi:10.1186/alzrt105

Fenstemaker SB, Kiorpes L, Movshon JA (2001) Effects of experimental strabismus on the architecture of macaque monkey striate cortex. J Comp Neurol 438:300–317

Friederici AD, Pfeifer E, Hahne A (1993) Event-related brain potentials during natural speech processing: effects of semantic, morphological and syntactic violations. Brain Res Cogn Brain Res 1:183–192

Galuske RA, Schlote W, Bratzke H, Singer W (2000) Interhemispheric asymmetries of the modular structure in human temporal cortex. Science 289:1946–1949

Gardenfors P (2000) Conceptual spaces: the geometry of thought. MIT Press, Cambridge, MA

Geschwind N, Levitsky W (1968) Human brain: left-right asymmetries in temporal speech region. Science 161:186–187

Geula C (1998) Abnormalities of neural circuitry in Alzheimer's disease: hippocampus and cortical cholinergic innervation. Neurology 51:S18–S29; discussion S65–67

Goate A, Chartier-Harlin MC, Mullan M, Brown J, Crawford F, Fidani L, Giuffra L, Haynes A, Irving N, James L et al (1991) Segregation of a missense mutation in the amyloid precursor protein gene with familial Alzheimer's disease. Nature 349:704–706

Goh JOS (2011) Functional dedifferentiation and altered connectivity in older adults: neural accounts of cognitive aging. Aging Dis 2(1):30–48

Goldstone RL, Barsalou LW (1998) Reuniting perception and conception. Cognition 65:231–262

Greene JR, Radenahmad N, Wilcock GK, Neal JW, Pearson RC (2001) Accumulation of calbindin in cortical pyramidal cells with ageing; a putative protective mechanism which fails in Alzheimer's disease. Neuropathol Appl Neurobiol 27:339–342

Guillery RW (2005) Is postnatal neocortical maturation hierarchical? Trends Neurosci 28:512–517

Hayes TL, Lewis DA (1993) Hemispheric differences in layer III pyramidal neurons of the anterior language area. Arch Neurol 50:501–505

Hensch TK, Stryker MP (2004) Columnar architecture sculpted by GABA circuits in developing cat visual cortex. Science 303:1678–1681

Hiorns RW, Neal JW, Pearson RC, Powell TP (1991) Clustering of ipsilateral cortico-cortical projection neurons to area 7 in the rhesus monkey. Proc Biol Sci 246:1–9

Hof PR, Morrison JH (2004) The aging brain: morphomolecular senescence of cortical circuits. Trends Neurosci 27:607–613

Hof PR, Cox K, Morrison JH (1990) Quantitative analysis of a vulnerable subset of pyramidal neurons in Alzheimer's disease: I. Superior frontal and inferior temporal cortex. J Comp Neurol 301:44–54

Honea R, Crow TJ, Passingham D, Mackay CE (2005) Regional deficits in brain volume in schizophrenia: a meta-analysis of voxel-based morphometry studies. Am J Psychiatry 162 (12):2233–2245

Hutsler JJ (2003) The specialized structure of human language cortex: pyramidal cell size asymmetries within auditory and language-associated regions of the temporal lobes. Brain Lang 86(2):226–242

Iritani S, Niizato K, Emson PC (2001) Relationship of calbindin D28K-immunoreactive cells and neuropathological changes in the hippocampal formation of Alzheimer's disease. Neuropathology 21:162–167

Jamison HL, Watkins KE, Bishop DV, Matthews PM (2006) Hemispheric specialization for processing auditory nonspeech stimuli. Cereb Cortex 16:1266–1275

Johnston PJ, Stojanov W, Devir H, Schall U (2005) Functional MRI of facial emotion recognition deficits in schizophrenia and their electrophysiological correlates. Eur J Neurosci 22 (5):1221–1232

Jones LB, Johnson N, Byne W (2002) Alterations in MAP2 immunocytochemistry in areas 9 and 32 of schizophrenic prefrontal cortex. Psychiatry Res 114(3):137–148

Jung-Beeman M (2005) Bilateral brain processes for comprehending natural language. Trends Cogn Sci 9:512–518

Karmiloff-Smith A (2009) Nativism versus neuroconstructivism: rethinking the study of developmental disorders. Dev Psychol 45(1):56–63

Kosik KS, Rogers J, Kowall NW (1987) Senile plaques are located between apical dendritic clusters. J Neuropathol Exp Neurol 46:1–11

Kretschmann HJ, Schleicher A, Wingert F, Zilles K, Loblich HJ (1979) Human brain growth in the 19th and 20th century. J Neurol Sci 40:169–188

Law AJ, Harrison PJ (2003) The distribution and morphology of prefrontal cortex pyramidal neurons identified using anti-neurofilament antibodies SMI32, N200 and FNP7. Normative data and a comparison in subjects with schizophrenia, bipolar disorder or major depression. J Psychiatr Res 37(6):487–499

Lindenberger U, Baltes PB (1994) Sensory functioning and intelligence in old age: a strong connection. Psychol Aging 9:339–355

McDonald B, Highley JR, Walker MA, Herron BM, Cooper SJ, Esiri MM, Crow TJ (2000) Anomalous asymmetry of fusiform and parahippocampal gyrus gray matter in schizophrenia: a post-mortem study. Am J Psychiatr 157(1):40–47

Meng M, Cherian T, Singal G, Sinha P (2012) Lateralization of face processing in the human brain. Proc R Soc B. doi:10.1098/rspb.2011.1784

Moore JK, Guan YL (2001) Cytoarchitectural and axonal maturation in human auditory cortex. J Assoc Res Otolaryngol 2:297–311

Morrison JH, Lewis DA, Campbell MJ, Huntley GW, Benson DL, Bouras C (1987) A monoclonal antibody to non-phosphorylated neurofilament protein marks the vulnerable cortical neurons in Alzheimer's disease. Brain Res 416:331–336

Nagy Z, Esiri MM, Jobst KA, Morris JH, King EM, McDonald B, Litchfield S, Barnetson L (1996) Clustering of pathological features in Alzheimer's disease: clinical and neuroanatomical aspects. Dementia 7:121–127

Ono K, Nakamura A, Yoshiyama K, Kinkori T, Bundo M, Kato T, Ito K (2011) The effect of musical experience on hemispheric lateralization in musical feature processing. Neurosci Lett 496(2):141–145

Opris I, Hampson RE, Gerhardt GA, Berger TW, Deadwyler SA (2012) Columnar processing in primate pFC: evidence for executive control microcircuits. J Cogn Neurosci 24(12):2334–2347

Paulsen JS, Romero R, Chan A, Davis AV, Heaton RK, Jeste DV (1996) Impairment in the semantic network in schizophrenia. Psychiatry Res 63:109–121

Pearlson GD, Barta PE, Powers RE, Menon RR, Richards SS, Aylward EH et al (1997) Ziskind-Somerfeld research award 1996. Medial and superior temporal gyral volumes and cerebral asymmetry in schizophrenia versus bipolar disorder. Biol Psychiatry 41:1–14

Pujol J, Vendrell P, Junque C, Marti-Vilalta JL, Capdevila A (1993) When does human brain development end? Evidence of corpus callosum growth up to adulthood. Ann Neurol 34:71–75

Radenahmad N, Neal JW, Wilcock GW, Pearson RCA (2003) A neurofilament antibody recognizes a subset of pyramidal cells in the human neocortex that are preserved in Alzheimer's disease. Neuropathol Appl Neurobiol 29:316–320

Rakic P (1995) A small step for the cell, a giant leap for mankind: a hypothesis of neocortical expansion during evolution. Trends Neurosci 18:383–388

Ramoa AS, Mower AF, Liao D, Jafri SI (2001) Suppression of cortical NMDA receptor function prevents development of orientation selectivity in the primary visual cortex. J Neurosci 21:4299–4309

Robson H, Keidel JL, Ralph MA, Sage K (2012) Revealing and quantifying the impaired phonological analysis underpinning impaired comprehension in Wernicke's aphasia. Neuropsychologia 50:276–288

Rojas DC, Teale P, Sheeder J, Simon J, Reite M (1997) Sex-specific expression of Heschl's gyrus functional and structural abnormalities in paranoid schizophrenia. Am J Psychiatry 154:1655–1662

Ropohl A, Sperling W, Elstner S, Tomandl B, Reulbach U, Kaltenhauser M et al (2004) Cortical activity associated with auditory hallucinations. Neuroreport 15:523–526

Rossell SL, Rabe-Hesketh S, Shapleske J, David AS (1999) Is semantic fluency differentially impaired in schizophrenic patients with delusions? J Clin Exp Neuropsychol 21(5):629–642

Rossion B, Dricot L, Devolder A, Bodart J-M, Crommelinck M, de Gelder B, Zoontjes R (2000) Hemispheric asymmetries for whole-based and part-based face processing in the human fusiform gyrus. J Cogn Neurosci 12(5):793–802

Seldon HL (1981a) Structure of human auditory cortex: I. Cytoarchitectonics and dendritic distributions. Brain Res 229:277–294

Seldon HL (1981b) Structure of human auditory cortex: II. Axon distributions and morphological correlates of speech perception. Brain Res 229:295–310

Shepard RN (1987) Toward a universal law of generalization for psychological science. Science 237(4820):1317–1323

Shergill SS, Brammer MJ, Williams SC, Murray RM, McGuire PK (2000) Mapping auditory hallucinations in schizophrenia using functional magnetic resonance imaging. Arch Gen Psychiatry 57:1033–1038

Simper R, Walker MA, Black G, Di Rosa E, Crow TJ, Chance SA (2011) The relationship between callosal axons and cortical neurons in the planum temporale: alterations in schizophrenia. Neurosci Res 71(2011):405–410

Spironelli C, Angrilli A, Stegagno L (2008) Failure in language lateralization in schizophrenia patients: an ERP study on early linguistic components. J Psychiatry Neurosci 33(3):235–243

Steinerman JR (2010) Minding the aging brain: technology-enabled cognitive training for healthy elders. Curr Neurol Neurosci Rep 10:374–380

Stephan KE, Baldeweg T, Friston KJ (2006) Synaptic plasticity and dysconnection in schizophrenia. Biol Psychiatry 59(10):929–939

Stern Y (2009) Cognitive reserve. Neuropsychologia 47:2015–2028

Sweet RA, Pierri JN, Auh S, Sampson AR, Lewis DA (2003) Reduced pyramidal cell somal volume in auditory association cortex of subjects with schizophrenia. Neuropsychopharmacology 28:599–609

Taubert J, Parr LA (2010) Geometric distortions affect face recognition in chimpanzees (Pan troglodytes) and monkeys (Macaca mulatta). Anim Cogn 14:35–43

Taylor K, Brugger P, Weniger D, Regard M (1999) Qualitative hemispheric differences in semantic category matching. Brain Lang 70:119–131

Toga AW, Thompson PM, Sowell ER (2006) Mapping brain maturation. Trends Neurosci 29:148–159

Tremblay P, Deschamps I, Gracco VL (2013) Regional heterogeneity in the processing and the production of speech in the human planum temporale. Cortex 49:143–157

Tucker AM, Stern Y (2011) Cognitive reserve in aging. Curr Alzheimer Res 8(4):354–360

Van Veluw SJ, Sawyer EK, Clover L, Cousijn H, De Jager C, Esiri MM, Chance SA (2012) Prefrontal cortex cytoarchitecture in normal aging and Alzheimer's disease: a relationship with IQ. Brain Struct Funct 217(4):797–808

Wang HY, Lee DH, Davis CB, Shank RP (2000) Amyloid peptide Abeta(1–42) binds selectively and with picomolar affinity to alpha7 nicotinic acetylcholine receptors. J Neurochem 75:1155–1161

Weisbrod M, Maier S, Harig S, Himmelsbach U, Spitzer M (1998) Lateralized semantic and indirect semantic priming effects in people with schizophrenia. Br J Psychiatry 172:142–146

Wynn ZJ, Cummings JL (2004) Cholinesterase inhibitor therapies and neuropsychiatric manifestations of Alzheimer's disease. Dement Geriatr Cogn Disord 17:100–108

Xiang Z, Huguenard JR, Prince DA (1998) Cholinergic switching within neocortical inhibitory networks. Science 281:985–988

Zhu Q, Song Y, Hu S, Li X, Tian M, Zhen Z, Dong Q, Kanwisher N, Liu J (2010) Heritability of the specific cognitive ability of face perception. Curr Biol 20(2):137–142

Chapter 12
The Relevance of Subplate Modifications to Connectivity in the Cerebral Cortex of Individuals with Autism Spectrum Disorders

Jeffrey J. Hutsler and Thomas Avino

Abstract Autism spectrum disorders (ASDs) have been widely described as a neurodevelopmental condition characterized by connectional changes within the cerebral cortex. These alterations to connectivity are thought to underlie many of the behavioral and neurophysiological findings that characterize the disorder. Suggestions of altered connectivity are not unique to ASD, but have also been hypothesized as being a key component in epilepsy, Down syndrome, and schizophrenia. Here we discuss other disorders characterized by disconnectivity and ASD's neuropathological relationship to these disconnectivity syndromes. One potentially important understudied contributor to disconnectivity in ASD is the cortical subplate. The subplate is a neurodevelopmental compartment that precedes the establishment of the cerebral cortex and that has been shown to be responsible for guiding a multitude of long-range connections within the brain. Although largely transitory during development, some subplate neurons persist into adulthood where they may play a role in gating activity within the overlying cortex and establishing synchronized activity between cortical regions. Findings in ASD suggest that the subplate may be an important contributor to alterations in cortical organization, but the complete nature of this contribution has yet to be explored.

Keywords Autism spectrum disorders • Neuropathology • Disconnectivity • Subplate

Autism spectrum disorders (ASDs) have been widely described as a neurodevelopmental condition that is characterized by connectional changes within the cerebral cortex (see Kana et al. 2011 for a review). These alterations are thought to be the basis for many of the behavioral and physiological findings that characterize the disorder. Suggestions of altered connectivity are not unique to ASD, but

---

J.J. Hutsler (✉) • T. Avino
Department of Psychology, Program in Neuroscience, University of Nevada, MS 296 1664 N Virginia Ave, Reno, NV 89557-0296, USA
e-mail: jhutsler@unr.edu; tavino@unr.edu

M.F. Casanova, I. Opris (eds.), *Recent Advances on the Modular Organization of the Cortex*, DOI 10.1007/978-94-017-9900-3_12

have also been hypothesized as a key component in epilepsy, Down syndrome, and schizophrenia (Penzes et al. 2011). In each of these conditions altered connectivity has been proposed to underlie changes in the computational architecture of the cortex and these altered computations produce the atypical motor, sensory and cognitive behaviors that are such a prominent characteristic of these conditions. Connectional change in autism is thought to take two simple forms: in the first, theories of long-range *functional* underconnectivity are meant to account for a lack of synchrony, coherence and cooperation between distant cortical regions, while in the second, theories of short-range overconnectivity are an attempt to explain behavioral phenomenon such as weak central coherence, hypersensitivity to sensory stimuli and superior performance in certain types of cognitive tasks, such as visual search. In theory, these two phenomenon can co-exist, as overconnectivity within local regions of cortex could reduce the influence of long range connections either by altering the balance of inputs onto individual neural units, or creating excess noise (Fellous et al. 2003).

Despite the popularity of these theories, the structural and neurodevelopmental basis for these hypotheses are underspecified. Here we will focus on neuroanatomical evidence that supports theories of alterations in both long-range and short-range connections with a focus on postmortem tissue findings from our own lab demonstrating structural alterations to cortical synapses in ASD. We will then focus on one potential neurodevelopmental mechanism for these changes: alterations to the cortical subplate. This focus is based on (1) observations of alterations to the subplate zone in ASD subjects, (2) the critical role of the subplate for the development of cortical connectivity, and (3) the presence of subplate abnormalities in other disorders that have been characterized as involving connectional change.

12.1 ASD as a Disconnectivity Syndrome

Functional underconnectivity between distant cortical locations has been discussed extensively as a defining characteristic of ASD (for example, Cherkassky et al. 2006; Courchesne and Pierce 2005; Frith 2004; Just et al. 2004, 2007, 2012), and a variety of studies have been cited in support of long-range connectional changes within the cortex of ASD individuals. Functional imaging studies have shown a reduction in coordinated activity between distant cortical regions (Just et al. 2004) and an absence of the top-down modulation of early sensory processing that is found in typically developing individuals (Frith 2004). Electrophysiological recordings show reduced coherence across electrode sites (Murias et al. 2007; Coben et al. 2008; Dinstein et al. 2011) and, behaviorally, long-range connectional changes may be linked to difficulties with integration of information into its wider context for higher-level meaning, a problem characteristic of ASD (Frith 2004; Geschwind and Levitt 2007). According to Just et al. (2004), functional underconnectivity is measured by the time correlation, or synchrony, of cortical network activity, and the core deficits of autism are those that require a high degree of coordination between distant cortical regions.

Disconnectivity, as a syndrome within the cerebral cortex, has been most intensely studied in callosotomy patients (Gazzaniga 2005) where partial or complete disconnection between the cerebral hemispheres is acquired as an adult and can have the beneficial effect of reducing seizure frequency (Asadi-Pooya et al. 2008). In contrast, disconnectivity in developmental disorders likely involves multiple populations of long-range connections, including both inter- and intrahemispheric connections, as well as connections between the cortex and subcortical locations. Rather than being imposed on the adult neuroarchitecture, connectivity changes that arise during neurodevelopment may forever alter the nature of neurobiological networks. One clear example of disconnectivity that arises in development is agenesis of the corpus callosum (ACC), a rare condition that only occurs in approximately two out of every 10,000 live births (Gośćik and Kulak 2011).

Interestingly, children with ACC and children with ASD show similar social deficits, that include problems maintaining conversations, as well as difficulties in both the development of peer relationships and comprehension of social etiquette (Badaruddin et al. 2007). Based on narratives of pictures depicting people in various social situations, individuals with ACC also show deficiencies imagining and inferring the mental, emotional, and social functioning of others, suggesting impairments in the cognitive construct known as theory of mind (Turk et al. 2010). Communication abilities are also abnormal in individuals with ACC (O'Brien 1994). Eighty-six percent of ACC subjects who do have speech, exhibit echolalia and the automatic, meaningless repetition of words and phrases. These symptoms are also found frequently in ASD (Badaruddin et al. 2007). Finally, problems with accurately and rapidly shifting attention have also been shown to occur in both developmental conditions (Reuter-Lorenz et al. 1990; Courchesne et al. 1994).

Although the corpus callosum and white matter are largely intact in ASD subjects, magnetic resonance imaging (MRI) studies of the corpus callosum in ASD have demonstrated alterations to its size, its shape, and its level of anisotropy (Chung et al. 2004; Casanova et al. 2011; Hardan et al. 2000; Piven et al. 1997). In high-functioning individuals with ASD, the anterior portions of the CC are significantly smaller relative to neurotypical subjects (Hardan et al. 2000). This regionally specific finding corresponds well with studies suggesting that cognitive abilities associated with the frontal lobes are specifically impaired in individuals with ASD (Courchesne et al. 2004). These include deficits in working memory (Luna et al. 2007), executive functioning (O'Hearn et al. 2008), theory of mind abilities (Sabbagh 2004), and the inability to suppress inappropriate responses (Minshew et al. 1999).

A wealth of recent diffusion tensor imaging (DTI) evidence also indicates that the fiber content within the corpus callosum is significantly altered in ASD. DTI studies have revealed smaller CC volumes, low anisotropy, and increased radial diffusivity (Shukla et al. 2010; Alexander et al. 2007; Barnea-Goraly et al. 2004). These measures may reflect underlying changes to myelin and/or the density and size of axons (Hong et al. 2011). Interestingly, decreased anisotropy is related to social impairments in high functioning children with ASD (Noriuchi et al. 2010), and these alterations are somewhat regionally specific, with the most notable changes being in the anterior third of the corpus callosum (Hong et al. 2011). Magnetization transfer imaging, a MRI technique sensitive to the presence of myelin, has also revealed abnormal myelin development in the corpus callosum, which may also contribute to altered interhemispheric connectivity in autism (Gozzi et al. 2012). Together, these results provide structural evidence for disrupted cortical connectivity in autism, particularly between the two cerebral hemispheres. Despite the clearly documented presence of these gross morphological alterations, little is known about how specific populations of fibers within the callosum are affected. Changes to cross-sectional area can occur without changes to total fiber numbers either by altering the diameter or the spacing of individual fibers. In addition, the presence of both large and small diameter fibers within the CC makes it difficult to establish a clear relationship between cross sectional area and overall fiber number (Aboitiz et al. 1992; also see Lamantia and Rakic 1990).

In typically developing subjects, the corpus callosum contains approximately 200 million interhemispheric axons that are distributed amongst a range of diameters and that can be either myelinated or unmyelinated (Aboitiz et al. 1992). These subpopulations of axons include a small number of large, myelinated, fast-conducting fibers, that likely participate in bringing distant cortical regions into synchronization (Uhlhaas et al. 2009), as well as larger populations of medium and slow-communicating fibers that likely support detailed information transfer (Singer and Gray 1995). In addition, because the CC is organized topographically from anterior to posterior, alterations in specific regions of the callosum may indicate that specific cortical regions are preferentially impacted by alterations in interhemispheric communication (Aboitiz et al. 1992; Piven et al. 1997).

The behavioral effects of acquired callosotomy have been documented extensively and have demonstrated a number of impairments and unique abilities in this population (Gazzaniga 2005). For example, callosotomy patients have been shown to process bilateral stimuli at the same level of performance as unilaterally presented objects due to a lack of interference between the hemispheres. Other disorders characterized by disconnectivity could also show reduced effects of hemispheric interference, and we have recently demonstrated that high-functioning children with ASD show behavioral patterns that are remarkably similar to those documented in adult callosotomy patients for both a simple redundancy gain task, and the simultaneous evaluation of lateralized moving patterns. In the latter task, subjects must observe a moving pattern in each visual field and then, after a brief delay, determine whether a single, lateralized test pattern matches the pattern that was previously presented in that location. The initial bilateral patterns are either the same or different. Neurotypical subjects exhibit a large cost when evaluating bilateral patterns that differ, relative to bilateral patterns that are the same. In contrast, both callosotomy patients (Holtzman and Gazzaniga 1985) and ASD subjects (Jung and Hutsler, in submission), although showing poorer performance when the patterns are the same, show almost no additional costs when evaluating patterns that differ (see Fig. 12.1). In callosotomy patients this pattern of results has been attributed to the independence of each disconnected cerebral hemisphere (Holtzman and Gazzaniga 1985).

Although the axonal content of the corpus callosum has not been directly examined in ASD, postmortem studies of frontal lobe white matter have shown both alterations to myelin and changes in axon populations (Zikopoulos and Barbas 2010). These studies point directly to reductions in the largest myelinated fibers, which likely correspond to long-range connections. Although further studies of the microanatomy of the white matter in ASD are needed to fully evaluate the potential structural bases of functional underconnectivity, the corpus callosum (CC) provides an ideal location to examine this issue because of its predictable fiber orientation, topographic organization, and lack of axons that form local connections.

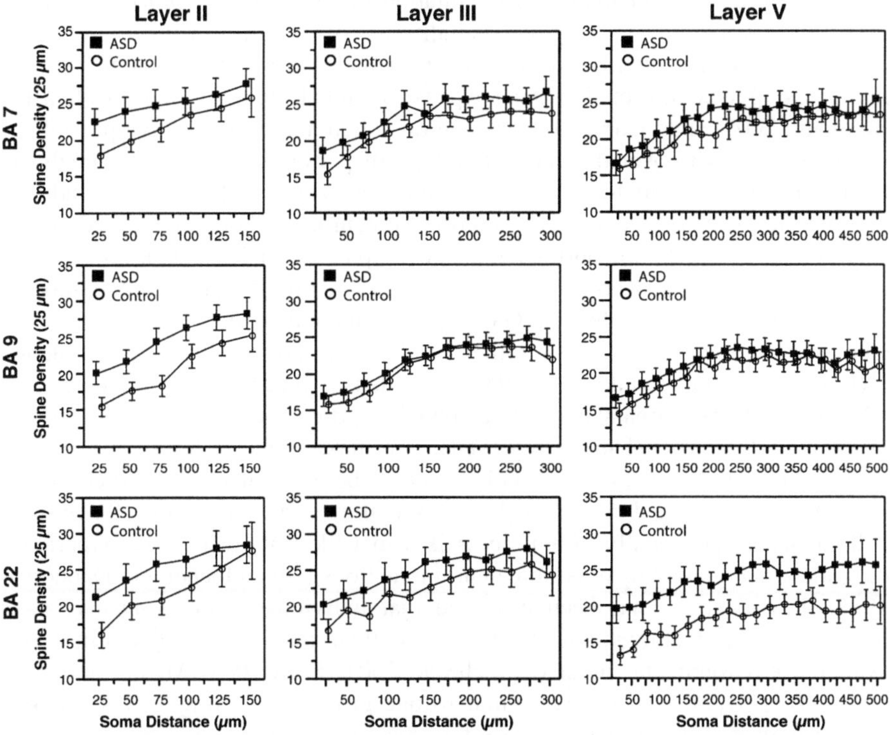

Fig. 12.1 Averaged spine densities on pyramidal cell apical dendrites as a function of distance from the cell body, cortical region, and cortical layer for both ASD and control subjects. Error bars indicate the standard error of the mean for each value. ASD subjects showed consistently elevated values within superficial layer II of parietal (BA 7), frontal (BA 9) and temporal (BA 21) regions (Reprinted from Hutsler and Zhang (2010) with permission from Elsevier)

12.2 Overconnectivity in ASD

While long-range functional underconnectivity has received a great deal of focus in ASD, it has also been suggested that ASD individuals may demonstrate short-range overconnectivity (Courchesne and Pierce 2005; Hughes 2007). Structural imaging studies have reported an increased volume in white matter compartments associated with short-range connections (Herbert et al. 2003, 2004).

In addition, this type of connectional change has been proposed to co-occur alongside of deficiencies in long-range cortico-cortical connections in ASD and include increases in short-range connections, as well as increased connections between subcortical areas and the cortex (Mizuno et al. 2006). This pattern of connectional change is partially supported by functional imaging studies showing a disconnection between distant cortical regions and studies demonstrating increased activity in regions associated with early cortical processing (Just et al. 2004; Koshino et al. 2005). It has been proposed that this connectional architecture may result in the absence of top-down modulation of early sensory processing (Frith 2004) and an inability to integrate spatially distant information (Frith 2004; Just et al. 2004). In addition, local overconnectivity has also been tied to improvements in processing local information at the expense of global structure (Baron-Cohen and Belmonte 2005) and theories of weak central coherence (Frith and Happé 1994).

The spines that cover the dendritic arbors of pyramidal cells in mammals are thought to support excitatory connections and have been widely used as an index of presynaptic excitatory connectivity onto these neurons in both animal and human studies (Fiala et al. 2002). During human development, dendritic spines first appear on neurons within the deep cortical layers in the late prenatal period (Huttenlocher and Dabholkar 1997; Michel and Garey 1984). In contrast, neurons in the superficial layers do not express spines until several months postnatally (Koenderink and Uylings 1995; Koenderink et al. 1994). Spine density reaches a maximum between 12 and 36 months postnatal depending upon the cortical location (Huttenlocher and Dabholkar 1997; Michel and Garey 1984). This maximum is followed by a gradual decline in spine numbers, which is believed to be associated with the culling of unused connections and the establishment of mature cortical networks. Since spine culling occurs postnatally, this process is also widely thought to be experience dependent and guided by environmental interactions (Huttenlocher and Dabholkar 1997). Previously, it has been proposed that individuals with ASD may have impaired synapse elimination attributable to their inability to fully utilize this available environmental input (Courchesne 2004; Frith 2004; Mundy and Neal 2001). This hypothesis is very similar to one proposed for fragile X syndrome, another pervasive developmental disorder where over-expression of synaptic spines has been demonstrated using Golgi methods in postmortem human tissue (Irwin et al. 2000, 2001; Sabaratnam 2000).

In ASD, we examined synaptic spines on pyramidal neurons located within the superficial and deep cortical layers of frontal, temporal, and parietal eulaminate isocortex (Brodmann's Areas 7, 9, and 21; Hutsler and Zhang 2010). Spine densities were averaged across the distance from the cell body to directly compare three cortical layers (II, III, and V), and three morphological types of dendrites (apical, basilar, and oblique). Using these highly averaged values, ASD subjects demonstrated higher average spine densities when compared to age-matched controls. Both layer and dendrite type also had a significant effect on spine density. Pyramidal cells from the superficial layers (II and III) had greater spine densities than those from the deeper layer, and spine densities were greatest on small, oblique dendrites as compared to basilar, but not apical dendrite densities.

Spine densities as a function of distance from the pyramidal cell soma were evaluated exclusively on apical dendrites, since these can often be followed for some distance from the cell body. As expected, densities varied according to distances from the cell body, but there was no interaction between region and distance. Spine densities were again found to be higher in ASD subjects as compared to controls, and this effect did not interact with distance. Within layer II, ASD and control subjects differed from each other in spine density, and these differences were consistent across regions and at varying distances from the cell body (see Fig. 12.2). The only other layer region combination that showed increased spine densities in ASD subjects was layer V of the temporal lobe.

In our sample, density differences between neurotypical and ASD subjects were driven largely by a subgroup of cases. Only seven of the ten ASD cases showed increased averaged spine densities relative to age-matched control cases, and in one instance the magnitude of this difference was very small. Available case histories allowed the assessment of seizure disorders, secondary medical conditions, medications, educational history, level of cognitive functioning, and the severity of specific domains of diagnosis as assessed by a postmortem ADI-R (Lord et al. 1994).

When level of cognitive functioning was considered, each of the four lowest-functioning cases showed increased spine densities relative to their age-matched control. Of the cases identified as either not retarded or mildly retarded, only one had greater spine densities. In two cases, the level of cognitive functioning was unknown, and one showed higher spine densities while the other did not. Many developmental disorders, such as fetal alcohol syndrome (Ferrer and Galofré 1987), severe infant protein-calorie malnutrition (Benítez-Bribiesca et al. 1999), infant brain damage (Dietzmann and von Bossanyi 1994), and Down syndrome (Suetsugu and Mehraein 1980), show spine loss rather than spine increases. These spine reductions are presumed to be associated with connectional loss, abnormal cortical circuits and impaired cognitive abilities (Fiala et al. 2002; Halpain et al. 2005). There are only a few conditions associated with mental retardation where an increase in spine densities has been shown, including fragile X syndrome (Irwin et al. 2001) and hemi-megalencephaly (Takashima et al. 1991). Fragile X and ASD show some notable similarities. Like ASD, fragile X is more common in males than females (Demark et al. 2003), and it shares several cognitive and behavioral traits (Kaufmann et al. 2004). Overall brain size is also larger in individuals with fragile X (Sabaratnam 2000). Estimates of the co-occurrence of fragile X in ASD subjects range from 1.6 to 16 % (Demark et al. 2003), but none of the ASD subjects in our

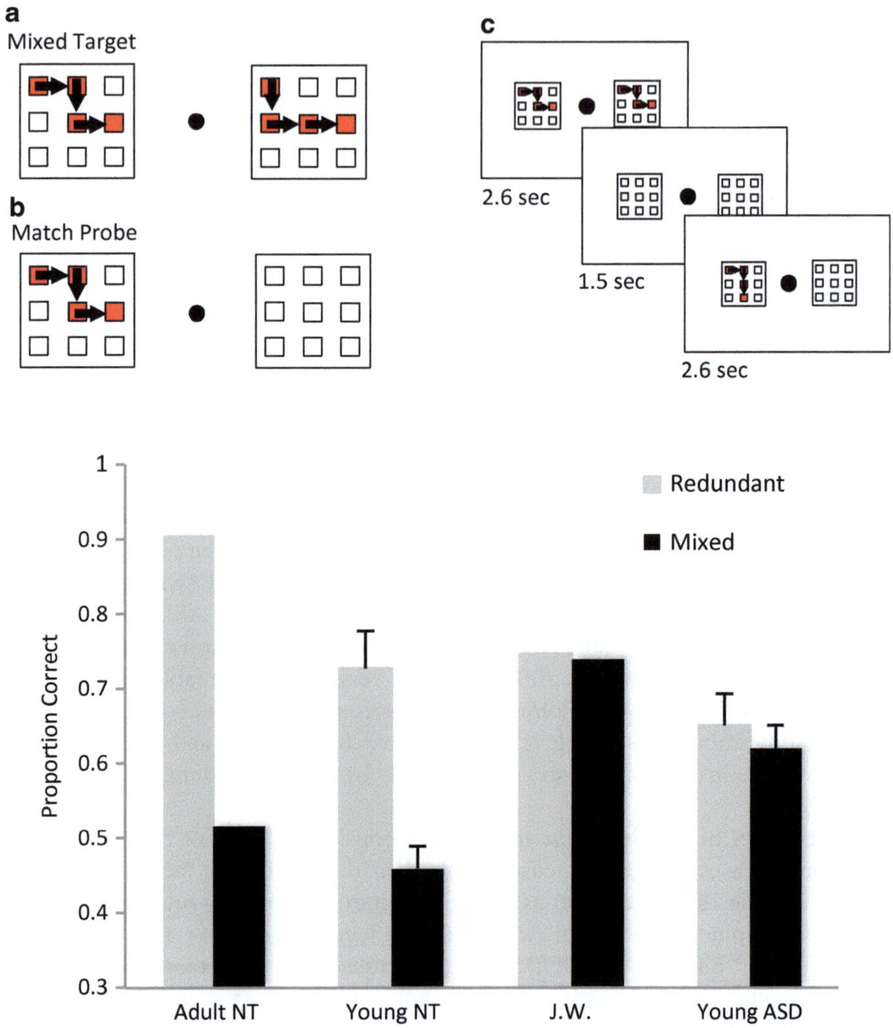

Fig. 12.2 *Top:* Subjects are presented with bilateral 3 × 3 targets that are either mixed or the same, and then, following a 1,500 msec delay, subjects are shown a single match probe and asked to determine it matches the previously presented target. *Bottom:* Comparisons of percent correct according to whether the initial target was mixed or the same (redundant). Data for NT adults and callosotomy patient JW (From Holtzman and Gazzaniga 1985, error bars are unavailable). Neurotypical adult performance drops to chance when the targets are mixed and perform close to ceiling when they are the same. Patient JW gets fewer correct overall, but shows almost no cost when the targets are mixed relative to when they are the same. Young ASD subjects and age-matched NTs, show a pattern that is very similar to that demonstrated in callosotomy patient JW

sample carried a comorbid diagnosis of fragile X. Although fragile X and ASD may not share similar causative antecedents, it is interesting that two developmental disorders sharing greater spine densities also share similar behavioral phenotypes and can coexist in the same individual (Demark et al. 2003).

Greater spine densities were also loosely associated with smaller brain sizes, while in our control subjects there was no relationship between spine density and brain size. Although average brain sizes in ASD populations are consistently enlarged in younger individuals between the ages of 1 and 7 years (Courchesne et al. 2003), the adult ASD population is characterized by high variance in brain size, which includes a 20 % rate of megencephaly (Bailey et al. 1993; Filipek et al. 1992; Piven et al. 1995) and average values that do not differ significantly from control groups (Courchesne et al. 1999). As ASD subjects age there are significant reductions in total brain volume (Aylward et al. 2002) and accelerated morphometric changes (Hardan et al. 2004).

Because spines are plastic, many additional factors could contribute to alterations in their density. For example, in both the ASD and control group spine densities decreased with age, although this reduction was not significant. Interestingly, higher spine density tended to be associated with the smallest brains in our ASD group, but not in neurotypical subjects. Spine densities also decreased slightly with increasing PMI time in our control group, but increased slightly in our ASD group and these relationships were nonsignificant.

Spine densities can be reduced by the presence of seizure activity (Multani et al. 1994; Swann et al. 2000), as well as by the neuroleptics used in the treatment of epilepsy (Benes et al. 1985; Garey et al. 1998). In our sample, four of the ten ASD subjects had a history of seizures early in development that were treated with anti-convulsant medications (phenytoin, divalproex sodium, lorazepam, carbamazepine, or phenobarbital). Two members of this group showed greater spine densities relative to their age-matched control, while the other two showed spine densities similar to controls. Although it is unclear whether a relationship exists between spine density, seizure frequency, seizure type, or the medications utilized to maintain seizure control, it is interesting to note that half of the epilepsy cases did not show higher spine densities. Whether medication may have masked an effect in these cases is unknown.

The temporal lobes showed a region-specific elevation in spine density within layer V (Fig. 12.2), a finding that supports a variety of studies implicating the temporal lobes as one of the primary loci of impairment in ASD. Finally, alterations in spine densities within the ASD group did not appear to be associated with a history of epileptic seizure activity, although a weak relationship may exist between high spine densities, the smallest brain sizes, and lower levels of cognitive functioning.

Alterations in spine densities within the deep layers of the temporal lobe suggest that this region may have additional impairments in ASD subjects. Although no single neural structure is associated with ASD, both limbic areas and mesial temporal lobe structures have been identified as having significant neuroanatomical abnormalities in ASD groups (Bauman and Kempter 1985; Bauman 1991, 1996;

Bauman and Kemper 1994). These regions are heavily interconnected with adjacent temporal regions (Amaral and Price 1984), and abnormalities in temporal-limbic circuits might be associated with deficits in facial and emotional recognition, as well as associated social-emotional functioning (Critchley et al. 2000; Schultz et al. 2000). These regions also show abnormal activation in auditory studies, theory of mind tasks (Baron-Cohen et al. 1999), and the processing of simple (Gage et al. 2003) and complex speech-related sounds (Just et al. 2004; Boddaert et al. 2004; Gervais et al. 2004). Finally, volumetric and morphometric alterations to temporal lobe structures have also been reported (Boddaert et al. 2004; Bailey et al. 1998; Bigler et al. 2003). In aggregate, these findings point to substantial deficits in temporal lobe circuits that underlie some of the behavioral symptoms that characterize ASD patients.

In sum, ASD is among a small group of developmental disabilities involving mental retardation where there is no loss of dendritic spines. Instead, a substantial subgroup of ASD individuals show increased spine densities relative to age-matched controls. This alteration in cortical circuitry organization is not directly associated with the occurrence of epilepsy but is associated with decreased brain size. In addition, higher spine densities are most often found in the lowest functioning subset of cases examined. The presence of supernumerary spines and increased local white matter volumes, in combination with a lack of functional coherence between distant cortical regions, suggests a model of synapse formation in ASD where experience-dependent strengthening and weakening of neuronal interconnections during the postnatal period is impaired (Mundy and Neal 2001; Courchesne 2004; Frith 2004). Finally, these alterations to the connectional patterns that are established postnatally in the developing human brain may have additional implications for the cognitive strengths and impairments demonstrated in ASD (Happe and Frith 2006) and the success of early behavioral intervention programs (Dawson et al. 2000; Goldstein et al. 2002) that specifically impact the time frame in which the process of synaptic culling is the most active.

Because almost all excitatory neurotransmission in the cortex occurs at synaptic spines, alterations to the number and/or morphology of these spines could disrupt the balance between excitation and inhibition. An increased number of immature spines (much like that of FXS) could cause decreased excitation across networks leading to hypoconnectivity. In fact, an imbalance in the excitation inhibition ratio has been hypothesized to underlie ASD and altered connectivity (LeBlanc and Fagiolini 2011). Animal models for altered neuroligin expression have also been shown to disrupt the balance between excitation and inhibition, which has important implications for synaptic transmission and the motor stereotypies seen in ASD (Hines et al. 2008; Blundell et al. 2010).

Alterations to synaptic density and/or morphology have been found in a number of neurological disorders, including Alzheimer's disease (AD), schizophrenia, Fragile X syndrome (FXS), and autism spectrum disorders. In AD, for example, researchers note prominent synaptic loss in the cortex of these patients (Baloyannis et al. 2007), which has even been shown to correlate with cognitive severity (DeKosky and Scheff 1990). Furthermore, researchers have also implicated synapse loss within the hippocampus at an early stage of disease progression in individuals with mild AD (Scheff et al. 2007). Similar results have been found in the schizophrenic brain where experimenters have identified a marked decrease in dendritic spines in both the auditory cortex (Sweet et al. 2009) and the prefrontal cortex (Glantz and Lewis 2000). As Penzes et al. (2011) note, Alzheimer's disease and schizophrenia have disruptions at the synapse in which synaptic loss is consistently found; however FXS and ASD are characterized by a marked increase in dendritic spines. FXS, a disorder highly comorbid with ASD, has been shown to result in an overall increase of dendritic spines as well as a larger proportion of immature spines (Irwin et al. 2001).

Animal models and molecular studies further implicate the synapse as being highly involved in the disruption seen in ASDs. Namely, the neuroligin-neurexin pairing is responsible for the maintenance and function of the synapse with neuroligins presenting at the post-synaptic site and neurexins presenting at the pre-synaptic site. These cell adhesion molecules have been shown to be abnormally expressed in human autistic genetic studies (Jamain et al. 2003; Laumonnier et al. 2004). Furthermore, knockout studies of neuroligins in mice provide a neuroanatomical basis for some of the behavioral phenotypes seen in ASDs with some researchers claiming increased inhibitory synaptic transmission for neuroligin-3 (Tabuchi et al. 2007) and decreased excitatory transmission for neuroligin-1 deletion (Blundell et al. 2010). Other synapse-specific molecules have been implicated in the disorder, such as Shank3 (Durand et al. 2007). These studies provide converging evidence for a disruption at the synapse as being a potentially important factor in the development of ASD, however alterations at these specific markers may be present in only a small subset of the larger ASD population (O'Roak et al. 2012).

12.3 Synaptic Morphology

Although synaptic spines are increased in ASD, the presence of this alteration does not indicate the nature of the pre- and postsynaptic relationship. Spines possess morphological characteristics that have been associated with specific functional roles. For example, variation in shape and length, and the presence or absence of a spine head, have been used to inform the nature of the pre- and postsynaptic relationship at individual dendritic spines. The size of the spine head appears to have a direct relationship with the strength of the synapse such that large heads indicate a strong, consolidated synapse (Yuste and Majewska 2001). Characteristics such as the thickness of the spine neck may underlie calcium compartmentalization allowing spines to independently control biochemical fluxes in and out of the dendrite. Lastly, spine length decreases follow prolonged stimulation and the associated induction of long-term potentiation (Yuste 2011). This previous work informs the stability of the synapse such that large, thick spines are thought to reflect stable synaptic connections, whereas long, thin spines are indicative of immature or malleable connections (Woolfrey et al. 2009; Holtmaat et al. 2005; Irwin et al. 2001).

Importantly, both width and length of the spine control calcium by creating a barrier from the dendrite through which chemical compartmentalization can be independently controlled. These characteristics of the spine can be experience dependent (Nikonenko et al. 2005), are relatable to maturity of the spine (Bourne and Harris 2007; Irwin et al. 2001), and have even been implicated in several neurological disorders (Penzes et al. 2011).

Dendritic spines were evaluated in ASD for length and morphological characteristics, such as thickness and the presence of a spine head. Previous research (Irwin et al. 2001) has subdivided the morphology of dendritic spines into eight different types, including several morphologies that were rarely found in our tissue samples. Most of these spine types are based upon the morphological characteristics of spine thickness and the presence of a synaptic head. In order to encompass most of the variations in these previous systems, we used a three-part classification scheme: length, presence or absence of a spine head, and whether the spine was thick or thin. Using this method we collected morphological data on over 20,000 synaptic spines from three cortical regions and from pyramidal cells located in three cortical layers. Our results revealed that the morphology of apical dendritic spines in ASD subjects differ from neurotypical subjects in two key respects. First, although spines length varied tremendously in both groups, in the ASD cases were longer on average and the distribution of their lengths was positively skewed indicating a greater frequency of long, immature morphologies (see Fig. 12.3).

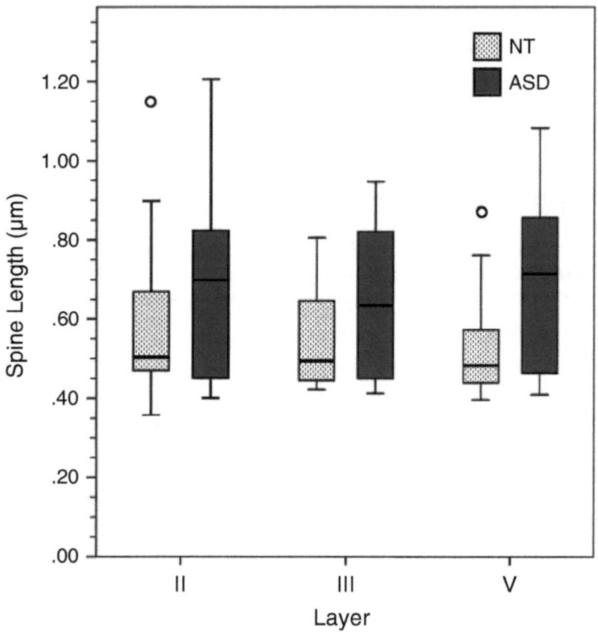

Fig. 12.3 A comparison of spine lengths as a function of cortical layer for NT subjects and individuals with ASD. ASD subjects show positively skewed distributions and overall greater spine lengths relative to neurotypical subjects. *Boxes* are the interquartile range and median values, while the whiskers show the 5th and 95th percentiles

Second, along with the increased lengths, we encountered a greater preponderance of spines with heads, which has also been described as a characteristic of immature spine morphologies (Avino and Hutsler, in prep.).

Spines were reliably longer in the ASD group and immature spine types were found more frequently in ASD subjects relative to neurotypical controls. Given what is known about the relationship of spine morphology to physiology, these results suggest that, along with the previously reported density increase, there is an overall reduction in the proportion of stable, strengthened spines in the autistic brain. A failure of proper synaptic spine maturation may underlie the deviation from neurotypical spine morphology. Like spine density changes, these disruptions likely reflect abnormal neurodevelopmental maturation and/or abnormal experience-dependent processes. Given the strong link between spine structure and function, both an increase in spine density with concurrent dysmorphology likely contribute to the connectional changes seen in the disorder and point towards the synapse as a major factor in the autistic phenotype.

12.4 Cortical Connectivity and the Subplate

The cortical subplate appears prior to the cortical plate during development and plays a critical role in guiding cortical connectivity. Although many of the neurons within the cortical plate undergo apoptosis after they have served their developmental function, in primates many are retained and appear as a diffuse band of neurons located directly subjacent to cortical layer VI. These neurons maintain connections with the overlying cortex and are integrated into the mature cortical circuitry.

Within 2 months of gestation, subplate neurons are co-generated with cells that will become the Cajal-Retzius cells of layer I, making them amongst the earliest born neurons of the cerebral cortex (Allendoerfer and Shatz 1994). Subplate and Cajal-Retzius cells are also the earliest active neurons in the presumptive cortex with endogenous spontaneous activity that begins around 8 weeks gestation and extends until 18 weeks of gestation. The two cell groups are split into an upper (layer I) and a lower (subplate) compartment by migrating neuroblasts that form the early cortical plate.

Between 13 and 18 weeks gestation the subplate expands in volume due to axonal ingrowth from a variety of fiber types. These include catecholaminergic and serotoninergic afferents originating from the brainstem, cholinergic afferents from the basal forebrain, glutamatergic afferents from the thalamus, and glutamatergic afferents from other cortical locations. Between 19 and 23 weeks of gestation thalamocortical fibers begin to drive activity in the subplate population. It is not until between 23 and 24 weeks of gestation that these afferents will penetrate the overlying cortical plate (Shatz and Luskin 1986). As such, during this time period subplate neurons exhibit mature morphologies and physiological properties, while the layers of the developing cortical plate are still immature (McAllister 1999; Luhmann et al. 2009).

In addition to the suplate's role in guiding thalamocortical axons, these neurons guide efferent connections from the cerebral cortex. Anterograde DiI labeling has demonstrated in the cat that cortical plate axons project to the thalamus at a relatively late stage in development, well after subplate neurons have made projections into the thalamus. Studies examining the influence of subplate neurons on 'pioneering' the corticothalamic pathway from visual cortex have utilized kainic acid injections into the subplate zone followed by H-leucine labeling. In 50 % of the animals that received these lesions, pathways from the visual cortex to subcortical targets including the thalamus were disrupted (McConnell et al. 1994).

Excessive neuronal profiles within the subcortical white matter have been casually reported in neuropathological examinations of ASD cortex for some time (Simms et al. 2009; Bailey et al. 1998; Hutsler et al. 2007). Subplate neuron numbers and densities are difficult to quantitatively assess for two reasons. First, the density of neuronal profiles within the subplate zone is low which creates difficulties for widely accepted stereological sampling methods. Second, the boundary between layer VI of the cerebral cortex and the subplate is often unclear, making it difficult to reliably draw a boundary around the region. Trained raters may disagree upon the boundary placement depending upon their susceptibility to the perceptual pull of excess profiles and their distribution within the white matter. In cases where the boundary is indistinct the problem is exacerbated (Hutsler and Avino 2013). To circumvent these difficulties we quantified the changes is density, as a function of depth, in images centered on the gray-white matter boundary in both ASD and neurotypical subjects. To quantify the transition we fit sigmoid functions to these density profiles (see Fig. 12.4), and found that ASD subjects had indistinct transition zones relative to a NT control group. This finding was apparent in all of the regions examined, which included eulaminate isocortex from frontal, temporal, and parietal locations (Avino and Hutsler 2010). Currently, we are attempting to directly count the number of profiles using computer-generated sigmoid based boundaries in combination with immunuhistochemical labeling for Nuen and several additional antibodies known to morphologically and neurochemical identify this cell population.

Excess subplate neurons are not unique to neuropathological descriptions of ASD. Supernumerary profiles have also been found in the dorsolateral prefrontal cortex, parahippocampal gyrus, and the superior temporal gyrus of individuals with schizophrenia (Eastwood and Harrison 2003, 2005). Friedlander and Torres-Reveron (2009) describe inhibitory subplate neurons in the mature brain as 'gate-keepers' for cortical signals. An excess of these inhibitory cells in schizophrenia may result in functional disconnectivity between frontal and limbic areas (Kostovic et al. 2011).

Epilepsy, which is commonly comorbid with autism, also shows heterotopic white matter neurons as well as other focal cortical dysplasias (Hildebrandt et al. 2005; Emery et al. 1997). This characteristic in individuals with epilepsy suggests that persistent white matter neurons in autism could be a contributor to

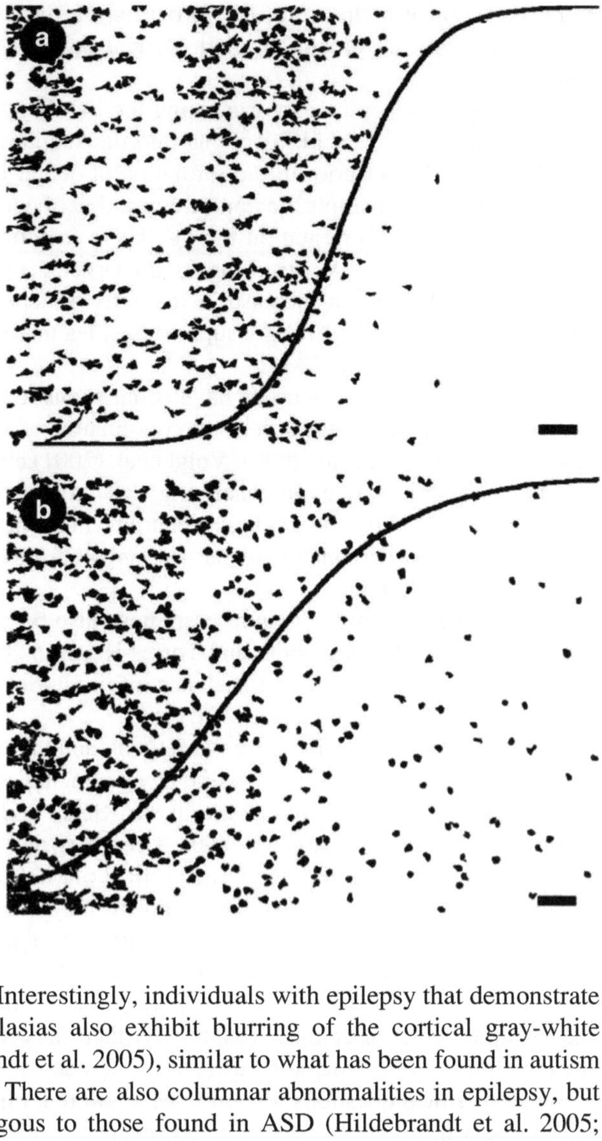

Fig. 12.4 Quantification of the transition zone between *cortical gray* (*left-hand side*) and *white* matter (*right-hand side*) using overlaid sigmoid functions in binary images. Neurotypical subject boundaries (**a**) were typically more distinct than those found in ASD subjects (**b**). Scale bar = 100 μm (Reprinted from Avino and Hutsler 2010 with permission from Elsevier)

comorbid seizure activity. Interestingly, individuals with epilepsy that demonstrate type I focal cortical dysplasias also exhibit blurring of the cortical gray-white matter boundary (Hildebrandt et al. 2005), similar to what has been found in autism (Avino and Hutsler 2010). There are also columnar abnormalities in epilepsy, but these may not be homologous to those found in ASD (Hildebrandt et al. 2005; Casanova et al. 2002).

Subplate neurons serve as an active hub where axons destined for the cerebral cortex accumulate prior to invading the overlying cortical layers (Shatz and Luskin 1986). They are an intermediary through which thalamocortical axons activate the cortex and, as development proceeds, many subplate-layer IV connections are dissolved as thalamic axons begin to make their connections to layer IV directly. In line with this sequence of events, computational modeling studies within the visual system suggest that crude visual maps first develop within the subplate via the retinogeniculate-subplate pathway (Grossberg and Seidman 2006). Activity from the subplate orchestrates the formation of visual maps within layer IV, and then these maps are taught to other cortical layers until most of the subplate is dissolved.

The subplate's key role in organizing the overlying cortex has also been demonstrated using kainic acid injections into the subplate after thalamic axons have invaded layer IV, but prior to the formation of ocular dominance columns. Kainic acid excitotoxicity disrupts the normal formation of ocular dominance columns and results in the loss of orientation tuning. These injections also disrupt the laminar organization of layer IV, resulting in mistargeted thalamic terminations within layers II and III, and an increase in the activity-dependent expression of BDNF and GAD. These results strongly support the idea that the subplate is a regulator of cortical activity (Kanold 2004).

In the adult, subplate neurons have connections to the overlying cortical plate, and may participate as "amplifiers" or coordinators of afferent signals (Suarez-Sola et al. 2009; Luhmann et al. 2009). Voigt et al. (2001) elaborated a functional role of subplate neurons as one of generating synchronous oscillatory activity that is dependent upon GABAergic subplate neurons. In the newborn rat, calcium-imaging studies have also demonstrated that the subplate mediates synchronous activity (Hanganu et al. 2009).

Subplate neurons in the mature brain have also been described as 'gatekeepers' for cortical signals (Friedlander and Torres-Reveron 2009) and regulators of afferent signals that modulate the balance of inhibition and excitation (Kostovic et al. 2011; LeBlanc and Fagiolini 2011).

In sum, experimentation in a variety of animal models has shown that the subplate plays a critical role in several developmental processes, including guiding thalamocortical and corticothalamic axons (Shatz and Luskin 1986), influencing columnar structure (Kanold et al. 2003; Grossberg and Seitz 2003), and generating synchronous activity in the overlying cortical plate (Luhmann et al. 2009; Suarez-Sola et al. 2009; Voigt 1989). The subplate may also play an important role in regulating connectivity in the mature brain and abnormalities of this neuronal compartment may drive disruptions in long-range connections. Because abnormalities of the cortical subplate in autism may provide a microanatomical substrate for the disrupted cortical connectivity theory, there has recently been an increase in interest in this embryonic compartment (McFadden and Minshew 2013).

Several critical questions remain to be answered. First, are the neuronal profiles within the white matter truly subplate neurons, or are they vestiges of cortical neurons that failed to fully migrate? A partial answer to this question is already available, as Golgi staining in the ASD brain demonstrates that many of these profiles have multistellar and horizontally fusiform morphologies, a characteristic of the neurotypical subplate population (Avino and Hutsler, in prep.). Second, what proportion of the subplate population in ASD is GABAergic? Immunohistochemistry for several antigens that are known subplate markers can readily provide this information, and this is an especially important question given previously proposed theories of disconnectivity in the schizophrenic brain that are dependent upon an increased prevalence of inhibitory subplate neurons. Finally, how does this excess of white matter neurons arise during development? Are they a product of failed prenatal apoptosis, or are cells overproduced prior to the formation of the cortical plate and subsequent apoptosis? This question is, of course, difficult to answer in human postmortem material and will require a valid animal model of ASD to properly assess.

Although an early pathology in ASD that may underlie subsequent deviations from neurotypical development is certainly of interest, there is also a practical aspect to further elucidation of the role of the subplate in autism. The subplate region can already be distinguished from the intermediate zone using sonographic techniques (Pugash et al. 2012), and the future possibility of prenatal identification of subplate abnormalities could provide one of the earliest chances to identify children at risk for ASD.

References

Aboitiz F, Scheibel AB, Fisher RS, Zaidel E (1992) Fiber composition of the human corpus-callosum. Brain Res 598:143–153. doi:10.1016/0006-8993(92)90178-C

Alexander AL, Lee JE, Lazar M, Boudos R, DuBray MB, Oakes TR, Miller JN, Lu J, Jeong EK, McMahon WM, Bigler ED, Lainhart JE (2007) Diffusion tensor imaging of the corpus callosum in autism. NeuroImage 34:61–73. doi:10.1016/j.neuroimage.2006.08.032

Allendoerfer KL, Shatz CJ (1994) The subplate, a transient neocortical structure – its role in the development of connections between thalamus and cortex. Annu Rev Neurosci 17:185–218. doi:10.1146/annurev.neuro.17.1.185

Amaral DG, Price JL (1984) Amygdalo-cortical projections in the monkey (Macaca fascicularis). J Comp Neurol 230:465–496. doi:10.1002/cne.902300402

Asadi-Pooya AA, Sharan A, Nei M, Sperling MR (2008) Corpus callosotomy. Epilepsy Behav 13:271–278. doi:10.1016/j.yebeh.2008.04.020

Avino TA, Hutsler JJ (2010) Abnormal cell patterning at the cortical gray-white matter boundary in autism spectrum disorders. Brain Res 1360:138–146. doi:10.1016/j.brainres.2010.08.091

Aylward EH, Minshew NJ, Field K, Sparks BF, Singh N (2002) Effects of age on brain volume and head circumference in autism. Neurology 59:175–183

Badaruddin DH, Andrews GL, Bolte S, Schilmoeller KJ, Schilmoeller G, Paul LK, Brown WS (2007) Social and behavioral problems of children with agenesis of the corpus callosum. Child Psychiatry Hum Dev 38:287–302. doi:10.1007/s10578-007-0065-6

Bailey A, Luthert P, Bolton P, Le Couteur A, Rutter M, Harding B (1993) Autism and megalencephaly. Lancet 341:1225–1226

Bailey A, Luthert P, Dean A, Harding B, Janota I, Montgomery M, Rutter M, Lantos P (1998) A clinicopathological study of autism. Brain 121:889–905. doi:10.1093/brain/121.5.889

Baloyannis SJ, Costa V, Mauroudis I, Psaroulis D, Manolides SL, Manolides LS (2007) Dendritic and spinal pathology in the acoustic cortex in Alzheimer's disease: morphological and morphometric estimation by Golgi technique and electron microscopy. Acta Otolaryngol 127:351–354. doi:10.1080/00016480601126986

Barnea-Goraly N, Kwon H, Menon V, Eliez S, Lotspeich L, Reiss AL (2004) White matter structure in autism: preliminary evidence from diffusion tensor imaging. Biol Psychiatry 55:323–326. doi:10.1016/j.biopsych.2003.10.022

Baron-Cohen S, Belmonte MK (2005) Autism: a window onto the development of the social and the analytic brain. Annu Rev Neurosci 28:109–126. doi:10.1146/annurev.neuro.27.070203.144137

Baron-Cohen S, Ring HA, Wheelwright S, Bullmore ET, Brammer MJ, Simmons A, Williams SCR (1999) Social intelligence in the normal and autistic brain: an fMRI study. Eur J Neurosci 11:1891–1898
Bauman ML (1991) Microscopic neuroanatomic abnormalities in autism. Pediatrics 87:791–796
Bauman ML (1996) Brief report: neuroanatomic observations of the brain in pervasive developmental disorders. J Autism Dev Disord 26:199–203
Bauman ML, Kemper TL (1994) Neuroanatomic observations of the brain in autism. In: Bauman ML, Kemper TL (eds) The neurobiology of autism. Johns Hopkins University Press, Baltimore, pp 119–145
Bauman M, Kempter TL (1985) Histoanatomic observations of the brain in early infantile autism. Neurology 35:866–874. doi:10.1212/WNL.35.6.866
Benes FM, Paskevich PA, Davidson J, Domesick VB (1985) Synaptic rearrangements in medial prefrontal cortex of haloperidol-treated rats. Brain Res 348:15–20
Benítez-Bribiesca L, De la Rosa-Alvarez I, Mansilla-Olivares A (1999) Dendritic spine pathology in infants with severe protein-calorie malnutrition. Pediatrics 104:e21
Bigler ED, Tate DF, Neeley ES, Wolfson LJ, Miller MJ, Rice SA, Cleavinger H, Anderson C, Coon H, Ozonoff S, Johnson M, Dinh E, Lu J, McMahon W, Lainhart JE (2003) Temporal lobe, autism, and macrocephaly. Am J Neuroradiol 24:2066–2076
Blundell J, Blaiss CA, Etherton MR, Espinosa F, Tabuchi K, Walz C, Bolliger MF, Sudhof TC, Powell CM (2010) Neuroligin-1 deletion results in impaired spatial memory and increased repetitive behavior. J Neurosci 30:2115–2129. doi:10.1523/JNEUROSCI. 4517-09.2010
Boddaert N, Chabane N, Gervais H, Good CD, Bourgeois M, Plumet MH, Barthelemy C, Mouren MC, Artiges E, Samson Y, Brunelle F, Frackowiak RSJ, Zilbovicius M (2004) Superior temporal sulcus anatomical abnormalities in childhood autism: a voxel-based morphometry MRI study. NeuroImage 23:364–369. doi:10.1016/j.neuroimage.2004.06.016
Bourne J, Harris KM (2007) Do thin spines learn to be mushroom spines that remember? Curr Opin Neurobiol 17:381–386. doi:10.1016/j.conb.2007.04.009
Casanova MF, Buxhoeveden DP, Switala AE, Roy E (2002) Minicolumnar pathology in autism. Neurology 58:428–432
Casanova MF, El-Baz A, Elnakib A, Switala AE, Williams EL, Williams DL, Minshew NJ, Conturo TE (2011) Quantitative analysis of the shape of the corpus callosum in patients with autism and comparison individuals. Autism 15:223–238. doi:10.1177/1362361310386506
Cherkassky VL, Kana RK, Keller TA, Just MA (2006) Functional connectivity in a baseline resting-state network in autism. Neuroreport 17:1687–1690. doi:10.1097/01.wnr.0000239956.45448.4c
Chung MK, Dalton KM, Alexander AL, Davidson RJ (2004) Less white matter concentration in autism: 2D voxel-based morphometry. NeuroImage 23:242–251. doi:10.1016/j.neuroimage.2004.04.037
Coben R, Clarke AR, Hudspeth W, Barry RJ (2008) EEG power and coherence in autistic spectrum disorder. Clin Neurophysiol 119:1002–1009. doi:10.1016/j.clinph.2008.01.013
Courchesne E (2004) Brain development in autism: early overgrowth followed by premature arrest of growth. Ment Retard Dev Disabil Res Rev 10:106–111. doi:10.1002/mrdd.20020
Courchesne E, Pierce K (2005) Why the frontal cortex in autism might be talking only to itself: local over-connectivity but long-distance disconnection. Curr Opin Neurobiol 15:225–230. doi:10.1016/j.conb.2005.03.001
Courchesne E, Townsend J, Akshoomoff NA, Saitoh O, Yeung-Couchesne R, Lincoln AJ, James HE, Haas RH, Schreibman L, Lau L (1994) Impairment in shifting attention in autistic and cerebellar patients. Behav Neurosci 108:848–865. doi:10.1037//0735-7044.108.5.848
Courchesne E, Müller RA, Saitoh O (1999) Brain weight in autism: normal in the majority of cases, megalencephalic in rare cases. Neurology 52:1057–1059
Courchesne E, Carper R, Akshoomoff N (2003) Evidence of brain overgrowth in the first year of life in autism. JAMA 290:337–344. doi:10.1001/jama.290.3.337

Courchesne E, Redcay E, Kennedy DP (2004) The autistic brain: birth through adulthood. Curr Opin Neurol 17:489–496

Critchley HD, Daly EM, Bullmore ET, Williams SCR, Van Amelsvoort T, Robertson DM, Rowe A, Phillips M, McAlonan G, Howlin P, Murphy DGM (2000) The functional neuroanatomy of social behaviour: changes in cerebral blood flow when people with autistic disorder process facial expressions. Brain 123:2203–2212

Dawson G, Ashman SB, Carver LJ (2000) The role of early experience in shaping behavioral and brain development and its implications for social policy. Dev Psychopathol 12:695–712. doi:10.1017/S0954579400004089

DeKosky ST, Scheff SW (1990) Synapse loss in frontal cortex biopsies in Alzheimer's disease: correlation with cognitive severity. Ann Neurol 27:457–464. doi:10.1002/ana.410270502

Demark JL, Feldman MA, Holden JJA (2003) Behavioral relationship between autism and fragile x syndrome. Am J Ment Retard 108:314–326. doi:10.1352/0895-8017(2003)108<314:BRBAAF>2.0.CO;2

Dietzmann K, von Bossanyi P (1994) Dendritic spines and immunoreactivity of synaptophysin in the frontal cortex of humans with infantile brain damage: a correlative study. Clin Neuropathol 13:127–133

Dinstein I, Pierce K, Eyler L, Solso S, Malach R, Behrmann M, Courchesne E (2011) Disrupted neural synchronization in toddlers with autism. Neuron 70:1218–1225. doi:10.1016/j.neuron.2011.04.018

Durand CM, Betancur C, Boeckers TM, Bockmann J, Chaste P, Fauchereau F, Nygren G, Rastam M, Gillberg IC, Anckarsäter H, Sponheim E, Goubran-Botros H, Delorme R, Chabane N, Mouren-Simeoni M-C, de Mas P, Bieth E, Rogé B, Héron D, Burglen L, Gillberg C, Leboyer M, Bourgeron T (2007) Mutations in the gene encoding the synaptic scaffolding protein SHANK3 are associated with autism spectrum disorders. Nat Genet 39:25–27. doi:10.1038/ng1933

Eastwood SL, Harrison PJ (2003) Interstitial white matter neurons express less reelin and are abnormally distributed in schizophrenia: towards an integration of molecular and morphologic aspects of the neurodevelopmental hypothesis. Mol Psychiatry 8:821–831. doi:10.1038/sj.mp.4001371

Eastwood SL, Harrison PJ (2005) Interstitial white matter neuron density in the dorsolateral prefrontal cortex and parahippocampal gyrus in schizophrenia. Schizophr Res 79:181–188. doi:10.1016/j.schres.2005.07.001

Emery JA, Roper SN, Rojiani AM (1997) White matter neuronal heterotopia in temporal lobe epilepsy: a morphometric and immunohistochemical study. J Neuropathol Exp Neurol 56:1276–1282

Fellous J-M, Rudolph M, Destexhe A, Sejnowski TJ (2003) Synaptic background noise controls the input/output characteristics of single cells in an in vitro model of in vivo activity. Neuroscience 122:811–829

Ferrer I, Galofré E (1987) Dendritic spine anomalies in fetal alcohol syndrome. Neuropediatrics 18:161–163. doi:10.1055/s-2008-1052472

Fiala JC, Spacek J, Harris KM (2002) Dendritic spine pathology: cause or consequence of neurological disorders? Brain Res Rev 39:29–54

Filipek PA, Richelme C, Kennedy DN, Rademacher J, Pitcher DA, Zidel S, Caviness VS (1992) Morphometric analysis of the brain in developmental language disorders and autism. Ann Neurol 32:475

Friedlander MJ, Torres-Reveron J (2009) The changing roles of neurons in the cortical subplate. Front Neuroanat 3:15. doi:10.3389/neuro.05.015.2009

Frith C (2004) Is autism a disconnection disorder? Lancet Neurol 3:577. doi:10.1016/S1474-4422(04)00875-0

Frith U, Happé F (1994) Autism: beyond "theory of mind". Cognition 50:115–132

Gage NM, Siegel B, Callen M, Roberts TPL (2003) Cortical sound processing in children with autism disorder: an MEG investigation. Neuroreport 14:2047–2051. doi:10.1097/01.wnr.0000090030.460874a

Garey LJ, Ong WY, Patel TS, Kanani M, Davis A, Mortimer AM, Barnes TR, Hirsch SR (1998) Reduced dendritic spine density on cerebral cortical pyramidal neurons in schizophrenia. J Neurol Neurosurg Psychiatry 65:446–453

Gazzaniga MS (2005) Forty-five years of split-brain research and still going strong. Nat Rev Neurosci 6:653–659. doi:10.1038/nrn1723

Gervais H, Belin P, Boddaert N, Leboyer M, Coez A, Sfaello I, Barthelemy C, Brunelle F, Samson Y, Zilbovicius M (2004) Abnormal cortical voice processing in autism. Nat Neurosci 7:801–802. doi:10.1038/nn1291

Geschwind DH, Levitt P (2007) Autism spectrum disorders: developmental disconnection syndromes. Curr Opin Neurobiol 17:103–111. doi:10.1016/j.conb.2007.01.009

Glantz LA, Lewis DA (2000) Decreased dendritic spine density on prefrontal cortical pyramidal neurons in schizophrenia. Arch Gen Psychiatry 57:65–73

Goldstein G, Minshew NJ, Allen DN, Seaton BE (2002) High-functioning autism and schizophrenia: a comparison of an early and late onset neurodevelopmental disorder. Arch Clin Neuropsychol 17:461–475

Gościk E, Kulak P (2011) Agenesis of corpus callosum: genetics, epidemiology and neuroimaging findings. Prog Health Sci 1:126–131

Gozzi M, Nielson DM, Lenroot RK, Ostuni JL, Luckenbaugh DA, Thurm AE, Giedd JN, Swedo SE (2012) A magnetization transfer imaging study of corpus callosum myelination in young children with autism. Biol Psychiatry 72:215–220. doi:10.1016/j.biopsych.2012.01.026

Grossberg S, Seidman D (2006) Neural dynamics of autistic behaviors: cognitive, emotional, and timing substrates. Psychol Rev 113:483–525. doi:10.1037/0033-295x.113.3.483

Grossberg S, Seitz A (2003) Laminar development of receptive fields, maps and columns in visual cortex: the coordinating role of the subplate. Cereb Cortex 13:852–863

Halpain S, Spencer K, Graber S (2005) Dynamics and pathology of dendritic spines. Prog Brain Res 147:29–37. doi:10.1016/S0079-6123(04)47003-4

Hanganu IL, Okabe A, Lessmann V, Luhmann HJ (2009) Cellular mechanisms of subplate-driven and cholinergic input-dependent network activity in the neonatal rat somatosensory cortex. Cereb Cortex 19:89–105. doi:10.1093/cercor/bhn061

Happe F, Frith U (2006) The weak coherence account: detail-focused cognitive style in autism spectrum disorders. J Autism Dev Disord 36:5–25. doi:10.1007/s10803-005-0039-0

Hardan AY, Minshew NJ, Keshavan MS (2000) Corpus callosum size in autism. Neurology 55:1033–1036

Hardan AY, Jou RJ, Keshavan MS, Varma R, Minshew NJ (2004) Increased frontal cortical folding in autism: a preliminary MRI study. Psychiatry Res 131:263–268. doi:10.1016/j.pscychresns.2004.06.001

Herbert MR, Ziegler DA, Deutsch CK, O'Brien LM, Lange N, Bakardjiev A, Hodgson J, Adrien KT, Steele S, Makris N, Kennedy D, Harris GJ, Caviness VS (2003) Dissociations of cerebral cortex, subcortical and cerebral white matter volumes in autistic boys. Brain 126:1182–1192. doi:10.1093/brain/awg110

Herbert MR, Ziegler DA, Makris N, Filipek PA, Kemper TL, Normandin JJ, Sanders HA, Kennedy DN, Caviness VS (2004) Localization of white matter volume increase in autism and developmental language disorder. Ann Neurol 55:530–540. doi:10.1002/ana.20032

Hildebrandt M, Pieper T, Winkler P, Kolodziejczyk D, Holthausen H, Blumcke I (2005) Neuropathological spectrum of cortical dysplasia in children with severe focal epilepsies. Acta Neuropathol 110:1–11. doi:10.1007/s00401-005-1016-6

Hines RM, Wu L, Hines DJ, Steenland H, Mansour S, Dahlhaus R, Singaraja RR, Cao X, Sammler E, Hormuzdi SG, Zhuo M, El-Husseini A (2008) Synaptic imbalance, stereotypies, and impaired social interactions in mice with altered neuroligin 2 expression. J Neurosci 28:6055–6067. doi:10.1523/JNEUROSCI.0032-08.2008

Holtmaat A, Trachtenberg JT, Wilbrecht L, Shepherd GM, Zhang XQ, Knott GW, Svoboda K (2005) Transient and persistent dendritic spines in the neocortex in vivo. Neuron 45:279–291. doi:10.1016/j.neuron.2005.01.003

Holtzman JD, Gazzaniga MS (1985) Enhanced dual task performance following corpus commissurotomy in humans. Neuropsychologia 23:315–321

Hong SS, Ke XY, Tang TY, Hang YY, Chu KK, Huang HQ, Ruan ZC, Lu ZH, Tao GT, Liu YJ (2011) Detecting abnormalities of corpus callosum connectivity in autism using magnetic resonance imaging and diffusion tensor tractography. Psychiatry Res 194:333–339. doi:10.1016/j.pscychresns.2011.03.009

Hughes JR (2007) Autism: the first firm finding = underconnectivity? Epilepsy Behav 11:20–24. doi:10.1016/j.yebeh.2007.03.010

Hutsler JJ, Avino TA (2013) Sigmoid fits to locate and characterize cortical boundaries in human cerebral cortex. J Neurosci Methods 212:242–246. doi:10.1016/j.jneumeth.2012.10.007

Hutsler JJ, Zhang H (2010) Increased dendritic spine densities on cortical projection neurons in autism spectrum disorders. Brain Res 1309:83–94. doi:10.1016/j.brainres.2009.09.120

Hutsler JJ, Love T, Zhang H (2007) Histological and magnetic resonance imaging assessment of cortical layering and thickness in autism spectrum disorders. Biol Psychiatry 61:449–457. doi:10.1016/j.biopsych.2006.01.015

Huttenlocher PR, Dabholkar AS (1997) Regional differences in synaptogenesis in human cerebral cortex. J Comp Neurol 387:167–178

Irwin SA, Galvez R, Greenough WT (2000) Dendritic spine structural anomalies in fragile-X mental retardation syndrome. Cereb Cortex 10:1038–1044

Irwin SA, Patel B, Idupulapati M, Harris JB, Crisostomo RA, Larsen BP, Kooy F, Willems PJ, Cras P, Kozlowski PB, Swain RA, Weiler IJ, Greenough WT (2001) Abnormal dendritic spine characteristics in the temporal and visual cortices of patients with fragile-X syndrome: a quantitative examination. Am J Med Genet 98:161–167. doi:10.1002/1096-8628(20010115)98:2<161::aid-ajmg1025>3.0.co;2-b

Jamain S, Quach H, Betancur C, Rastam M, Colineaux C, Gillberg IC, Soderstrom H, Giros B, Leboyer M, Gillberg C, Bourgeron T (2003) Mutations of the X-linked genes encoding neuroligins NLGN3 and NLGN4 are associated with autism. Nat Genet 34:27–29. doi:10.1038/ng1136

Just MA, Cherkassky VL, Keller TA, Minshew NJ (2004) Cortical activation and synchronization during sentence comprehension in high-functioning autism: evidence of underconnectivity. Brain 127:1811–1821. doi:10.1093/brain/awh199

Just MA, Cherkassky VL, Keller TA, Kana RK, Minshew NJ (2007) Functional and anatomical cortical underconnectivity in autism: evidence from an fMRI study of an executive function task and corpus callosum morphometry. Cereb Cortex 17:951–961. doi:10.1093/cercor/bhl006

Just MA, Keller TA, Malave VL, Kana RK, Varma S (2012) Autism as a neural systems disorder: a theory of frontal-posterior underconnectivity. Neurosci Biobehav Rev 36:1292–1313. doi:10.1016/j.neubiorev.2012.02.007

Kana RK, Libero LE, Moore MS (2011) Disrupted cortical connectivity theory as an explanatory model for autism spectrum disorders. Phys Life Rev 8:410–437. doi:10.1016/j.plrev.2011.10.001

Kanold PO (2004) Transient microcircuits formed by subplate neurons and their role in functional development of thalamocortical connections. Neuroreport 15:2149–2153

Kanold PO, Kara P, Reid RC, Shatz CJ (2003) Role of subplate neurons in functional maturation of visual cortical columns. Science 301:521–525. doi:10.1126/science.1084152

Kaufmann WE, Cortell R, Kau ASM, Bukelis I, Tierney E, Gray RM, Cox C, Capone GT, Stanard P (2004) Autism spectrum disorder in fragile X syndrome: communication, social interaction, and specific behaviors. Am J Med Genet A 129A:225–234. doi:10.1002/ajmg.a.30229

Koenderink MJ, Uylings HB (1995) Postnatal maturation of layer V pyramidal neurons in the human prefrontal cortex: a quantitative Golgi analysis. Brain Res 678:233–243

Koenderink MJ, Uylings HB, Mrzljak L (1994) Postnatal maturation of the layer III pyramidal neurons in the human prefrontal cortex: a quantitative Golgi analysis. Brain Res 653:173–182
Koshino H, Carpenter PA, Minshew NJ, Cherkassky VL, Keller TA, Just MA (2005) Functional connectivity in an fMRI working memory task in high-functioning autism. NeuroImage 24:810–821. doi:10.1016/j.neuroimage.2004.09.028
Kostovic I, Judas M, Sedmak G (2011) Developmental history of the subplate zone, subplate neurons and interstitial white matter neurons: relevance for schizophrenia. Int J Dev Neurosci 29:193–205. doi:10.1016/j.ijdevneu.2010.09.005
Lamantia AS, Rakic P (1990) Axon overproduction and elimination in the corpus-callosum of the developing rhesus monkey. J Neurosci 10:2156–2175
Laumonnier F, Bonnet-Brilhault F, Gomot M, Blanc R, David A, Moizard MP, Raynaud M, Ronce N, Lemonnier E, Calvas P, Laudier B, Chelly J, Fryns JP, Ropers HH, Hamel BCJ, Andres C, Barthelemy C, Moraine C, Briault S (2004) X-linked mental retardation and autism are associated with a mutation in the NLGN4 gene, a member of the neuroligin family. Am J Hum Genet 74:552–557. doi:10.1086/382137
LeBlanc JJ, Fagiolini M (2011) Autism: a "critical period" disorder? Neural Plast 2011:921680. doi:10.1155/2011/921680
Lord C, Rutter M, Le Couteur A (1994) Autism diagnostic interview-revised: a revised version of a diagnostic interview for caregivers of individuals with possible pervasive developmental disorders. J Autism Dev Disord 24:659–685
Luhmann HJ, Kilb W, Hanganu-Opatz IL (2009) Subplate cells: amplifiers of neuronal activity in the developing cerebral cortex. Front Neuroanat 3:19. doi:10.3389/neuro.05.019.2009
Luna B, Doll SK, Hegedus SJ, Minshew NJ, Sweeney JA (2007) Maturation of executive function in autism. Biol Psychiatry 61:474–481. doi:10.1016/j.biopsych.2006.02.030
McAllister AK (1999) Subplate neurons: a missing link among neurotrophins, activity, and ocular dominance plasticity? Proc Natl Acad Sci U S A 96:13600–13602
McConnell SK, Ghosh A, Shatz CJ (1994) Subplate pioneers and the formation of descending connections from cerebral cortex. J Neurosci 14:1892–1907
McFadden K, Minshew NJ (2013) Evidence for dysregulation of axonal growth and guidance in the etiology of ASD. Front Hum Neurosci 7:671–681. doi:10.3389/fnhum.2013.00671
Michel AE, Garey LJ (1984) The development of dendritic spines in the human visual cortex. Hum Neurobiol 3:223–227
Minshew NJ, Luna B, Sweeney JA (1999) Oculomotor evidence for neocortical systems but not cerebellar dysfunction in autism. Neurology 52:917–922. doi:10.1212/WNL.52.5.917
Mizuno A, Villalobos ME, Davies MM, Dahl BC, Muller RA (2006) Partially enhanced thalamocortical functional connectivity in autism. Brain Res 1104:160–174. doi:10.1016/j.brainres.2006.05.064
Multani P, Myers RH, Blume HW, Schomer DL, Sotrel A (1994) Neocortical dendritic pathology in human partial epilepsy: a quantitative Golgi study. Epilepsia 35:728–736
Mundy P, Neal R (2001) Neural plasticity, joint attention and autistic developmental pathology. Int Rev Res Ment Ret 23:139–168
Murias M, Webb SJ, Greenson J, Dawson G (2007) Resting state cortical connectivity reflected in EEG coherence in individuals with autism. Biol Psychiatry 62:270–273. doi:10.1016/j.biopsych.2006.11.012
Nikonenko I, Boda B, Alberi S, Muller D (2005) Application of photoconversion technique for correlated confocal and ultrastructural studies in organotypic slice cultures. Microsc Res Tech 68:90–96. doi:10.1002/jemt.20239
Noriuchi M, Kikuchi Y, Yoshiura T, Kira R, Shigeto H, Hara T, Tobimatsu S, Kamio Y (2010) Altered white matter fractional anisotropy and social impairment in children with autism spectrum disorder. Brain Res 1362:141–149. doi:10.1016/j.brainres.2010.09.051
O'Brien G (1994) The behavioral and developmental consequences of corpus callosal agenesis and Aicardi syndrome. In: Lassonde M, Jeeves M (eds) Callosal agenesis. Springer, New York, pp 235–246

O'Hearn K, Asato M, Ordaz S, Luna B (2008) Neurodevelopment and executive function in autism. Dev Psychopathol 20:1103–1132. doi:10.1017/s0954579408000527

O'Roak BJ, Vives L, Fu WQ, Egertson JD, Stanaway IB, Phelps IG, Carvill G, Kumar A, Lee C, Ankenman K, Munson J, Hiatt JB, Turner EH, Levy R, O'Day DR, Krumm N, Coe BP, Martin BK, Borenstein E, Nickerson DA, Mefford HC, Doherty D, Akey JM, Bernier R, Eichler EE, Shendure J (2012) Multiplex targeted sequencing identifies recurrently mutated genes in autism spectrum disorders. Science 338:1619–1622. doi:10.1126/science.1227764

Penzes P, Cahill ME, Jones KA, VanLeeuwen JE, Woolfrey KM (2011) Dendritic spine pathology in neuropsychiatric disorders. Nat Neurosci 14:285–293. doi:10.1038/nn.2741

Piven J, Arndt S, Bailey J, Havercamp S, Andreasen NC, Palmer P (1995) An MRI study of brain size in autism. Am J Psychiatry 152:1145–1149

Piven J, Bailey J, Ranson BJ, Arndt S (1997) An MRI study of the corpus callosum in autism. Am J Psychiatr 154:1051–1056

Pugash D, Hendson G, Dunham CP, Dewar K, Money DM, Prayer D (2012) Sonographic assessment of normal and abnormal patterns of fetal cerebral lamination. Ultrasound Obstet Gynecol 40:642–651. doi:10.1002/uog.11164

Reuter-Lorenz PA, Kinsbourne M, Moscovitch M (1990) Hemispheric control of spatial attention. Brain Cogn 12:240–266. doi:10.1016/0278-2626(90)90018-J

Sabaratnam M (2000) Pathological and neuropathological findings in two males with fragile-X syndrome. J Intellect Disabil Res 44(Pt 1):81–85

Sabbagh MA (2004) Understanding orbitofrontal contributions to theory-of-mind reasoning: implications for autism. Brain Cogn 55:209–219. doi:10.1016/j.banc.2003.04.002

Scheff SW, Price DA, Schmitt FA, DeKosky ST, Mufson EJ (2007) Synaptic alterations in CA1 in mild Alzheimer disease and mild cognitive impairment. Neurology 68:1501–1508. doi:10.1212/01.wnl.0000260698.46517.8f

Schultz RT, Gauthier I, Klin A, Fulbright RK, Anderson AW, Volkmar F, Skudlarski P, Lacadic C, Cohen DJ, Gore JC (2000) Abnormal ventral temporal cortical activity during face discrimination among individuals with autism and Asperger syndrome. Arch Gen Psychiatry 57:331–340

Shatz CJ, Luskin MB (1986) The relationship between the geniculocortical afferents and their cortical target cells during development of the cat's primary visual cortex. J Neurosci 6:3655–3668

Shukla DK, Keehn B, Lincoln AJ, Muller RA (2010) White matter compromise of callosal and subcortical fiber tracts in children with autism spectrum disorder: a diffusion tensor imaging study. J Am Acad Child Adolesc Psychiatry 49:1269–1278. doi:10.1016/j.jaac.2010.08.018

Simms ML, Kemper TL, Timbie CM, Bauman ML, Blatt GJ (2009) The anterior cingulate cortex in autism: heterogeneity of qualitative and quantitative cytoarchitectonic features suggests possible subgroups. Acta Neuropathol 118:673–684. doi:10.1007/s00401-009-0568-2

Singer W, Gray CM (1995) Visual feature integration and the temporal correlation hypothesis. Annu Rev Neurosci 18:555–586

Suarez-Sola ML, Gonzalez-Delgado FJ, Pueyo-Morlans M, Medina-Bolivar OC, Hernandez-Acosta NC, Gonzalez-Gomez M, Meyer G (2009) Neurons in the white matter of the adult human neocortex. Front Neuroanat 3:7. doi:10.3389/neuro.05.007.2009

Suetsugu M, Mehraein P (1980) Spine distribution along the apical dendrites of the pyramidal neurons in Down's syndrome: a quantitative Golgi study. Acta Neuropathol 50:207–210

Swann JW, Al-Noori S, Jiang M, Lee CL (2000) Spine loss and other dendritic abnormalities in epilepsy. Hippocampus 10:617–625. doi:10.1002/1098-1063(2000)10:5<617::AID-HIPO13>3.0.CO;2-R

Sweet RA, Henteleff RA, Zhang W, Sampson AR, Lewis DA (2009) Reduced dendritic spine density in auditory cortex of subjects with schizophrenia. Neuropsychopharmacology 34:374–389. doi:10.1038/npp.2008.67

Tabuchi K, Blundell J, Etherton MR, Hammer RE, Liu XR, Powell CM, Sudhof TC (2007) Neuroligin-3 mutation implicated in autism increases inhibitory synaptic transmission in mice. Science 318:71–76. doi:10.1126/science.1146221

Takashima S, Chan F, Becker LE, Kuruta H (1991) Aberrant neuronal development in hemimegalencephaly: immunohistochemical and Golgi studies. Pediatr Neurol 7:275–280

Turk AA, Brown WS, Symington M, Paul LK (2010) Social narratives in agenesis of the corpus callosum: linguistic analysis of the Thematic Apperception Test. Neuropsychologia 48:43–50. doi:10.1016/j.neuropsychologia.2009.08.009

Uhlhaas PJ, Roux F, Singer W, Haenschel C, Sireteanu R, Rodriguez E (2009) The development of neural synchrony reflects late maturation and restructuring of functional networks in humans. Proc Natl Acad Sci U S A 106:9866–9871. doi:10.1073/pnas.0900390106

Voigt T (1989) Development of glial cells in the cerebral wall of ferrets. J Comp Neurol 289:74–88

Voigt T, Opitz T, de Lima AD (2001) Synchronous oscillatory activity in immature cortical network is driven by GABAergic preplate neurons. J Neurosci 21:8895–8905

Woolfrey KM, Srivastava DP, Photowala H, Yamashita M, Barbolina MV, Cahill ME, Xie Z, Jones KA, Quilliam LA, Prakriya M, Penzes P (2009) Epac2 induces synapse remodeling and depression and its disease-associated forms alter spines. Nat Neurosci 12:1275-U96. doi:10.1038/nn.2386

Yuste R (2011) Dendritic spines and distributed circuits. Neuron 71:772–781. doi:10.1016/j.neuron.2011.07.024

Yuste R, Majewska A (2001) On the function of dendritic spines. Neuroscientist 7:387–395

Zikopoulos B, Barbas H (2010) Changes in prefrontal axons may disrupt the network in autism. J Neurosci 30:14595–14609. doi:10.1523/jneurosci. 2257-10.2010

Chapter 13
The Minicolumnopathy of Autism

Manuel F. Casanova

Abstract The organization of the cerebral cortex is centered around a modular construct. The smallest module capable of information processing is the minicolumn. Recent studies on minicolumnar morphometry using either pyramidal cell arrays or the gray level index (GLI) suggest distinct abnormalities of this structure in several psychiatric conditions including schizophrenia, dyslexia, and autism. More specifically, minicolumns in autism as compared to controls seem thinner and more numerous. An increased number of minicolumns indicates the supernumerary division of periventricular germinal cells. A reduction in size of pyramidal cell somas within affected minicolumns suggests a bias towards shorter corticocortical connectivity. Compartmentalization of the minicolumn indicates that the majority of the deficit is found within the peripheral neuropil space. This compartment includes, among other things, many of the inhibitory elements of the cerebral cortex and provides the so-called "shower curtain of inhibition" to the minicolumn. Laminae studies indicate that in autism the peripheral neuropil defect extends the width of the cerebral cortex. A possible explanation to the above described minicolumnar abnormalities is the heterochronic division of germinal cells. Neuroblasts generated from heterochronic divisions of germinal cells can give rise to heterotopias and dysplastic cortical lesions. Once the radially migrating neuroblasts (future pyramidal cells) reach the cortex they develop asynchronously from tangentially dividing neuronal elements (interneurons). The resultant excitatory/inhibitory imbalance may provide for seizures and sensorimotor abnormalities

Keywords Minicolumns • Cerebral cortex • Autism spectrum disorders • Interneurons • Pyramidal cells

The known neuropathology of the different entities comprising syndromic autism strongly suggest an onset during brain development. The percentage of individuals showing an autism phenotype in many of these conditions is exceedingly high. Different series suggest that 25–50 % of all patients with tuberous sclerosis meet

M.F. Casanova, M.D. (✉)
Department of Psychiatry, University of Louisville, Louisville, KY, USA
e-mail: manuel.casanova@louisville.edu

M.F. Casanova, I. Opris (eds.), *Recent Advances on the Modular Organization of the Cortex*, DOI 10.1007/978-94-017-9900-3_13

criteria for an autism spectrum disorder (ASD). According to some series the frequency is even larger for the Lujan Fryns syndrome (mental retardation, X-linked, Marfanoid body habitus) and the Smith-Lemli-Opitz syndrome (mental retardation and an inborn error of cholesterol synthesis). The commonality among many syndromic cases appears to be a migratory abnormality where neuroblasts fated to form part of either cortex or brainstem either fail to reach their normal target areas or reach the same and are desynchronized in their development from other cellular elements already in place. Syndromic cases with migratory abnormalities exhibit subependymal nodular heterotopias, cluster of neuronal cells within the white matter, and a variety of cortical malformations. Although the above mentioned cases are the result of genetic conditions (including *de novo* mutations), exogenous factors such as viral infections (e.g., cytomegalovirus) and even cocaine during gestation may alter cellular divisions of the germinal matrix thus providing for similarities in both neuropathology and clinical presentations.

Abnormalities of neuronal migration are intimately related to malformations of the cerebral cortex. Malformations during brain development, also called dysplastic changes, are usually manifested as cortical thickening or thinning, blurring of the gray-whiter matter junction and a variety of histological findings, e.g. effacement of laminar patters, minicolumnar morphometric abnormalities, malpositioned neurons (supernumerary neurons in the molecular layer and white matter). In this regard the pathology observed in the cerebral cortex is secondary to processes impinging on its formation, e.g. germinal cell divisions, migration of neuroblasts to the cortical plate. All of the previously observed dysplastic changes have been reported in both syndromic and idiopathic autism (Casanova et al. 2013). The significance of the findings lays in their capacity to explain some of the phenomenology in autism spectrum disorders, e.g., medically refractory seizures and sensory abnormalities.

This chapter will focus on certain aspects of cortical dysplasia in ASD as revealed by minicolumnar morphometry. Vertical aggregates of cells forming closed circuits were first described anatomically by Lorente de Nó (1938): “All the elements of the cortex are represented in it, and therefore it may be called an elementary unit, in which, theoretically, the whole process of transmission of impulses from the afferent fiber to the efferent axon may be accomplished”. According to Mountcastle, the minicolumn is the smallest element of information processing within the cerebral cortex (Mountcastle 1998). Recent electrophysiological studies reveal that sensory processing at the level of the minicolumn modulates the expression of higher cognitive functions (Opris et al. 2013). Differences in minicolumnar morphometry between hemispheres have been used to explain cerebral dominance for the language regions of the brain (Buxhoeveden et al. 2001). Since this difference appears to be specific to our species (not present in non-human primates), it may be a putative speciation event relating language to cytoarchitectonic features of the brain (Buxhoeveden et al. 2001). Defects in minicolumnar morphometry may therefore be present in psychiatric conditions such as ASD which are defined by language abnormalities.

13.1 Minicolumnar Morphometry in Autism Spectrum Disorders

The first study on minicolumnar morphometry in autism spectrum disorders sampled nine cases and an equal number of age-matched controls (Casanova et al. 2002a, b, c). All of the specimens were collected and processed by Drs. Thomas Kemper and Margaret Bauman. There were no abnormalities for any of the cases at gross examination. Four of the nine autistic cases were macrocephalic with brain weights more than two standard deviations above the means for their corresponding ages. The medical records detailed little information as to whether the abnormal brain weights were due to postmortem swelling, an artifact, or a real phenomenon. These cases had been cut midsagittally, freezing one hemisphere and leaving the other for microscopic examination. In order to avoid shrinkage artifacts, brains were embedded in celloidin. Sections were cut at 35 μm and Nissl stained. Every 20th section was collected and for each 100th section and adjacent slide was stained for myelin using the Loyez method. Previous reports on the pathology of these cases used a Zeiss comparison microscope and offered subjective assessments of findings when viewing simultaneously side-by-side anatomically matched slides. In this patient population, Bauman and Kemper indicated the presence of reduced neuronal size and increased cell density in the amygdala, hippocampal complex, subiculum, entorhinal cortex, medial septal nuclei, and mammillary bodies (Bauman and Kemper 2005).

Bauman and Kemper reported few cortical findings in this series. Most prominent among these were an indistinct pattern of lamination in the anterior cingulate gyrus and a minor malformation in one case of the orbitofrontal cortex (Bauman and Kemper 1987; Kemper and Bauman 1998). Subjective appraisal of the slides (whole hemisphere coronal sections) made it difficult to draw firm conclusions about the cerebral cortex. It is in this regard that we attempted relative quantitation by using computerized image analysis and implementing an algorithm based on the Euclidian minimum spanning tree. Images from lamina III in three cortical regions (Brodmann areas 9, 21, and 22) were digitized and cells reduced to points using lines such that the total length of all lines was minimized. Minicolumns were identified as vertical clusters of large neurons delimited on either side by cell-sparse areas. Imaginary lines through the sparse areas partitioned the field into polygonal regions thus defining minicolumnar segments. Descriptive statistic of minicolumnar morphometry included columnar width (CW), peripheral neuropil space (NS), interneuronal distance (MCS), and compactness (RDR). Results from this study revealed that minicolumns in ASD were narrower with a mean CW of 46.8 *vs.* 52.8 μm in the control brain. Both the peripheral neuropil space and compactness were reduced in autism.

Results from this first study revealed that minicolumns in ASD are of reduced width with a significant portion of the abnormality accounted by a diminution of the peripheral neuropil space. The lateral compartment of the minicolumns, the peripheral neuropil space, is the conduit, among other things, for inhibitory circuit projections. Previous researchers have called this compartment of the minicolumns: "a shower curtain of inhibition" or a "strong vertically directed stream of inhibition" (Szentágothai 1978; Mountcastle 1997). This inhibitory compartment sharpens the functional borders of the minicolumn and increases their definition or discreteness (DeFelipe 1999; Favorov and Kelly 1994a, b; Szentágothai 1978). The primary source of this inhibitory effect stems from the action of double-bouquet and basket cells. Double-bouquet cell axons arrange themselves in repeating vertical patterns between 15 and 30 μm apart depending on cortical area examined (DeFelipe 1999). It is thought that the vertical descent of double bouquet cells across laminae acts as a buffer by inhibiting dendritic terminals belonging to excitatory (pyramidal) cells of neighboring minicolumns. A defect in these inhibitory elements could cause the signals being processed in the core of the minicolumn to suffuse into adjacent minicolumns. The resultant avalanche of activity could help explain the significant prevalence of seizures in ASD.

An interesting possibility worth considering is what would happen in ASD if thalamic fields retain the same area dimension while minicolumns are smaller? The result for ASD would be an increased number of minicolumns innervated per thalamic afferent than in the normal brain. Alternatively, the failure to assimilate additional minicolumns into a thalamic field would impel these processing units to establish connections with functionally dissimilar sets of thalamic neurons (Favorov and Kelly 1994a, 1994b). In the later instance, the failure of the system to assimilate the additional minicolumns would result in cortical noise; that is, supernumerary units of activity that overtax the system (Casanova et al. 2002a, b, c).

Individual cells when depolarized cannot produce gamma activity. This fast oscillation, in the gamma range, is responsible for creating the unity of conscious perception. Only thalamic modulation of neuronal networks can provide for intracortical gamma activity (Sukov and Barth 2001; Macdonald et al. 1998). Because interneurons can help synchronize neuronal discharges and some exhibit fast spiking activity, they are important mediators of gamma activity. Lack of integration among separate specialized local networks (a binding failure) may account for a weak central coherence in autism spectrum disorders. Several laboratories have recently reported a disorder of binding related gamma EEG oscillatory activity in autism spectrum disorders (Grice et al. 2001; Sokhadze et al. 2009; Baruth et al. 2010).

The same patients and regions of interests reported by Casanova et al. (2002a) were examined for descriptive parameters of the Gray Level Index (Casanova et al. 2002b; Schlaug et al. 1995). According to this method a fractional area of Nissl-stained objects was computed (profiles of 11 pixels 110 um wide). Images were smoothed with the resultant profiles leading to the analysis of its peaks and troughs. The method produced measurements of peak width distance and a verticality index. The verticality index quantified the degree of columnar organization relative to the mean of the entire sample. The overall statistical test revealed significant differences between autistic patients and controls and between hemispheres. There was no significant age dependency between factors. Follow-up univariate tests showed significant diagnosis-dependent effects in feature distance. No significant differences were evident in overall verticality.

Results from the latter two studies using different algorithms indicate the presence of vertical cellular structures in the cerebral cortex of individuals with autism spectrum disorders that are packed more closely together and more regularly spaced than in controls (Casanova et al. 2002a, b). The results equally applied to all areas examined (Brodmann areas 9, 21, 22) in both hemispheres. The findings suggest a developmental defect of the nervous system.

A third study on minicolumnar morphometry in ASD was carried in an independent sample as an international effort sponsored by the Autism Tissue Program (Casanova et al. 2006a, b). Different laboratories were charged with collecting samples, constructing photomicrograph montages, and performing the computerized image analysis. The results were made available to all of the participating researchers before the codes were broken and the data analyzed statistically. The patient population consisted of six age-matched pairs of ASD subjects (DSM-IV-TR and ADI-R diagnosed). Tissue specimens consisted of full coronal sections embedded in celloidin and cut at 200 μm. Gallocyanin-stained sections were used to identify cortical areas M1, V1, frontal association cortex, and S1 (Brodmann areas 4, 17, 9, and 3b). Sections were delineated with a stereology workstation and photographed with a digital camera at 40X. The Virtual Slice module of the StereoInvestigator was used to assemble digitized photomicrographs into one mosaic. Computerized image analysis of minicolumnar morphometry in Laminae II through VI was performed with algorithms previously described in the literature (Casanova and Switala 2005). A threshold function eliminated the smaller cellular elements (interneurons) and allowed the researchers to concentrate on measurements provided by pyramidal cells. Minicolumnar width was reduced in autism spectrum disorder individuals as compared to controls. Mean neuron (soma) and nucleolar size was reduced in ASD, while neuron density (based on a point process model) in autism exceeded the control group by 23 %.

The findings reported in the international study on minicolumnar morphometry in ASD reproduced previous findings (Casanova et al. 2006a). In this study the feature extraction properties of the program were corrected for minicolumnar fragments, curvature of the tissue section, and were adapted to 3D proportions (stereological modeling). The smaller minicolumns per given brain region translates into increased numbers when corrected for 3D quantitation. The smaller minicolumns and their overall increase in total numbers translated into an increased neuronal density. However, the basis for the increase cellular density remained conjectural, e.g., the presence of smaller supernumerary minicolumns, an increase in the total number of cells per minicolumn, or both. A subsequent analysis based on a Delaunay triangulation addressed the aforementioned concerns.

The Delaunay triangulation parcellates points by joining them with edges forming triangles. The edges satisfy the criteria of an "empty circle" property where for each edge we can find a circle containing the edges' endpoints but no other points within the same. Owing to the clustering of cells within vertical arrangements, the Delaunay triangulation demonstrates a bimodal distribution of edges between cells in the same minicolumn (intracluster) and edges between cells in neighboring minicolumns (interclusters). The Delaunay triangulation indicated a significant reduction in the edges of intercluster distances but not within intracluster differences. Increased neuronal density in this series was due to the presence of supernumerary minicolumns, otherwise the total number of neurons (pyramidal cells) per minicolumn was normal (Casanova et al. 2006a).

Minicolumns in autism, although smaller in size, are increased in total numbers as compared to neurotypicals. Furthermore within each minicolumn a reduction of both neuronal soma and nucleolar size biases connectivity towards shorter connections. Casanova et al. (2006a) concluded that, "Just et al. (2004) have subsumed the evidence for a lower degree of information integration in autism under the rubric of the "underconnectivity theory". However, the term may prove to be a misnomer when applied to shorter intra-areal connections (arcuate or u fibers). In autism, an increase in the total number of minicolumns requires a scale increase (roughly a 3/2 power law) in white matter to maintain modular interconnectivity (Hofman 1985). This additional white matter takes the form of short-range connections which makes up the bulk of intracortical connections." Dysfunction of long projections translates into complex abnormalities within widely distribute networks. Some authors have suggested that a "dysexecutive syndrome" in autism could possibly result from the frontal lobe's complex pattern of connectivity. The widely distributed network of connectivity of the frontal lobes accounts for the phenomenon of diaschisis where executive cognitive deficits may become apparent in lesions distant to the anterior cortical region (Mesulam 2002; Casanova et al. 2006b).

The international group that produced the Casanova et al. (2006a) study also expanded on previous findings by examining the topographical distribution of the autistic minicolumnopathy (Casanova et al. 2006b). Tissue embedding and processing was identical as before but the number of sampled regions of interest increased to include Brodmann areas: 10 (frontopolar), 11 (orbitofrontal), 9 (dorsolateral prefrontal), 4 (primary motor), 3b (primary sensory), 43 (frontoinsular), 44 (ventrolateral), 24 (anterior cingulate), and 17 (primary visual). The sampling included areas of the cortex representing paralimbic, heteromodal association, unimodal association, and primary areas. The study found an interaction of diagnosis and region for peripheral neuropil space. *Post hoc* analysis revealed significant differences for the frontopolar region (area 10) and the anterior cingulate gyrus (area 24). The role of the frontopolar cortex in executive functions and of the anterior cingulate gyrus in the analysis of socially salient information suggests that involvement of these areas may provide a correlate to some of the more salient clinical manifestations of autism.

The initial results of a minicolumnopathy in autism has been reproduced in an independent sample by Buxhoeveden et al. (2006). The study included two autistic subjects with an average age of 22 years and 5 controls whose average age was 35 years of age. Two of the controls were embedded in paraffin, cut at 20 μm and stained with a silver impregnation method. The other subjects within this study were cryoprotected, cut at 80 μm and Nissl stained. The significance of this study is arguable given the salient limitations imposed by the limited number of participants, lack of age matched controls, and differences in processing of tissue as well as staining. Despite all of this, the study reproduced previous findings. The average minicolumnar width reported in Casanova et al. (2002a) study was 46.8 μm for 9 autistic subjects and 52.8 μm for 9 controls, compared to results in Buxhoeveden et al. (2006) of 45.5 μm and 56.2 μm in autistic and controls respectively.

Many of the aforementioned studies on minicolumnar morphometry in ASD focused on elucidating changes in lamina III as the columnar nature of pyramidal cell arrays is easily discernible in supragranular layers. In deeper laminae the arrangement of pyramidal cells in relation to the central axis of the minicolumns is more variable. For this reason photomicrograph mosaics from the Casanova et al. (2006b) study were used to analyze with computerized imaging methods the minicolumnar width at the supragranular, granular, and infragranular levels. Images were smoothed with a Markov-Gibbs random field and used a Gibbs energy function to find the gray level that minimized the difference in intensity between each pixel and its surrounding neighbors. We then optimized the gray level value to increase the contrast between neurons and background. Images obtained by the previous steps were segmented by thresholding to produce binary images. The individual neurons were then identified using a region growing algorithm. Minicolumns were recognized with a line tracing method that grouped cells into columnar structures by finding the shortest cell-to-cell paths from one end of the layer to the next. Results of the study corroborated again previous findings of reduced minicolumnar width in ASD but expanded the same to be inclusive of supragranular, granular and infragranular layers. The reduction was accounted by findings within the peripheral neuropil space.

The peripheral neuropil space of minicolumns contains, among other anatomical elements, the interneurons and inhibitory projections that help frame the activation of modules by an inhibitory surround (Buxhoeveden and Casanova 2002). A diminution of the peripheral neuropil space across the different cortical layers probably reflects involvement of a shared anatomical element. It seems possible that the abnormality belongs to inhibitory elements suggesting that individuals with ASD have a defect in the inhibitory surround of minicolumns. Proof for this supposition has been obtained from studies of tactile resolution and habituation to stimuli (Tommerdahl et al. 2007; Tannan et al. 2008) and others using a lateral masking paradigm to assess visuospatial processing information (Keita et al. 2011).

Surround or lateral inhibition is a common feature of cortical modules. They are easily recognized in sensory regions where an excitatory receptive field is surrounded by an inhibitory area or areas. Probably the best known examples of cells organized in center surround fashion are located in the retina and the lateral geniculate nucleus. The presence of surround inhibition allows for contrast detection. Neurons with classical surround inhibition respond best to a stimulus applied to its central receptive field but less well as the stimulus activates the surrounding areas. An abnormality of lateral inhibition in the brains of autistic individuals would affect how the individual processes information, specifically why do they seem to examine individual elements of a figure but fail to acquire the complete picture of the same.

Besides a minicolumnopathy, the brains of autistic individuals evidence an effacement of the normal lamination pattern, variation in neuronal density, and gyral malformations (Schmitz and Rezaie 2008). The described defects suggest an abnormality during brain development. These malformations or dysplasias define a disturbance of cellular proliferation, migration, and cortical organization that starts long before a person is born. A recent study used cortical width as a proxy measure of dysplasia (Hustler et al. 2007). The study reviewed the histologic findings in 8 ASD individuals and an equal number of controls in three different regions of eulaminate cortex (BA7, 9, and 21). Findings revealed an increased number of cells in both lamina I and subplate region of ASD individuals. A later study using the same patient population evaluated the boundary of the gray and white matter by computerized image analysis (Avino and Hustler 2010). The results indicated the presence of supernumerary cells beneath the cortical plate probably the result of a defect of cell migration or failed apoptosis within the subplate region. Studies by Wegiel and colleagues (2010) suggest that defects of neurogenesis and neuronal migration account for described dysplastic changes.

A recent study by our group attempted to identify the nature of the dysplastic process (Casanova et al. 2013). The study analyzed celloidin-embedded and Nissl stained full coronal sections of 7 autistic (ADI-R diagnosed) and an equal number of age/sex matched controls. Sections were scanned and manually segmented. Digitized images were analyzed with an algorithm using Laplace's equation to measure cortical width. Results of the study revealed multiple circumscribed regions of dysplastic cortex distributed throughout the whole brain of ASD individuals. Described defects varied greatly in overall size and location but were most abundant within the frontal lobes. Microscopic assessment of dysplastic regions revealed and absence of dysmorphic or balloon neurons, and no evidence of gliosis. Affected gyri were not mushroom shaped, nor did they acquire the shape of tubers. There was no evidence of ulegyria. Neuronal morphometry suggested the presence of smaller pyramidal cells and a total reduction in the number of interneurons. The authors concluded that supernumerary minicolumns in ASD are the product of heterochronic divisions of periventricular germinal cells. When these cells are forced to divide and migrate to the cortex they are uncoupled from those that migrate tangentially (interneurons) thus creating an excitatory-inhibitory imbalance.

13.2 Discussion

During corticogenesis progenitor cells located within the ventricular zone provide for symmetric divisions wherein the total number of postmitotic cells will help define the future number of minicolumns within the cerebral cortex. A subsequent wave of asymmetric divisions (following the fortieth embryonic day) provides for daughter cells that migrate along glial fascicles to the cerebral cortex. During mammalian evolution this process of symmetrical and asymmetrical divisions has resulted in a 1,000-fold increase in cortical surface area (considering the dimensions of the brains of mice and humans) but only a two or three-fold increase in cortical thickness. Putative factors affecting the total number of cells and minicolumns within the cerebral cortex include: (1) the number of founder cells, (2) the duration of the cell-division cycle, (3) the number of successive cell cycles during the period of neurogenesis, (4) the modes of cell division, and (5) selective cell death. The minicolumnar findings from previous studies (Casanova et al. 2002a, b, 2006a) suggest the presence of supernumerary symmetrical cell divisions of periventricular germinal cells providing an increased number of radial units (minicolumns) within the cerebral cortex of individuals with autism spectrum disorders.

Some researchers believe that if neuronal connectivity is kept at a fixed percentage during cortical expansion, then a significant portion of the growth would be devoted to maintaining wiring. This would mean that bigger brains would require increasing axonal lengths and a consequent reduction in neural computational speed. It is this consideration that has driven some researchers to maintain that our brains are almost at their limit for biologic intelligence (Hofman 2001). Encephalization has therefore provided for increased cortical gray with disproportionate growth in white matter. It has been suggested that each minicolumn is connected to on the order of 1,000 other modules (Casanova 2004). Findings of increased number of cortical minicolumns in ASD would help explain why MRIs studies have reported enlarged gray matter volumes with an even bigger increase in white matter (Herbert et al. 2004).

The increased white matter seen with supernumerary minicolumns takes primarily the form of short-range fibers. The total layout of the connectivity within the brain minimizes total connections costs. Long-range connections incur the penalty of increases in both conduction times and metabolism. A small-world network helps to reach neighboring and distant modules in a small number of steps. A restricted number of longer projections, may help explain why despite larger brains, on average, ASD individuals have a reduction in size of their corpus callosum. Indeed, long commissural projections may serve as a better index of macrocolumnar rather than microcolumnar connectivity (Casanova et al. 2004).

The presence of a minicolumnopathy in autism spectrum disorders distances the condition from classic neuropathology by emphasizing abnormalities in cell assemblies rather than single cells. During development these cell assemblies organize into columnar aggregates. Columns will be active when a given set of features is present in the input. Objects with similar features activate columns in proximity to each other. The unequal distribution of synaptic weights among neurons allow for variability in the output. According to Gustafsson (1997) a wide column that contains many neurons has a broader range of synaptic weights than a narrow one. This feature allows for a greater variability in the processing of objects that activate columns. Gustafsson (1997) believes that narrow columns with fewer cells facilitates discrimination, whereas wider columns facilitate generalization.

The morphometric variability of minicolumns in autism spectrum disorders is very specific. Multiple studies have corroborated that diminished minicolumnar width is associated with major reductions in their peripheral neuropil space. While development of the core compartment of the minicolumn is constrained by the process of radial cell migration and its attendant radially oriented projections, anatomical elements within the peripheral neuropil space is more heterogeneous as to their sources, modes of migration, morphogenesis and synaptogenesis. The large variability of the peripheral neuropil space in humans as compared to other species suggests that genetic and epigenetic influences may act primarily in this compartment (Casanova et al. 2009). This heterogeneity may provide the basis for adapting minicolumns to function within specific networks during both evolution and development.

Variability among the multiple components of the minicolumn may contribute to the fault tolerance among larger networks such as macrocolumns. Variability may allow individual minicolumns to compute different functions within the same module that can be used to their advantage with respect to sensitivity to error. This variability in otherwise redundant circuits is the basis for majority voting circuits (Stroud 1994). In effect plasticity is associate not only with the tuning of synaptic activity states but also with the optimal selection among alternate sub-networks of microcircuits developing within a given context. Competition among networks allow for circuit optimization. Morphometric variability of minicolumns, representing variations in their microcircuitry, may provide the substrate of this competition and the basis for adapting learned behavior to context (Casanova 2008).

It has been said that, "During development, neurogenetic programs interact with epigenetic factors to regulate formation of cortical microcircuit templates, which are then shaped and pruned by differential patterns of sensory activity. Incipient behavioral patterns in turn constrain selection for mechanisms of plasticity, establishing a dynamic, mutually informing loop, a process referred to as the Baldwin effect" (Casanova 2008, p. 357). In this regard the variability in minicolumnar shapes and microcircuitry gives rise to a greater potential for combinatorial activity with overlapping or neighboring networks. For these reasons minicolumnar variability may be a phenotypic character under selection in the truest Darwinian sense.

The constellation of minicolumnar findings reported in this chapter along with other neuropathological findings described in the literature (e.g., abnormalities of gyrification and lamination, heterotopias) indicate that many cases diagnosed as ASD are linked through a cascading chain of events propitiated by disordered periventricular cell divisions. In this regard ASD should be considered a sequence rather than a syndrome. In a syndrome the various relationships between pathological findings are not understood. In ASD one primary defect can account for defects in neurogenesis, cellular migration, and corticogenesis. Like other sequences (e.g., Pierre Robin) there may be multiple known causes for ASD. Nevertheless, neuropathological studies of idiopathic and syndromic autism (e.g., tuberous sclerosis,

Ehlers-Danlos syndrome, congenital cytomegalovirus) all evidence an injury of the germinal cell matrix (Casanova et al. 2013).

The presence of a minicolumnopathy and corresponding defects in surround inhibition has promoted a possible intervention based on repetitive transcranial magnetic stimulation (rTMS). This technique offers a non-invasive method of altering patterns of brain activity. TMS operates based on Faraday's Law of induction that describes how a changing magnetic field procreates the flow of electrical current in a nearby conductor. Studies have shown that low frequency rTMS increases the activation of inhibitory circuits probably through the mechanisms of long-term depotentiation (Hoffman and Cavus 2002). We have hypothesized that the presence of double bouquet cells that bear a perpendicular geometrical relationship to the cortex makes them likely targets for induction by low frequency rTMS (Sokhadze et al. 2009). "Slow" rTMS in this regard would help restore the inhibitory tone of the cerebral cortex that my group has tested as a reduction in excess gamma band activity. We have also reported on the positive effects of rTMS on clinical and behavioral questionnaires and a visual attention task employing Kanizsa illusory figures (Sokhadze et al. 2009). Thus far we have found that in individuals with ASD gamma activity does not differentiates between target and non-target stimuli. Following rTMS, individuals with ASD showed significant improvement in their discriminatory activity between relevant and irrelevant stimuli. We have also noted significant reductions in irritability, repetitive behaviors, and response to error monitoring and post-error correction as a result of rTMS (Sokhadze et al. 2012).

References

Avino TA, Hustler JJ (2010) Abnormal cell patterning at the cortical gray–white matter boundary in autism spectrum disorders. Brain Res 1360:138–146

Baruth JM, Casanova MF, El-Baz A, Horrell T, Mathai G, Sears L, Sokhadze E (2010) Low frequency repetitive transcranial magnetic stimulation (rTMS) modulates evoked-gamma frequency oscillations in autism spectrum disorder (ASD). J Neurother 14(3):179–194

Bauman ML, Kemper TL (1987) Limbic involvement in a second case of early infantile autism. Neurology 37(Suppl 1):147

Bauman ML, Kemper TL (2005) Structural brain anatomy in autism: what is the evidence? In: Bauman ML, Kemper TL (eds) The neurobiology of autism, 2nd edn. Johns Hopkins University Press, Baltimore, pp 121–135

Buxhoeveden DP, Casanova MF (2002) The minicolumnar hypothesis in neurosciences. Brain 125:935–951

Buxhoeveden DP, Switala AE, Litaker M, Roy E, Casanova MF (2001) Lateralization of minicolumns in human planum temporale is absent in nonhuman primate cortex. Brain Behav Evol 57(6):349–358

Buxhoeveden DP, Semendeferi K, Buckwalter J, Schneker N, Switzer R, Courchesne E (2006) Reduced minicolumns in the frontal cortex of patients with autism. Neuropathol Appl Neurobiol 32:483–491

Casanova MF (2004) White matter volume increase and minicolumns in autism. Ann Neurol 56 (3):453

Casanova MF (2008) The significance of minicolumnar size variability in autism: a perspective from comparative anatomy (ch. 16). In: Zimmerman A (ed) Autism: current theories and evidence. Current clinical neurology. The Humana Press, Totowa, pp 349–360
Casanova MF (2013) The minicolumnopathy of autism spectrum disorders (ch. 3.7). In: Buxbaum JD, Hof PR (eds) The neuroscience of autism spectrum disorders. Academic, Oxford, pp 327–333
Casanova MF, Switala AE (2005) Minicolumnar morphometry: computerized Image analysis. In: Casanova MF (ed) Neocortical modularity and the cell minicolumn. Nova Biomedical, New York, pp 161–180
Casanova MF, Buxhoeveden DP, Switala AE, Roy E (2002a) Minicolumnar pathology in autism. Neurology 58:428–432
Casanova MF, Buxhoeveden DP, Switala AE, Roy E (2002b) Neuronal density and architecture (Gray Level Index) in the brains of autistic patients. J Child Neurol 17:515–521
Casanova MF, Buxhoeveden DP, Brown C (2002c) Clinical and macroscopic correlates of minicolumnar pathology in autism. J Child Neurol 17:692–695
Casanova MF, Araque J, Giedd J, Rumsey JM (2004) Reduced brain size and gyrification in the brains of dyslexic patients. J Child Neurol 19:275–281
Casanova MF, Van Kooten IAJ, Switala AE, Van Engeland H, Heinsen H, Steinbusch HWM, Hof PR, Trippe J, Stone J, Schmitz C (2006a) Minicolumnar abnormalities in autism. Acta Neuropathol 112:287–303
Casanova MF, Van Kooten IAJ, Switala AE, Van Engeland H, Heinsen H, Steinbusch HWM, Hof PR, Schmitz C (2006b) Abnormalities of cortical minicolumnar organization in the prefrontal lobes of autistic patients. Clin Neurosci Res 6:127–133
Casanova MF, Trippe J, Tillquist C, Switala AE (2009) Morphometric variability of minicolumns in the striate cortex of Homo sapiens, Macaca mulatta, and Pan troglodytes. J Anat 214 (2):226–234
Casanova MF, El-Baz AS, Kamat SS, Dombroski BA, Khalifa F, Elnakib A, Soliman A, Allison-McNutt A, Switala AE (2013) Focal cortical displasias in autism spectrum disorders. Acta Neuropathol Commun 1:67. doi:10.1186/2051-5960-1-67
DeFelipe J (1999) Chandelier cells and epilepsy. Brain 122(10):1807–1822
Favorov OV, Kelly DG (1994a) Minicolumnar organization within somatosensory cortical segregates, I: development of afferent connections. Cereb Cortex 4(4):408–427
Favorov OV, Kelly DG (1994b) Minicolumnar organization within somatosensory cortical segregates, II: emergent functional properties. Cereb Cortex 4(4):428–442
Grice SJ, Spratling MW, Karmiloff-Smith A et al (2001) Disordered visual processing and oscillatory brain activity in autism and Williams syndrome. Neuroreport 12:2697–2700
Gustafsson L (1997) Inadequate cortical feature maps: a neural circuit theory of autism. Biol Psychiatry 42:1138–1147
Herbert MR, Ziegler DA, Makris N, Filipek PA, Kemper TL, Normandin JJ, Sanders HA, Kennedy DN, Caviness VS Jr (2004) Localization of white matter volume increase in autism and developmental language disorder. Ann Neurol 55(4):530–540
Hofman MA (1985) Neuronal correlates of corticalization in mammals: a theory. J Theor Biol 112(1):77–95
Hofman MA (2001) Brain evolution in hominids: are we at the end of the road? In: Falk D, Gibson KR (eds) Evolutionary anatomy of the primate cerebral cortex. Cambridge University Press, Cambridge, pp 113–127
Hoffman RE, Cavus I (2002) Slow transcranial magnetic stimulation, long-term depotentiation, and brain hyperexcitability disorders. Am J Psychiatr 159:1093–1102
Hustler JJ, Love T, Zhang H (2007) Histological and magnetic resonance imaging assessment of cortical layering and thickness in autism spectrum disorders. Biol Psychiatry 61:449–457
Just MA, Cherkassky VL, Keller TA, Minshew NJ (2004) Cortical activation and synchronization during sentence comprehension in high-functioning autism: evidence of underconnectivity. Brain 127(8):1811–1821

Keita L, Mottron L, Dawson M, Bertone A (2011) Atypical lateral connectivity: a neural basis for altered visuospatial processing in autism. Biol Psychiatry 70(9):806–811
Kemper TL, Bauman ML (1998) Neuropathology of infantile autism. J Neuropathol Exp Neurol 57(7):645–652
Lorente de Nó (1938) Cerebral cortex: architecture, intracortical connections, motor projections. In: Fulton JF (ed) Physiology of the nervous system. Oxford University Press, Oxford
Macdonald KD, Fifkova E, Jones MS, Barth DS (1998) Focal stimulation of the thalamic reticular nucleus induces focal gamma waves in the cortex. J Neurophysiol 79:474–477
Mesulam M-M (2002) The human frontal lobes: transcending the default mode through contingent encoding (ch. 2). In: Stuss DT, Knight RT (eds) Principles of frontal lobe function. Oxford University press, Oxford, pp 8–30
Mountcastle VB (1997) The columnar organization of the neocortex. Brain 120(4):701–722
Mountcastle VB (1998) Perceptual neuroscience: the cerebral cortex. Harvard University Press, Cambridge, MA
Opris I, Santos L, Gerhardt GA, Song D, Berger TW, Hampson RE, Deadwyler SA (2013) Prefrontal cortical microcircuits bind perception to executive control. Sci Rep 3:2285
Schlaug G, Schleicher A, Zilles K (1995) Quantitative analysis of the columnar arrangement of neurons in the human cingulate cortex. J Comp Neurol 351(3):441–452
Schmitz C, Rezaie P (2008) The neuropathology of autism: where do we stand? Neuropathol Appl Neurobiol 34:4–11
Sokhadze EM, El-Baz AS, Baruth J, Mathai G, Sears L, Casanova MF (2009) Effects of low frequency repetitive transcranial magnetic stimulation (rTMS) on gamma frequency oscillations and event-related potentials during processing of illusory figures in autism. J Autism Dev Disord 39(4):619–634
Sokhadze EM, Baruth JM, Sears L, Sokhadze GE, El-Baz AS, Casanova MF (2012) Prefrontal neuromodulation using rTMS improves error monitoring and correction function in autism. Appl Psychophysiol Biofeedback 37(2):91–102
Stroud CE (1994) Reliability of majority voting based VLSI fault-tolerant circuits. IEEE Trans VLSI Syst 2:516–521
Sukov W, Barth DS (2001) Cellular mechanisms of thalamically evoked gamma oscillations in auditory cortex. J Neurophysiol 85:1235–1245
Szentágothai J (1978) The Ferrier Lecture, 1977: the neuron network of the cerebral cortex: a functional interpretation. Proc R Soc Lond B 201:219–248
Tannan V, Holden JK, Zhang Z, Baranek GT, Tommerdahl M (2008) Perceptual metrics of individuals with autism provide evidence for disinhibition. Autism Res 1:223–230
Tommerdahl M, Tannan V, Cascio CJ, Baraneck GT, Whistle BL (2007) Vibrotactile adaptation fails to enhance spatial location in adults with autism. Brain Res 1154:116–123
Wegiel J, Kuchna I, Nowicki K, Imaki H, Wegiel J, Marchi E, Ma SY, Chauhan A, Chauhan V, Wierzba Bobrowicz T et al (2010) The neuropathology of autism: defects of neurogenesis and neuronal migration, and dysplastic changes. Acta Neuropathol 119:755–770

Chapter 14
Clinical Applications of Electrophysiological Approaches Based on Cortical Modularity in Autism

Estate M. Sokhadze, Lonnie Sears, Ayman S. El-Baz, Allan Tasman, and Manuel F. Casanova

Abstract Autism spectrum disorder (ASD) consists of a set of pervasive developmental problems marked by measurable deficits in social interaction and communication, often coupled with specific and repetitive patterns of behavior. Featured restrictions in the capability to communicate and remain attentive can directly relate to the individual's ability to interact with others within societal norms. Comparing ASD subjects to neurotypical (NT) controls, and other developmental disorders such attention deficit/hyperactivity disorder (ADHD), previous investigations had shown that evoked electroencephalographic (EEG) gamma oscillations and event-related potentials (ERPs) to sensory stimuli do display certain aberrations in latency or amplitude in the ASD individuals. To investigate the aforementioned phenomena the series of studies by our group employed EEG recording technology while subjects participated in several oddball-paradigm reaction time tests. The paper reports on the differences in behavioral reactions as well as variances in amplitude and latency of ERP in autistic individuals and age-matched matched NT controls and children with ADHD. Subjects were evaluated using various ERP components as well as power of EEG gamma oscillations recorded at fronto-central and parietal sites. Findings of our studies suggest that the irregularities arise from deficits in the perception, integration and cognitive processing of sensory inputs. Previous research investigating the neuropathology of autism has identified

E.M. Sokhadze, Ph.D. (✉) • A. Tasman, M.D.
Department of Psychiatry & Behavioral Sciences, University of Louisville,
401 E. Chestnut Street, #600, Louisville, KY 40202, USA
e-mail: tato.sokhadze@louisville.edu; allan.tasman@louisville.edu

L. Sears, Ph.D.
Department of Pediatrics, University of Louisville, Louisville, KY, USA
e-mail: lonnie.sears@louisville.edu

A.S. El-Baz, Ph.D.
Department of Bioengineering, University of Louisville, Louisville, KY, USA
e-mail: aselba01@louisville.edu

M.F. Casanova, M.D.
Department of Psychiatry, University of Louisville, Louisville, KY, USA
e-mail: manuel.casanova@louisville.edu

M.F. Casanova, I. Opris (eds.), *Recent Advances on the Modular Organization of the Cortex*, DOI 10.1007/978-94-017-9900-3_14

abnormalities in the structure, number and activity of the cortical minicolumns. The minicolumns of ASD individuals appear in greater number coupled with increased neuronal density due to a reduction in the volume of peripheral neuropil space and neuronal cell bodies. Such a cortical and cellular arrangement favors the formation of short intralobular connections between neurons at the expense of longer interlobular fibers. Our studies propose that aberrations in sensory processing and functional cortical binding, as evidenced by EEG recordings related to the tasks, further reflect the underlying abnormalities of minicolumns in ASD individuals. Thus, the results of our studies suggest that dysfunction of sensory information processing by way of minicolumn irregularity may in turn lead to symptoms commonly associated with ASD.

Keywords Autism spectrum disorder • Gamma oscillations • Electroencephalography • Event-related potential • Minicolumn

14.1 Introduction

Autism is a severe developmental disorder characterized by deficits in social interaction and communication, and by stereotyped, restrictive and repetitive behavior and interests (DSM-IV TR 2000). Arguably, two-thirds of autistic subjects are mentally retarded. The rate of occurrence is reported to be increasing (up to 1 in 150, current data claiming up to 1 in 100). According to the Autism Society of America the current cost to the national economy for autism treatment, education, and services is around $90 billion a year. More than 1.5 million Americans today are believed to have some form of autism spectrum disorder (ASD).

Informed clinical consensus defines autism as a behavioral syndrome characterized by pervasive impairment in several areas of development: social interaction, communication skills, and repertoire of interests and activities. Thus far there have been no neuropathological findings nor laboratory/performance based measures providing construct validity to the diagnosis. It is not surprising that given the complexity of the clinical symptoms researchers have claimed abnormalities in widely divergent areas of the brain (Palmen et al. 2004; Bauman and Kemper 2005). Generalized deficits have been inferred from recent neuropathological studies suggesting disturbances of brain size, cortical lamination and columnarity (Bailey et al. 1998; Casanova et al. 2002a, b; Lainhart 2006).

In the absence of pathognomonic abnormalities research in autism has been guided by a variety of ideologies and epistemological assumptions each contributing to the development of explanatory models or "theories": executive functions (Ozonoff et al. 1991), "weak central coherence" (Frith and Happe 1994), complex information processing (Minshew et al. 1997), theory-of-mind (Baron-Cohen et al. 1985), and empathy (Baron-Cohen et al. 2002). Recent attempts at deriving such an overarching meta-theory have focused on a basic abnormality of neural connectivity (Belmonte et al. 2004a, b). This model is empirically based on lack of coordinated brain activity (functional imaging) and abnormal "binding" (EEG tracings) in the brains of autistic patients (Brock et al. 2002; Just et al. 2004; Brown 2005; Koshino et al. 2005).

By themselves these theories are incapable of accounting for all of the developmental, social, cognitive and affective variables which define autistic psychopathology. Autism is thought to be influenced by multiple genes as well as environmental factors thus providing for multifactorial inheritance. One more current theory of autism take "minicolumnar" perspective and is based on neuropathology of autism (Casanova et al. 2002a; Casanova 2006).

14.1.1 Neuropathology of Autism

In recent years, neuropathological studies of autism have revealed abnormalities in several brain regions. The studies of Bauman and Kemper (2005) demonstrated reduced cell size and increased cell-packing density in the amygdala, entorhinal cortex, subiculum, mammillary bodies, and septum. The above related changes suggest that the underlying pathology in autism consists of widely distributed histological abnormalities. The available neuropathological and structural imaging data suggest that autism is the result of a developmental lesion capable of affecting brain growth. One possible explanation for this is the recent finding of minicolumnar abnormalities in autism (i.e., minicolumns of reduced size and increased numbers) (Casanova et al. 2002a). In this initial study measures of minicolumnar morphometry were obtained relative to pyramidal cell arrays in nine autistic cases and an equal number of controls. The feature extraction properties of the algorithms were corrected for minicolumnar fragments, curvature of the tissue section, and 3D proportions, so called stereological modeling (Casanova and Switala 2005). Later on, the same patient population was used to confirm the presence of cortical radial abnormalities in a study using the Grey Level Index (GLI) ratio, i.e., area covered by Nissl-stained to unstained elements in postmortem samples (Casanova et al. 2002b).

14.1.2 Minicolumns and Inhibition

Minicolumns consist of radially arranged cell arrays along with some of their afferent, efferent and interneuronal connections (Casanova et al. 2003). Minicolumns are considered a basic anatomical and physiological unit of the cortex (Mountcastle 2003). An increase in the number of minicolumns is thought to underlie the neocortical expansion accompanying human encephalization, i.e., the process by which the brain has increased in size to a degree greater than expected when taking body size into account (Rakic 1995). Empirical evidence and theoretical models indicate that local circuit neurons increase in number, complexity and proportion relative to projection neurons during primate encephalization (Hofman

1985; Rakic 1975). The interaction of diverse projection and interneuron types within the developing ontogenetic cell column gives rise to characteristic architectonic components and response properties within the adult cell minicolumn (Kozloski et al. 2001; Casanova et al. 2003). Four of these features have been investigated in studies of minicolumnar morphometry: apical dendritic bundles, double bouquet axon bundles, pyramidal cell arrays, and myelinated axon bundles of projection neurons (del Río and DeFelipe 1997; Ong and Garey 1990; Peters and Sethares 1991, 1996; Seldon 1981a, b; Viebahn 1990; von Bonin and Mehler 1971). Each of these features provides complementary information, and study of each entails specific experimental and analytic challenges.

Double-bouquet cells in the peripheral neuropil space of minicolumns provide a "vertical stream of negative inhibition" (Mountcastle 1997) surrounding the minicolumnar core. Other GABAergic cells in the minicolumn, having collateral projections extending hundreds of microns tangentially, provide lateral inhibition of surrounding minicolumns on a macrocolumnar scale. The value of each minicolumn's output is insulated to a greater or lesser degree from the activity of its neighbors by GABAergic inhibition in its peripheral neuropil space (see double bouquet cells above). This allows for gradations in amplitude of excitatory activity across a minicolumnar field. Rubenstein and Merzenich (2003) have posited that reductions in GABAergic inhibitory activity may explain some symptomatology of autism, including increased incidence of seizures and auditory-tactile hypersensitivity (see also Casanova et al. 2003).

This hypothesis is consistent with findings of reduced minicolumnar peripheral neuropil space in the neocortex of autistics relative to controls (Casanova et al. 2002a). In this model, reduced peripheral neuropil would result in smaller minicolumns which would coalesce into discrete, isolated islands of coordinated excitatory activity. These islands could serve as potential ictal foci. Moreover, their autonomous activity would hinder the binding of associated cortical areas, arguably promoting focus on particulars as opposed to general features. We proposed that these neurodevelopmental "minicolumnar" abnormalities (compact minicolumns, deficient lateral inhibition, high cortical excitation/inhibition ratio, etc.) will be manifested in electrophysiological measures as well.

14.2 Electrophysiological Studies on Autism

14.2.1 General Introduction to EEG and Gamma-Band

The electroencephalogram (EEG) is typically subdivided into several frequency bands: delta (0.5–4 Hz), theta (4–8 Hz), alpha (8–13 Hz), beta (13–30 Hz), and gamma (30–80 Hz, which is often limited only to 40 Hz centered activity in 30–50 Hz range). Spectral analysis of the waking EEG in autism have shown increased slow-wave activity, reduced EEG power in frontal and temporal regions, decreased variability, and decreased inter- and intra-hemispheric asymmetries relative to healthy individuals (Cantor et al. 1986; Dawson et al. 1995).

Oscillatory activity in the gamma-band (30–80 Hz) of the EEG has been related to Gestalt perception and to cognitive functions such as attention, learning, and memory (Kaiser 2003). Gamma frequency oscillations have been implicated in a variety of verbal and nonverbal cognitive tasks. Gamma activity increase has been most widely associated with top-down attentional processing and object perception (Gruber and Muller 2005; Gruber et al. 1999). Rodriguez et al. (1999), and was shown to be higher in a delayed match-to-sample task with visual (Tallon-Baudry et al. 2005), auditory (Kaiser et al. 2005), and audio-visual (Kaiser et al. 2005) stimuli, as well as in working memory tasks (Howard et al. 2003). Electrophysiological studies on animal models show strong evidence that synchronized cortical activity in the gamma frequency range could be a correlate of feature binding (Herrmann and Knight 2001). In particular, neurons in the animal brain, which oscillate at gamma frequencies (~40 Hz) are believed to represent the binding of different features of one object to form a single coherent percept (Engel et al. 1992). Human experiments also found induced gamma activity to correlate with binding mechanism (Gray et al. 1989). Binding of widely spread cell assemblies by synchronizing their gamma frequency activity is thought to underlie cohesive stimulus representation in the human brain (Pavlova et al. 2006). According to this assumption, changes in gamma EEG activity have been considered as indicators of processing Gestalt-like patterns (Herrmann and Mecklinger 2000; Keil et al. 1999; Rodriguez et al. 1999).

14.2.2 Abnormalities in Induced Gamma Activity in Autism

It was proposed that the weak "central coherence" (Frith and Happe 1994) in autism could result from a reduction in the integration of specialized local networks in the brain caused by a deficit in temporal binding (Brock et al. 2002). Visual and auditory perception anomalies associated with weak central coherence may be attributed to a reduction in synchronization of gamma frequency activity between networks processing local features. Furthermore, temporal binding deficits can explain some of the features of language processing deficits, executive dysfunctions, and other deficits in social communication in autism.

In one study, a group of adults with autism and a group of control adults viewed pictures of upright or inverted faces while their EEG was recorded. In the control group, upright faces induced gamma activity over the frontal regions, which was larger than the inverted face condition. In autistic patients by contrast, both upright and inverted faces generated the same level of gamma therefore giving no useful information in order to discriminate between upright and inverted faces (Grice et al. 2001). This result shows that whereas in normal subjects attention was greater to upright faces, in those with autism equal attention was paid to both upright and inverted faces. Brown (2005) tested adolescents with autism in an experiment which presented either shapes with visual illusions (Kanizsa figures, see Kanizsa 1976) or random pictures. In these perceptual tasks gamma bursts have been reliably recorded over the posterior attention areas in unimpaired adults at both around 300 ms (Tallon-Baudry et al. 1996, 2005), and at around 100 ms (Herrmann et al. 1999). Whereas both groups showed almost identical activity in the Kanizsa-random condition they were very different in the Kanizsa-shape condition. When perceiving the shape, the group with autism differ from the control group in three ways: (1) overall gamma power is much higher than in the control group; (2) the group with autism shows a very early burst of gamma activity between 80 and 120 ms; and (3) the later gamma component in the group with autism is more powerful and starts much earlier. The inability to reduce gamma activity according to Brown (2005) would lead to the inability to decide which event requires paying attention to when there is a choice of two or more. This may be the underlying cause of an inability to direct attention either internally or externally.

Therefore, excessive gamma activity can be linked to a reduction in the ability to focus attention. In autism, this impairment of focus might be related to an inability to switch-off the activity of irrelevant neural circuits. Powerful and uninhibited gamma activity in autism suggests that none of the circuits in the brain can come to dominance because too many are active at the same time (Brown et al. 2005). A recently proposed "temporal binding deficit" hypothesis of autism (Rippon et al. 2007) suggests that many features of autism, such as superiority in processing of detail (local processing) and disadvantage in global processing, necessitating integration of information either over space (e.g., visuo-spatial perception), time, or context (integration of words into meaningful sentences) can be explained by a failure of binding between cortical areas. Increased and earlier burst of gamma activity in autism might be linked to increased ability to perform tasks where ignoring the contextual information is an advantage. The disadvantages of such an overactive system however are in an inflexible re-focusing of the brain system onto events rapidly occurring in sequence. For non-autistic people it is thought that the gamma binding signal allows the rapid formation of thought networks, which can be updated very quickly from both external and internally generated information (e.g. memory or feelings) (Singer 1999). Another disadvantage of an overactive gamma system is the increased likelihood of epilepsy or epileptic-like absence that is apparent in a significant percentage of autistic patients (Tuchman and Rapin 2002).

According to the "representational" hypothesis of gamma activity by Tallon-Baudry (2003), induced gamma oscillations are observed not only in response to a coherent object, but are also present during activation or rehearsal of an internal representation. This suggests that induced gamma oscillations may also underlie the activation of the neural representation of the object in mind. In addition, the topography of gamma oscillations is highly task-dependent, suggesting that depending on the task to perform, different functional areas are dynamically and flexibly bound by oscillatory synchrony. Thus in a broader framework, oscillatory synchrony should not be considered as a correlate of a particular cognitive process, but rather as a mechanism that could bind together the sensory and cognitive properties of an object into the experienced entity. Abnormalities of gamma oscillatory synchrony in autism can therefore result in significant perceptual and cognitive deficits.

14.2.3 ERP in Selective Attention Tests in Autism

The event-related potentials (ERP) reflect activation of neural structures in sensory cortex, association cortical areas, and brain areas related to higher order cognitive processes. They can be divided into short-latency (exogenous, e.g., N100) and long-latency (endogenous, e.g., N200, P300) ERPs. The endogenous ERP components reflect more cognitive, higher-level, less modality-specific processing, while the exogenous ERP reflect more basic, modality-specific processing. Neural generators of ERPs for auditory and visual stimuli are respectively located in the auditory cortex of the temporal lobe and in the occipital cortex. The most extensively studied endogenous ERP is the P300 (300–500 ms post-stimulus). The P300 is usually obtained in oddball paradigm, wherein two stimuli are presented in a random order, one of them frequent (*standard*) and another one rare (*target*) (Polich and Herbst 2000; Polich 2003; Pritchard et al. 2004).

Several studies have shown that children and adolescent with autism show abnormalities in ERPs (reviewed in Kemner et al. 1999; Bomba and Pang 2004). In one study the fronto-central N100 component has been found to be shorter in latency and higher in amplitude in an oddball task in children with autism than in normal children (Oades et al. 1988). Other studies have reported a normal N1b in autistic children (Lincoln et al. 1993; Kemner et al. 1995), while still others believe the same to be reduced and delayed (Bruneau et al. 1999). Autistic children have been found to differ from typical children mainly with respect to the P300 in standard oddball tasks. Kemner et al. (1995, 1999) have reported abnormally small occipital P300 in response to visual stimuli. Courchesne et al. (1989) also found a smaller frontal N450 to visual stimuli. Studies using a simple visual target detection have found no differences in the P3b to targets in autistic children compared to controls, but abnormalities were present in dissociations of frontal (delayed) and posterior P300 (intact) in spatial visual attention tasks (Townsend et al. 2001). Deficits in central auditory processing in autism, as indexed by ERPs, have been described by different authors (Courchesne et al. 1989; Ciesielski et al. 1995; Ferri et al. 2003; Kemner et al. 1999; Lincoln et al. 1993; Oades et al. 1988). In children with autism the most consistent and frequently reported abnormality is P3b amplitude attenuation with auditory stimulation (Bomba and Pang 2004; Bruneau et al. 1999; Seri et al. 1999), while P3b latency is spared (Courchesne et al. 1989; Lincoln et al. 1993; Oades et al. 1988). Reduced P3b is explained as reflecting deficiencies in allocation of attentional resources to stimuli (Dawson et al. 1988; Holcomb et al. 1985). It should be noted that P3b attenuation in autism is more specific to the auditory rather than visual modality (Bomba and Pang 2004).

An autistic deficit in rapid shifting of attention has been observed in behavioral tasks during shifts between sensory modalities, between spatial locations, and between object features (Belmonte and Yurgelun-Todd 2003a, b). Inefficient selective attention is manifested as well in deficits during engaging visual attention in the presence of distracters (Burack 1994), and in dividing attention between visual attributes (Casey et al. 1993). These behavioral observations of autism are complemented by ERP results which reveal that even when people with autism produce normal behavioral output, they tend to do so by abnormal physiological means. Frontal negativities associated with sustained attention are reduced or absent in the autistic brain (Ciesielski et al. 1995), the frontal late positive component to peripheral visual stimuli is delayed, and the visual P3b is highly variable (Courchesne et al. 1989) with a somewhat low average amplitude (Townsend et al. 2001). In addition to these failures of normal modulation of ERP peaks, neural systems in the autistic brain are often inappropriately activated. The visual N2b to novel stimuli is larger when a person with autism is performing a task in a passive observation mode, i.e., even when these novel stimuli are not relevant to the task in question (Kemner et al. 1994). This inappropriate activation occurs across modalities, for instance, when a response is required to an auditory stimulus autistic children manifest an enhanced P3b at occipital sites overlying visual processing areas (Kemner et al. 1995). In general, perceptual filtering in autism seems to occur in an all-or-none manner, with little specificity for the location of the stimulus, for the behavioral relevance of the stimulus, or even for the sensory modality in which the stimulus appears. Autistic attention, it seems, is founded more on the coarse control of general arousal than on selective activation of specific perceptual systems.

14.3 Testing Minicolumnar Hypotheses of Autism in Psychological Tasks with Physiological Recordings

The combination of local sensory hyper-arousal and low-level over-processing of incoming sensory stimuli, and at the same time abnormalities in attention selectivity and focus, according to Baron-Cohen and Belmonte (2005) may tap at the over-connected low-level processing neural networks in autism spectrum disorders. In such over-wired networks signal is insufficiently differentiated from noise or task-irrelevant information and as result information capacity is drastically reduced (Belmonte et al. 2004a, b; Rubenstein and Merzenich 2003). Higher-than-normal noise in cortical processes also affects normal development of differentiated representations, because cortical response selectivity in space and time is a product of balanced inhibitory and excitatory processes (Casanova 2005, 2006). Such over-representation by non-differentiated systems could plausibly account, for example, for the strong aversive reactions to auditory, tactile and visual stimuli that are commonly recorded in autistic individuals. Long-range abnormal neural connectivity model is suggested to explain dysfunctions deficits in high-level complex information processing functions where rapid and integrated operation of many separate neural systems is required (Brock et al. 2002; Minshew et al. 1997; Welchew et al. 2005). In the autistic brain, high local connectivity may develop along with deficient low long-range connectivity (Just et al. 2004; Belmonte et al. 2004a, b; Courchesne and Pierce 2005).

The modular arrangement of the cortex is based on the cell minicolumn: a self-contained ecosystem of neurons and their afferent, efferent, and interneuronal connections. Our preliminary studies indicate that minicolumns in the brains of autistic patients are narrower, with an altered internal organization. More specifically, their minicolumns reveal less space for inhibitory local circuit projections. Based on the descriptions given thus far, it is possible to propose a disruption of the normal balance between excitation and inhibition in the columnar organization of autistic patients. Computer modeling suggests that such an imbalance biases the information processing system towards more signal or discrimination. In this regard, a series of noteworthy studies report that both children and adults with autism were superior to a control group in their ability to discriminate novel, highly similar stimuli (O'Riordan et al. 2001). Autistic children also have superior ability to discriminate display items in visual search tasks (Plaisted et al. 2003). The authors suggested states that enhanced discrimination in autism results from low-level perceptual processing of incoming stimuli or what is called the bottom-up approach. Networks of inhibitory interneurons acting as GABA gated pacemakers are critically involved in gamma oscillations (Bragin et al. 1995; Traub et al. 1996; Bartos et al. 2007; Grothe and Klump 2000; Porjesz et al. 2002). Abnormalities in these mechanisms have been associated with binding problems (the co-activation of neural assemblies), which may be present in both autism and schizophrenia (Brock et al. 2002; Grice et al. 2001; Lee et al. 2003; Shergill et al. 2000).

The increased number of minicolumns reported in autism (Casanova et al. 2002a, b) therefore suggests a disruption during the earlier stages of gestation. The proposed timing correlates well with the observation of a high incidence of pervasive developmental disorders in children with prenatal exposure to thalidomide (Rodier et al. 1996, 1997) and the concurrence of structural and functional brainstem abnormalities in autistic children (Hashimoto et al. 1992; Zikopoulos and Barbas 2010). This early insult may well interfere with some of the unfolding capabilities of the developing brain of an autistic patient. Furthermore, a minicolumnar abnormality may translate difficulties in the integration of information into a delay in language acquisition. In all, minicolumnar abnormalities may incapacitate a patient as a social being by distorting elements of the child's biopsychological experience.

14.3.1 Perceptual, Attentional and Cognitive Abnormalities in Autism in Visual Oddball Task with Illusory Kanizsa Figures

Processing Kanizsa figures in autism is different and has distinct electrocortical reactivity profile as it was noted above. In autism, the increased gamma activity indicates that activity induced by perceptual process starts earlier and continues longer because neural networks subserving cognitive processes involved in combining information processing are not functioning normally (Brown 2005; Casanova et al. 2013; Sokhadze et al. 2009b). Reduced ability to decrease gamma oscillations may reflects as well inhibitory deficit, and will result in difficulties in directing attention. Earlier onset and enhanced gamma activity can be used by some individuals to improve their atypical perception abilities and demonstrate well-known islets of superior local processing of details (Brown 2005) at the expense of reduced global processing abilities. We proposed that in autism the pattern of evoked and induced gamma activity demonstrate abnormal functioning of anterior top-down processes controlling matching of perceived complex stimulus with a template held in working memory along with excessive activation and deficits in inhibition of bottom-up processes in the sensory-specific posterior areas. Effects predicted for gamma oscillations should be observed as well in event-related potentials (ERP) recorded during visual three-stimuli oddball task with Kanizsa figures.

Our study was designed to evaluate the cortical responses to Kanizsa visual stimuli evoked at short latencies (below 200 ms post-stimulus) over frontal and parieto-occipital regions-of-interest (ROI) in both children with ASD and typical age-matched controls. Kanizsa stimuli consist of inducer disks of a shape feature and either constitute an illusory figure (square, triangle) or not (colinearity feature). For the purposes of this study the stimuli consisted of Kanizsa targets, Kanizsa non-targets, and non-Kanizsa stimuli. We focused our analysis on short-latency field event-related potentials (ERP) such as P50, N100, and P200, with the intent of gaining insight into early-stage visual processing abnormalities associated with autism. We included the visually associated, parieto-occipital region-of-interest (ROI) in order to better capture early-stage, visual activity associated with extrastriate areas. The frontal ROI was included as it is associated with working memory, executive function, and selective attention. We hypothesized that individuals with ASD will manifest deficits in early-stage visual processing shown by an augmentation of evoked potentials elicited by task-irrelevant distracter stimuli in early stages of visual processing, and this will consequently disrupt stimulus discrimination as compared to the control group.

In the task used in the study subjects were required to respond with a button-press to rare (25 % probability) Kanizsa squares (targets) among Kanizsa triangles (rare non-target distracters, 25 % probability) and non-Kanizsa figures (standards, 50 % probability). This task is a classic three-stimuli 'oddball' with rare target and distracter stimuli presented among frequent standards. The non-target Kanizsa triangle was introduced to differentiate processing of Kanizsa figures and targets. The stimuli consist of either three or four inducer disks which are considered the shape feature, and they either constitute an illusory figure (square, triangle) or not (collinearity feature).

In this study, the mean age of 15 participants enrolled in the ASD group was 13.9 years, while the mean age of the Control (CNT) group ($N = 15$) was 15.5 years. Reaction time to targets was not significantly different in the ASD group compared to the typical control group, but a difference in total error rate was significantly higher (14.8 % vs. 1.4 %, $F = 8.43$, $p = 0.007$). Amplitude of the parieto-occipital P50 in the ASD group was bilaterally higher to non-target Kanizsa stimuli as compared to controls (2.25 μV vs. 0.94 μV; $F = 4.49$, $p = 0.043$). A *Stimulus* (target Kanizsa, non-target Kanizsa, no-Kanizsa standard) X *Group* (ASD, CNT) interaction reached significance ($F = 3.43$, $p = 0.048$) over the left hemisphere and can be described as higher P50 amplitude to both target and non-target Kanizsa figures but not to standard stimuli in the ASD group. Latency of P50 at the right ROI to non-target Kanizsa figures was some 14 ms shorter in the ASD group compared to controls ($F = 4.87$, $p = 0.035$). Latency of the early N100 component in the ASD group as compared to controls was prolonged to targets across all parieto-occipital areas (136.2 vs. 120.8 ms, $F = 5.10$, $p = 0.032$), and was also significant at the right hemisphere ($F = 4.30$, $p = 0.048$). Repeated measures analysis revealed a *Stimulus* X *Group* interaction ($F = 3.43$, $p = 0.04$) across both hemispheres; this effect was expressed as a significantly longer latency to target stimuli in the ASD group with a relatively longer latency to non-target stimuli without any between group differences to standard stimuli. Amplitude of the parieto-occipital P200 in the ASD group was higher to target stimuli and yielded a *Stimulus* (target, standard) X *Group* interaction at the left hemisphere ($F = 7.48$, $p = 0.011$). Comparison of P200 latency between target and non-target Kanizsa stimuli revealed a *Stimulus X Group* interaction ($F = 5.32$, $p = 0.029$) with longer latency to non-targets in the ASD group. Amplitude of the midline frontal P50 in the ASD group compared to controls was significantly higher to non-target Kanizsa figures ($F = 6.24$, $p = 0.019$). Latency of the midline frontal P50 to targets was marginally longer in ASD compared to at the left hemisphere (62.3 ms vs. 48.1 ms, $F = 7.05$, $p = 0.013$). Amplitude of the frontal N100 to targets was more negative in the ASD group compared to controls ($F = 8.63$, $p = 0.007$). A *Stimulus* (target Kanizsa, standard non-Kanizsa, non-target Kanizsa) X *Group* (ASD, CNT) interaction was significant ($F = 7.52$, $p = 0.003$) with the ASD group exhibiting comparable amplitudes to each non-target category of stimuli, whereas the ASD group had a more negative amplitude to targets. Latency of the frontal N100 over both hemispheres showed a *Stimulus* (target Kanizsa, non-Kanizsa standard) X *Group* interaction ($F = 5.80$, $p = 0.023$) with shorter latency to standards in the ASD group. Amplitude of the left frontal P200 was significantly higher in the ASD group compared to controls. It should be noted that amplitude of the frontal P200 in the ASD group was indiscriminative and comparably high in all three conditions.

Over parieto-occipital ROI we found P50 amplitudes to be significantly more positive to non-target Kanizsa stimuli in the ASD group compared to the control group. P50 latency over parieto-occipital ROI was also significantly reduced in the ASD group to non-target Kanizsa figures at the right ROI compared to the control group. The early P50 potential in visual tasks is associated with the sensory processing of attended stimuli and is generally larger to attended stimuli (Hillyard et al. 1995). These results may point to sensory over reactivity in individuals with ASD in early stages of visual processing especially to task irrelevant stimuli. Our earlier study (Sokhadze et al. 2009a) found similar results in a traditional visual three-stimuli oddball task over frontal electrode sites: ASD patients had significantly increased amplitudes of this early positivity to irrelevant distracter stimuli compared to controls. As altered inhibitory control of sensory intake (Khalfa et al. 2004), sensory overload (Ratey and Johnson 1997), and hypersensitivity (Charman 2008) have all been associated with ASD, these results at early stages of visual processing are not surprising.

Also over parieto-occipital ROI the latency of N100 was significantly prolonged to target stimuli in the ASD group relative to the control group. These results are similar to our earlier study (Sokhadze et al. 2009a) where we found a prolonged N100 latency to targets in ASD patients compared to controls over centro-parietal ROI. Delayed negativity in this range has been associated with prolonged memory comparison processes during stimulus discrimination (Casanova et al. 2002a, b). It is plausible that sensory hyper-reactivity at early stages of visual processing may be delaying stimulus discrimination processes at the stage of the N100. Comparison of P200 latencies between target and non-target Kanizsa stimuli over parieto-occipital ROI revealed a *Stimulus X Group* interaction with prolonged latencies to non-targets relative to target stimuli in the ASD group. Since the P200 has been associated with visual categorization processes, these results may point to compromised attentive orientating in ASD patients due to excessive sensitivity in early visual processing stages (Fig. 14.1).

Over frontal ROI P50 amplitudes were significantly more positive to non-target Kanizsa and non-Kanizsa stimuli in individuals with ASD, while P50 latency was prolonged to target stimuli. These results are similar to our findings over parieto-occipital ROI as well as our previous study (Sokhadze et al. 2009b) where ASD patients had excessive frontal positivity to distracter stimuli compared to controls. Again since this early positivity is associated with early categorization and recognition processes and is generally higher to attended stimuli, these results suggest ASD patients are abnormally orientating to task irrelevant stimuli. An exaggerated response to sensory inputs may result in a global inundation of higher level integrative centers with task-irrelevant information during early stages of visual processing.

N100 amplitude over frontal ROI was significantly more negative to target stimuli in the ASD group relative to the control group. Also, a *Group X Stimulus* interaction for N100 latency indicated a significantly reduced latency in the ASD group to non-Kanizsa stimuli compared to target Kanizsa stimuli. These results corroborate with our previous study where we found an augmented N100 amplitude to target stimuli over frontal ROI as well as a prolonged N100 latency to targets over centroparietal ROI in ASD patients. Amplified and delayed responses to rare,

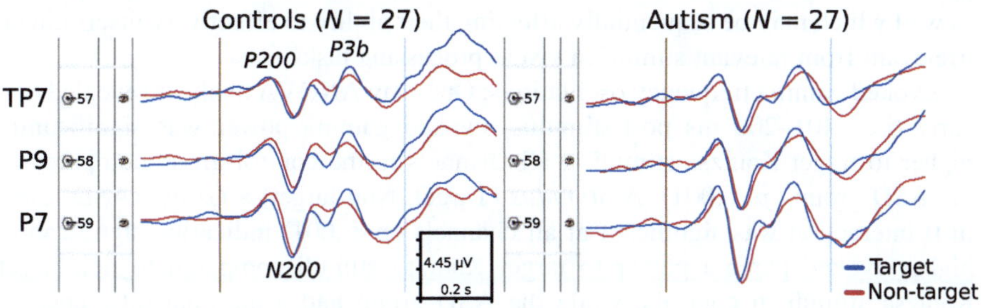

Fig. 14.1 Grandaverage left parietal and parieto-temporal ERP to target and rare non-target Kanizsa illusory stimuli in ASD and control groups (N = 27/per group). The ASD group shows more negative N200 component both to targets and non-targets as compared to controls. The ASD group presents higher and delayed P3b peaks both to target and non-target stimuli as compared to controls

target stimuli in the ASD group may point to visual hypersensitivity and increased general arousal relative to controls, and this may disrupt and delay the processing of target stimuli. Courchesne et al. (1985) reported larger N100 amplitudes at the Cz electrode to all stimuli (Target, Novel, Standard) in autistic patients compared to controls, but the group differences were not found to be significant. These findings may also point to augmented visual responses in ASD during early stages of processing.

P200 amplitude over frontal ROI was found to be equally more positive to all stimuli in the ASD group with a lack of stimulus discrimination; where P200 amplitudes over frontal ROI were indiscernible between target and distracter stimuli in the ASD group, where in the control group P200 amplitude was more positive to targets. The P200 over frontal ROI has been associated with the hierarchal selection of task-relevant features (Kenemans et al. 1993). In ASD globally augmented cortical responses, especially to irrelevant stimuli at early stages of visual processing may be complicating stimulus discrimination processes at the stage of the P200. At behavioral stages, responses of ASD patients did not differ from the control group in reaction time although they had a significantly higher rate or error. The significantly higher rate of error in ASD patients may be a manifestation of early-stage visual sensitivity and consequently a disruption in selective attention and executive function.

Our results show that individuals with ASD have abnormally large cortical responses to task irrelevant stimuli over both parieto-occipital and frontal ROI during early stages of visual processing compared to the control group. Also, ASD patients showed signs of an overall disruption in stimulus discrimination compared to the control group as evidenced by ERPs and a significantly higher rate of motor response errors. Sensory hyper-reactivity has been well documented in the auditory domain (see for review Gomes et al. 2008) but not during visual tasks. Patients with ASD may have sensitivities at early-stages of visual processing as well which may be sequentially affecting their ability to effectively discriminate irrelevant from relevant stimuli in visual processing tasks.

Evoked gamma frequency oscillations: One-way ANOVA analysis revealed that early (i.e., 401–200 ms post-stimulus) evoked gamma power was significantly higher to target Kanizsa stimuli at all channels in the control group compared to the ASD group ($p < .001$). A *Stimulus* (Target, Non-target) × *Group* (ASD, Control) interaction was significant at all channels ($p < .001$) indicating significantly higher evoked gamma power to target Kanizsa stimuli compared to non-target Kanizsa stimuli in controls; while the ASD group had a minimal difference in evoked gamma power between target and non-target Kanizsa stimuli actually demonstrating more gamma power to non-targets. An analysis of differences in evoked gamma power between anterior and posterior regions revealed a *Topography* (Anterior, Posterior) × *Group* (ASD, Control) interaction over the left hemisphere to all stimuli where controls had higher evoked gamma power over frontal (F7) compared to posterior (P7) regions ($F = 5.48$, $p = 0.024$); while the ASD group showed a negligible difference with slightly higher evoked gamma power over posterior (P7) regions. There were no significant hemispheric differences elucidated between ASD and control groups in evoked gamma power during baseline analysis (Figs. 14.2 and 14.3).

Additional investigations characterizing early-stage visual processing deficits using similar 'oddball' paradigms maintain a large amount of significance for future ASD research and treatment. These visual tasks are capable of detecting difficulty in filtering irrelevant sensory stimuli in early stages of visual processing, and could potentially play an important role in identifying sensory processing characteristic of the autistic disorder.

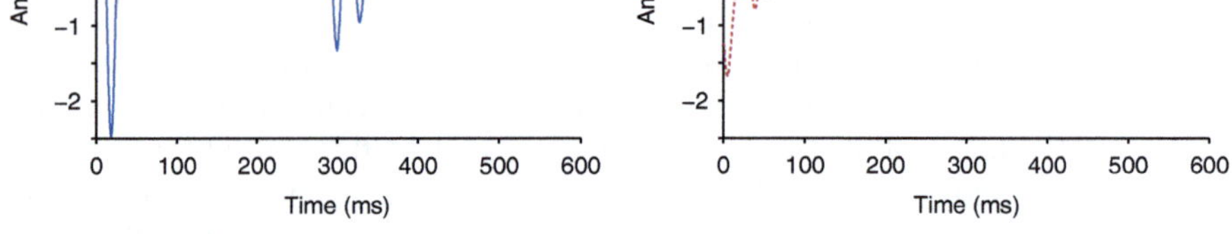

Fig. 14.2 Frontal evoked (peak around 50 ms) and induced (peak around 300 ms) gamma responses in children with autism spectrum disorder (ASD, N = 15) and age-matched neurotypical controls (N = 15) in Kanizsa oddball task. The control group shows higher amplitude of gamma burst to target stimuli, while the ASD group shows higher gamma response to non-target Kanizsa figures

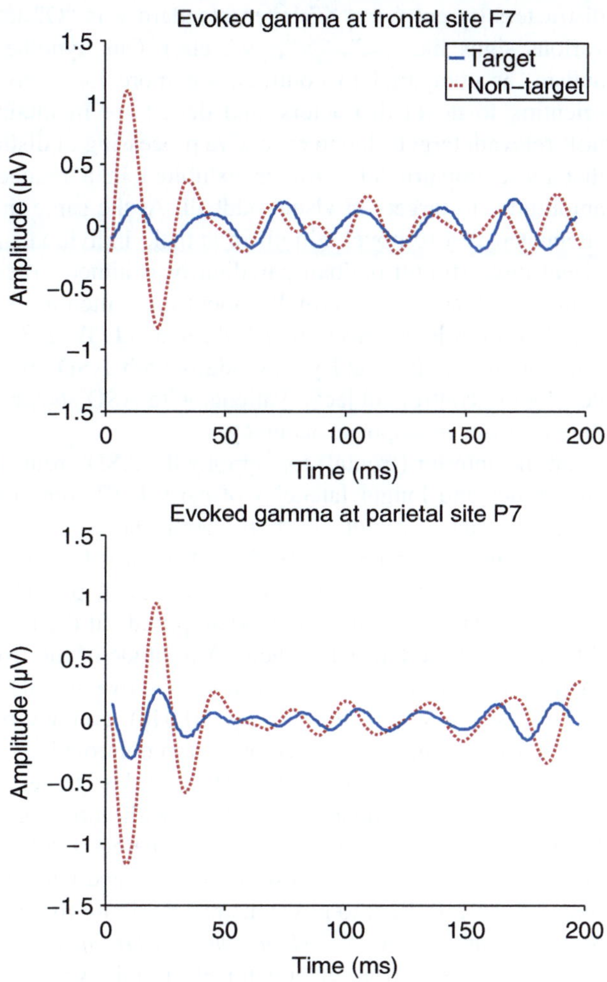

Fig. 14.3 Early evoked gamma oscillations to target and non-target rare Kanizsa figures at the left lateral frontal (F7) and parietal (P7) sites in a group of children with autism (N = 16). Evoked gamma to non-target stimuli is comparable and even larger in amplitude than gamma response to target stimuli

14.4 Other ERP Outcomes in Children with Autism

14.4.1 Processing Novelty in Visual Oddball Task

In order to understand better the cognitive basis of visual processing abnormalities in autism, and in particular deficit of attentive orienting to novel signals, we investigated (Sokhadze et al. 2009a) the novelty-related anterior-frontal P3a component, which is recorded at the frontal sites as a response to unexpected novel event occurring in a sequence of repetitive stimuli, and other posterior ERP indices of orienting and sustained attention (N200, P3b). We investigated amplitude and latency of EPR components in a three-condition oddball task in visual modality to assess reactivity to the novelty. In this experiment we used task with variable novel distracters (target was “X” letter, standard was “O” letter, while distracters were various signs like “<”, “>”, “v”, etc.). Our hypothesis was that children with autism, as compared to controls, are more likely to show deficits in attentive orienting to novel distracters, and deficits in maintaining sustained attention to task-relevant targets due to excessive processing of distracter stimuli. We predicted that these impairments will be exhibited in a reduced parietal N200 and P3b amplitudes to targets in visual oddball. At the same time we predict delayed and attenuated P3a to novel stimuli in autistic individuals compared to controls. The visual three-stimuli oddball paradigm was aimed to test the hypothesis that individuals with autism abnormally orient their attention to novel distracters as compared to controls. A dense-array 128 channel EGI EEG system was used on 11 high-functioning children and young adults with ASD and 11 age-matched, typically developing control subjects. Patients with ASD showed slower reaction times but did not differ in response accuracy.

At the anterior (frontal) topography the ASD group showed significantly higher amplitudes and longer latencies of early ERP components (e.g., P100, N100) to novel distracter stimuli in both hemispheres (e.g., to novel distracters 4.17 vs. 2.09 μV, $F = 5.65$, $p = 0.027$). A *Hemisphere* (left, right) X *Group* (ASD, CNT) interaction effect was found ($F = 5.54$, $p = 0.028$) with the autism group showing significantly higher P100 amplitude at the right hemisphere, without any differences at the left hemisphere. Amplitude of the frontal N100 to targets at the right hemisphere was marginally more negative in ASD group compared to controls (−3.12 vs. −1.69 μV, $F = 4.54$, $p = 0.046$). Latency of the N100 was prolonged bilaterally in ASD group as compared to controls in response to novels (136.2 ms vs. 123.9 ms, $F = 4.79$, $p = 0.043$). *Stimulus* (target, standard, novel) X *Group* interaction was significant ($p = 0.019$) with autism group exhibiting comparable latencies to each category of stimuli, whereas CNT group had faster latency to targets. The ASD group also showed prolonged latencies of late ERP components (e.g., P2a, N200, P3a) to novel distracter stimuli in both hemispheres. There was a *Stimulus* (target, novel) X *Hemisphere* X *Group* interaction ($F = 5.78$, $p = 0.029$) when responses to target and novel stimuli were compared. In particular, the anterior-frontal P2a was similar both to targets and novels in the ASD group and this effect was more visible in the right hemisphere. The only difference in the frontal N200 was found for latency, which was prolonged to novels in the ASD group as compared to the CNT group at the left hemisphere ($F = 4.50$, $p = 0.048$). The latency comparisons for P3a in response to targets and novels showed a *Stimulus* (target, novel) X *Group* interaction ($F = 5.30$, $p = 0.04$), where the ASD group had delayed P3a latency to novels but not targets, whereas the CNT group had longer latency to targets but not novels. However, differences were more profound in the right hemisphere for both early and late ERP components. Our results indicated augmented and prolonged early frontal potentials and a delayed P3a component to novel stimuli, which suggest low selectivity in pre-processing and later-stage under-activation of integrative regions in the prefrontal cortices.

At the centro-parietal topography the ASD group showed significantly prolonged N100 latencies both to standard and novel stimuli ($F = 5.22$, $p = 0.035$), and reduced amplitudes of the N200 component to target stimuli.

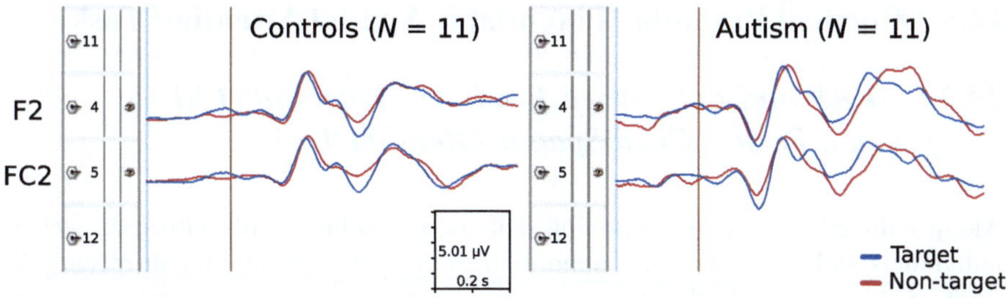

Fig. 14.4 Frontal (F2, FC2) ERP to target and target and novel distracters in visual oddball task in ASD and typically developing children (N = 11/per group). Children with autism show higher magnitude of N100 and N200 components and delayed P3a to novel stimuli

Amplitude of the cento-parietal N200 was globally lower at the right hemisphere in ASD group as compared to controls, however, a between group difference was significant only at the right hemisphere in response to novel distracters ($F = 5.30$, $p = 0.034$). We found a trend in the *Stimulus* (target, standard, novel) X *Group* interaction, which can be described as a more negative N200 to standards and novel distracters, but not to targets in the ASD group as compared to the CNT group ($F = 3.62$, $p = 0.049$). The latency of P3b to novel distracters was delayed in the ASD group ($F = 4.96$, $p = 0.041$), which was more pronounced in the right hemisphere (410 ms vs. 356 ms, $F = 8.09$, $p = 0.012$). In general, the autistic group showed prolonged latencies to novel stimuli especially in the right hemisphere. These results suggest that individuals with autism over-process information needed for the successful differentiation of target and novel stimuli (Fig. 14.4).

Neural systems in the brains of autistic patients are often inappropriately activated (Belmonte and Yurgelun-Todd 2003a; Brown 2005). In particular, abnormally enhanced sensory responses have been reported, and associated with this are deficits in orienting attention and transferring information to higher levels of processing (Townsend et al. 1996, 1999). According to Belmonte and Yurgelun-Todd (2003a, b) perceptual filtering selectivity in autism occurs in an all-or-none manner with little specificity for the task relevance of the stimulus. These authors suggest that perceptual filtering primarily depends on the control of general arousal rather than the activation of a specific perceptual system. Since in many tasks requiring attention persons with autism perform at close to normal levels (Belmonte 2000) despite generally high arousal and low selectivity, some compensatory mechanisms may be operating at a higher stage of processing to sort out the relevant stimuli from poorly discriminated background. One candidate mechanism has been suggested as the active inhibition of irrelevant distracters having passed through earlier filtering (Belmonte and Yurgelun-Todd 2003b). In general, the autistic group showed prolonged latencies to novel stimuli especially in the right hemisphere. These results suggest that autistic subjects over-process information needed for the successful differentiation of target and novel stimuli. The examination of ERP measures in novelty task holds promising potential for contributing to our knowledge of autism.

14.5 Cortical Premotor Potential in Spatial Attention Task

14.5.1 ERP and Lateralized Readiness Potential (LRP) in a Posner Cued Spatial Attention Task

Along with other executive function deficits, individuals with autism as well as individuals with ADHD present abnormalities of spatial attention. Our recent pilot study (Sokhadze et al. 2014) was aimed to understand the abnormal neural and functional mechanisms underlying attention abnormalities in autism and in ADHD by incorporating ERP and behavioral measures of spatial attention. The participants for the study were 24 high-functioning ASD and 14 individuals with ADHD who complied with ERP task requirements and tolerated EEG recording. Mean age of patients was 14.2 ± 3.4 years. The contrast group consisted of 24 age-matched (15.2 ± 3.1 year) typically developing children.

The spatial attention task was a modification of a cued Posner spatial attention task and had 2 blocks – one with horizontal, while the second with diagonal windows where target appeared either at the left or the right side. The analysis included comparison of behavioral performance (RT, accuracy) and ERP measures. In addition to the second cue stimulus (S2) locked ERPs, we analyzed also lateralized readiness potential (LRP) recorded as a difference wave between responses at motor strip (C3/C4) starting from the first cue (pre-cue S1). ERP data set was analyzed using ANOVA with within subject factors *Cue Position* (Horizontal, Diagonal), S1 *Cue Congruence* (valid, invalid), and *Hemisphere* (left, right) and between group factor *Group* (ASD, ADHD, Controls).

Reaction time (RT) analysis for ASD and typical children (CNT) showed a *Congruence* X *Group* effect ($F = 7.14$, $p = 0.011$), in particular the ASD group had similarly slower RT both in valid and invalid pre-cued conditions, while controls responded faster to correctly prompted targets. Accuracy of responses was lower in the ASD group, mostly due to more omission error rate ($F = 6.17$, $p = 0.017$). Comparison of ERPs across 3 groups (ASD, ADHD, CNT) yielded following results. Midline frontal N100 showed a target *Position* (horizontal, diagonal) *X Pre-cue Congruence* (valid, invalid) *X Group* (ASD, ADHD, CNT) interaction ($F = 13.42$, $p = 0.001$), where ASD group had more negative N100 amplitude during diagonal target condition regardless of congruence of cues, as pre-cue congruence had less effect in this group. In the ADHD group latency of N100 was prolonged, e.g., in a diagonal condition ADHD group had longer latency than ASD. Frontal N200 showed a *Position X Congruence X Hemisphere X Group* (ADHD, ASD, CNT) interaction effect ($F = 4.11$, $p = 0.023$). Amplitude was higher in ASD, more in horizontal target position, and effect was better expressed at the right hemisphere. Latency of the frontal N200 was showed significant group differences, namely shorter latency, along with lower amplitude in the ADHD (ADHD vs. ASD, $F = 2.42$, $p = 0.018$). The P3b component showed differences between ASD and CNT groups at the midline ($F = 5.38$, $p = 0.026$) in invalidly cued diagonal target condition and was significantly prolonged in the ASD group.

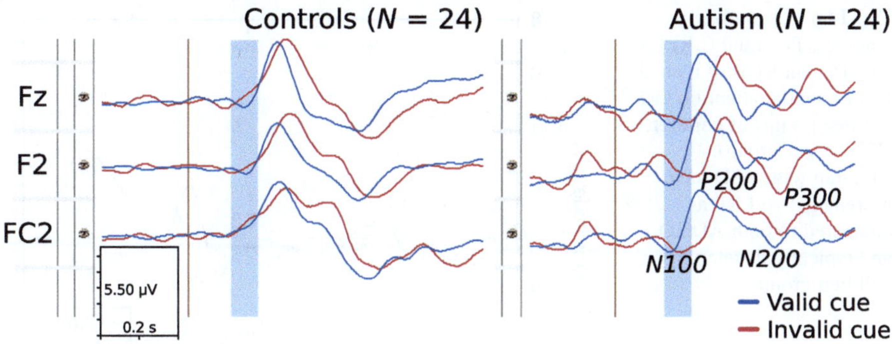

Fig. 14.5 Frontal (Fz, F2, FC2) ERP to imperative (S2) stimuli in cued Posner spatial attention task in ASD and neurotypical (NT) children groups (N = 24/per group). Children with autism show delayed latencies of N100 and P200 components in incongruent trials of the test

Most of ERP differences between controls and ADHD and ASD were observed at the frontal sites thus pointing at the possible frontal executive deficits both in autism and ADHD. Of particular interest are frontal hemispheric differences present at the pre-attentive early processing stages (N100), and less discrimination between correctly and incorrectly cued targets at the later stages of processing in autism, and globally lower magnitude and delayed latency of early frontal component in the ADHD (Fig. 14.5).

The lateralized readiness potential (LRP) is an index of motor processes and it is assumed that this brain potential is generated by a source within the MI. The lateralized readiness potential (LRP, reviewed in Coles 1989; Eimer 1998; Leuthold et al. 2004; Leuthold and Jentzsch 2001) reflects the response-specific involvement of the left and right cortices of the brain. This LRP has mostly been recorded by just one pair of electrodes located at scalp locations overlying the motor cortex (C3/C4 or C3′/C4′). The LRP enables the determination of the point in time at which the activation of the motor cortex controlling one hand surpasses the activation of the motor cortex controlling the other side. On the basis of this feature, the LRP has been advocated as a tool for psychophysiological studies that can be used to index the time of response selection and the onset of hand-specific movement preparation (Coles 1989; Eimer 1998). Hand-specific movement preparation, as reflected in the LRP, can be detected, for instance, in the foreperiod of a warned RT task such as cued Posner task, when a subject knows which hand has to be moved (Leuthold et al. 2004). Moreover, it has been shown that the LRP is sensitive to covert aspects of movement preparation (Eimer 1998). In conjunction with RT (or EMG) measures, this sensitivity of the LRP to covert response tendencies has proven to be very valuable in studies on stimulus–response translation. Given the contralateral dominance and central distribution of movement-related potentials preceding self-paced and externally cued movements, it seems very likely that the different types of movement-related potentials at least share the primary motor cortex (MI) as a generator (Praamstra et al. 1999). That MI has a major part in the generation of the readiness potential is indeed well established by recordings of

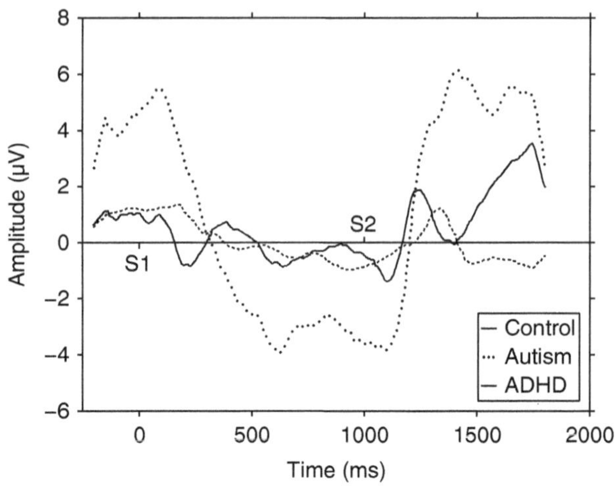

Fig. 14.6 Lateralized readiness Potential (LRP) in incongruent trials of a cued Posner spatial attention task in three groups of children (N = 24/per group). Children with ASD have more negative LRP as compared to both ADHD and typically developing children groups

cortical field potentials, topographical EEG, and dipole source analyses (Cui et al. 2000; Praamstra et al. 1999).

The LRP is assumed to be related to selective response activation. It captures the asymmetric portion of the late BP preceding hand or foot movements. The LRP is computer on the basis of ERP potentials recorded prior to and during the execution of a motor response over the left and right motor cortices (C/C4 or C3′/C4′-later located 1 cm anterior to C3/C4 sites by 10–20 EEG system). The lateralization negativities are made explicit with the double substraction method, when at 1st stage C3-C4 amplitude differences are computed separately for the left and right hand responses. The differences for the right hand responses are then subtracted from difference waves for left-hand responses. As a result of these two subtractions, lateralized negativities of motor cortex are unrelated to the side of the response. There are known other methods of LRP extraction (e.g., Coles 1989, averaging method). In all methods LRP waveform reflects the lateralization of slow motor ERP activity observed prior to movement onset that is assumed to be related to a central activation of an unimanual response (Eimer 1998) Furthermore, it is important for the mental chronometry that LRP helps to determine exact point in time when sensory information starts affect motor processing and response execution. The LRP can be computed on the basis of stimulus-locked average waveforms or as a response-locked averages. In our study (cued Posner attention task) we used stimulus-locked LRP method and since S1-S2 interval was set on 1 s, we computed mean LRP and integrated LRP values for 2 windows (early 600–800 ms post S1 stimulus; and late – 800–1,200 ms post-stimulus). In our sample of ASD, ADHD and NT controls, differences at the early stages were not significant at the early stage, but became significant at the late stage of the LRP yielding significant *Time* (early, late) X *Group* interaction ($F = 6.77$, $p = 0.012$). Post-hoc analysis showed group differences at 1,000–1,200 ms post stimulus window ($F = 4.81$, $p = 0.033$) between autism and control groups. Autism group also showed more pronounced differences of LRPs in incongruent trials (Fig. 14.6).

14.5.2 Audio-Visual Selective Attention Test

The purpose of our another study (Kiser et al. 2012) was to employ electrophysiological measures to study and provide observable evidence of atypical neurological multimodal sensory processing in ASD. This study used the audio-visual selective attention test and a measure of ERP and evoked gamma oscillations. It is believed that aberrations of the fundamental cortical structures, i.e., minicolnopathy of ASD subjects create a deficit in essential processing and integration of sensory modalities. We predicted that ASD subjects will display deficits in the ability to attend to and respond to rare, combined audio-visual target stimuli. It was hypothesized that because of a dysfunction in target discrimination ASD subjects will display impaired cognitive inhibition, thus exhibit hyper-excitable responses to non-targets. Furthermore, it was expected that lower ability to selectively discriminate between targets and non-targets will create delayed latency in the ERP components and induced EEG gamma responses of in the subjects with ASD.

This study was based on the premise that the previously described neuroanatomical differences in ASD – the hypotheses that ASD subjects have hyper-connected, hyper-excitable cortical regions – will be prominently observable through EEG recordings in combined audio-visual target detection task. Because of the aforementioned anatomical differences, ASD subjects were anticipated to display exaggerated responses to both target and non-targets alike eliciting early and late stage differences.

If there are deficits in single modality stimulus processing (as it was described above for visual target detection tasks), one would naturally expect there to be deficits in tasks when the individual is required to integrate information from multiple modalities. The anticipated deficiencies associated with ASD individuals may be reflections of the failure to successfully attend to and/or process multiple modalities, e.g. visual, auditory, or concurrently (O'Neill and Jones 1997). It is expected that the multimodal audiovisual tasks will more readily elicit observable processional sensory deficits (Marco et al. 2011). In accordance with the previously mentioned neuroanatomical observations linked to ASD, it was also likely that abnormal ERPs and gamma oscillations associated with multimodal sensory processing might be the results of altered minicolumn morphology and decreased inhibition (Baruth et al. 2010a, b; Sokhadze et al. 2009a, b, 2010a, b). The aim of this study was to establish any possible differences in concurrent audio-visual processing between in individuals with autism and typically developing control group.

Twelve ASD subjects, 12 control subjects, completed the tasks of this protocol. The ASD group was comprised of 4 female and 8 male participants while the control group consisted of 3 female and 9 male participants. Each participant performed 5 target detection tasks during a single ERP recording session. Total Task time lasted approximately 20 min. Each task consisted of a block of (100) trials with a break every (50) trials. Students were instructed to press a key for the specific target in each block. Stimuli were presented pseudo randomly with a target to standard ratio of (20:80). Stimuli had (150 ms) duration with a random inter-stimulus interval between 1,000 and 1,250 ms. In combined audiovisual task subjects were presented with concurrent visual and auditory stimuli, and were instructed to only respond when the visual target "X" and auditory target "low tone" were presented simultaneously.

Tailored algorithms generated in MATLAB were used to extract measures of gamma frequency from the EEG recordings. The extracted data was then processed using SPSS to assess between group differences for power (μV^2), hemispheric activity, response to visual, auditory, and combined audiovisual target conditions (Baruth et al. 2010a; Baruth et al. 2011; Gross et al. 2012). ANOVA was used to analyze the following factors within all participants: (1) Modality (Visual, Auditory, Combined), (2) Stimulus (target, non-target), (3) Hemisphere (left, right), (4) Group, (ASC, control). Statistically significant findings were found in the latency of the N100 in the frontal region in response to the combined audiovisual non-target: the ASD group showed significantly shorter N100 latencies to a non-target stimulus ($F = 8.93$, $p < 0.01$). In the tempo-parietal region there was a significant difference in the amplitude of the N200 component in response to the audiovisual non-targets ($F = 4.36$, $p = 0.049$).

The parietal P300 showed a significant difference in amplitude response to audio-visual targets ($F = 7.66$, $p = 0.01$), manifested as higher P300 in autism. Upon collapsing the activities measured at frontal left and frontal right as well as the responses to targets and non-targets of the auditory and audiovisual stimuli, a significant hemispheric difference was found in gamma activity between ASD subjects and neurotypical controls. The individuals with autism displayed a significant bias towards right frontal hemispheric gamma activity ($F = 4.93$, $p = 0.04$). The ASD group shows increases in gamma activity within 0–50 ms and as well as the 150–200 ms across all modalities in the parietal region additionally showing frontal increases in activity for the combined condition (Fig. 14.7).

14.6 Discussion and Conclusions

For high functioning ASC individuals it generally appears that higher cognitive inhibitory control remains intact, however it also appears that the parietal and frontal regions of these patients display increased activity related to standard and novel stimuli (Baruth et al. 2010b; Sokhadze et al. 2009a). It has been conjectured that there could be two reasons for the observed cortical hyperactivity: ASD subjects may have atypical neuroanatomical development, or ASD subjects may employ unconventional compensatory cognitive processing techniques that require more cortical activity (Sokhadze et al. 2009a). Tannan et al. (2008) found that in a sensory discrimination task, ASD individuals failed to adapt to the changes in the stimulus while controls did, and the authors suggested that the lack of adaptation is

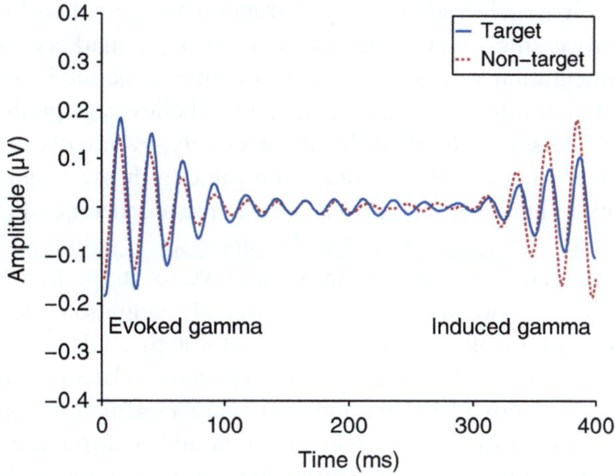

Fig. 14.7 Gamma frequency oscillations in response to combined audio-visual stimuli condition in selective attention task. Children with autism (N = 12) show similar power of the early (evoked) and late (induced) gamma oscillations (35–45 Hz range) in response to both target and non-target stimuli

indicative of the hyper-excited network based on ineffective GABAergic interneuron network mediation.

Cortical hyperexcitability associated with anatomical minicolumn pathology may be the basic anomaly affecting a subject's attention span as well as the individual's mitigation of sensory arousal (Casanova 2010). The exogenous overstimulation of an ASD individual's cortex would cause improper functioning due to the excess "cortical noise," further affecting how the individual relates to stimuli as well as other people (Ratey and Johnson 1997).

Hypotheses concerning exaggerated responses and local hyper-connectivity are bolstered by neuroanatomical findings of increased numbers of smaller and denser minicolumns – minicolumn cells – within frontal and temporal lobes (Casanova et al. 2006a, b). The GABAergic interneurons are responsible for the inhibitory surround of ASC minicolumns and are found at a decreased ratio to the excitatory pyramidal neurons (Casanova et al. 2006a, b). The coupling of decreased spatial distribution of pyramidal cells and disruption in the balance of excitation/inhibition can promote more localized connections and have a global effect on interregional connectivity (Casanova et al. 2006a, b). The increases in local intraregional excitation would serve to decrease stimuli specificity and functioning of interregional cortical networks (Casanova et al. 2006a, b). The excess accumulation of localized networks in the frontal and parietal regions could create a scenario where those regions are functioning in isolation at the expense of network integration (Kana et al. 2011). Corroborating evidence has been suggested through displays of the incongruously activated cortices associated with ASD subjects (Sokhadze et al. 2009a, b). Specifically there have been demonstrations of atypical exaggerated responses to sensory stimuli, coupled with deficits related to attention orientation as well as indices of dysfunction concerning downstream higher cognitive level processing (Sokhadze et al. 2009a).

It is believed that any disruption to the networks associated with "top-down processing" would be observed through markers indicating hampered sensory integration accompanied by a system of disjointed cognitive processing. A cortical system affected in such a manner is believed to predispose the ASD individual to processing individual details separately, eventually leading to sensory information overwhelming the regions responsible for higher cognitive processes and an inability to integrate the various details into a coherent whole (Kana et al. 2011). Brock et al. (2002), propose that the processes of neural network integration along with regional specialization are imperative for normal anatomical and cognitive development. The balance between neural integration and cortico-region function specialization development persists through adolescence, and a disruption of such could be tied to ASD cognitive symptoms (Brock et al. 2002; Rippon et al. 2007).

Collective ERP and induced gamma oscillation research suggests that ASD is a condition of neurobiological origin and with possible genetic contributors. More confounding to understanding the cause is the notion that research has found several brain regions, cortical and subcortical, to be either atypical in their cytoarchitectural arrangement or with their overall functioning (Kana et al. 2011). Many of the typical demonstrative symptoms of ASD are associated with impairments in social interaction, communication, and repetitive behaviors. It can be reasonably assumed that the observable symptoms of autism are related to decreased integrative capabilities between specialized intraregional neural networks, reflected in a decrease in frequency oscillation coupling and coherence, between cortical regions.

Findings of increased connectivity within posterior regions in more severe ASD traits could be evidence of the formation of compensatory mechanisms, or possibly evidence of the prevalence of more short connection fibers due to minicolumn pathology (Casanova et al. 2006a, b; Schipul et al. 2011). Long distance anatomical connections, as evidenced by white matter tracts have suggested that greater behavioral disturbances arise with decreases in anatomical white matter connectivity between regions (Schipul et al. 2011). Converging results from several studies fit the theme of under-connectivity, showing that diminished white matter integrity and connectivity appear to be underlying many of the observable symptoms of ASD (Schipul et al. 2011).

While many other studies attempting to measure executive dysfunction in ASD individuals have shown reduced performance in task completion, particularly on protocols that affect attentional focus (Frith 1989; Brock et al. 2002; Sokhadze et al. 2013). The studies suggest that tasks of attention orienting, sensory perception and integration, and motor response processes require a well-coordinated system of cortical modulation. A reduction in synchronized trans-cortical activity, as evidenced by excessive unsynchronized gamma activity, elicits the consequences of diminished capacity to integrate and discriminate sensory stimuli, affecting executive control over attentional and responsive processes.

It is possible to suggest that normal cognitive development is predicated fundamentally on the correct migration, cytoarchitectural arrangement and development of neural progenitor cells in the cortex (Casanova et al. 2006a, b; Casanova 2007, 2010). Any disruption of such, as previously described in the neuropathology of autism, may stymie any further developmental milestones related to higher order cognition and cognitive processes. It is believed that normal cortical and intellectual development is not only predicated on the gradual elaboration and specialization of cortical regions, but additionally on the formation of integrative connections between the specialized regions. Cytoarchitectural development dysfunction coupled with a gradient decrease in GABAergic neuron inhibition and prevalence of short excitatory connection fibers could in turn be the basis of impaired interregional connections, temporal binding and coherence between regions (Casanova et al. 2006a, b; Brock et al. 2002). Such anatomical aberrations could be responsible for excess oscillatory and hyperactivity recorded by EEG and ERPs above specific cortical regions. A combinatorial mechanism of excess activity, low temporal binding and interregional coherence could give segue to more difficult target discrimination as evidenced by increased ERPs and higher error rate in ASD subjects (Sokhadze et al. 2009a).

The collective results of the studies showing ERP irregularities, differences in coherence and gamma activity during active sensory processing, suggest that ASD individuals are equipped with cognitive mechanics that differ from neurotypical individuals. One would suggest that underlying biological differences correlate with EEG and behavioral changes. It is the opinion of the authors that changes in the cytoarchitecture of the basic minicolumn is inherently responsible for the changes in cortical activity and connectivity and represents the basis of atypical cortical sensory processing in ASD individuals.

Our results are indicative that oddball tasks are efficient at revealing some encumbrances ASD subjects have with sensory filtration of irrelevant stimuli. It is further believed that such protocols may be used to make strong correlative connections between the behavioral characteristics of ASD and electrophysiological activity. With such foundations in mind, the use of cortical measurements of electrophysiological differences may be instrumental in elucidating the underlying mechanisms of ASD. The use of electrophysiological research may be indicated as part of the process of establishing the quantitative EEG-based biomarkers of ASD, that would afford clinicians and researchers the ability to forecast the development of the atypical social features associated with ASD. Future research should improve the diagnostic capability of ERPs and induced EEG oscillations, and may contribute to earlier diagnosis and intervention in ASD, a disorder where timely intervention is critical.

References

American Psychiatric Association (2000) Diagnostic and statistical manual of mental disorders (DSM-IV-TR). American Psychiatric Association, Washington, DC

Bailey A, Luthert P, Dean A, Harding B, Janota I, Montgomery M, Rutter M, Lantos P (1998) A clinicopathological study of autism. Brain 121(5):889–905

Baron-Cohen S, Belmonte MK (2005) Autism: a window onto the development of the social and the analytic brain. Annu Rev Neurosci 28:109–126

Baron-Cohen S, Leslie AM, Frith U (1985) Does the autistic child have a 'theory of mind'? Cognition 21:37–46
Baron-Cohen S, Wheelwright S, Griffin R, Lawson J, Hill J (2002) The exact mind: empathizing and systemizing in autism spectrum conditions. In: Goswami U (ed) Handbook of cognitive development. Blackwell, Oxford, pp 491–508
Bartos M, Vida I, Jonas P (2007) Synaptic mechanisms of synchronized gamma oscillations in inhibitory interneuron networks. Nat Rev Neuroscience 8(1):45–56
Baruth JM, Casanova MF, El-Baz A, Horrell T, Mathai G, Sears L, Sokhadze E (2010a) Low-frequency repetitive transcranial magnetic stimulation (rTMS) modulates evoked-gamma oscillations in autism spectrum disorder (ASD). J Neurother 14(3):179–194
Baruth JM, Casanova M, Sears L, Sokhadze E (2010b) Early-stage visual processing abnormalities in autism spectrum disorder (ASD). Transl Neurosci 1:177–187
Baruth J, Williams E, Sokhadze E, El-Baz A, Sears L, Casanova MF (2011) Repetitive transcranial stimulation (rTMS) improves electroencephalographic and behavioral outcome measures in autism spectrum disorders (ASD). Autism Sci Dig J AutismOne 1(1):52–57
Bauman ML, Kemper TL (2005) Structural brain anatomy in autism: what is the evidence? In: Bauman ML, Kemper TL (eds) The neurobiology of autism, 2nd edn. John Hopkins University Press, Baltimore, pp 121–135
Belmonte MK (2000) Abnormal attention in autism shown by steady-state visual evoked potentials. Autism 4:269–285
Belmonte MK, Yurgelun-Todd DA (2003a) Functional anatomy of impaired selective attention and compensatory processing in autism. Cogn Brain Res 17:651–664
Belmonte MK, Yurgelun-Todd DA (2003b) Anatomic dissociation of selective and suppressive processes in visual attention. NeuroImage 19:180–189
Belmonte MK, Allen G, Beckel-Mitchener A, Boulanger LM, Carper RA, Webb SJ (2004a) Autism and abnormal development of brain connectivity. J Neurosci 24(42):9228–9231
Belmonte MK, Cook EH, Anderson GM, Rubenstein JL, Greenhough WT, Beckel-Mitchener A, Courchesne E, Boulanger LM, Powell SB, Levitt PR, Perry EK, Jiang YH, DeLorey TM, Tierney E (2004b) Autism as a disorder of neural information processing: directions for research and targets for therapy. Mol Psychiatry 9(7):646–663
Bomba MD, Pang EW (2004) Cortical auditory evoked potentials in autism: a review. Int J Psychophysiol 53(3):161–169
Bragin A, Jando G, Nadasdy Z, Hetke J, Wise K, Buzsaki G (1995) Gamma (40–100 Hz) oscillation in the hippocampus of the behaving rat. J Neurosci 15(1):47–60
Brock J, Brown CC, Boucher J, Rippon G (2002) The temporal binding deficit hypothesis of autism. Dev Psychopathol 14(2):209–224
Brown C (2005) EEG in autism: is there just too much going on in there? In: Casanova MF (ed) Recent developments in autism research. Nova, New York, pp 109–126
Brown C, Gruber T, Boucher J, Rippon G, Brock J (2005) Gamma abnormalities during perception of illusory figures in autism. Cortex 41:364–376
Bruneau N, Roux S, Adrien JL, Bathelemy C (1999) Auditory associative cortex dysfunction in children with autism: evidence from late auditory evoked potentials (N 1 wave- T Complex). Clin Neurophysiol 110:1927–1934
Burack JA (1994) Selective attention deficits in persons with autism: preliminary evidence for inefficient attentional lens. J Abnorm Psychol 103:515–543
Cantor DS, Thatcher RW, Hrybyk M, Kaye H (1986) Computerized EEG analysis of autistic children. J Autism Dev Disord 16:169–187
Casanova MF (2005) Minicolumnar pathology in autism. In: Casanova MF (ed) Recent developments in autism research. Nova Biomedical Books, New York, pp 133–144
Casanova MF (2006) Neuropathological and genetic findings in autism: the significance of a putative minicolumnopathy. Neuroscientist 12(5):435–441
Casanova MF (2007) The neuropathology of autism. Brain Pathol 17:422–433

Casanova MF (2010) Cortical organization: anatomical findings based on systems theory. Transl Neurosci 1(1):62–71
Casanova MF, Switala AE (2005) Minicolumnar morphometry: computerized image analysis. In: Casanova MF (ed) Neocortical modularity and the cell minicolumn. Nova Biomedical Books, New York, pp 161–180
Casanova MF, Buxhoeveden DP, Switala AE, Roy E (2002a) Minicolumnar pathology in autism. Neurology 58:428–432
Casanova MF, Buxhoeveden DP, Switala AE, Roy E (2002b) Neuronal density and architecture (gray level index) in the brains of autistic patients. J Child Neurol 17(7):515–521
Casanova MF, Buxhoeveden D, Gomez J (2003) Disruption in the inhibitory architecture of the cell minicolumn: implications for autism. Neuroscientist 9:496–507
Casanova MF, van Kooten I, van Engeland H, Heinsen H, Steinbursch HW, Hof PR, Trippe J, Stone J, Schmitz C (2006a) Minicolumnar abnormalities in autism II. Neuronal size and number. Acta Neuropathol 112(3):287–303
Casanova MF, van Kooten I, Switala AE, van England H, Heinsen H, Steinbuch HW, Hof PR, Schmitz C (2006b) Abnormalities of cortical minicolumnar organization in the prefrontal lobes of autistic patients. Clin Neurosci Res 6(3–4):127–133
Casanova MF, Baruth J, El-Baz AS, Sokhadze GE, Hensley M, Sokhadze EM (2013) Evoked and induced gamma-frequency oscillation in autism. In: Casanova MF, El-Baz AS, Suri JS (eds) Imaging the brain in autism. Springer, New York, pp 87–106
Casey BJ, Gordon CT, Mannheim GB, Rumsey JM (1993) Dysfunctional attention in autistic savants. J Clin Exp Neuropsychol 15(6):933–946
Charman T (2008) Autism spectrum disorders. Psychiatry 7:331–334
Ciesielski KT, Knoght JE, Prince RJ, Harris RJ, Handmaker SD (1995) Event-related potentials in cross-modal divided attention in autism. Neuropsychologia 33:225–246
Coles MG (1989) Modern mini-brain reading: psychophysiology, physiology, and cognition. Psychophysiology 26(3):251–269
Courchesne E, Pierce K (2005) Brain overgrowth in autism during a critical time in development: implications for frontal pyramidal neuron and interneuron development and connectivity. Int J Dev Neurosci 23(2–3):153–170
Courchesne E, Lincoln AJ, Kilman BA, Galambos R (1985) Event-related brain potential correlates of the processing of novel visual and auditory information in autism. J Autism Dev Disord 15(1):55–76
Courchesne E, Lincoln AJ, Yeung-Courchesne R, Elmasian R, Grillon C (1989) Pathophysiologic findings in nonretarded autism and receptive developmental disorder. J Autism Dev Disord 19 (1):1–17
Cui RQ, Huter D, Egkher A, Lang W, Lindinger G, Deecke L (2000) High resolution DC-EEG mapping of the Bereitschaftspotential preceding simple or complex bimanual sequential finger movement. Exp Brain Res 134(1):49–57
Dawson G, Finley C, Phillips S, Galpert L, Lewy A (1988) Reduced P 3 amplitude of the event-related brain potential: its relationship to language ability in autism. J Autism Dev Disord 18 (4):493–504
Dawson G, Klinger LG, Panagiotides H, Lewy A, Catelloe P (1995) Subgroups of autistic children based on social behavior display distinct patterns of brain activity. J Abnorm Child Psychol 23 (5):569–583
Del Rio MR, DeFelipe J (1997) Double bouquet cell axons in the human temporal neocortex: relationship to bundles of myelinated axons and colocalization of calretinin and calbindin D-28k immunoreactivities. J Chem Anat 13(4):243–251
Eimer M (1998) The lateralized readiness potential as an on-line measure of central response activation processes. Behav Res Methods Instrum Comput 30:146–156
Engel AK, Konig P, Kreiter AK, Schillen TB, Singer W (1992) Temporal coding in the visual cortex: new vistas on integration in the nervous system. Trends Neurosci 15(6):218–226

Ferri R, Elia M, Agarwal N, Lanuzza B, Musumeci SA, Pennisi G (2003) The mismatch negativity and the P3a components of the auditory event-related potentials in autistic low-functioning subjects. Clin Neurophysiol 114:1671–1680
Frith U (1989) Autism: explaining the enigma. Blackwell, Oxford
Frith U, Happé F (1994) Autism: beyond "theory of mind". Cognition 50(1–3):115–132
Gomes E, Pedroso FS, Wagner MB (2008) Auditory hypersensitivity in the autistic spectrum disorder. Pro Fono 20(4):279–284
Gray CM, Konig P, Engel AK, Singer W (1989) Oscillatory responses in cat visual cortex exhibit inter-columnar synchronization which reflects global stimulus properties. Nature 338 (6213):334–337
Grice SJ, Spratling MW, Karmiloff-Smith A, Halit H, Csibra G, de Haan M, Johnson MH (2001) Disordered visual processing and oscillatory brain activity in autism and Williams syndrome. Neuroreport 12(12):2697–2700
Gross E, El-Baz A, Sokhadze G, Sears L, Casanova M, Sokhadze E (2012) Induced EEG gamma oscillation alignment improves differentiation between autism and ADHD group responses in a facial categorization task. J Neurother 16(2):78–91
Grothe B, Klump GM (2000) Temporal processing in sensory systems. Curr Opin Neurobiol 10 (4):467–473
Gruber T, Muller MM (2005) Oscillatory brain activity dissociates between associative stimulus content in a repetition priming task in the human EEG. Cereb Cortex 15(1):109–116
Gruber T, Muller MM, Keil A, Elbert T (1999) Selective visual-spatial attention alters induced gamma band responses in the human EEG. Clin Neurophysiol 110(12):2074–2085
Hashimoto T, Tayama M, Miyazaki M, Sakurama N, Yoshimoto T, Murakawa K, Kuroda Y (1992) Reduced brainstem size in children with autism. Brain and Development 14(2):94–97
Herrmann CS, Knight RT (2001) Mechanisms of human attention: event related potentials and oscillations. Neurosci Biobehav Rev 25(6):465–476
Herrmann CS, Mecklinger A (2000) Magnetoencephalographic responses to illusory figures: early evoked gamma is affected by processing of stimulus features. Int J Psychophysiol 38 (3):265–281
Herrmann CS, Mecklinger A, Pfeifer E (1999) Gamma responses and ERPs in a visual classification task. Clin Neurophysiol 110(4):636–642
Hillyard SA, Mangun GR, Woldorff MG, Luck SJ (1995) Neural mechanisms mediating selective attention. In: Gazzaniga MS (ed) The cognitive neurosciences. MIT Press, Cambridge, MA
Hofman MA (1985) Size and shape of the cerebral cortex in mammals. I. The cortical surface. Brain Behav Evol 27(1):28–40
Holcomb PJ, Ackerman PT, Dykman RA (1985) Cognitive event-related brain potentials in children with attention and reading deficits. Psychophysiology 22(6):656–667
Howard MW, Rizzuto DS, Caplan JB, Madsen JR, Lisman J, Aschenbrenner-Scheibe R, Schulze-Bonhage A, Kahana MJ (2003) Gamma oscillations correlate with working memory load in humans. Cereb Cortex 13(12):1369–1374
Just MA, Cherkassky V, Keller TA, Minshew NJ (2004) Cortical activation and synchronization during sequence comprehension in high-functioning autism: evidence of underconnectivity. Brain 127(8):1811–1821
Kaiser J (2003) Induced gamma-band activity and human brain function. Neuroscientist 9 (6):475–484
Kaiser J, Hertrich I, Ackermann H, Mathiak K, Lutzenberger W (2005) Hearing lips: gamma-band activity during audiovisual speech perception. Cereb Cortex 15(5):646–653
Kana RK, Libero LE, Moore MS (2011) Disrupted cortical connectivity theory as an explanatory model for autism spectrum disorder. Phys Life Rev 8(4):410–437
Kanizsa G (1976) Subjective contours. Sci Am 234(4):48–52
Keil A, Muller MM, Ray WJ, Gruber T, Elbert T (1999) Human gamma band activity and perception of a gestalt. J Neurosci 19(16):7152–7161

Kemner C, Verbaten MN, Cuperus JM, Camfferman G, Van Engeland H (1994) Visual and somatosensory event-related brain potentials in autistic children and three different control groups. Electroencephalogr Clin Neurophysiol 92(3):225–237
Kemner C, Verbaten MN, Cuperus JM, Camfferman G, Van Engeland H (1995) Auditory event-related potentials in autistic children and three different control groups. Biol Psychiatry 38 (3):150–165
Kemner C, van der Gaag RJ, Verbaten M, van Engeland H (1999) ERP differences among subtypes of pervasive developmental disorders. Biol Psychiatry 46(6):781–789
Kenemans JL, Kok A, Smulders FT (1993) Event-related potentials to conjunctions of spatial nd orientation as a function of stimulus parameters and response requirements. Electroencephalogr Clin Neurophysiol 88(1):51–63
Khalfa S, Bruneau N, Rogé B, Georgieff N, Veuillet E, Adrien JL, Barthelemy C, Collet L (2004) Increased perception of loudness in autism. Hear Res 198(1–2):87–92
Kiser RM, Sokhadze G, Gross E, El-Baz A, Sokhadze E, Casanova M (2012) Selective attention and audiovisual integration in children with autism. Appl Psychophysiol Biofeedback 37:307
Koshino H, Carpenter PA, Minshew NJ, Cherkassky VL, Keller TA, Just MA (2005) Functional connectivity in an fMRI working memory task in high-functioning autism. Neuroimage 24 (3):810–821
Kozloski J, Hamzei-Sichani F, Yust R (2001) Stereotyped position of local synaptic targets in neocortex. Science 293(5531):868–872
Lainhart JE (2006) Advances in autism neuroimaging research for the clinician and genetics. Am J Med Genet Semin Med Genet 142C(1):33–39
Lee KH, Williams LM, Breakspear M, Gordon E (2003) Synchronous gamma activity: a review and contribution to an integrative neuroscience model of schizophrenia. Brain Res Brain Res Rev 41(1):57–78
Leuthold H, Jentzsch I (2001) Neural correlates of advance movement preparation: a dipole source analysis approach. Brain Res Cogn Brain Res 12(2):207–224
Leuthold H, Sommer W, Ulrich R (2004) Preparing for action: inferences from CNV and LRP. J Psychophysiol 18(2–3):77–88
Lincoln AJ, Courchesne E, Harms L, Allen M (1993) Contextual probability evaluation in autistic, receptive developmental disorder and control children: event-related potential evidence. J Autism Dev Disord 23(1):37–58
Marco EJ, Hinkley LB, Hill SS, Nagarajan SS (2011) Sensory processing in autism: a review of neurophysiologic findings. Pediatr Res 69(5):48–54
Minshew NJ, Goldstein G, Siegel DJ (1997) Neuropsychologic functioning in autism: profile of a complex information processing disorder. J Int Neuropsychol Soc 3(4):303–316
Mountcastle VB (1997) The columnar organization of the neocortex. Brain 120(4):701–722
Mountcastle VB (2003) Introduction. Computation in cortical columns. Cereb Cortex 13(1):2–4
O'Neill M, Jones RS (1997) Sensory-perceptual abnormalities in autism: a case for more research? J Autism Dev Disord 27(3):283–293
O'Riordan MA, Plaisted KC, Driver J, Baron-Cohen S (2001) Superior visual search in autism. J Exp Psychol Hum Percept Perform 27(3):719–730
Oades RD, Walker MK, Geffen LB, Stern LM (1988) Event-related potentials in autistic and healthy children on an auditory choice reaction time task. Int J Psychophysiol 6(1):25–37
Ong WY, Garey LJ (1990) Pyramidal neurons are immunopositive for peptides, but not GABA, in the temporal cortex of the macaque monkey (Macaca fascicularis). Brain Res 533(1):24–41
Ozonoff S, Pennington BF, Rogers SJ (1991) Executive function deficits in high-functioning autistic individuals: relationship to theory of mind. J Child Psychol Psychiatry Allied Discip 32(7):1081–1105
Palmen SJ, van Engeland H, Hof PR, Schmitz C (2004) Neuropathological findings in autism. Brain 127(12):2572–2583
Pavlova M, Birbaumer N, Sokolov A (2006) Attentional modulation of cortical neuromagnetic gamma response to biological movement. Cereb Cortex 16(3):321–327

Peters A, Sethares C (1991) Organization of pyramidal neurons in area 17 of monkey visual cortex. J Comp Neurol 306(1):1–23

Peters A, Sethares C (1996) Myelinated axons and the pyramidal cell modules in monkey primary visual cortex. J Comp Neurol 365(2):232–255

Plaisted K, Saksida L, Alcantara JI, Weisblatt E (2003) Towards an understanding of the mechanisms of weak central coherence effects: experiments in visual configural learning and auditory perception. Philos Trans R Soc Lond B Biol Sci 358(1430):375–386

Polich J (2003) Theoretical overview of P3a and P3b. In: Polich J (ed) Detection of change: event-related potential and fMRI findings. Kluwer Academic Press, Boston, pp 83–98

Polich J, Herbst KL (2000) P300 as a clinical assay: rationale, evaluation, and findings. Int J Psychophysiol 38(1):3–19

Porjesz B, Almasy L, Edenberg HJ, Wang K, Chorlian DB, Foroud T, Goate A, Rice JP, O'Connor SJ, Rohrbaugh J, Kuperman S, Bauer LO, Crowe RR, Schuckit MA, Hesselbrock V, Conneally PM, Tischfield JA, Li TK, Reich T, Begleiter H (2002) Linkage disequilibrium between the beta frequency of the human EEG and a GABAA receptor gene locus. Proc Natl Acad Sci U S A 99(6):3729–3733

Praamstra GF, Schmitz F, Freund HJ, Schnitzler A (1999) Magneto-encephalographic correlates of the laterized readiness potential. Brain Res Cogn Brain Res 8(2):77–85

Pritchard W, Sokhadze E, Houlihan M (2004) Effects of nicotine and smoking on event-related potentials. Nicotine Tob Res 6(6):961–984

Rakic P (1975) Timing of major ontogenetic events in the visual cortex of the rhesus monkey. In: Buchwald NA, Brazier M (eds) Brain mechanisms in mental retardation. Academic, New York, pp 3–40

Rakic P (1995) Radial versus tangential migration of neuronal clones in the developing cerebral cortex. Proc Natl Acad Sci U S A 92(25):11323–11327

Ratey JJ, Johnson C (1997) Shadow syndromes: the mild forms of major mental disorders that sabotage us. Bantam Books, New York

Rippon G, Brock J, Brown C, Boucher J (2007) Disordered connectivity in the autistic brain: challenges for the 'new psychophysiology'. Int J Psychophysiol 63(2):164–172

Rodier PM, Ingram JL, Tisdale B, Nelson S, Romano J (1996) Embryological origin for autism: developmental anomalies of the cranial nerve motor nuclei. J Comp Neurol 370(2):247–261

Rodier PM, Ingram JL, Tisdale B, Croog VJ (1997) Linking etiologies in humans and animal models: studies of autism. Reprod Toxicol 11(2–3):417–422

Rodriguez E, George N, Lachaux JP, Martinerie J, Renault B, Varela FJ (1999) Perception's shadow: long distance synchronization of human brain activity. Nature 397:430–433

Rubenstein JL, Merzenich MM (2003) Model of autism: increased ratio of excitation/inhibition in key neural systems. Genes Brain Behav 2(5):255–267

Schipul SE, Keller TA, Just MA (2011) Inter-regional brain communication and its disturbance in autism. Front Syst Neurosci 5:10. doi:10.3389/fnsys.2011.00010.eCollection 2011

Seldon HL (1981a) Structure of human auditory cortex. I. Cytoarchitectonics and dendritic distributions. Brain Res 229(2):277–294

Seldon HL (1981b) Structure of human auditory cortex. II. Axon distributions and morphological correlates of speech perception. Brain Res 229(2):295–310

Seri S, Cerquiglini A, Pisani F, Curatolo P (1999) Autism in tuberous sclerosis: evoked potential evidence for a deficit in auditory sensory processing. Clin Neurophysiol 110(10):1825–1830

Shergill SS, Brammer MJ, Williams SC, Murray RM, McGuire PK (2000) Mapping auditory hallucinations in schizophrenia using functional magnetic resonance imaging. Arch Gen Psychiatry 57(11):1033–1038

Singer W (1999) Neuronal synchrony: a versatile code for the definition of relations. Neuron 24 (1):49–65

Sokhadze E, Baruth J, Tasman A, Sears L, Mathai G, El-Baz A, Casanova M (2009a) Event-related potential study of novelty processing abnormalities in autism. Appl Psychophysiol Biofeedback 34(1):37–51

Sokhadze EM, Singh S, El-Baz A, Baruth J, Mathai G, Sears L, Casanova M (2009b) Effects of a low frequency repetitive transcranial magnetic stimulation (rTMS) on gamma frequency oscillations and event-related potentials during processing of illusory figures in autism. J Autism Dev Disord 39(4):619–634

Sokhadze E, Baruth J, Tasman A, Mansoor M, Ramaswamy R, Sears L, Mathai G, El-Baz A, Casanova MF (2010a) Low-frequency repetitive transcranial magnetic stimulation (rTMS) affects event-related potential measures of novelty processing in autism. Appl Psychophysiol Biofeedback 35(2):147–161

Sokhadze E, Baruth J, El-Baz A, Horrell T, Sokhadze G, Carroll T, Tasman A, Sears L, Casanova M (2010b) Impaired error monitoring and correction function in autism. J Neurother 14 (2):79–95

Sokhadze EM, Baruth J, Tasman A, Casanova MF (2013) Event-related potential studies of cognitive processing abnormalities in autism. In: Casanova MF, El-Baz A, Suri JS (eds) Imaging the brain in autism. Springer, New York, pp 61–86

Sokhadze E, Edelson S, Casanova MF, Sokhadze E (2014) Motor response preparation deficits in a cued spatial attention task in autism. Psychophysiology 41(Supplement):S61

Tallon-Baudry C (2003) Oscillatory synchrony and human visual cognition. J Physiol Paris 97 (2–3):355–363

Tallon-Baudry C, Bertrand O, Delpuech C, Pernier J (1996) Stimulus specificity of phase-locked and non-phase-locked 40 Hz visual responses in human. J Neurosci 16(13):4240–4249

Tallon-Baudry C, Bertrand O, Henaff MA, Isnard J, Fischer C (2005) Attention modulates gamma-band oscillations differently in the human lateral occipital cortex and fusiform gyrus. Cereb Cortex 15(5):654–662

Tannan V, Holden JK, Zhang Z, Baranek GT, Tommerdahl MA (2008) Perceptual metrics of individuals with autism provide evidence for disinhibition. Autism Res Int Soc Autism Res 1 (4):223–230

Townsend J, Courchesne E, Egaas B (1996) Slowed orienting of covert visual-spatial attention in autism: specific deficits associated with cerebellar and parietal abnormality. Dev Psychopathol 8:503–584

Townsend J, Courchesne E, Covington J, Westerfield M, Harris NS, Lyden P, Lowry T, Press GA (1999) Spatial attention deficits in patients with acquired or developmental cerebellar abnormality. J Neurosci 19(13):5632–5643

Townsend J, Westerfield M, Leaver E, Makeig S, Jung T, Pierce K, Courchesne E (2001) Event-related brain response abnormalities in autism: evidence for impaired cerebello-frontal spatial attention networks. Brain Res Cogn Brain Res 11(1):127–145

Traub RD, Whittington MA, Stanford IM, Jefferys JG (1996) A mechanism for generation of long-range synchronous fast oscillations in the cortex. Nature 383(6601):621–624

Tuchman R, Rapin I (2002) Epilepsy in autism. Lancet Neurol 2(6):352–358

Viebahn C (1990) Correlation between differences in the structure of dendrite bundles and cytoarchitectonic patterns in the cerebral cortex of the rabbit. J Hirnforsch 31(5):645–652

Von Bonin G, Mehler WR (1971) On columnar arrangement of nerve cells in cerebral cortex. Brain Res 27(1):1–9

Welchew DE, Ashwin C, Berkouk K, Salvador R, Suckling J, Baron-Cohen S, Bullmore E (2005) Functional disconnectivity of the medial temporal lobe in Asperger's syndrome. Biol Psychiatry 57(9):991–998

Zikopoulos B, Barbas H (2010) Changes in prefrontal axons may disrupt the network in autism. J Neurosci 30(44):14595–14609

Chapter 15
Modular Signatures and Neural Avalanches in Epileptic Brain Networks

Ana Ciurea, Ioana Mîndruţă, Mihai Dragos Maliiă, Alexe Ciurea, Jean Ciurea, Andrei Barborică, Cristian Donos, Manuel F. Casanova, and Ioan Opris

Abstract Epileptic seizures are characterized by a rich dynamic spectrum consisting of excessive, abnormal and synchronized firing of neuron ensembles. Such abnormal firing has been quantitatively characterized via power laws in neural avalanches. The term "neural avalanche" has been used to illustrate the excessively amplified neural firing patterns that lead to epileptic seizures. The pattern of

---

A. Ciurea
Neurosurgery Department, Bagdasar-Arseni Hospital; and Physics Department, University of Bucharest, Bucureşti, Romania
e-mail: ana.ciurea@bagdasar-arseni.ro

I. Mîndruţă
Neurology Department, Carol Davila University of Medicine and Pharmacy; and Neurology Department, University Emergency Hospital, Bucureşti, Romania
e-mail: ioanamindruta@me.com

M.D. Maliiă
Neurology Department, University Emergency Hospital, Bucureşti, Romania
e-mail: mihaidragosh@yahoo.com

A. Ciurea
Applied Electronics and Information Engineering Department, Medical Electronics and Computing, Bucureşti, Romania
e-mail: alexe.ciurea@elmed.pub.ro

J. Ciurea
Neurosurgery Department, Bagdasar-Arseni Hospital, Bucureşti, Romania
e-mail: ciureaj@bagdasar-arseni.ro

A. Barborică • C. Donos
Physics Department, University of Bucharest, Bucureşti, Romania
e-mail: andrei.barborica@fizica.unibuc.ro; cristian.donos@g.unibuc.ro

M.F. Casanova, M.D.
Department of Psychiatry, University of Louisville, Louisville, KY, USA
e-mail: manuel.casanova@louisville.edu

I. Opris, Ph.D. (✉)
Department of Physiology and Pharmacology, Wake Forest University Health Sciences, Winston-Salem, NC 27103, USA
e-mail: ioanopris.phd@gmail.com

M.F. Casanova, I. Opris (eds.), *Recent Advances on the Modular Organization of the Cortex*, DOI 10.1007/978-94-017-9900-3_15

amplified firing in neural avalanches betrays a modular signature in the spread of activation across cortical minicolumns. According to this modular approach of epilepsy, the excessive amplification of neural firing in a cortical minicolumn results from a defect within the "inhibitory curtain" surrounding the pyramidal cells. The functional basis of this approach provides insights into potential clinical interventions.

Keywords Epilepsy seizure • Power law • Neural avalanche • Cortical microcircuits • Minicolumns

15.1 Introduction

Anatomical connectivity patterns of brain complexity have revealed a wiring organization that can be captured by neural modules that are neither regular nor random (Mountcastle 1997; Opris and Casanova 2014). In many instances the emergence of behaviors and pathological states are accompanied by various changes in brain connectivity patterns (Varela et al. 2001). Recently, it has been discovered that functional connectivity patterns obtained from electrophysiological recordings, functional magnetic resonance imaging (fMRI) and electroencephalography (EEG) signals during different pathological and cognitive brain states (including epilepsy) display structural and functional signatures of modular properties (Vaessen et al. 2014). Experimental studies have also shown that brain functions are coordinated via scattered specialized modules, forming a "brainweb" of neural ensembles (Varela et al. 2001). Modularity is a basic concept in the complex networks of cells endowed with columnar laminar structure (Buxhoeveden and Casanova 2002; Mountcastle 1997). A module can be defined as a group of cells within a network, in which connections between cells in the group are denser than connections of the cells with the rest of the network (Mountcastle 1997). Modularity appears to play a key role in the stability, flexibility and robustness of the network (Sole and Valverde 2008). Modular architecture of the brain might play a key role in the integration of microcircuit and of macro-scale brain activities (Opris et al. 2011, 2012a, b, 2013). Occurrence of rhythmic spike wave discharges of high amplitude that synchronizes over multiple cortical areas, suddenly emerging from a normal activity background, is a main characteristic of epilepsy.

Dysfunction of microcircuit and of macro-scale brain activity facilitates the emergence of epileptic discharges. While brain connectivity studies mainly focus on the onset, evolution and onset of epileptic discharges, in this chapter we study the modular signature of brain microcircuits and networks extracted from signals of epileptic patients compared to healthy subjects. Epileptic seizures are pathological activity states in the human brain having a rich dynamic spectrum consisting of excessive, abnormal and synchronized firing of neuronal ensembles. Such abnormal firing has been quantitatively characterized via power laws that are characteristic to avalanche processes (Wu et al. 2014).

The term "neural avalanche" illustrates excessively amplified neural firing patterns. The pattern of amplified firing in neural avalanches betrays a modular signature in the spread of activation across cortical minicolumns and large-scale networks. According to this modular view of epilepsy, one could posit that excessive amplification of neural firing in a cortical minicolum may be the outcome of pyramidal cells crosstalk that pierce the inhibitory wall surrounding the minicolumns.

15.2 Neural Avalanches

15.2.1 Definition of Neural Avalanches

Neuronal populations exhibit a type of activity named neuronal avalanches. A neuronal avalanche is a cascade of bursts of activity in neuronal networks whose size distribution can be approximated by a power law, as illustrated in critical sand pile models (Bak et al. 1987). The size distribution of neuronal avalanches in cortical networks of the brain has been reported to follow a power law distribution with an exponent close to $-3/2$, which reflect the long range spatial correlations in spontaneous activity (Klaus et al. 2011).

Neuronal avalanche has been demonstrated in cortical seizure-like activity in the anterior cingulate cortex (ACC) induced by 4-aminopyridine (4AP) and bicuculline (Bic) antagonists of selected voltage-gated potassium channels and γ-aminobutyric acid-A (GABAA) receptors, respectively (Wu et al. 2014). To illustrate the concept of neuronal avalanche induced by the application of 4AP and Bic, a layered brain slice together with the recording position of the multi-electrode array (MEA) are depicted in Fig. 15.1a. To detect the active neuronal responses, a threshold was set on the high-pass filtered sweep of negative local field potentials (nLFPs). These nLFPs have been correlated with neuronal spikes (Wu et al. 2014).

Seizure-like activity consists of ictal and tonic bursts (Fig. 15.1b, red line, upper panel), followed by clonic bursts (Fig. 15.1b, gray line, upper panel). The time-points at which each nLFPs exceeded the specific threshold is marked as a unit raster in the lower panel of Fig. 15.1b. The time units were summed together in a timescale-binned plot to calculate the avalanche size and lifetime. Figure 15.1c shows the cumulated time units framed by a time window (Δt, gray regions) from eight-channel recording. According to Wu et al. 2014, the avalanche is defined as "a series of activity separated by a blank at the beginning and end of the events". The avalanche size was calculated as the total number of electrodes with active units, while the lifetime was calculated as the summation of the total time frame in each avalanche event. The distribution of the avalanche size with its probability P (s) were plotted on a log-log scale. A neuronal avalanche that has a fitted straight-line slope of α value indicates a power-law relationship and the event's dependence. The 4AP-Bic group (red solid line) showed a power-law distribution

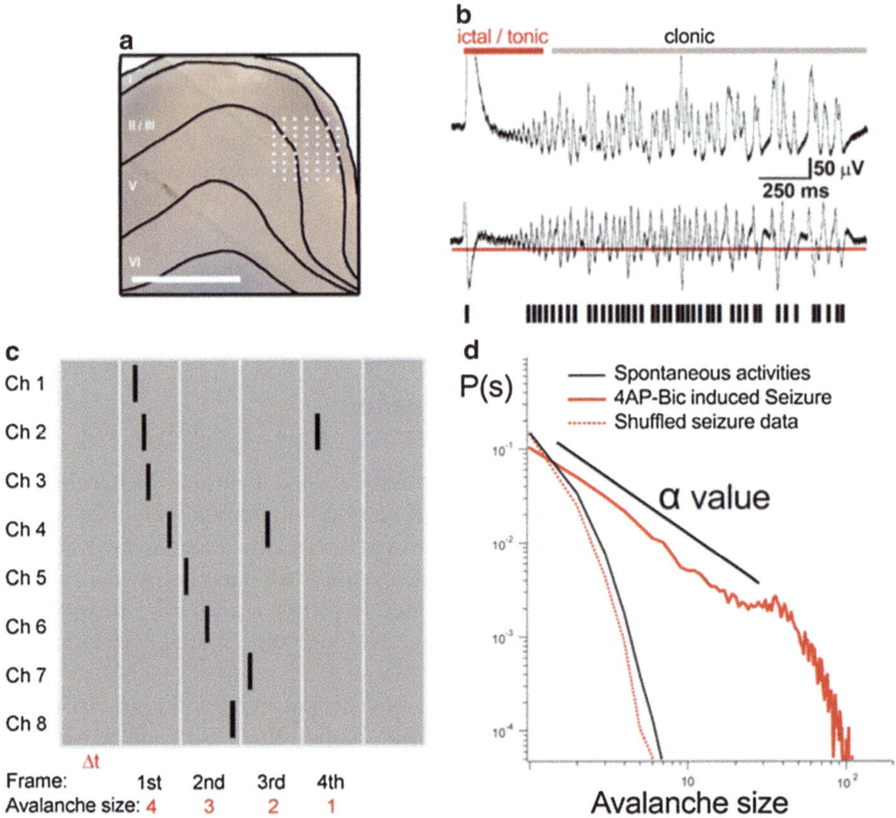

Fig. 15.1 Definition of the neuronal avalanche. (**a**) Brain slice with the recording site. (**b**) Comparison of seizure-like activity of ictal events, followed by tonic and clonic bursts with spontaneous activity. (**c**) Example of the collective time step, which is framed in 4 ms time bins from eight channels. The definition of an avalanche is separated by blank activity at the beginning and end of the events. The activated electrodes are counted as the avalanche size, and each event's lifetime is the summation of the total time frames. (**d**) Distribution of different avalanche sizes plotted on a log-log scale. The neuronal avalanche could follow the power-law distribution, and its slope could be calculated as the α value. The original data were shuffled in order to disturb their spatiotemporal arrangement. Scale bar = 1 mm in A (Adapted with permission from Wu et al. 2014 BMC Neuroscience)

with the α value around −1.5 (Fig. 15.1d). An important test was performed to assess whether event dependence is essential for the power-law relationship. Thus, the event dependence was disturbed, by randomly shuffling the data with regard to the order of temporal sequence and spatial arrangement of the events. Both the shuffling data (dashed red line) and spontaneous activity data (black line) showed an exponential distribution, which is a type of Poisson distribution in which the events occur independently (Fig. 15.1d).

15.2.2 Power Law Distributions

Power law distributions of event sizes are often seen in complex phenomena including semiconductor devices (Levinshtein et al. 2005), forest fires, earthquakes, phase transitions, financial market fluctuations, snow avalanches and many other instances (Bak 1996). For example, earthquake models integrate local rules in which forces at one site are distributed to nearest neighbors without dissipation. This conservation of forces is similar to the conservation of synaptic strengths (Royer and Pare 2003) and it could be a mechanism responsible for maintaining a network near the critical point. Computer simulations indicate that networks can be kept nearly critical levels when the total sum of synaptic strengths soars near a constant value (Hsu and Beggs 2006). This could be accomplished through a mechanism like synaptic scaling (Turrigiano and Nelson 2000), which has been observed experimentally. Finally, recently "burned" areas in forest fire models are refractory, while unburned areas are more likely to ignite. This balance of refractoriness and excitability combines to maintain the system near the critical point. Recent models of neuronal avalanches (Levina et al. 2005) have suggested that short-term synaptic depression and facilitation may also serve to drive neuronal networks toward the critical point where avalanches occur. Thus, an understanding of power laws in diverse complex systems can suggest mechanisms that might underlie criticality in neuronal networks. Computational models have also explored the potential relationship between neuronal avalanches and epilepsy (Beggs 2008; Hsu and Beggs 2006; Hsu et al. 2007, 2008).

Although the normal brain firing activity can be moderately-synchronous, in epileptic seizures, a group of neurons begin firing in an abnormal, excessive, and synchronized manner (McPhee and Hammer 2010; National Institute for Health and Clinical Excellence 2012; Plenz 2012; Yang et al. 2012). This excessive firing may occur due to structural or functional anomalies within the epileptic brain. Such abnormal activity could be produced either by hyper excited neurons acting independently or it could involve abnormal interactions among many neurons. Many forms of collective activity including waves, spirals, oscillations, synchrony, and neuronal avalanches have been identified in abnormal epileptic firing. All these emergent activity patterns have been hypothesized to show pathologic signatures associated with epilepsy (Hobbs et al. 2010). From this perspective, epileptic activity would occur when regulatory mechanisms failed and the network entered a super critical regimen. Operating at the critical point depends on the appropriate balance between excitation and inhibition implies a structured activity that is far from random. In addition, the activity in one neuron would, on average, lead to activity in other neurons, amplifying activity excessively and possibly leading to seizures (Hobbs et al. 2010). Therefore, when network activity is randomly shuffled, it no longer follows a power law distribution characteristic of avalanches (Beggs 2008).

Hobbs et al. (2010) examined neural activity from human and rat cortical tissue in local cortical networks using 60 channel multielectrode arrays to record local field potentials in brain slices removed from the most active epileptogenic area (as identified by intraoperative electrocorticography). In the human cortical tissue they found periods of pronounced hyperexcitability and lack of a clear power law in avalanche size distributions. Analysis showed that during these periods, there was a significant positive correlation between the branching parameter and the firing rate, suggesting a positive feedback loop. This aspect was not present in the activity examined in rat tissue. These results indicate that cortical tissue removed from pediatric epilepsy patients produces aberrant neuronal avalanches (Hobbs et al. 2010).

15.2.3 Self-Organized Criticality

Self-organized criticality (SOC) represents a property of complex dynamic systems that evolve to a critical state, capable of producing scale-free energy fluctuations. A characteristic feature of dynamical systems exhibiting SOC is the power-law probability distributions that describe the dynamics of energy release. Worrell and colleagues in their study investigated the probability distribution *of in vivo* pathological energy fluctuations in human epileptic hippocampus by analyzing data from seven consecutive patients with temporal lobe epilepsy who required depth electrode iEEG monitoring during evaluation for epilepsy surgery (Worrell et al. 2002). Contacts that recorded the earliest clear seizure onset on iEEG delineated the seizure onset zone that was determined by visual inspection as located within the mesial (middle) temporal lobe in each patient. Typical waveforms of interest were epileptiform spikes, sharp waves, and sharp and slow wave complexes. They concluded that the probability densities of interictal epileptic energy fluctuations and the quiescent time between successive fluctuations exhibit power-law scaling, which provides evidence for SOC in human epileptic hippocampus. They hypothesized that interictal epileptiform discharges are a mechanism for energy release within epileptic brain, and that these events may provide a method for identifying the network involved in seizure generation and they even may assure a physiological mechanism for preventing seizures (Worrell et al. 2002).

15.3 Epileptic Seizures

15.3.1 Definitions, Cause, Symptoms and Features of Epileptic Seizures

Epilepsy has been defined as a chronic, complex neurological disorder associated with abnormal electrical activity in the brain, marked by sudden recurrent episodes of sensory/motor disturbance, changes in behavior, with or without loss of consciousness, and/or convulsions (Nunes et al. 2012). This excessive activity can be produced by hyperexcited neurons acting independently or involve abnormal interactions among many neurons (Hobbs et al. 2010). Epilepsy is one of the most common chronic neurological affections, with more than 50 million patients worldwide and approximate 2 million new cases each year. Epilepsy responds to treatment in about 70 % of cases. Not all cases of epilepsy are lifelong, a substantial number of people improve to the point that medication is no longer needed. One seizure does not signal epilepsy (up to 10 % of people worldwide have one seizure during their lifetimes) (Chang and Lowenstein 2003). Epilepsy has been defined since 2005 by two or more unprovoked seizures, >24 h apart (Fisher et al. 2005). The definition of epilepsy has been recently revised in order to consider the diagnosis even after occurrence of a single unprovoked seizure, providing that we can demonstrate an enduring predisposition for recurrence, similar to the general recurrence risk after two seizures (more than 60 %) over the next 10 years. For instance a patient can have a single unprovoked seizure after trauma, a stroke, or central nervous system infection. A patient with such brain disorders has a high risk of developing epilepsy after a single unprovoked seizure.

The tendency to respond to particular stimuli with seizures also meets the conceptual definition of epilepsy. Reflex epilepsies are also associated with an enduring abnormal predisposition to have recurrent seizures.

The document also mentions as a general agreement that epilepsy should no longer be considered a disorder but a disease, the term implying a more substantial, long lasting derangement of neuronal functionality (Fisher et al. 2014).

15.3.1.1 Classification of Seizures and Epilepsies

Generalized epileptic seizures are defined as originating within bilaterally spread networks that, at some point, are rapidly involved. These networks do not necessary include the entire cortex and could involve both cortical and subcortical structures. Generalized seizures can be asymmetric in clinical appearance. Focal epileptic seizures are defined as originating within networks restricted to one hemisphere. They may be localized in a small area or have a wider spread. Depending on the seizure type, ictal onset ("area of cortex that initiates clinical seizures" (Rosenow and Luders 2001) is consistent from one seizure to another), with distinctive propagation patterns that can sometimes imply the contralateral hemisphere. Focal seizures might provoke alteration of consciousness or awareness and could evolve into secondary tonic-clonic generalization. Auras are purely subjective clinical manifestation usually occurring at the very onset of a focal seizure as a warning event. The epileptic nature is sometimes difficult to prove if the aura symptoms do not evolve into a more objective clinical pattern (Luders et al. 1998). Focal seizures might develop abruptly without any warning symptoms.

15.3.1.2 Cause Types (Etiology)

The International League Against Epilepsy (ILAE) Commission on Classification and Terminology has simplified the classification of seizures and they proposed the following three concepts:

1. Genetic epilepsy represents the result of a known or supposed genetic error in which seizures are the main symptom of the disorder. In recent years, some types of epilepsy have been correlated to mutations in genes, mostly involving ion channels, assumptions derived from specific molecular genetic studies or from suitable adapted family studies. For example a mutation in the GABA1 gene has been detected in some members of a family with juvenile myoclonic epilepsy (Cossette et al. 2002). No definitive conclusions can be made because there is no knowledge regarding specific surrounding influences as causes of or factors that contribute to these forms of epilepsy.
2. Structural or metabolic: This type of epilepsy refers to a condition or disease with a metabolic or structural background associated with a high risk of developing epilepsy. Structural lesions are frequently associated with focal seizures. They include cerebral changes resulting from, trauma, stroke, perinatal brain damage, malformations, brain tumors, and infections. These epilepsies are characterized by the presence in epileptogenic foci of residual neurons with no afferents and less dendritic spines, destroyed probably by infections, trauma, stroke, or other lesion. Structural lesions of the cerebral cortex can disrupt, inhibitory GABAergic interneurons, thereby minimizing the inhibition that controls large pyramidal cells (Menzler et al. 2011).
3. "Unknown cause": There are several cases wherein the underlying pathophysiology can't be identified. These are usually encountered in children and young adults but can occur at any age in persons who have a family history of epilepsy or seizures. This epilepsy may have a genetic defect at its origin or it may be the result of a distinct and yet unknown disorder (Berg et al. 2010).

15.3.2 Power Law Distributions in Epilepsy

Diseases of central nervous system are often associated with altered brain dynamics (Expert et al. 2010). It has been hypothesized that the dynamical properties characterizing a critical state may be considered as an important marker of brain well-being in both health and disease (Plenz 2012). During epileptic seizures the distribution of phase-locking intervals (PLI) is providing additional evidence for the criticality hypothesis. Furthermore, deriving the distribution of PLI from electrocorticogram (ECoG) data as an indicator of critical brain dynamics has shown that the system deviates from scale-free behavior during seizures. All scales closely follow a power-law probability distribution during pre-ictal time intervals with the exponent between 22 and 23.5 (see Fig. 15.2). The apparent robustness of the

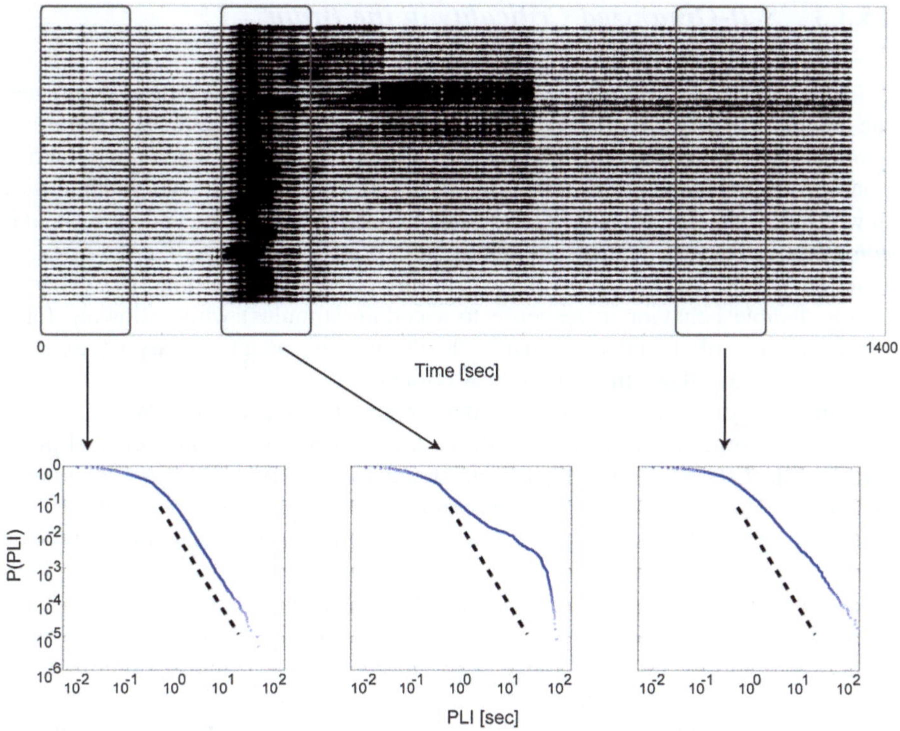

Fig. 15.2 The distribution of phase-locking intervals deviates from a power-law during epileptic seizures. *Top*: The electrocorticogram (ECoG) recording shows the onset of a focal epileptic seizure attack around 300 s time. *Bottom*: Cumulative distributions of phase-locking intervals (PLI) are obtained during three time intervals of 150 s: preictal (*left*), ictal (*middle*) and postictal (*right*). *Dashed lines* indicate a power-law with exponent 23.1. While the distribution appears to follow a power-law during the pre-ictal period, intervals of increased phase-locking disturb this characteristic distribution with the onset of seizure activity. Data shown are from a patient scale 3, corresponding to the frequency band 25–12.5 Hz (Adapted with permission from Meisel et al. 2012, PLOS Computational Biology)

power-law against exact conditions (different anatomical regions with varying number of channels) strengthens the hypothesis of the relevance of a critical state in human brain dynamics. While the PLI distribution followed a power-law in time intervals preceding the seizure onset, a deviation from power-law behavior was observed in intervals containing the seizure attack.

Figure 15.2 shows the distribution of PLI derived from a pre-ictal, an ictal and a postictal time interval. The probability to find longer PLI increased during attacks thereby destroying the scale-free property of the original distribution. After the seizure this distribution slowly relaxed back to a power-law. In Fig. 15.2 this relaxation is not yet complete in the postictal time interval as there is still some residual seizure dynamics in the ECoG recording.

15.3.3 Self-Organized Criticality in the Brain

Critical systems have been defined as systems that are close to a critical point, near the boundary of an order-disorder phase transition. At criticality, these systems can avoid being trapped in one of two extreme cases: a disordered state (when interactions are too weak and the system is dominated by noise) or a globally ordered state in which all elements are locked (when interactions are too strong and the system is completely static). A dualism is essential for a complex system, like the brain, to function: it must maintain some order to ensure coherent functioning (i.e., generate a reproducible behavior in response to a certain stimulus) while allowing for a certain degree of disorder to enable flexibility (i.e., adapt to varying external conditions). Such dualism is possible at criticality.

While many degrees of order/disorder are possible, the subtle balance between order and disorder at criticality manifests itself in certain general statistical properties: critical systems exhibit spatial and temporal correlations that are long range (i.e., on scales that are larger than those on which mutual interactions take effect) and follow power-law distributions. Recent research has shown that brain networks, by being in the critical state, optimize their response to inputs and maximize their information processing ability (Shew and Plentz 2013).

Haimovici and colleagues have presented a simple brain model that, if tuned to criticality, explains the broad range of experimental observations of human brain activity, in particular, reproducing key findings obtained with functional magnetic resonance imaging (fMRI) (Haimovici et al. 2013). fMRI research has been able to deliver an important observation: the human brain at rest exhibits a large-scale spatiotemporal organization into distinct functional networks—so-called resting state networks (RSN) (Fox and Raichle 2007). RSNs are areas of the resting brain—measured in subjects performing any cognitive, or motor tasks—in which fluctuations of neural activity are correlated, as revealed by the fact that BOLD signal fluctuations within the same network are synchronous. Each RSN can be related to a specific set of cortical areas associated with certain functions: cognitive, sensory (visual, auditory), and motor RSNs, for instance, have been identified (see Fig. 15.3).

What Haimovici and colleagues found is that such a model can lead to activity clusters similar to those found for RSN activity (Haimovici et al. 2013). But for the model to match the experimental data, the activation threshold had to be set exactly at the level at which their model becomes critical, as illustrated in Fig. 15.3. At criticality, their model predicts a number of statistical properties that are consistent with experiments: the brain forms activity clusters whose size follows a power law with slope of $-3/2$, the hallmark of neuronal avalanches (Beggs and Plenz 2003), corresponding to a peak in the size of the second-largest cluster, as found in percolation models (Margolina et al. 1982); the correlation length (the distance at which two points in the system behave independently) and its fluctuations diverge and match those seen in human brain data.

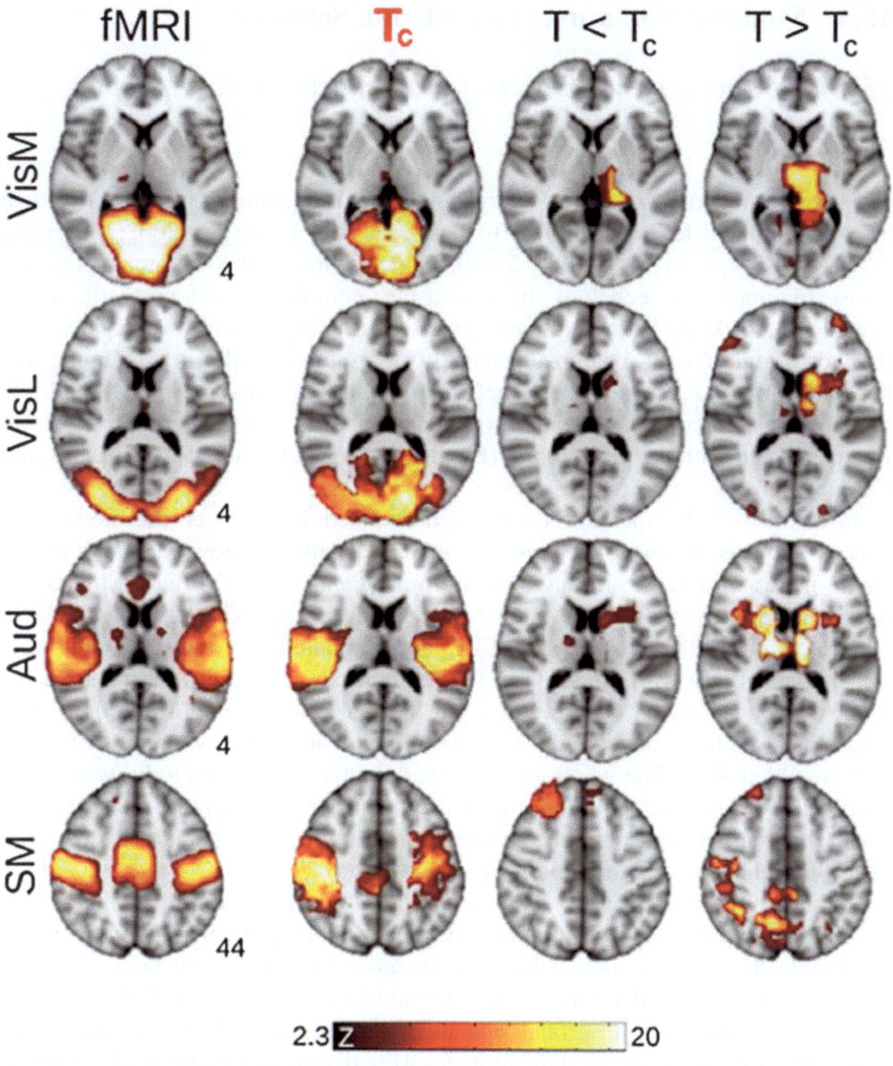

Fig. 15.3 Functional magnetic resonance imaging (fMRI) experiments have revealed that the brain at rest is organized into several areas in which fluctuations of brain activities are correlated, so-called resting state networks (RSN). From *top* to *bottom*: medial visual (VisM), lateral visual (VisL), auditory (Aud), and sensory-motor (SM) RSNs. (Right columns) Results from the work of Haimovici et al. 2013 show that a simple model can reproduce the statistical properties of RSNs only if the model is tuned to criticality (at TC) (Adapted with permission from Haimovici et al. 2013, Phys Rev Letters)

15.4 Modular Signatures in Epileptic Seizures

15.4.1 Functional vs. Structural Modularity of the Epileptic Brain

The human neocortex consists of a large number of minicolumns in parallel vertical arrays (Mountcastle 1957, 1997; Buxhoeveden and Casanova 2002; Casanova et al. 2007; Shepherd and Grillner 2010). Minicolumns are the first step in a nested series of nodes or echelons of increasing complexity (Mountcastle 1997; Szentagothai 1975). Within minicolumns, cortical neurons are aggregated into five horizontal layers (or laminae), namely two supra-granular, one granular and two infra-granular layers. Other levels of modular organization include multiple minicolumns, macrocolumns, and large-scale networks of macrocolumns that are interconnected with the entire brain (Buxhoeveden and Casanova 2002; Opris and Casanova 2014).

In contrast to the sparse but organized connectivity between the modules of control subjects, brain connectivity of epileptic patients shows a configuration where nodes in a functional module are more connected to different functional modules. Recently, Vaessen et al. (2014) examined connectivity of whole brains of children with frontal lobe epilepsy (FLE) and compared their structural and functional connectivity with the same in healthy controls. Their measurements of the functional connectivity was derived from the dynamic fluctuations of the fMRI, while the structural connectivity was determined from fiber tractograms of diffusion weighted MRI. The whole brain network patterns of connectivity were characterized with graph theoretical metrics and further decomposed into modules. Then, the graph metrics with the extracted connectivity within and between modules were related to cognitive performance.

As shown in Fig. 15.4a, the modularity algorithm extracted four modules from the averaged functional connectivity matrix over all subjects. Spatial organization of module #1 was considered the "default mode network" with network nodes distributed in the frontal, temporal and parietal lobes (Vaessen et al. 2014). Module #2 was distributed over the frontal and subcortical regions. Module #3 was localized in the occipital lobe. Module #4 nodes were distributed over frontal, temporal and occipital regions. All four modules seemed to be symmetric to the interhemispheric fissure. As shown in Figs. 15.4a, b when the structural connections were organized according to the modularity index of the frontal cortex, the structural modularity revealed bilateral structural sub networks. For the structural connectivity matrix the modularity algorithm determined only two modules, which were highly symmetric over the two hemispheres. After functional connections were organized according to these structural connectivity modules, the sub organization vanished (Fig. 15.4c, d). An interactive view of functional connectivity and structural connectivity was displayed as between-module, within-module averaged over all modules and individual within-module connections (Fig. 15.4e, f). It was further shown that functional "disturbances" of epileptic children were

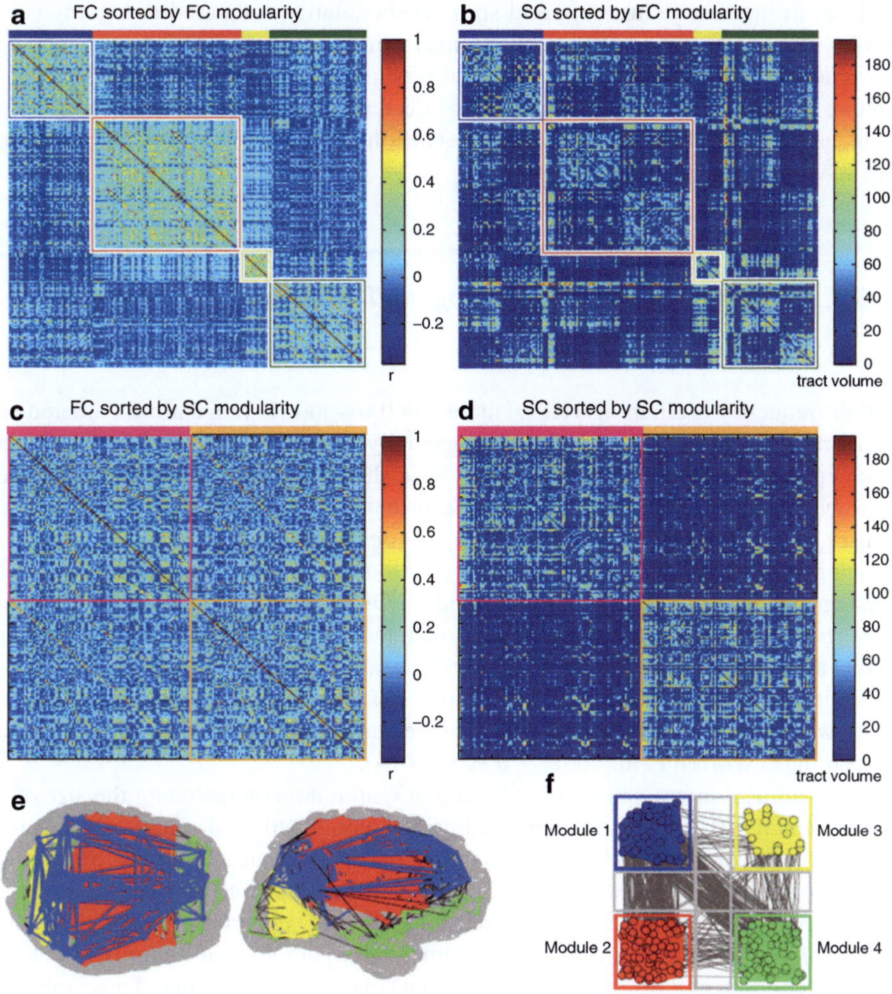

Fig. 15.4 Functional *vs*. structural brain modularity. (**a**) Functional connectivity showing the average connection matrices sorted by module. Colored rectangles indicate the modules. High within-module connectivity is illustrated by the higher values (more hot colors), while between-module connectivity is shown by more sparse (more cold colors). (**b**) Structural connectivity was sorted by functional modules. The functional modules are organized bilaterally, while the structural connectivity has strong inter-hemispheric connectivity and low intra-hemispheric connectivity clearly visible in the block patterns. (**c**) Functional connectivity matrix was sorted by the modular organization derived from the structural connectivity. The two structural connectivity modules represent the *left* and *right* hemisphere. Functional connectivity shows that strong inter-hemispheric connections are present within the two modules. (**d**) The structural connectivity was sorted by structural connectivity modularity. Strong intra-hemispheric connections are obvious, while inter-hemispheric connections (and thus between-module connections) are weaker. (**e**) Illustration of modular organization of functional connectivity. Within-module connections are colored as in panel A. (**f**) An alternative presentation of the modular organization. The nodes of each separate module are depicted spatially segregated. The *gray lines* indicate the between-module connections (Adapted with permission from Vaessen et al. 2014, PLOS one)

related to increased clustering and stronger modularity compared to healthy controls, which was accompanied by stronger within- and weaker between-module functional connectivity. While structural modularity increased with stronger cognitive impairment, it was concluded that decreased coupling between large-scale functional network modules may represent a hallmark for impaired cognition in childhood FLE.

15.4.2 Spatial Scale of Epileptogenicity Biomarkers Matches Minicolumns Size

High-frequency oscillations (HFO) in the 80(100)–500 Hz range are considered an important biomarker of cortical epileptogenicity. Although HFO may be present in non-epilepogenic brain structures, it has been shown that the ripple (<250 Hz) and particularly fast ripples (>250 Hz) subbands have are highly specific to the seizure onset zone (Zijlmans et al. 2012). Using dense 2D microelectrode arrays (MEA) having a spacing of 0.4 mm between electrodes, Schevon and colleagues have shown that about ~90 % of the recorded HFOs were limited to a single channel (Schevon et al. 2009). This is consistent with the idea that most of the spontaneous HFO events are confined within a minicolumn, and spreading the activity requires a recruitment process of several minicolumns, possibly through an avalanche process following a power-law, as suggested by several other studies using intracranial recordings (Worrell et al. 2002; Wu et al. 2014).

Sub-clinical micro-seizures confined to a spatial domain matching the size of a minicolumn (~0.5 mm) were observed using a combination of micro- and macro-contact cortical grids and depth electrodes on domains in a study by Stead and colleagues (Stead et al. 2010). They hypothesize that 'relatively sparse pathological cortical columns are the anatomical substrate of focal neocortical epilepsy', or 'the sick column hypothesis'. Such minicolumns with pathological neurons or pathological connectivity (intra- or inter-columnar) may be the initiators of a seizure, as epileptiform discharges have been detected early during the ictal onset in microcontacts, before being detected on the macrocontacts (Stead et al. 2010; Zijlmans et al. 2012).

15.4.3 Does Cortical Columnar Firing in Layer 2/3 Get Amplified Across Neighboring Minicolumns?

A great deal of structural and functional implications emerged with the decreased coupling between large-scale functional network modules resulting in impaired cognition in epilepsy. It is intriguing to know what happens at the minicolumn and microcircuit level during epileptic seizures. Does cortical columnar firing in layer

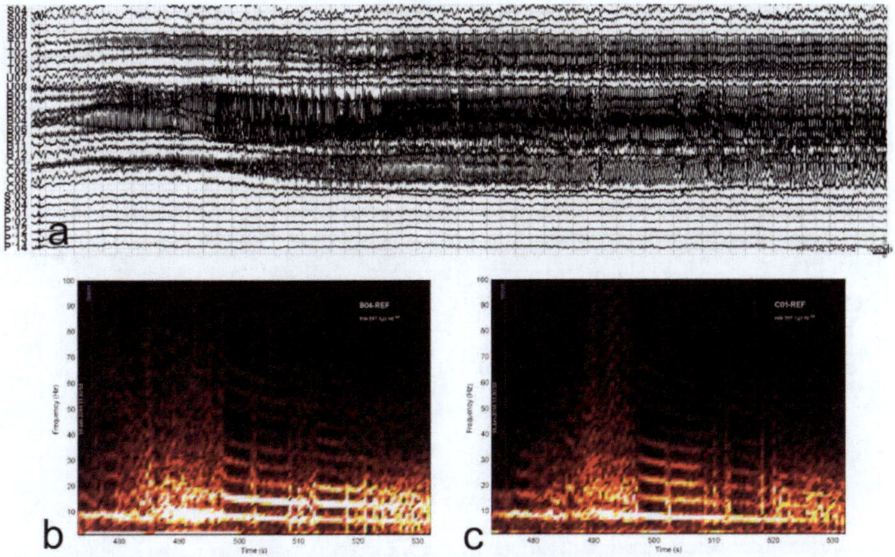

Fig. 15.5 Illustration of synchronized burst in epilepsy. (**a**) EEG data from intracranial depth electrodes showing seizure initiation on contacts B03-B04 and propagation on contacts B01-B02, B06-B07, C01-C02 and T01-T06; (**b**) and (**c**) Time-frequency maps for contacts B04(b) and C01 (c) showing sustained 8 Hz activity and multiple harmonics

2/3 (Wu et al. 2014; Giresh and Plenz 2008) get amplified across neighboring minicolumns?

During "neural avalanche" the firing pattern of pyramidal cells gets excessively amplified. Such synchronized amplification is also reflected by the 8 Hz multiple harmonics (Fig. 15.5a–c). According to such modular view of epilepsy, one hypothesis posits that such excessive amplification of neural firing in a cortical minicolumn may be the outcome of pyramidal cells crosstalk that pierce the minicolumnar inhibitory wall surrounding the pyramidal cells. Indeed if that would be the case layer 2 and 3 cells that have pyramidal cells extend cortico-cortical projections intra- and inter-hemisphere and the excitation is thus spread across many areas (Fig. 15.6). This excitation is then sent through the minicolumns top-down to the subcortical structures that deal with the behavior, resulting in massive seizures with uncoordinated movements. The minicolumn hypothesis fits also with local generation of epileptic seizures in cortical microcircuits.

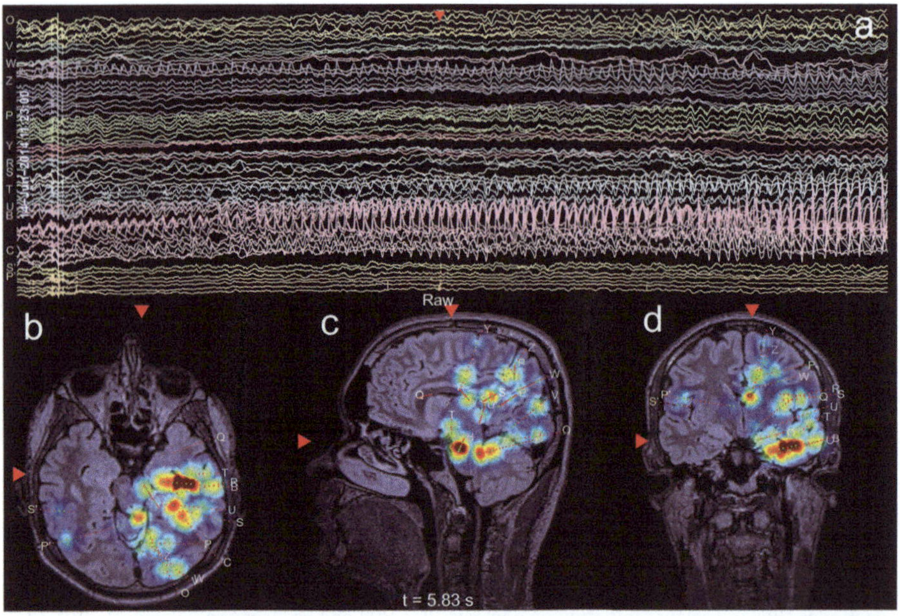

Fig. 15.6 Instantaneous spatial extent of the epileptic seizure. (**a**) Raw (unfiltered) EEG data from intracranial depth electrodes showing seizure initiation on electrode B and propagation on electrodes C and T. The red arrow represent the timestamp at which the EEG voltage maps are created. Maximum intensity projection voltage maps in axial (**b**), sagittal (**c**) and coronal (**d**) views, where *green dots* are contacts included in the montage and *red dots* are contacts not included in the montage (The EEG data is represented in color code: *red* for higher EEG amplitude (voltage), *blue* for lower EEG amplitude (voltage))

15.5 Future Directions

15.5.1 Future Directions in Treating Epilepsy

In contrast to the rapid advances in other therapies, epilepsy medical therapy was usually limited to anti-epileptic drugs. Yet, approximately 34 % of the patients suffering from epilepsy are described as medically intractable epilepsy patients which still suffer from sustained frequent seizures in defiance of receiving adequate treatment with anti-epileptic drugs of sufficient duration (Kwan and Brodie 2000). Recent studies show that the notable advance in the field of inflammation and immunology are estimated to be important elements of the pathobiology of epilepsy that may offer future directions in treatment of this disease. Inflammatory processes have been noticed in the brain tissue of both experimental animal models and subjects with epilepsy. Anti-inflammatory and immunotherapies demonstrated considerable anticonvulsant properties both in experimental and in clinical findings. These arguments denote the fact that modulation of inflammatory processes and immunity could represent a new particular objective to accomplish eventual anticonvulsant effects for the subjects with epilepsy, especially for those suffering of medically intractable epilepsy (Yua et al. 2013). For modern neurosurgery, the aim is to find a procedure which is minimally invasive and does not involve a large craniotomy thus avoiding cognitive morbidity and decreasing hospital stay.

15.5.1.1 Responsive Neurostimulation (RNS)

An additional treatment option for those suffering from medically intractable epilepsy, consists in direct brain stimulation with a system that provides responsive focal cortical stimulation (RNS System). The aim is to interrupt epileptiform abnormalities before the seizure starts. The RNS System detects epileptiform activity with electrodes placed in the brain. A programmable brain implanted neurostimulator provides control on seizure focus. Subdural cortical strip leads and/or one or two brain depth electrodes are surgically implanted, according to the seizure onset zone. These electrodes are connected to the stimulator. This device permanently detects and analyze brain electrical activity and is programmed to select specific electrical patterns and then to generate pulses to control seizures in a closed loop feed-back. Heck and coworkers studied the safety and effectiveness of this method in 191 patients with medically intractable partial onset seizures emerging from one or two foci. After a 2 years of post-implant follow-up they demonstrated that responsive stimulation of the seizure onset zone, decreased the frequency of seizures and revealed seizure reduction over time, being acceptably safe and well tolerated (Heck et al. 2014).

15.5.1.2 Stereotactic Laser Ablation

Stereotactic laser ablation of the epileptogenic structures may be an alternative to standard epilepsy surgery. Laser ablation leaves areas situated in the pathway to the target largely intact. This technique could be correlated with fewer deficits by avoiding important network regions and critical white matter pathways. Particularly when guided by MRI it allows real time thermal monitoring of the resection and also provides feedback for the laser energy delivery (Curry et al. 2012; Tovar-Spinoza et al. 2013).

15.5.1.3 Vagus Nerve Stimulation

Vagus nerve stimulation can be used to prevent epileptic seizures by sending regular, mild electrical pulses to the brain via the vagus nerve. These pulses are supplied by a pacemaker-like device. The pulse generator is surgically placed in the upper part of the chest, under the skin. A connecting wire passes from the stimulator to an electrode that is adjacent to the vagus nerve, which is accessible through a small incision in the cervical area. After it is implanted, depending on the patient's tolerance, the device may be programmed to stimulate the nerve at regular intervals. The settings on the stimulator are adjustable, and the electrical pulses are gradually raised as the subject's tolerance to the pulses increases. A certain number of patients suffering from medically intractable epilepsy have undergone implantation of the left vagus nerve. Less than half of the subjects reported more than 50 % diminution in seizure frequency and few patients became seizure free (Ramsey et al. 1991).

AspireSR generator is a new technology which provides automatic stimulation in response to seizure detection. This technology is based on the fact that seizures are most of the time being accompanied by ictal tachycardia. This generator can analyze relative heart rate changes in order to adapt to seizures control. The stimulation can be activated manually when the patient anticipate the onset of a seizure, by using a hand-held magnet. This kind of stimulation has been proven to reduce seizure severity, shorten or stop a seizure, and ameliorate or reduce the postictal recovery period (Cyberonics 2014).

15.5.1.4 Transcranial Magnetic Stimulation

Repetitive transcranial magnetic stimulation (rTMS) may be considered a new therapeutic tool in treating epilepsy. It can be used to control seizures even though its safety and tolerability among the patients with epilepsy are still discussed in the literature. Hsu et al. (2011) did a meta-analysis on 164 participants in previous studies during twenty years period (1990–2010) to estimate the antiepileptic efficacy of rTMS (low frequency repetitive transcranial magnetic stimulation) in medically intractable epilepsy. They concluded the fact that low frequency rTMS has a favorable effect on seizure control, particularly evident in subjects with cortical dysplasia or neocortical epilepsy. These findings require confirmation in extensive studies.

15.5.1.5 Biomarkers for Seizure Detection

Electrophysiological Biomarkers

It is necessary to identify and validate electrophysiological biomarkers of epilepsy in order to diagnose, prevent, treat and cure epilepsy. The aim is to identify the epileptogenic zone and obtain functional mapping that could be used to determine if epilepsy surgery can be performed and to define risk /benefit ratio. Tailored resections are recommended. Electrophysiological biomarkers of epilepsy may be high-frequency oscillations which consist of ripples 80(100)–250 Hz and fast ripples >250 Hz. High-frequency oscillations (HFO) could be present in normal brain, for instance in area CA1 of the hippocampus (Buzsaki et al. 1992). Sensory-evoked high-frequency oscillations could also be present in CA3, subiculum and the entorhinal cortex (Chrobak and Buzsaki 1996, Csicsvari et al. 1999). Interictal fast ripples (250–600 Hz) are strongly associated with brain areas capable of generating spontaneous seizures (Zijlmans et al. 2012). Ripples in the dentate gyrus and the neocortex should be considered pathological HFOs. The association between pathological HFOs and epileptogenicity suggests pathological HFOs could support the localization of the seizure onset zone and might identify the epileptogenic zone more accurately. Interictal spikes represent a good spatial biomarker for the seizure onset zone and the irritative zone. However, there is little evidence that interictal spikes predict seizure frequency or the disease's severity. The functional role of interictal spikes in epilepsy is not known, but some interictal spikes might reduce ictal discharges. In a different view, the presence and clustering of interictal spikes after induced status epilepticus in rats could predict the subsequent appearance of spontaneous seizure (Staba et al. 2014).

Cortical Excitability (CE)

Transcranial magnetic stimulation (TMS) is a noninvasive and safe method to explore cortical excitability (CE). Bauer et al. 2014 reviewed fifty studies that determined CE in patients with epilepsy and most of them revealed cortical hyperexcitability, which can be improved with anti-epileptic drug treatment. Other studies state that reduction of CE after resection (epilepsy surgery) is an important predictor of good seizure outcome. Cortical direct electrical stimulation (DES) studies have indeed confirmed an increased excitability in the seizure onset zone (Iwasaki et al. 2010; Enatsu et al 2012). Cortical excitability may therefore be a feasible biomarker for epilepsy.

Molecular Biomarkers

Notable advance has been made recently concerning biomarkers of epilepsy. Analysis on the genomic, proteomic and mRNA levels were made from samples of brain tissue or cerebrospinal fluid, blood, plasma, or serum. The purpose is to discover specific molecular biomarkers. Similar to the role of molecular biomarkers for brain tumors, a temporary dysfunction of the blood–brain barrier may contribute to epileptogenesis. Combining molecular analysis with other procedures, such as imaging and electrophysiological biomarkers could be beneficial for understanding and treatment of epileptic patients (Lukasiuk and Becker 2014).

15.5.2 Control of Epilepsy

Device controlled epilepsy is an option that has been developed recently to help patients with drug resistant epilepsy. Among the emerging applications are anti-epilepsy neural implants and the nanostructures such as carbon nanowire/nanotubes that allow the drug to reach the critical location in the brain (Opris 2013).

15.5.2.1 Anti-epilepsy Neural Implants

An anti-seizure neural implant should operate on the brain signals that cause the epileptic seizures in order to maintain its proper functioning. Anderson et al. (2013) has shown high-frequency spectral changes in the prefrontal cortex of primates with potential use in neuoroprosthetics. In another study (Johnson et al. 2012), 5 human subjects were implanted with electrocorticography (ECoG) arrays for long-term epilepsy monitoring. Subjects were trained on a brain-computer interface to control a cursor on a computer screen by modulating the spindle activity involved in learning. The fact that the pattern of increased spindle activity was locally modulated by training on a brain-computer interface, supports the idea of future neural implants to control epilepsy.

15.5.2.2 Nanotechnological Applications to Control Epilepsy

The nanomaterials play a critical role in controlling the multitude of events that determine the composition and structure of the elements for brain machine interface. A number of approaches that allow the control at the nano-level (metals, carbon nanowires/nanotubes and organic conducting polymers) have revealed a critical relationship between nanostructure features and cellular behavior (Wallace et al. 2012). In this regard, epilepsy related drugs, delivery systems and devices, seem to operate successfully based on nanotechnology (Bennewitz and Saltzman 2009; Pathan et al. 2010). Thus, the nanotechnology for delivery of drugs to the brain for epilepsy treatment use drug loaded nanoparticles with improved efficiency.

15.5.2.3 Nano-devices for Epilepsy Control

The possibility to use small scale, monolithic integrated circuits, allows the easy and efficient implantation of a whole energy efficient system. Neuroprosthetics for Parkinson's patients send electrical pulses to the deep brain structures, with the aim of reducing or eliminating symptoms. Still, to have a more complete system, one could take into consideration the feedback-loop design. In such a case, the device would pick up the signals related to the intention of the user and format them back to the brain in such a way that the effects of neurodegenerative diseases are reduced.

The implementation of such as device should be as minimally invasive and precise as possible. Battery supply of such devices can be either long-term, depending on energy consumption or rechargeable using induction coils. One major obstacle for Brain-Machine Interfaces (BMI) today represents the lack of upgrade-ability of the hardware of a device, once implanted. For each physical change, the patient must be operated again. A proposed solution to this question is the "Neural Dust" (Seo et al. 2013), based on micrometre size, ultra-low power CMOS "node" sensors used for detection and reporting of extracellular data and a central control unit. The device aims to replace the electrode implantation that is in use today. The free-floating sensors are powered ultrasonically by the transmitters and use backscatter communication. In detail, the main control unit is powered by a standalone radio frequency (RF) transmitter on the skin of the skull, completely wireless. This type of power and communication transmission reduces significantly the risk of infection associated with standard wire implants.

Ultrasound EM (electromagnetic waves) frequency was chosen for power and communication transfer because the tissue attenuation is far less when compared to standard micro-wave frequencies used for wireless today. The sensors are using piezo-electric transducers, since they seem to be best suited for the application. The software of the system is based on the function of the device: feedback-loops will allow the control of Parkinson's Disease and Epilepsy. Currently solutions are employed with varying success on breaking down the neural electrical code. Sending the right electrical signals to the epileptic tissue should stop its propagation. But being able to recognise the beginning of an ictal phase is possible only using learning algorithms such as support vector machines, since the patients signals are different. The algorithm itself is used as an early-warning system, based on the analysis of electroencephalogram recordings (EEG) that enables the controlling part of the device to enable suppressing signals to the affected tissue. Current best-performing algorithms allow around 70 % accuracy. Improvement in this field will produce an automated stimulator for the patients (Wang 2013).

In conclusion, as recent results point out, the pattern of amplified firing in neural avalanches betrays a modular signature in the spread of activation across cortical minicolumns. According to such modular approach of epilepsy, it is likely that the excessive amplification of neural firing in a cortical minicolum results from a defect within the "inhibitory curtain" surrounding the pyramidal cells (Opris and Casanova 2014). The functional basis of this approach provides valuable insights into potential clinical interventions.

Acknowledgments Ana Ciurea was supported by the strategic grant POSDRU/159/1.5/S/133652, "Integrated system to improve the quality of doctoral and postdoctoral research in Romania and promotion of the role of science in society" cofinanced by the European Social Found within the Sectorial Operational Program Human Resources Development 2007–2013.

References

Anderson NR, Blakely T, Brunner P et al (2013) High-frequency spectral changes in Dorsolateral Prefrontal Cortex for potential neuoroprosthetics. In: Conference proceedings of the IEEE engineering in medicine and biology society 2013, pp 2247–2250

Bak P (1996) How nature works: the science of self-organized criticality. Copernicus, New York

Bak P, Tang C, Wiesenfeld K (1987) Self-organized criticality: an explanation of the 1/f noise. Phys Rev Lett 59:381–384

Bauer PR, Kalitzin S, Zijlmans M et al (2014) Cortical excitability as a potential clinical marker of epilepsy: a review of the clinical application of transcranial magnetic stimulation. Int J Neural Syst 24(2):1430001. doi:10.1142/S0129065714300010

Beggs JM (2008) The criticality hypothesis: how local cortical networks might optimize information processing. Phil Trans R Soc A 366(1864):329–343

Beggs JM, Plenz D (2003) Neuronal avalanches in neocortical circuits. J Neurosci 23:11167–11177

Bennewitz MF, Saltzman WM (2009) Nanotechnology for delivery of drugs to the brain for epilepsy. Neurotherapeutics 6(2):323–336

Berg AT, Brkovic SF, Brodie MJ et al (2010) Revised terminology and concepts for organization of seizures and epilepsies: report of the ILAE commission on classification and terminology, 2005–2009. Epilepsia 51(4):676–685

Blackspear M, Roberts JA, Terry JR et al (2006) A unifying explanation of primary generalized seizures through nonlinear brain modeling and bifurcation analysis. Cereb Cortex 16 (9):1296–1313

Buxhoeveden D, Casanova M (2002) The minicolumn Hypothesis in neuroscience. Brain 125:935–951

Buzsaki G, Horvath Z, Urioste R et al (1992) High-frequency network oscillation in the hippocampus. Science 256:1025–1027

Casanova MF, Trippe J, Switala A (2007) A temporal continuity to the vertical organization of the human neocortex. Cereb Cortex 17(1):130–137

Chang BS, Lowenstein DH (2003) Epilepsy. N Engl J Med 349(13):1257–1266. doi:10.1056/NEJMra022308, PMID 14507951

Chrobak JJ, Buzsaki G (1996) High-frequency oscillations in the output of the hippocampal–entorhinal axis of the freely behaving rat. J Neurosci 16:3056–3066

Cossette P, Liu L, Brisebois K et al (2002) Mutation of GABRA1 in an autosomal dominant form of juvenile myoclonic epilepsy. Nat Genet 31(2):184–189

Csicsvari J, Hirase H, Czurko A et al (1999) Oscillatory coupling of hippocampal pyramidal cells and interneurons in the behaving rat. J Neurosci 19(1):274–287

Curry DJ, Gowda A, McNichols RJ et al (2012) MR-guided stereotactic laser ablation of epileptogenic foci in children. Epilepsy Behav 24(4):408–414

Cyberonics (2014) Cyberonics announces CE mark approval of the aspireSR™ generator for VNS therapy®. http://ir.cyberonics.com/releasedetail.cfm?ReleaseID=826981 Accessed 22 July 2014

Enatsu R, Piao Z, O'Connor T et al (2012) Cortical excitability varies upon ictal onset patterns in neocortical epilepsy: a cortico-cortical evoked potential study. Clin Neurophysiol 123:252–260

Expert P, Lambiotte R, Chialvo DR et al (2010) Self-similar correlation function in brain resting-state functional magnetic resonance imaging. J R Soc Interface 8(57):472–479

Fisher RS, van Emde BW, Blume W et al (2005) Epileptic seizures and epilepsy: definitions proposed by the International League Against Epilepsy (ILAE) and the International Bureau for Epilepsy (IBE). Epilepsia 46:470–472

Fisher RS, Acevedo C, Arzimanoglou A et al (2014) A practical clinical definition of epilepsy. Epilepsia 55(4):475–482

Fox MD, Raichle ME (2007) Spontaneous fluctuations in brain activity observed with functional magnetic resonance imaging. Nat Rev Neurosci 8(9):700–711

Giresh E, Plenz D (2008) Neuronal avalanches organize as nested theta and beta/gamma-oscillations during development of cortical layer 2/3. Proc Natl Acad Sci U S A 105(21):7576–7581

Haimovici A, Tagliazucchi E, Balenzuela P et al (2013) Brain organization into resting state networks emerges at criticality on a model of the human connectome. Phys Rev Lett 110:178101
Heck CN, King-Stephens D, Massey AD et al (2014) Two-year seizure reduction in adults with medically intractable partial onset epilepsy treated with responsive neurostimulation: final results of the RNS System Pivotal trial. Epilepsia 55(3):432–441
Hobbs JP, Smith JL, Beggs JM (2010) Aberrant neuronal avalanches in cortical tissue removed from juvenile epilepsy patients. J Clin Neurophysiol 27:380–386
Hsu D, Beggs JM (2006) Neuronal avalanches and criticality: a dynamical model for homeostasis. Neurocomputing 69:1134–1136
Hsu D, Tang A, Hsu M et al (2007) Simple spontaneously active Hebbian learning model: homeostasis of activity and connectivity, and consequences for learning and epileptogenesis. Phys Rev E Stat Nonlin Soft Matter Phys 76(4 Pt 1):041909
Hsu D, Chen W, Hsu M et al (2008) An open hypothesis: is epilepsy learned, and can it be unlearned? Epilepsy Behav 13(3):511–522
Hsu WY, Cheng CH, Lin MW et al (2011) Antiepileptic effects of low frequency repetitive transcranial magnetic stimulation: a meta-analysis. Epilepsy Res 96(3):231–240
Iwasaki M, Enatsu R, Matsumoto R et al (2010) Accentuated cortico-cortical evoked potentials in neocortical epilepsy in areas of ictal onset. Epileptic Disord 12(4):292–302. doi:10.1684/epd.2010.0334
Johnson LA, Blakely T, Hermes D et al (2012) Sleep spindles are locally modulated by training on a brain-computer interface. Proc Natl Acad Sci U S A 109(45):18583–18588
Klaus A, Yu S, Plenz D (2011) Statistical analyses support power law distributions found in neuronal avalanches. PLoS ONE 6(5):e19779. doi:10.1371/journal.pone.0019779
Kwan P, Brodie MJ (2000) Early identification of refractory epilepsy. N Engl J Med 342:314–319
Levina A, Herrmann JM, Geisel T (2005) Dynamical synapses give rise to a power law distribution of neuronal avalanches. Neural Information Processing Systems (NIPS). Adv Neural Inf Process Syst 18:771–778
Levinshtein M, Kostamovaara J, Vainshtein S (2005) Breakdown phenomena in semiconductors and semiconductor devices. World Scientific, Singapore
Lüders H, Acharya J, Baumgartner C et al (1998) Semiological seizure classification. Epilepsia 39 (9):1006–1013
Lukasiuk K, Becker AJ (2014) Molecular biomarkers of epileptogenesis. Neurotherapeutics 11 (2):319–323
Margolina A, Herrmann HJ, Stauffer D (1982) Size of largest and second-largest cluster in random percolation. Phys Lett 49A:73–75
McPhee SJ, Hammer GD (2010) Pathophysiology of disease: an introduction to clinical medicine, 6th edn. McGraw-Hill Medical, New York
Meisel C, Storch A, Hellmeyer-Elgner S et al (2012) Failure of adaptive self-organized criticality during epileptic seizure attacks. PLoS Comput Biol 8(1):1–8
Menzler K, Thiel P, Hermsen A et al (2011) The role of underlying structural cause for epilepsy classification: clinical features and prognosis in mesial temporal lobe epilepsy caused by hippocampal sclerosis versus cavernoma. Epilepsia 52(4):707–711
Mountcastle VB (1957) Modality and topographic properties of single neurons of cat's somatic sensory cortex. J Neurophysiol 20(4):408–434
Mountcastle VB (1997) The columnar organization of the neocortex. Brain 120:701–722
National Institute for Health and Clinical Excellence (2012) The epilepsies: the diagnosis and management of the epilepsies in adults and children in primary and secondary care. National Clinical Guideline Centre, London, pp 21–28
Nunes VD, Sawyer L, Neilson J et al (2012) Diagnosis and management of the epilepsies in adults and children: summary of BMJ. BMJ 344:e281. doi: 10.1136/bmj.e281

Opris I (2013) Inter-laminar microcircuits across the neocortex: repair and augmentation. frontiers. In: Lebedev MA, Opris I, Casanova MF (ed) Systems neuroscience research topic: "Augmentation of brain function: facts, fiction and controversy". doi:10.3389/fnsys.2013.00080
Opris I, Casanova MF (2014) Prefrontal cortical minicolumn: from executive control to disrupted cognitive processing. Brain. doi:10.1093/brain/awt359
Opris I, Hampson RE, Stanford TR et al (2011) Neural activity in frontal cortical cell layers: evidence for columnar sensorimotor processing. J Cogn Neurosci 23(6):1507–1521. doi:10.1162/jocn.2010.21534
Opris I, Fuqua JL, Huettl PF et al (2012a) Closing the loop in primate prefrontal cortex: inter-laminar processing. Front Neural Circuits 6:88. doi:10.3389/fncir.2012.00088
Opris I, Hampson RE, Gerhardt GA et al (2012b) Columnar processing in primate pFC: evidence for executive control microcircuits. J Cogn Neurosci 24(12):2334–2347
Opris I, Santos L, Gerhardt GA et al (2013) Prefrontal cortical microcircuits bind perception to executive control. Sci Rep 3:2285. doi:10.1038/srep02285
Pathan SA, Jain GK, Akhter S et al (2010) Insights into the novel three 'D's of epilepsy treatment: drugs, delivery systems and devices. Drug Discov Today 15(17–18):717–732
Plenz D (2012) Neuronal avalanches and coherence potentials. Eur Phys J Spec Top 205 (1):259–301
Ramsey RE, Uthman B, Ben-Menachem E et al (1991) Efficacy of vagal nerve stimulation in partial seizueres: double blind comparison of two stimulus paradigms. Epilepsia 32 (suppl):90–91
Rosenow F, Lüders HO (2001) Presurgical evaluation of epilepsy. Brain 124:1683–1700
Royer S, Pare D (2003) Conservation of total synaptic weight through balanced synaptic depression and potentiation. Nature 422(6931):518–522
Schevon CA, Trevelyan AJ, Schroeder CE et al (2009) Spatial characterization of interictal high frequency oscillations in epileptic neocortex. Brain 132(Pt 11):3047–3059
Seo D, Carmena JM, Rabaey JM et al (2013) Neural dust: an ultrasonic, low power solution for chronic brain-machine interfaces, Cornell University Library, arXiv:1307.2196 [q-bio.NC]
Shepherd G, Grillner S (2010) Handbook of brain microcircuits. Oxford University Press, Oxford
Shew W, Plentz D (2013) The functional benefits of criticality in the cortex. Neuroscientist 19:88–100
Sole RV, Valverde S (2008) Spontaneous emergence of modularity in cellular networks. J R Soc Interface 5:129–133
Staba RJ, Stead M, Worrell GA (2014) Electrophysiological biomarkers of epilepsy. Neurotherapeutics 11(2):334–346. doi:10.1007/s13311-014-0259-0
Stead M, Bower M, Brinkmann BH et al (2010) Microseizures and the spatiotemporal scales of human partial epilepsy. Brain 133(9):2789–2797. doi:10.1093/brain/awq190
Szentagothai J (1975) The module-concept in cerebral cortex architecture. Brain Res 95:475–496
Tovar-Spinoza Z, Carter D, Ferrone D et al (2013) The use of MRI-guided laser-induced thermal ablation for epilepsy. Childs Nerv Syst 29(11):2089–2094
Turrigiano GG, Nelson SB (2000) Hebb and homeostasis in neuronal plasticity. Curr Opin Neurobiol 10:358–364
Vaessen MJ, Jansen JFA, Braakman HMH et al (2014) Functional and structural network impairment in childhood frontal lobe epilepsy. PLoS ONE 9(3):e90068
Varela F, Lachaux JP, Rodriguez E et al (2001) The brainweb: phase synchronization and large-scale integration. Nat Rev Neurosci 2:229–239
Wallace GG, Higgins MJ, Moultonand SE et al (2012) Nanobionics: the impact of nanotechnology on implantable medical bionic devices. Nanoscale 4:4327–4347
Wang S (2013) Online seizure prediction using an adaptive learning approach. IEEE Trans Knowl Data Eng 25(12):2854–2866
Worrell GA, Cranstoun SD, Echauz J et al (2002) Evidence for self-organized criticality in human epileptic hippocampus. Neuroreport 13:2017–2021

Wu JJS, Chang WP, Shih HC et al (2014) Cingulate seizure-like activity reveals neuronal avalanche regulated by network excitability and thalamic inputs. BMC Neurosci 15(3):1–17

Yang H, Shew WL, Roy R et al (2012) Maximal variability of phase synchrony in cortical networks with neuronal avalanches. J Neurosci 32(3):1061–1072. doi:10.1523/JNEUROSCI.2771-11.2012

Yua N, Liub H, Dia Q (2013) Modulation of immunity and the inflammatory response: a new target for treating drug-resistant epilepsy. Curr Neuropharmacol 11:114–127

Zijlmans M, Jiruska P, Zelmann R et al (2012) High-frequency oscillations as a new biomarker in epilepsy. Ann Neurol 71(2):169–178

Chapter 16
Adaptive Coding in Visual Cortical Circuits

Bryan J. Hansen and Valentin Dragoi

Abstract Understanding the rules by which brain networks represent incoming stimuli in population activity to influence the accuracy of behavioral responses remains one of the deepest mysteries in neuroscience. We have embarked on a set of projects to investigate the real-time operation of multiple neuronal networks and their capacity to undergo adaptive changes and plasticity. What are the fundamental units of network computation and the principles that govern their relationship with behavior? By employing state-of-the-art electrophysiological techniques we were able to record from large pools of cells in the non-human primate brain while animals performed a fixation task. We found that spatio-temporal correlations between neurons could act as an active 'switch' to control network performance in real time by modulating the communication between neurons. We believe that 'cracking' the mysteries of the population code will offer unique insight into a network-based mechanistic explanation of behavior and new therapeutic solutions to cure brain dysfunction.

Keywords Neuronal network • Local networks • Population code • Adaptation • Local field potential

16.1 Introduction

A fundamental feature of cortical neurons is their ability to rapidly adapt to changes in incoming stimuli. Several lines of evidence indicate that cortical neurons dynamically change their responses and selectivity to match the changes in the statistics of the input stimuli. Whereas our understanding of information processing at the single neuron level has substantially improved over the past three decades, how populations of brain cells encode information remains largely unknown. In

B.J. Hansen
The Salk Institute of Biological Sciences, La Jolla, CA, USA
e-mail: bhansen@salk.edu

V. Dragoi (✉)
Department of Neurobiology and Anatomy, University of Texas Medical School, 6431 Fannin St, Ste 7.166, Houston, TX 77030, USA
e-mail: valentin.dragoi@uth.tmc.edu

M.F. Casanova, I. Opris (eds.), *Recent Advances on the Modular Organization of the Cortex*, DOI 10.1007/978-94-017-9900-3_16

particular, understanding how adaptation changes information processing by individual neurons and networks is essential for understanding the relationship between efficient sensory coding and behavior. We have recently performed neurophysiological investigations to examine how rapid adaptation at the time scale of visual fixation impacts information processing in networks of brain cells.

16.1.1 Rapid Adaptation

Adaptation is a ubiquitous phenomenon in sensory processing. The time-scale of adaption can range from hundreds of milliseconds to several days. In normal visual processing, adaptation occurs along many stimulus dimensions such as luminance, contrast, orientation, spatial frequency, direction of motion, color, and curvature and shape. The effects of pattern adaptation have been extensively explored in psychophysical and electrophysiological experiments. In the orientation domain, adaptation has been shown to influence the signaling capabilities of neurons in primary visual cortex (V1). For example, adaptation to an oriented grating induces a repulsive shift in the preferred orientation of individual V1 neurons and changes orientation selectivity when the adapting grating is similar to the cell's optimal orientation (Dragoi et al. 2002; Felsen et al. 2002). Examining rapid adaptation in visual cortex is important for understanding how individual neurons and local circuits change their coding properties in real-time.

An important perceptual effect of adaptation is the change in orientation discrimination performance. Both perceptual studies and physiological recordings in V1 have demonstrated that adaptation improves the capacity of neurons to signal small differences between stimuli such that orientation discrimination is improved (Muller et al. 1999; Dragoi et al. 2002; Dragoi and Sur 2006). However, it remains unclear whether and how the information encoded in population activity is influenced after adaptation. In particular, how the structure of correlations across a population of neurons is affected by adaptation, and how it influences the efficiency of coding is unknown.

16.1.2 Cortical Layers

A fundamental issue in our understanding of brain circuits is how networks in different layers of the cerebral cortex encode information. Cortical layers are ubiquitous structures throughout neocortex (Nassi and Callaway 2009; Hansen et al. 2011) that consist of highly recurrent local networks that communicate among each other to possibly influence the information encoded in population activity. In recent years, significant progress has been made in our understanding of the differences in response properties of neurons across cortical layers (Lakatos et al. 2009; Hansen and Dragoi 2011), yet there is still a great deal to learn about whether and how neuronal populations encode information in a layer-specific manner. A measure of the activity of a local population of neurons (Pesaran et al. 2002) is captured by local field potentials (LFPs), which are composed of low-frequency extracellular voltage fluctuations (Logothetis 2003) believed to originate from within 250–500 μm of the recording site (Kruse and Eckhorn 1996; Katzner et al. 2009).

16.1.3 Neural Synchronization

It has been proposed that one way in which networks of cells can efficiently process information about incoming stimuli is through synchronization between the spiking activity of multiple neurons and local oscillatory activity (measured as local field potentials, LFPs). In visual cortex, it has been found that neuronal groups exhibit strong responses in the gamma-band frequency (30–80 Hz) (Gray et al. 1989; Engel et al. 1991; Fries et al. 2001), and that single neurons synchronize their responses with the local population activity (Freiwald et al. 1995; Tsodyks et al. 1999). Synchronization in visual cortex, particularly in the gamma-band, has been found to be critically involved in sensory processing (Gray et al. 1989; Engel et al. 1991; Cardin et al. 2009), grouping (Gray et al. 1989; Engel et al. 1991), attention (Fries et al. 2001; Womelsdorf et al. 2006; Gregoriou et al. 2009; Chalk et al. 2010), working memory (Pesaran et al. 2002), and behavioral reaction times (Womelsdorf et al. 2006). Importantly, selective activation of fast-spiking interneurons and their phase relationship with excitatory pyramidal cell activity has been shown to enhance the gamma rhythm (Traub et al. 1996; Cardin et al. 2009; Sohal et al. 2009). This inhibition-based mechanism is also consistent with anatomical results indicating that both the density of interneurons and the distribution of GABAb receptors, known to be involved in gamma oscillations, favor superficial layers of cortex (Fitzpatrick et al. 1987; Whittington et al. 1995; Eickhoff et al. 2007).

In this chapter we will describe our work examining the possibility that the adaptive capacity of individual neurons may exhibit cortical layer dependency. Indeed, while gamma synchronization has been found to be involved in a variety of conditions (Gray and Singer 1989; Engel et al. 1991; Fries et al. 2001; Womelsdorf et al. 2006; Cardin et al. 2009; Gregoriou et al. 2009; Chalk et al. 2010) whether a fundamental feature of individual neurons, such as the capacity to exhibit adaptive changes or plasticity, is influenced by synchrony in the gamma frequency band remains unclear.

16.1.4 Interneural Correlations

Understanding how adaptation influences population coding requires an understanding of how adaptation changes the structure of interneuronal correlations across a network of cells. Indeed, during the past decade, it has become increasingly understood that the trial-by-trial variability in neuronal responses, or 'noise', is not independent, but it exhibits correlations (Kohn and Smith 2005; Gutnisky and Dragoi 2008; Hansen et al. 2012). This implies that the accuracy of the population code must depend on the distribution of noise correlations across the network. Theoretically, it has been proposed that adaptation would reduce neuronal correlations, and hence redundancy (Reich et al. 2001; Schneidman et al. 2003), to improve stimulus coding (Barlow 1961). We addressed this issue in the context of the macaque V1, where adaptation has been previously shown to induce changes in the response magnitude and selectivity of individual neurons (Muller et al. 1999; Dragoi et al. 2000; Dragoi et al. 2002; Felsen et al. 2002; Sharpee et al. 2006). We focused on a rapid form of adaptation that is believed to occur spontaneously during visual fixation when cortical cells are exposed to redundant information for hundreds of milliseconds (Muller et al. 1999; Dragoi et al. 2002). Our hypothesis is that rapid adaptation changes the structure of noise correlations in V1 and increases the amount of information in a population code in a way that is consistent with perceptual performance.

16.2 Experimental Methods

16.2.1 Identification of Cortical Layers

To identify cortical layers, we measured the evoked response potentials (ERPs) of LFPs across equally-spaced contacts (Fig. 16.1b left, inter-contact distance = 100 μm) in response to a full-field flashed stimulus. We then performed current-source density (CSD; Fig. 16.1b right) analysis of the LFP time-series to identify the polarity inversion accompanied by the sink-source configuration at the base of layer 4; the sink is inside layer 4, subsequently referred to as the granular layer (Lakatos et al. 2009; Nassi and Callaway 2009; Hansen and Dragoi 2011; Hansen et al. 2011, 2012). Current-source density analysis is useful because it provides an index of the location, direction, and density of transmembrane current flow, allowing us to accurately position electrodes to record from all layers in a single penetration.

We observed that cells recorded on laminar probes had highly overlapping receptive fields (Fig. 16.1c) and highly similar preferred orientations (PO) (e.g., the difference in PO for over 58 % of the pairs of neurons was within 0°–10° range, $p < 0.01$, Wilcoxon signed-rank test).

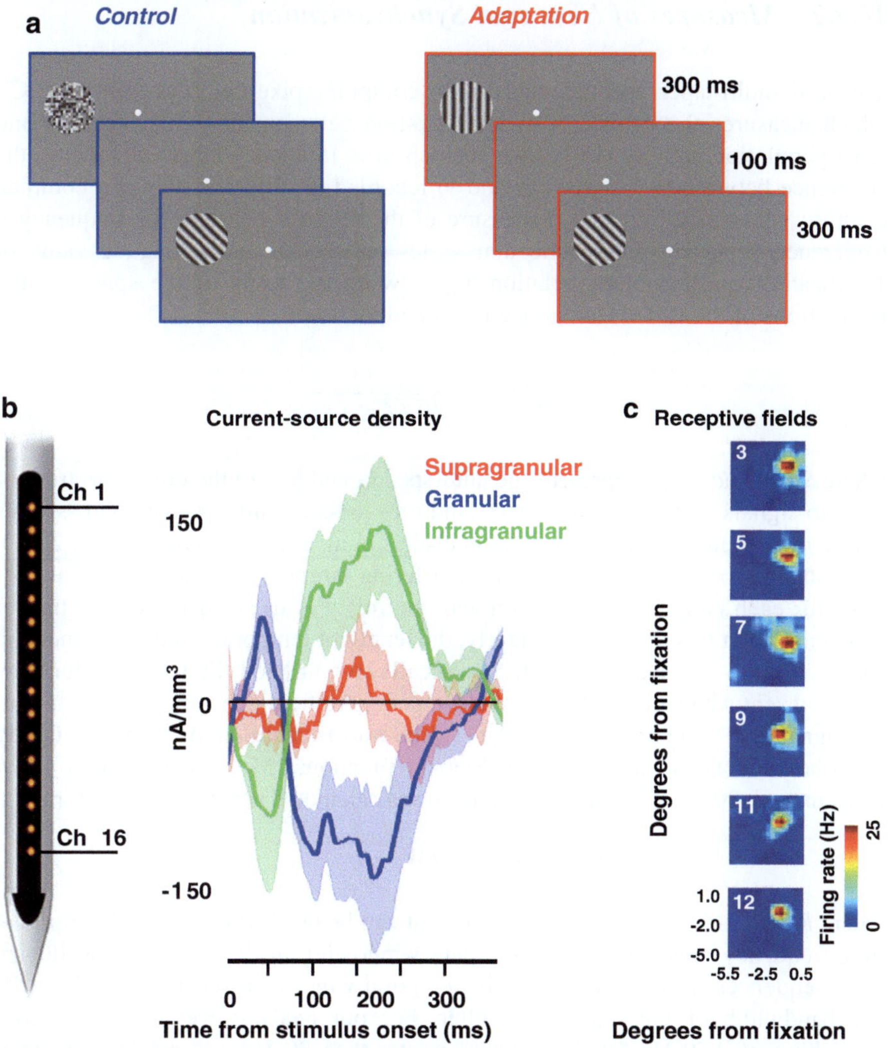

Fig. 16.1 Rapid adaptation across cortical layers in V1. (**a**) Schematic description of the orientation adaptation protocol. (**b**) Multi-contact laminar electrodes were used to record neuronal activity across cortical depth (*left*). Current-source density analysis (based on the second spatial derivate of the LFP time-series) was used to identify the polarity inversion accompanied by the sink-source configuration at the base of the granular layer. The CSD traces represent the average of those contacts assigned to a given layer. (**c**) Receptive fields across contacts were mapped using oriented stimuli presented in random patches. Firing rates for each neuron are calculated independently at 5 ms intervals and the maximum firing rates (shown as *red*) were used to computed the centroid for each time delay

16.2.2 Measures of Neuronal Synchronization

We used multi-taper spectral analysis to compute spike-field coherence (SFC), which measures the degree of synchronization between individual neurons and local population activity (LFPs) for each specific frequency band. In general, the coherence between two signals (x and y) recorded at different sites is a complex quantity whose magnitude is a measure of the phase synchrony for frequency f. Coherence is an absolute value that varies between 0 and 1 (e.g., a value of 1 indicates a perfect phase relationship between the firing of the spikes to the fluctuations of the LFP). Coherence is defined as

$$C_{xy}(f) = \frac{S_{yx}(f)}{\sqrt{S_x(f)S_y(f)}}$$

where $S_x(f)$ and $S_y(f)$ represent the auto-spectra and $S_{yx}(f)$ the cross-spectrum of the two signals x and y. Auto-spectra and cross-spectra are averaged across trials before the coherency calculation (Womelsdorf et al. 2006; Gregoriou et al. 2009). In an attempt to eliminate any bias from differing sample sizes, the same number of trials for each condition (adaptation and control) was used for the calculation of coherence for a given pair. Importantly, the length of temporal window included in each condition was also constant. Specifically, we utilized the Chronux function *coherencycpt* which computes the multi-taper spike-field coherence for a continuous signal (LFP) and point process data (spike-train) according to an optimal family of orthogonal tapers derived from Slepian functions (Mitra and Pesaran 1999; Pesaran et al. 2002). The number of tapers was calculated according to the formula:

$$K = 2*TW - 1$$

where K is the highest number of tapers that can be used while preserving optimal time-frequency concentration of the data windowing available from the Slepian taper sequences, T is the length of the temporal window in seconds, and W is the half-bandwidth of the multi-taper filter. For our analysis we applied spectral smoothing of ± 10 Hz for frequencies greater than 30 Hz and ± 4 Hz for lower frequencies.

16.2.3 Correlated Variability

Multiple single-unit recordings were performed from V1 of two fixating monkeys (*Macaca mulatta*). Stimuli were presented such as to cover the center of the neurons' receptive fields. In control trials, movie strips were presented for ~1.86 s (16 orientations × 7 repeats at 60 Hz; random spatial phase). In adaptation trials, movies were preceded by a 400-ms grating of fixed orientation. The Pearson correlation coefficient of spike counts, R_{sc}, of two cells is defined as:

$$R_{sc} = \frac{\sum_{i=1}^{N} \left(r_1^i - \overline{r}_1\right) \cdot \left(r_2^i - \overline{r}_2\right)}{\sigma_1 \cdot \sigma_2}$$

where N is the number of trials, r_j^i is the firing rate of cell j in trial i averaged over the entire stimulus sequence, and σ is standard deviation of the responses. Correlation coefficients after adaptation depend on three variables: the adapting orientation, ϕ_a, and the preferred orientation of the cells in a pair, θ_1 and θ_2. To ensure that the parameter space is adequately sampled, the distance between the adapting orientation and the preferred orientation of one of the cells was held constant while varying the relative difference between the two cells' preferred orientations ($\Delta\theta$).

16.3 Results

16.3.1 Adaptation Increases Spike-LFP Gamma Synchronization

We used multi-contact laminar electrodes (Plextrode® U-Probe, Plexon Inc.) to record neuronal activity at 20 V1 recording sites, each measured at 16 different depths, while two monkeys (W: 13; P: 7) performed a rapid adaptation fixation task (Fig. 16.1a). While animals fixated a white dot at the center of a screen, an adapting stimulus was flashed for 300 ms in the center of the neurons' receptive field. After a 100 ms blank, a test stimulus of random orientation (8 equally spaced orientations spanning 0–180°) was presented for 300 ms. The adapting stimulus was either a random dot patch (control condition) or a sine-wave grating with spatial characteristics identical to those of the test stimulus, but fixed in orientation within 45° of each cell's preferred orientation (adaptation condition).

In control trials, the presentation of the test stimulus led to pronounced synchronous activity across all cortical layers, with the most significant increase in gamma synchronization in the granular layers (Fig. 16.2a). However, after adaptation, there was a significant increase in gamma synchronization specifically in the supragranular layers (Fig. 16.2b). Our population analysis confirms these results – before adaptation we found spike-LFP gamma synchronization across all layers (mean + S.E.M; SG: 0.08 + 0.002; G: 0.12 + 0.007; IG: 0.08 + 0.006) with the largest SFC level in the granular layers (one-way ANOVA, F (2, 74) = 8.75, P = 0.0004; post-hoc multi-comparison, Tukey's Least Significant Difference).

We also calculated the percentage change in gamma SFC between adaptation and control (ΔSFC) across the entire frequency range and found a significant

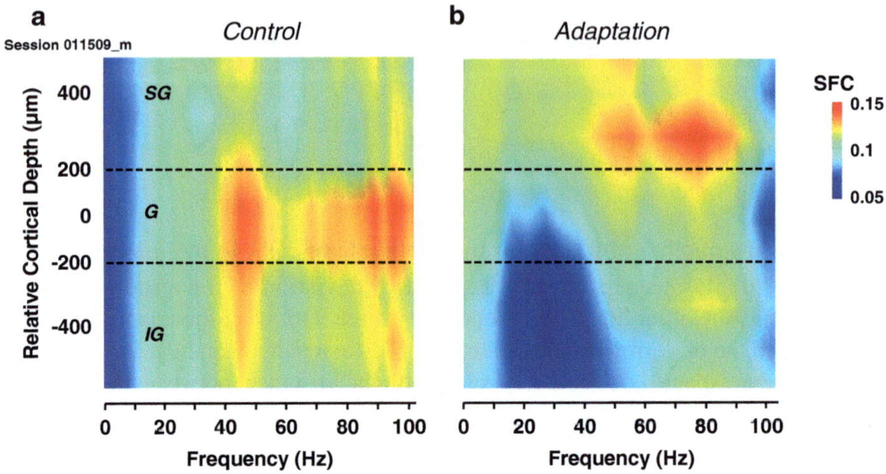

Fig. 16.2 Example of synchronization, measured as spike-field coherence as a function of frequency, across cortical layers during control and adaptation across cortical depth during control (*left*) and adaptation (*right*). During the presentation of the control stimuli there is an increase in gamma activity in the granular layer (Adaptation increases SFC across cortical layers with the largest increase in the supragranular layer. *Dashed lines* equal the granular layer)

increase in synchronization for the supragranular layer (68.91 % + 6.75 %; Fig. 16.3a, b; one-way ANOVA, F (2, 74) = 35.24, P = 1.77 10^{-11}; post-hoc multi-comparison, Tukey's Least Significant Difference). The post-adaptation increase in gamma SFC in the supragranular layers was observed only when both recording sites were stimulated with test stimuli within 45° of the cells' preferred orientation. Non-optimal test orientations reduced spike rates, LFP amplitudes, and SFC in the control condition; adaptation at these orientations did not result in a significant increase in gamma SFC (P = 0.20, Wilcoxon signed-rank test).

16.3.2 Relationship Between Gamma Synchronization and Neuronal Discrimination

We further investigated whether the ability of neurons in different cortical layers to discriminate stimulus orientation (Fig. 16.3c) is influenced by the post-adaptation change in synchronization between individual cells and their local population (Fig. 16.3d). We addressed this issue by examining the relationship between the post-adaptation change in gamma-band spike-field coherence and the change in neurons' capacity (d′) (Green and Swets 1966; Macmillan and Creelman 2005) to discriminate nearby orientations (22.5° apart). We found a significant correlation between the post-adaptation change in d′ and the corresponding change in SFC only for the recording sites in the supragranular layers (Fig. 16.2d; r = 0.38, P = 0.02,

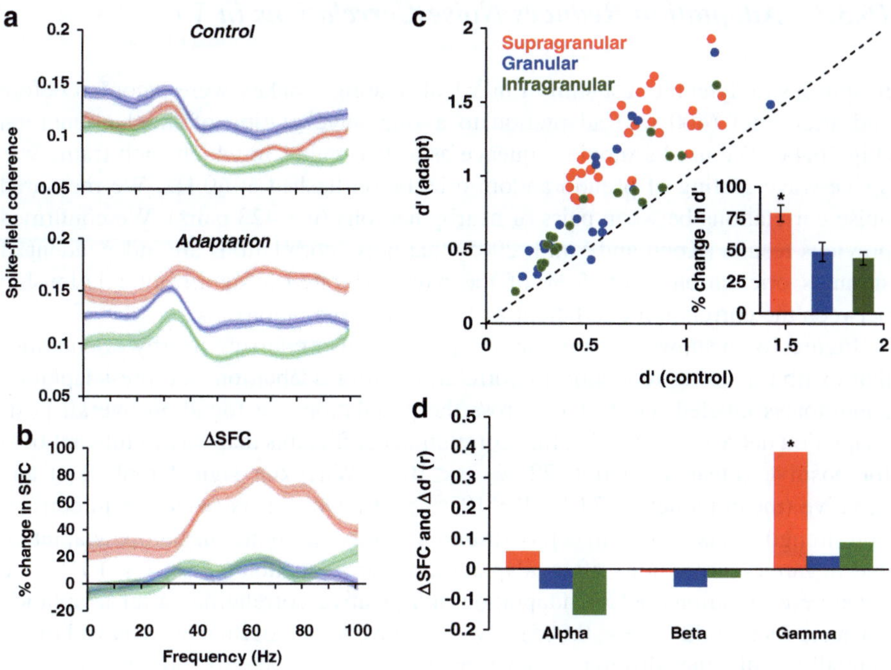

Fig. 16.3 Adaptation influences synchronization between individual neurons and local populations in a layer-specific manner. (**a**, *top*) Population analysis during the presentation of the control stimulus results in a significant increase in SFC between 30 and 80 Hz in the granular layer. (**a**, *bottom*) Adaptation increases SFC in the supragranular layer for all frequency bands between 0 and 80 Hz, with the largest increase in the gamma-band (30–80 Hz; shaded regions represent S.E.M.) (**b**) We calculated the percentage change between adaptation and control across the entire frequency range and observed a significant increase in gamma-band spike-field coherence for the supragranular layer (shaded regions represent S.E. for percentage change). (**c**) Scatter plot showing the effects of adaptation on neuronal discrimination performance (d′) at the population level across cortical layers. Each dot represents the mean d′ during control and adaptation, while the different colors indicate the layer in which the neuron was isolated. Across the total population of cells (n = 77), adaptation significantly increases orientation discriminability. Inset: Although adaptation significantly increases d′ across all cortical layers, the largest increase occurred in the supragranular layer. (**d**) There is a significant and positive correlation between the gamma-band spike-field coherence after adaptation and the change in d′ that is specific to the supragranular layers. The analysis of lower frequency bands (<30 Hz) does not reveal a statistically significant correlation between the post-adaptation change in SFC and the change in d′

Pearson correlation). In contrast, neurons in granular and infragranular layers exhibited post-adaptation changes in discriminability that were independent of the changes in gamma spike-field coherence (G: $r = 0.04$, $P = 0.84$; IG: $r = 0.08$, $P = 0.70$).

16.3.3 Adaptation Reduces Noise Correlations in V1

Responses to dynamic test stimuli in V1 of fixating monkey were recorded before and after brief (400 ms) adaptation to a sine-wave grating of fixed orientation (Fig 16.4a). We used a movie sequence as test stimulus, in which each frame was a sine-wave grating of pseudorandom orientation flashed at 60 Hz. We measured noise correlations between pairs of nearby neurons ($n = 423$ pairs). We confirmed previous results (Kohn and Smith 2005) that noise correlations are independent of stimulus orientation (only 5 % of the pairs exhibited a significant relationship between the correlation coefficient and stimulus orientation).

Figure 16.4b shows an example of a pair of cells preferring nearby orientations that exhibit a strong reduction in correlations after adaptation (the pre-adaptation condition is labeled 'control'). Across the population, we found an overall post-adaptation decrease in the absolute correlation coefficients that was significant both for positive (mean reduction 22 %, $P < 10^{-8}$, Wilcoxon signed rank test) and negative (mean reduction 74 %, $P < 10^{-5}$) coefficients. This reduction in correlation strength is also found in cells exhibiting positive correlations before adaptation and negative correlations after adaptation (mean reduction 73 %, $P < 10^{-6}$) and negative correlations before adaptation and positive correlations after adaptation (mean reduction 42 %, $P < 0.005$). Overall, correlation coefficients decayed exponentially with the difference ($\Delta\theta$ in the cells' preferred orientation (Ts'o et al. 1986). We further examined whether the decrease in correlations after adaptation could be due to the small (5.8 %), but significant ($P < 10^{-6}$), reduction in mean firing rates. However, we found no relationship between the mean changes in firing rates after adaptation ($P > 0.3$, Pearson correlation) and the changes in correlation coefficients (de la Rocha et al. 2007).

Since adaptation is an orientation-specific phenomenon (Dragoi et al. 2000, 2002; Felsen et al. 2002), we reasoned that the degree of decorrelation would depend on the relationship between the adapting stimulus and the preferred orientation of the cells in a pair. We thus selected the cell pairs preferring nearby orientations ($\Delta\theta < 30°$), and compared the mean correlation coefficients before and after adaptation to stimuli of different orientation (we defined as the minimum difference between the adapting stimulus and the preferred orientation of each cell in a pair. There is a strong reduction in correlations (Fig. 16.4c) for adapting stimuli near ($\Delta\phi \leq 30°$, 31 %, $P < 0.005$; Wilcoxon sum rank test) and far ($\Delta\phi > 60°$, 43 %, $P < 0.0005$) relative to the orientation of the cell pair, but intermediate adaptation ($30° < \Delta\phi \leq 60°$) was ineffective in reducing correlations (19 %; $P > 0.1$).

These results indicate that brief adaptation reduces the strength of correlations in an orientation-asymmetric manner. We quantified the orientation dependency of this decorrelation by estimating the probability density function (pdf) of correlations, before and after adaptation, as a function of $\Delta\theta$ and $\Delta\phi$ (using the kernel density estimation technique). Contrary to common belief that adaptation would only influence the responses of cells of similar preferred orientation (but consistent with Fig. 16.4b), we found a non-monotonic decorrelation profile. That is, cells of similar ($\Delta\theta < 30°$)

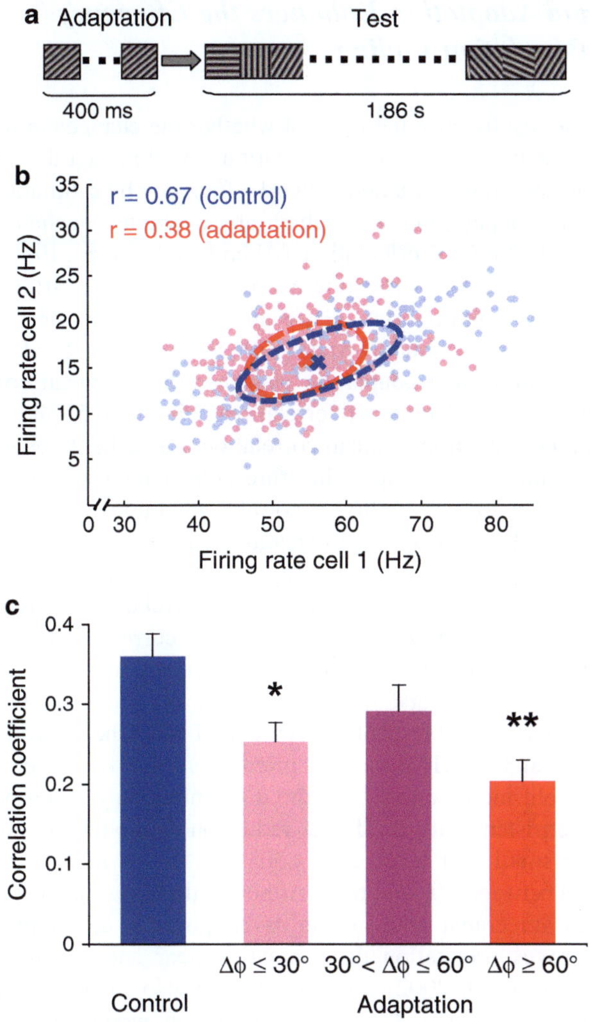

Fig. 16.4 Adaptation-induced response decorrelation in V1. (**a**) Schematic representation of the stimulus sequence: an adapting stimulus of fixed orientation was presented for 400 ms and was followed by a 60-Hz test stimulus of random orientation presented for 1.86 s. (**b**) Scatter plot showing the trial-by-trial responses of two cells recorded simultaneously. Each dot represents the firing rates of both cells in a given trial. The dotted ellipses represent the two-dimensional gaussian fits of the firing rate distributions during control and adaptation (crosses represent the means). 'r' represents the correlation coefficient. (**c**) The reduction in the mean correlation coefficients after adaptation depends on the adapting orientation (for pairs for which $\Delta\varphi < 30°$). All panels are based on the correlation analysis of $n = 423$ cell pairs ($\Delta\theta < 30°$, 223 pairs; $\Delta\theta > 30°$, 200 pairs). Error bars represent s.e.m. (*$P < 0.005$; **$P < 0.0005$)

and largely dissimilar orientation preference ($\Delta\theta > 60°$) exhibit significant decorrelation, whereas cells with $\Delta\theta$ between 30°–60° show only a weak decrease in correlations, both for near and far adaptation (using an adaptation decorrelation index, DI, as the magnitude of post-adaptation changes in absolute correlations).

16.3.4 Rapid Adaptation Enhances the Efficiency of Population Coding

Altogether, these results raise the issue of whether the changes in the strength and variability of noise correlations after adaptation would impact the efficiency of the population code. We thus computed network efficiency by estimating Fisher Information (FI) as the upper limit with which any decoding mechanism can extract information about stimulus orientation (Abbott and Dayan 1999; Sompolinsky et al. 2001). We assumed that the joint neuronal responses to stimulus orientation can be described by a multivariate Gaussian defined by the mean firing rate and covariance matrix (Abbott and Dayan 1999; Sompolinsky et al. 2001). FI was computed by assuming that adaptation changes (i) only mean correlations, and (ii) both the mean and variability of correlations. Figure 16.5a shows that whereas the post-adaptation reduction in mean correlations caused a 25 % improvement in the network orientation discriminability threshold, taking into accounts both the changes in the mean and variability of correlations improved the post-adaptation discrimination threshold by 40 %. Interestingly, for small populations, the post-adaptation network performance is slightly better than that of uncorrelated (independent) neurons, possibly due to a reduction in correlation variability.

The fact that adaptation changes interneuronal correlations in an orientation-asymmetric manner could cause the network efficiency to depend on the relationship between the adapting and test orientations. As shown in Fig. 16.5b, brief adaptation caused an almost fourfold increase in FI when the network discriminated stimuli of similar and largely dissimilar orientation relative to the adapting stimulus, and a threefold increase in FI for the discrimination of intermediate orientations. This is consistent with the larger reduction in the mean and variability of correlations for small and large $\Delta\theta$ ($<30°$ and $>60°$) relative to intermediate orientations ($\Delta\theta$ between 30° and 60°). Although these results may appear surprising, they are in agreement with human psychophysical data reporting that brief adaptation improves orientation discrimination near and far from the adapting orientation (Dragoi et al. 2002). Importantly, we also found that the increase in population coding efficiency through decorrelation would be equivalent to an overall post-adaptation increase in firing rates of approximately 55 %.

16.4 Discussion and Future Directions

Theoretically, adaptation has been proposed to reduce redundancy in sensory neurons, possibly by decorrelating responses, to improve coding efficiency (Barlow 1961; Sharpee et al. 2006). However, besides the lack of experimental support, theories proposing the decorrelation hypothesis were unable to predict the changes in correlations across the entire network engaged in sensory computations. We provide empirical evidence that adaptation causes both a selective reduction in the

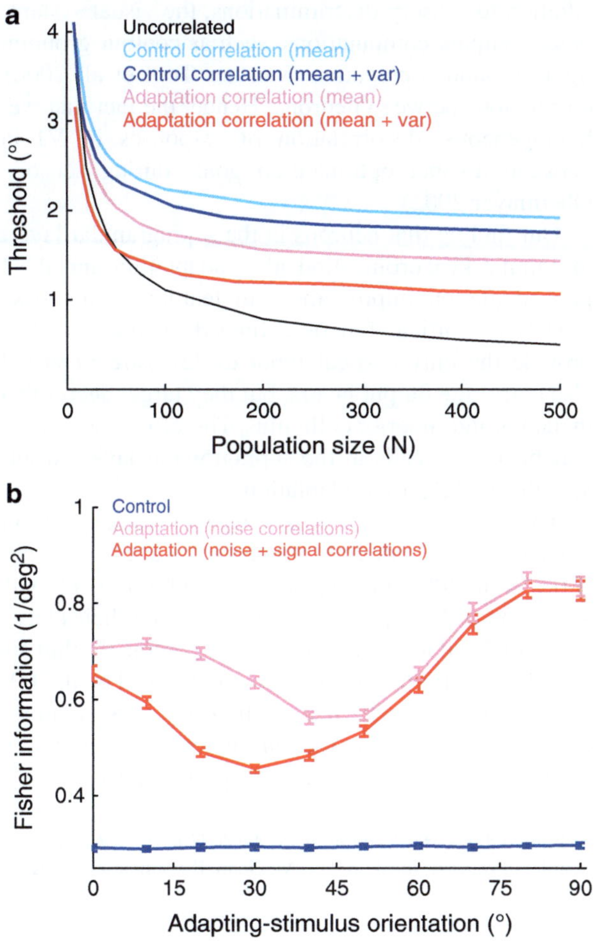

Fig. 16.5 Rapid adaptation enhances the efficiency of population coding. (**a**). *Light blue/red curves*: adaptation changes only the mean correlation coefficients; *Dark blue/red curves*: adaptation changes both the mean and variability of correlations. (**b**) The efficiency of population coding (FI) depends on $\Delta\theta$ and the adapting orientation (fixed at 0°; abscissa represents the test orientation). In control trials, FI is independent on the test orientation. Combining the effects of noise and signal correlations yields a more pronounced U-shape profile of FI

strength and variability of correlations and an improvement in population coding efficiency. These results are consistent with the 'efficient coding hypothesis' (Barlow 1961; Sharpee et al. 2006), i.e., sensory neurons are adapted to the statistical properties of the stimuli they are exposed to, and with psychophysical changes in human discrimination performance after adaptation. We further propose that adaptation takes advantage of the rapid sequence of fixations during natural viewing to optimize image discrimination performance in real time (Dragoi et al. 2002; Dragoi and Sur 2006).

Our results argue that the visual system employs a metabolically inexpensive solution (selective decorrelation) to adapt neural responses to the statistics of the input stimuli and improve coding efficiency. For instance, theoretical studies have suggested that temporally decorrelated inputs are transmitted less efficiently than correlated inputs (Salinas and Sejnowski 2000). Indeed, it is well known that in addition to sensory discriminations, the visual system is often required to perform other complex computations, such as contour grouping (Roelfsema et al. 2004) or figure-ground segregation (van der Togt et al. 2006), which may require strong correlations between neurons. Hence, the fact that we did not observe a complete, homogeneous, decorrelation of responses in V1 could constitute a trade-off between distinct optimization goals during sensory processing (Schwabe and Obermayer 2002).

Our finding that neurons in the supragranular layers exhibit the largest increase in gamma synchronization after adaptation and the highest correlation with the post-adaptation improvement in feature coding has functional implications for models of cortical function. Indeed, neurons in the supragranular layers of V1 provide the only cortical input to downstream visual areas. Infragranular layers also constitute output layers, but they target deep subcortical structures such as the thalamus and superior colliculus. Therefore, neurons in higher-order cortices would benefit most if cells in the supragranular layers would exhibit a large increase in stimulus coding after adaptation.

The possible relationship between gamma synchronization and neuronal performance has been indirectly suggested by attention studies in mid-level cortical areas (Fries et al. 2001; Gregoriou et al. 2009). Recent evidence indicates that selective activation of fast-spiking interneurons and their phase relationship with excitatory pyramidal cell activity enhances the gamma rhythm and controls sensory responses (Traub et al. 1996; Cardin et al. 2009; Sohal et al. 2009). This raises the possibility that an increase in local inhibition due to adaptation (Chelaru and Dragoi 2008) could subsequently cause an increase in gamma synchronization possibly to improve neuronal discrimination performance.

Acknowledgments Supported by the NIH EUREKA Program, National Eye Institute, the Pew Scholars Program, the James S. McDonnell Foundation (V.D.) and an NIH Vision Grant (B.J.H).

References

Abbott LF, Dayan P (1999) The effect of correlated variability on the accuracy of a population code. Neural Comput 11:91–101

Barlow HB (1961) Possible principles underlying the transformations of sensory messages. In: Rosenblith WA (ed) Sensory communication. MIT Press, Cambridge, MA, pp 217–234

Cardin JA, Carlen M, Meletis K, Knoblich U, Zhang F, Deisseroth K, Tsai LH, Moore CI (2009) Driving fast-spiking cells induces gamma rhythm and controls sensory responses. Nature 459:663–667

Chalk M, Herrero JL, Gieselmann MA, Delicato LS, Gotthardt S, Thiele A (2010) Attention reduces stimulus-driven gamma frequency oscillations and spike field coherence in V1. Neuron 66:114–125

Chelaru MI, Dragoi V (2008) Asymmetric synaptic depression in cortical networks. Cereb Cortex 18:771–788

de la Rocha J, Doiron B, Shea-Brown E, Josic K, Reyes A (2007) Correlation between neural spike trains increases with firing rate. Nature 448:802–806

Dragoi V, Sur M (2006) Image structure at the center of gaze during free viewing. J Cogn Neurosci 18:737–748

Dragoi V, Sharma J, Sur M (2000) Adaptation-induced plasticity of orientation tuning in adult visual cortex. Neuron 28:287–298

Dragoi V, Sharma J, Miller EK, Sur M (2002) Dynamics of neuronal sensitivity in visual cortex and local feature discrimination. Nat Neurosci 5:883–891

Eickhoff SB, Rottschy C, Zilles K (2007) Laminar distribution and co-distribution of neurotransmitter receptors in early human visual cortex. Brain Struct Funct 212:255–267

Engel AK, Konig P, Kreiter AK, Singer W (1991) Interhemispheric synchronization of oscillatory neuronal responses in cat visual cortex. Science 252:1177–1179

Felsen G, Shen YS, Yao H, Spor G, Li C, Dan Y (2002) Dynamic modification of cortical orientation tuning mediated by recurrent connections. Neuron 36:945–954

Fitzpatrick D, Lund JS, Schmechel DE, Towles AC (1987) Distribution of GABAergic neurons and axon terminals in the macaque striate cortex. J Comp Neurol 264:73–91

Freiwald WA, Kreiter AK, Singer W (1995) Stimulus dependent intercolumnar synchronization of single unit responses in cat area 17. Neuroreport 6:2348–2352

Fries P, Reynolds JH, Rorie AE, Desimone R (2001) Modulation of oscillatory neuronal synchronization by selective visual attention. Science 291:1560–1563

Gray CM, Singer W (1989) Stimulus-specific neuronal oscillations in orientation columns of cat visual cortex. Proc Natl Acad Sci U S A 86:1698–1702

Gray CM, Konig P, Engel AK, Singer W (1989) Oscillatory responses in cat visual cortex exhibit inter-columnar synchronization which reflects global stimulus properties. Nature 338:334–337

Green DM, Swets JA (1966) Signal detection and psychophysics. Wiley, New York

Gregoriou GG, Gotts SJ, Zhou H, Desimone R (2009) High-frequency, long-range coupling between prefrontal and visual cortex during attention. Science 324:1207–1210

Gutnisky DA, Dragoi V (2008) Adaptive coding of visual information in neural populations. Nature 452:220–224

Hansen BJ, Dragoi V (2011) Adaptation-induced synchronization in laminar cortical circuits. Proc Natl Acad Sci U S A 108:10720–10725

Hansen BJ, Eagleman S, Dragoi V (2011) Examining local network processing using multi-contact laminar electrode recording. J Vis Exp 55:2806

Hansen BJ, Chelaru MI, Dragoi V (2012) Correlated variability in laminar cortical circuits. Neuron 76:590–602

Katzner S, Nauhaus I, Benucci A, Bonin V, Ringach DL, Carandini M (2009) Local origin of field potentials in visual cortex. Neuron 61:35–41

Kohn A, Smith MA (2005) Stimulus dependence of neuronal correlation in primary visual cortex of the macaque. J Neurosci 25:3661–3673

Kruse W, Eckhorn R (1996) Inhibition of sustained gamma oscillations (35–80 Hz) by fast transient responses in cat visual cortex. Proc Natl Acad Sci U S A 93:6112–6117

Lakatos P, O'Connell MN, Barczak A, Mills A, Javitt DC, Schroeder CE (2009) The leading sense: supramodal control of neurophysiological context by attention. Neuron 64:419–430

Logothetis NK (2003) The underpinnings of the BOLD functional magnetic resonance imaging signal. J Neurosci 23:3963–3971

Macmillan NA, Creelman CD (2005) Detection theory: a user's guide. Lawrence Erlbaum Associates, Mahwah

Mitra PP, Pesaran B (1999) Analysis of dynamic brain imaging data. Biophys J 76:691–708

Muller JR, Metha AB, Krauskopf J, Lennie P (1999) Rapid adaptation in visual cortex to the structure of images. Science 285:1405–1408

Nassi JJ, Callaway EM (2009) Parallel processing strategies of the primate visual system. Nat Rev Neurosci 10:360–372

Pesaran B, Pezaris JS, Sahani M, Mitra PP, Andersen RA (2002) Temporal structure in neuronal activity during working memory in macaque parietal cortex. Nat Neurosci 5:805–811

Reich DS, Mechler F, Victor JD (2001) Independent and redundant information in nearby cortical neurons. Science 294:2566–2568

Roelfsema PR, Lamme VA, Spekreijse H (2004) Synchrony and covariation of firing rates in the primary visual cortex during contour grouping. Nat Neurosci 7:982–991

Salinas E, Sejnowski TJ (2000) Impact of correlated synaptic input on output firing rate and variability in simple neuronal models. J Neurosci 20:6193–6209

Schneidman E, Bialek W, Berry MJ 2nd (2003) Synergy, redundancy, and independence in population codes. J Neurosci 23:11539–11553

Schwabe L, Obermayer K (2002) Rapid adaptation and efficient coding. Biosystems 67:239–244

Sharpee TO, Sugihara H, Kurgansky AV, Rebrik SP, Stryker MP, Miller KD (2006) Adaptive filtering enhances information transmission in visual cortex. Nature 439:936–942

Sohal VS, Zhang F, Yizhar O, Deisseroth K (2009) Parvalbumin neurons and gamma rhythms enhance cortical circuit performance. Nature 459:698–702

Sompolinsky H, Yoon H, Kang K, Shamir M (2001) Population coding in neuronal systems with correlated noise. Phys Rev E Stat Nonlin Soft Matter Phys 64:051904

Traub RD, Whittington MA, Stanford IM, Jefferys JG (1996) A mechanism for generation of long-range synchronous fast oscillations in the cortex. Nature 383:621–624

Ts'o DY, Gilbert CD, Wiesel TN (1986) Relationships between horizontal interactions and functional architecture in cat striate cortex as revealed by cross-correlation analysis. J Neurosci 6:1160–1170

Tsodyks M, Kenet T, Grinvald A, Arieli A (1999) Linking spontaneous activity of single cortical neurons and the underlying functional architecture. Science 286:1943–1946

van der Togt C, Kalitzin S, Spekreijse H, Lamme VA, Super H (2006) Synchrony dynamics in monkey V1 predict success in visual detection. Cereb Cortex 16:136–148

Whittington MA, Traub RD, Jefferys JG (1995) Synchronized oscillations in interneuron networks driven by metabotropic glutamate receptor activation. Nature 373:612–615

Womelsdorf T, Fries P, Mitra PP, Desimone R (2006) Gamma-band synchronization in visual cortex predicts speed of change detection. Nature 439:733–736

Chapter 17
The Function of Cortical Microcircuits: Insights from Biomorphic Ceramic-Based Microelectrode Arrays

Greg A. Gerhardt, Ioan Opris, Jason J. Burmeister, Francois Pomerleau, Jorge E. Quintero, Peter Huettl, Robert E. Hampson, and Sam A. Deadwyler

Abstract Technological advancements in the manufacturing, design and use of biomorphic ceramic-based multi-electrode arrays have made it possible to study the function of the brain's microcircuits. Here we examine the literature on the fabrication, composition, design and use of biomorphic Microelectrode Arrays (MEAs) that were instrumental in understanding the function of cortical microcircuits. Recent findings highlight the importance of such MEAs for the study of cortical modularity from a broad range of perspectives such as electrophysiology, in vivo electrochemistry, optogenetics, and neuroprosthetics. In particular, biomorphic MEAs are a crucial milestone in the advancement of cortical modularity and have been used to simultaneously record neural activity from supra- and infra-granular layers along in adjacent cortical minicolumns. We have strived to develop MEAs that: (1) can be mass produced such that other laboratories can easily utilize the same recording technology, (2) are designed to be biomorphic to study multiple brain regions and neurotransmitters in various *in vivo* systems, (3) control online signal flow through multiple minicolumns and layers, and (4) can be used in the future in neuroprosthetics for patients with neurological and psychiatric disorders.

Author contributed equally with all other contributors

G.A. Gerhardt (✉) • J.J. Burmeister • F. Pomerleau • J.E. Quintero • P. Huettl
Department of Anatomy and Neurobiology, Parkinson's Disease Translational Research Center of Excellence, Center for Microelectrode Technology, University of Kentucky, Lexington, KY 40506, USA
e-mail: gregg@uky.edu; jason.burmeister@uky.edu; francois.pomerleau@uky.edu; george.quintero@uky.edu; peter.huetl@uky.edu

I. Opris, Ph.D.
Department of Physiology and Pharmacology, Wake Forest University Health Sciences, Winston-Salem, NC 27103, USA
e-mail: ioanopris.phd@gmail.com

R.E. Hampson • S.A. Deadwyler
Department of Physiology and Pharmacology, Wake Forest University Health Sciences, Medical Center Boulevard, Winston-Salem, NC 27157, USA
e-mail: rhampson@wakehealth.edu; sdeadwyl@wakehealth.edu

M.F. Casanova, I. Opris (eds.), *Recent Advances on the Modular Organization of the Cortex*, DOI 10.1007/978-94-017-9900-3_17

Keywords Amperometry • Cortical microcircuits • Microelectrode arrays • Minicolumn • Electrochemical recording • Glutamate • Optogenetics • Brain machine interface

17.1 Introduction

17.1.1 Brief History of Brain Recordings and Microelectrode Arrays

The neuronal electrical activity of the brain has been measured and studied since the 1930s. Of record, the first to implant electrodes into the diencephalon of the cat was Hess (1932). Fischer (1957) used various metals/insulators as single wire electrodes, while Delgado (1961) and Robinson and Johnson (1967) used handmade devices that reinforced histological findings. Next, Salcman and Bak (1973) began to record with parylene-coated microwires, and Chapin et al. (1980) developed multi-wire arrays. Wise and Angell (1970, 1975) used IC technology and Silicon (Si) wafers to develop the lithographically formed microelectrode arrays. Bement (1986) and coworkers developed the first multi-site electrode from Si (Michigan-style electrode) and Campbell et al. (1991) developed the first monolithic multi-shank electrode from Si, known as the Utah Electrode Array. Among these micro-electrode technologies, emerged our ceramic-based multisite electrode arrays for in vivo electrochemical recordings (Burmeister et al. 2000) and then chronic single-neuron recording (Moxon et al. 2004) that have been instrumental in recording both cortical and subcortical (including hippocampal) neurons in rodents and primates.

17.1.2 Electrochemical Recordings

Since the pioneering work of Adams and coworkers (Kissinger et al. 1973; McCreery et al. 1974; Wightman et al. 1976) in the 1970s, many research groups have strived to develop microelectrode technologies for the direct chemical measurements of neurotransmitters in the extracellular space of the brain (Wightman et al. 1976; Martin and Marsden 1987; Hu et al. 1994; Lowry et al. 1998; Burmeister et al. 2000; Dale et al. 2000; Philips et al. 2003; Heien et al. 2005; Oldenziel et al. 2006b; Wassum et al. 2008; Zesiewicz et al. 2013). Interestingly, chemical communication of the brain makes up approximately 90 % of brain communication. Over the last decade, our group has worked on the development of a more universal MEA technology for second-by-second measurements of neurotransmitters, neuromodulators, and markers of brain energy/metabolism (Burmeister et al. 2000, 2002; Dixon et al. 2002; Pomerleau et al. 2003; Hascup et al. 2008, 2010; Parikh et al. 2004, 2007; Parikh and Sarter 2006, 2008; Wilson and Gifford 2005). The most widely used microelectrode methods have been those employing carbon fiber-based microelectrodes. These have been developed and used for measuring rapid changes in CNS levels of dopamine and other electroactive monoamines in anesthetized, and more recently in awake, behaving animals (Phillips and Wightman 2004; Suaud-Chagny et al. 1993; Westerink 2004). These techniques, however, have had limited success for monitoring other non-electroactive neurotransmitters such as glutamate and acetylcholine mainly because of the inherent inability do directly oxidize or reduce these molecules without using an enzyme-based biosensor. Additional weaknesses have been the inability for self-referencing recording methods, poor limits of detection, and/or slow temporal resolution.

Our laboratory and others have made great strides toward developing biosensors capable of measuring these non-electroactive neurotransmitters (Burmeister et al. 2000; Burmeister and Gerhardt 2001; Hu et al. 1994; Oldenziel et al. 2006a; Tian et al. 2009). We have developed an advanced technology using enzyme-based ceramic MEAs for chronic neurotransmitter measurements in rodents and nonhuman primates (Rutherford et al. 2007; Hascup et al. 2008; Stephens et al. 2010). A variety of neurochemicals can be measured using the MEA technology in conjunction with constant voltage amperometry or chronoamperometry using the Fast Analytical Sensing Technology (FAST16mkIII; Quanteon, LLC, Nicholasville, KY) recording system. While the experiments and examples discussed in this chapter can be applied to other neurochemicals, we have the most extensive experience measuring the major excitatory neurotransmitter in the CNS, L-glutamate (Rutherford et al. 2007; Hascup et al. 2008, 2010). Therefore, in this chapter we focus on glutamatergic neurotransmission and what we have learned over the last decade about this important and elusive neurotransmitter.

17.1.3 Importance of MEAs for the Electrophysiological Analysis of Cortical Modularity

A question posed early on by Mountcastle (1957, 1997; Shepherd and Grillner 2010) and that has remained somewhat unresolved until recently is, "What are the transforming operations imposed in a local region of neocortex, a cortical column, upon its input, to produce its targeted outputs?" Previous anatomic and electrophysiological investigations (Buxhoeveden and Casanova 2002; Casanova et al. 2002a, b, 2003, 2011; Opris et al. 2011; González-Burgos et al. 2000; Kritzer and Goldman-Rakic 1995) have examined both the structural and the functional bases of this operation (de Kock et al. 2007; Sakurai and Takahashi 2006; Opris et al. 2005), but only a few studies have examined simultaneous firing patterns of cells in different cortical layers (Derdikman et al. 2006; Zhang and Alloway 2006; Krupa et al. 2004; Jung et al. 1998; Opris et al. 2012a, 2013; Takeuchi et al. 2011). However, in the last few years, a new approach for studying columnar/laminar structure and functionality of the brain has emerged (Opris et al. 2011, 2012a, b,

2013; Constantinople and Bruno 2013; Hansen et al. 2012; Takeuchi et al. 2011; Mahan and Georgopoulos 2013).

17.2 Biomorphic Microelectrode Arrays

17.2.1 MEA Composition

Over the last decade our group has used the basic concepts of photolithographic formation of microelectrodes on Si substrates (Wise 1976) to develop ceramic (Al_2O_3) substrate-based MEAs. Relatively inert FDA approved materials are used in fabrication including Al_2O_3, polyimide for encapsulation and platinum (Pt) recording (electrophysiological or chemical) sites and Pt/Ir stimulation sites. MEAs are mass fabricated using photolithographic techniques and cut out using computer-controlled diamond saw technology. Hundreds of MEAs with different or identical designs can be simultaneously patterned on a single ceramic wafer. Operationally we have defined biocompatibility as the ability to record single-unit activity of neurons or in vivo electrochemical signals of a test molecule such as peroxide and post-experimentation integrity of the MEAs out to 6 months *in vivo* in awake rats. High purity, non-conducting Al_2O_3 substrates (99.9 %, Coors, Golden, CO) are polished to 37.5–125 μm in thickness (2.5 × 2.5 cm or 5.0 × 5.0 cm wafers). The metal (Pt) recording sites or stimulation sites (Pt or Pt/Ir), connecting lines and bonding pads are patterned using a double mask design (Burmeister et al. 2000). One or both sides of the MEA are coated with up to a 1.25 μm polyimide layer to insulate the connecting lines. Pt or Pt/Ir (0.25 μm) can be additionally sputtered onto the recording or stimulation sites. MEAs with 4–16 recording sites have been fabricated with recording sites ranging from 10 × 10 μm to 15 × 330 μm. Various recording site sizes and arrangements have been selected to conform to specific brain layers, which we now term biomorphic (Burmeister et al. 2002; Moxon et al. 2004; Opris et al. 2011; Stephens et al. 2010). The front and back of the ceramic substrate can be patterned to increase recording site density and create isolated front and back site pairs. A practical advantage of multiple recording sites over a single MEA is that even if one site fails useful data can often be received from the other recording sites, thus redundancy for practical use.

17.2.2 MEA Fabrication

The MEAs developed by our laboratory are mass fabricated using photolithographic methods (Gerhardt and Burmeister 2006). There are several major advantages of photolithography over handmade electrodes. First, the technique allows routine production of reproducible recording surfaces as small as 10 × 10 μm.

Second, multiple MEAs are patterned onto a single fabrication substrate (usually 2.5 × 2.5 cm or 5.0 × 5.0 cm) allowing for increased fabrication number at a decreased cost. Third, photolithographic methods are used to manufacture numerous microelectrode designs with multiple recording sites in well-defined, highly reproducible geometrical configurations. This makes the method ideal for designing an MEA that can be easily, repetitively, and precisely positioned within a single or even multiple brain structures simultaneously with the aid of stereotaxic equipment.

Our biomorphic MEAs are currently constructed in conjunction with Hybrid Circuits, Sunnyvale, California). The fabrication process is also thoroughly explained in previous publications (Pomerleau et al. 2003; Nickell et al. 2005; Hascup et al. 2006; Gerhardt and Burmeister 2000; Burmeister and Gerhardt 2003). Initially microelectrode photographic masks are designed on a computer aided design (CAD) program where arrays of 4–16 recording sites are arranged on templates. A 2.5 cm × 2.5 cm or 5.0 cm × 5.0 cm × 125 μm thick polished, ceramic wafer (alumina, Al_2O_3, Coors Ceramic, Coors Superstrate 996) serves as a common substrate for the microelectrode arrays. Ceramic reduces the cross-talk from adjacent connecting lines. Additionally, ceramic is strong and rigid, which aids in precise stereotaxic placement into tissues with minimal flexing or breaking. The ceramic substrate may be polished or lapped down to achieve microelectrodes as thin as 37.5 μm (Burmeister and Gerhardt 2003). Following cleaning, photoresist is spun onto the ceramic wafer. Collimated light passing through the photo mask exposes the photoresist, thus transferring the microelectrode images onto the wafer. Locations of the recording sites, connecting lines and bonding pads are not exposed to light. Solvents are used to remove unexposed photoresist from the wafer. Next, an adhesion layer of titanium (Ti) is used to allow noble metals (Pt or Ir) to adhere to the ceramic substrate. Titanium (500 Å thick) is sputtered onto the developed photoresist covered ceramic wafer. Following the adhesion layer, the active recording metal layer, Pt for most of our work, is sputtered onto the substrate (usually 2,300 Å thick). Solvents are used to remove the developed photoresist and unwanted metals leaving the recording pads, connecting lines, and bonding pads.

Once the Pt recording sites, connecting lines, and bonding pads are in place, the connecting lines are coated with an insulator using another photolithographic step. The connecting lines act as wires to connect the bonding pads to the recording sites and must be insulated from aqueous environments. To accomplish this, the MEAs are again coated with photoresist. A second photomask is used to define the areas where polyimide is to be placed. Polyimide acts as an insulator to define the recording site active area as well as the bonding pads. It also reduces cross-talk between the connecting lines. After the photoresist is developed, polyimide is spun onto the wafer (1–1.25 μm thick). Once the insulating layer is applied, only the recording sites and bonding pads are exposed. The photoresist and excess polyimide are removed and the remaining polyimide is cured at 200 °C. The complete fabrication process is depicted in Fig. 17.1. Many 1 cm long microelectrodes (Hascup et al. 2006, 2008, 2009, 2010; Kulagina et al. 1999; Oldenziel et al. 2006a, b; Kinney et al. 1997; Drew et al. 2004; Clapp-Lilly et al. 1999;

Fig. 17.1 Photolithographic Process. Schematic representation of the photolithographic process used for MEA fabrication

Bungay et al. 2003; Yang et al. 1998; Borland et al. 2005; Marsden et al. 1988; Adams 1990; Michael and Wightman 1999; Gerhardt and Burmeister 2000, 2006; Burmeister and Gerhardt 2003; Stephens et al. 2011; Gerhardt et al. 1984; van Horne et al. 1992; Hebert et al. 1996; Hoffman et al. 1998; Friedemann et al. 1996) ; Lowry et al. 1998; Chen et al. 2002; Matsumoto et al. 2002; Photolithography 2001; MMMT 1984; Wire Bond 1998) can be patterned onto the wafer simultaneously to facilitate production and increase the number of microelectrodes that are made from the ceramic substrate (Fig. 17.2a).

After the formation of the MEAs on the ceramic wafer, a diamond saw or laser is used to form or "cut out" the individual MEAs. A major advantage of the diamond saw is that it produces highly polished edges for reduced tissue damage during implantation, but it can lead to small variations in tip sizes. As previously mentioned, the wafer thickness is usually polished to 125 μm. Once cut, the microelectrode tip is ~9,000 μm long and ~1,000 μm at the widest point furthest from the Pt

a
b
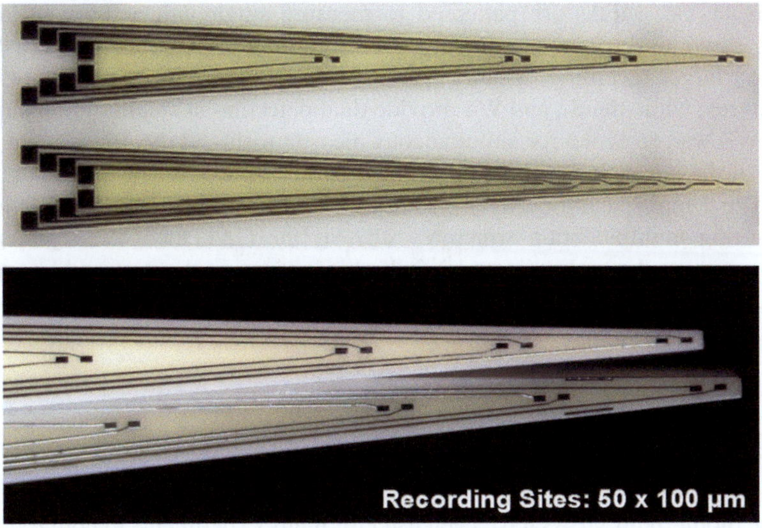

Fig. 17.2 (**a**) **Ceramic Wafer**. A ceramic wafer with two different recording site configurations patterned on its surface. Each biomorphic MEA contains 8 recording sites. Recording lines connect each recording site and run the length of the electrode tip for connection to the PCB. The length of each tip is approximately 1 cm. (**b**) **16-Site MEA.** The electrode tip of a 16 site MEA. Photolithographic techniques allow for patterning on the front and back of the ceramic wafer to increase surface recording site density. This photograph shows the Pt recording sites patterned on both sides by using a mirror to reflect the image from the underside of a microelectrode array

recording sites, where the tip is connected to the printed circuit board (see below). More importantly, the tapered tip of the microelectrode is considerably smaller and varies between 60 and 175 μm. We have found that the tapered design of the MEAs reduces tissue damage surrounding the recording sites.

The ceramic microelectrode tip with Pt recording sites is attached to a printed circuit board (PCB) holder for handling and connection to recording equipment. To connect the ceramic microelectrode to the PCB holder, each bonding pad is wire bonded to an individual Pt recording site on the ceramic microelectrode tip. The tips are epoxied onto the paddle for stability and to insulate the wire bonds. Cutting and assembly is currently performed in conjunction with Hybrid Circuits, Inc., (Sunnyvale, CA). The fully constructed MEAs for anesthetized and freely moving animal recordings are shown in Fig. 17.3a, b, respectively.

17.2.3 Types and Designs of MEAs

As previously mentioned, the major advantage of photolithography is the ability to design MEAs with well-defined, highly reproducible geometrical configurations. Our most commonly used MEAs have four or eight Pt recording sites and are shown in Fig. 17.3c. MEAs are patterned to measure single-unit activity or chemical analytes within layered structures or from multiple brain structures. The R1 provides a larger recording distance that is useful for large brain regions or layered structures, while the S2 and W4 provide dual detection in smaller brain structures. The W3 was designed for simultaneous measurements in the arcuate and median sulci of the non-human primate prefrontal cortex. One of the newest designs is the R2. The 1 mm spacing between pairs of sites makes this microelectrode ideal for measuring multiple brain structures such as the CA1 and CA3 regions of the hippocampus, the prelimbic and infralimbic regions of the prefrontal cortex, and the caudate/putamen and nucleus accumbens of the rat neostriatum. Additional designs are presented in (Gerhardt and Burmeister 2000).

17.2.4 Multiple Uses of MEAs

These biomorphic MEAs have been used for many experiments in both rodents and nonhuman primates and have been applied to chronic recordings of multiple single-unit neural activity, electrical stimulation and in vivo electrochemical recordings of neurotransmitters in awake rats, non-human primates and possibly in future applications to human neurosurgery (Day et al. 2006; Quintero et al. 2007; Dash et al. 2009; Zhang et al. 2009; Konradsson-Geuken et al. 2010; Stephens et al. 2010; Choi et al. 2012; Hampson et al. 2012a, b; Onifer et al. 2012; Opris and Casanova 2014; Howe et al. 2013; Zhou et al. 2013).

These arrays were capable of recording single neuron activity from each of their recording sites for at least 6 months during chronic implantation in the somatosensory cortex of rats (Moxon et al. 2004), and monkey hippocampus (Hampson et al. 2013a) as well as in the prefrontal cortex of nonhuman primates (Opris et al. 2011, 2012a, b). The vertical arrangement of the recording sites on these biomorphic MEAs is ideal for simultaneously recording across the different layers of brain areas such as the cerebral cortex and hippocampus in chronic preparations (Opris et al. 2011, 2012a, b, 2013; Hampson et al. 2012b, 2013a). Hippocampal pyramidal cells in the rat have been stimulated and recorded (Hampson et al. 2013a). Single-unit recordings from nonhuman primate frontal cortex and hippocampus in awake animals have been performed using a specially designed deep recording MEA (not shown) for the primate brain.

Neurochemical recordings of glutamate signaling in the motor and frontal cortex of anesthetized monkeys as well as recordings of glutamate levels in the prefrontal cortex, hippocampus and putamen of awake rodents and nonhuman primates have been performed (Burmeister et al. 2002; Pomerleau et al. 2003; Hascup et al. 2008, 2010; Opris et al. 2012b; Hampson et al. 2013b). New Pt/Ir surfaces formed on the more standard Pt surfaces have been characterized for improved electrical stimulation and enhanced recording site surface area. In conjunction with Ad-Tech® Medical Instrument Corporation, flexible shank electrodes have been developed for future studies in nonhuman primates and patients with epilepsy, brain tumors,

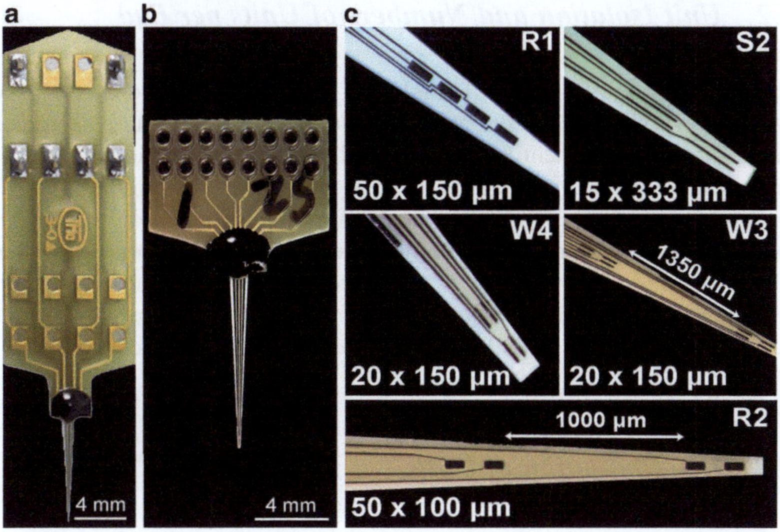

Fig. 17.3 Commonly used MEAs. Photographs of the fully fabricated MEAs for anesthetized (**a**) and freely moving (**b**) MEAs showing the PCB and wire bonded tip with black epoxy. (**c**) Magnified images of tips with several Pt recording sites patterned in unique geometrical configurations. The name of the type of MEA tip is shown in the *upper right*, while the size of the recording sites is shown in the *lower left*. Where applicable, distance between groups of recording sites is also labeled

Parkinson's disease deep brains stimulation (DBS) surgery or traumatic brain injury (TBI). In addition, a 4 shank electrode design with 32 active sites for recording and stimulation in the hippocampus of awake nonhuman primates is currently being used and further developed for electrochemical and electrophysiological studies in the hippocampus of awake nonhuman primates. Patterning can also be performed on the reverse side of the ceramic wafer to increase recording site density (Fig. 17.2b).

17.3 Electrophysiological Measures

17.3.1 Signal-to-Noise Ratio of Cell Recordings

Although the acceptable ratio of signal-to-noise is 2:1 for most recordings, the ratios for single and multiple neuron firing recordings with MEAs typically range between 3:1 and 10:1, in both rodent and nonhuman primates (Opris et al. 2011, 2012a, b). For in vivo electrochemical recording, MEA recordings achieve improved signal-to-noise by self-referencing noise subtraction (Burmeister and Gerhardt 2001), but can detect changes in glutamate as low as 0.1 μM/s.

17.3.2 Unit Isolation and Numbers of Units per Pad

The good signal-to-noise ratios of single cell recording achieved by the Pt MEAs allows for very clean isolation of cell waveforms. Extracellular neuron action potentials (spikes) were identified and isolated on-line by separation of extracellular action potential waveform duration and amplitude and confirmed via off-line analysis using principal components analysis and autocorrelation to confirm that (a) single neurons were isolated, (b) all spikes of a given neuron were identified within 95 % confidence limits, and (c) neurons detected simultaneously on adjacent sites were counted only once. In daily recording sessions we were able to isolate up to 4 units per pad (Opris et al. 2011), but 2 units per pad was most common.

17.3.3 Discrimination of Units

To discriminate neurons recorded simultaneously we usually employ a cluster separation method (using *Neuroexplorer* from www.Plexon.com), such as Principal Component Analysis (PCA) in real time or by offline sorting. We also perform peak-to-peak amplitude separations for cells at each pad where at least two well separated clusters of cells were recorded. All 8 pads discriminated the temporal pattern of neuron spiking from clusters of simultaneously recorded cells (Opris et al. 2011).

17.3.4 Durability and Longevity of the MEA

The durability of MEAs has continuously increased from 3 weeks (Moxon et al. 2004) to up to several years (Deadwyler et al. 2013). A good MEA used in chronic recording can provide reliable units for 6 months to 3 years. In acute recordings the MEAs can be reused many times with very high yield. For example a MEA with 8 pads can provide up to 32 reliable cells per session while a 4 shank MEA with 32 pads can yield up to 128 units in both acute (on a session base) and chronic (on long term base) recordings.

17.3.5 e. Recording Resolution

A good recording resolution comes from the distance between adjacent pads needed for isolation of separate units which is 30 μm. According to anatomic measures by Buxhoeveden and Casanova (2002) and quantitative analysis based on columnar neuron recordings by Mahan and Georgopoulos (2013), this is the ideal separation between pads for biomorphic minicolumnar recordings. Nevertheless, occasionally a cell may fall in between two adjacent pads, and thus one ends up recording the same unit with the adjacent pads. Neurons detected simultaneously on adjacent sites are isolated on only one pad and used for further analysis. On the chemical recording aspect, we have demonstrated that MEAs that are chemically modified for glutamate, acetylcholine, choline, lactate and glucose measures have fast temporal resolution (<1 s), excellent spatial resolution (microns), low detection limits (≤100 nM for all analytes) and cause minimal damage (50–100 μm) to surrounding brain tissue.

17.4 Electrophysiology of Minicolumns with Biomorphic MEAs

Biomorphic ceramic MEAs, designed collaboratively and manufactured in conjunction with Hybrid Circuits and the Center for Microelectrode Technology, University of Kentucky, consisted of eight patterned platinum recording pads on a ceramic substrate (Fig. 17.3c) specifically configured for recording single neuron activity (Hampson et al. 2004; Moxon et al. 2004). The model W3 version of the MEAs (Fig. 17.3c) was specially designed with 20 × 150-μm recording pads 2a–d located to record activity from supragranular (Layer 2/3) cells, whereas pads 5a–d simultaneously recorded neuron activity from infragranular Layer 5 neurons (Fig. 17.4b) in the nonhuman primate. The intra-pad distance between the edges of pads 2a–d and 5a–d was 1,350 μm, with a 30-μm separation between adjacent pads (i.e., 2a–c, 2b–d, etc.). Neural spike trains were analyzed with respect to firing rate during specific events within go/no-go trials (Opris et al. 2011).

17.4.1 Frontal Cortical Cell Activity Recorded with Biomorphic Ceramic MEAs

Neurons (Opris et al. 2011, 2012a, b, 2013) were recorded in Layer 2/3 and Layer 5 and from the dorsal bank of the arcuate sulcus in the dorsolateral prefrontal/premotor areas of prefrontal cortex (PFC). Figure 17.3c shows examples of biomorphic MEA recordings with the W3 MEAs in the dorsal premotor region during *go trials* of the go/no-go task in nonhuman primates. The waveforms recorded at the vertically separated (1,350 μm) recording sites on the MEA indicate individually isolated neuron firing exhibiting task-related discharges within different cortical cell layers, presumably supragranular Layer 2/3 cells shown at locations 2a–d and infragranular Layer 5 cells shown at MEA site locations 5a–d, 6a–f. Corresponding Post-Event Histograms (PEHs) in Fig. 17.4c illustrate the firing patterns of the same neurons, identified in terms of MEA recording pad location (i.e., 2a, 2b, 5a, 5b, 5a,

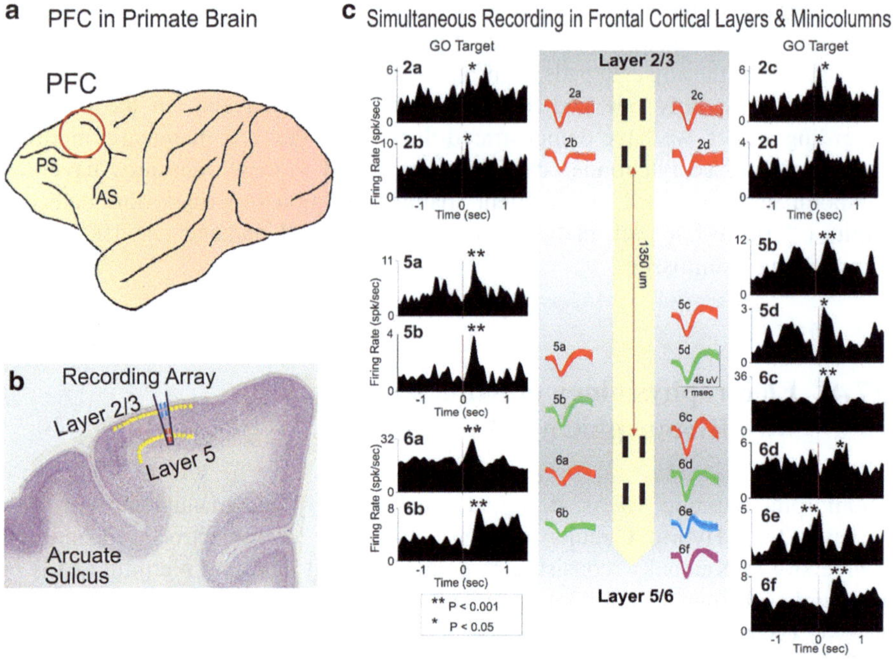

Fig. 17.4 Simultaneous Recording of Frontal Cortical Layers and Minicolumns. (**a**) The recording chamber located in the primate prefrontal cortex. (**b**) Coronal section in the primate prefrontal cortex with the MEA localized within the upper pad quartet in the supra-granular layers and with the lower pad quartet in the infra-granular layers. (**c**) The illustration of the MEA recording in prefrontal cortical layers and minicolumns (Adapted with permission from Opris et al. 2011)

6b, etc.) synchronized to *target* image presentations within the task. The MEA routinely recorded RS (define?) cells from both Layer 2/3 and Layer 5/6 simultaneously in the same session, and as indicated in Fig. 17.4c, most cells showed a significant either increase or decrease (noted by asterisks) in activity related to go/no-go task events.

17.4.2 Layer-Specific Cortical Activity Recorded with Biomorphic MEAs

Implementation of the newly applied MEA technology described here (Burmeister et al. 2000; Hampson et al. 2004; Moxon et al. 2004) allowed simultaneous recording from neurons at two different anatomically and physiologically distinct recording sites in the sensorimotor areas of PFC. The biomorphic design of the MEA allowed identification and designation of activity within a dual laminar arrangement of orthogonal MEA recording sites separated by 1,350 μm, the precise distance between Layer 2/3 and Layer 5 in the PFC Arcuate sulcus (Fig. 17.4a). This recording arrangement allowed unique comparisons of the simultaneous activity of cells from (i) each separate cortical layer and (ii) within the same layer but at adjacent physically separate locations on the MEA (Fig. 17.4b). The differential firing to task-related events of identified PFC cells in both Layer 2/3 and Layer 5 is described in detail in Figs. 17.3–17.5 in terms of individual and mean discharge patterns of simultaneously recorded neurons in each layer during either go or no-go trials.

Our recent results (Opris et al. 2011, 2012a, b, 2013) demonstrate many of the functional "operations" described above in which PFC cortical neurons in both the supragranular and the infragranular layers differentially encoded sensory stimuli relevant to performance of a simple go/no-go task. In addition, firing to task-related events was shown to be encoded between layers by synchronized firing of cell pairs comprising apparent functional "minicolumns" connecting the supragranular and the infragranular layers. This assessment was made possible by use of biomorphic MEAs designed to sample neuronal activity in a manner that would reveal the demonstrated functional columnar organization of the PFC. The results provide new insight into the manner in which underlying PFC microcircuitry integrates convergent sensory, cognitive and motor signals from other brain regions to select and control task-related behavioral response tendencies in primate brain (Opris et al. 2011, 2013).

17.5 Neurochemistry of Cortical Layers and Minicolumns

17.5.1 Amperometry

Amperometric measurements of L-glutamate using biomorphic MEAs provide spatially and temporally resolved measures of neuromolecules in the central nervous system of rats, mice and non-human primates (Talauliker et al. 2011). Although the functional capabilities of MEAs have been previously documented for both anesthetized and freely-moving paradigms, the performance enabling intrinsic physical properties of the MEA device have not heretofore been presented. In these studies, spectral analysis confirmed that the MEA recording sites were primarily composed of elemental platinum (Pt°). In keeping with the precision of the photolithographic process, scanning electron microscopy revealed that the Pt recording sites have unique microwell geometries postfabrication. Atomic force microscopy demonstrated that the recording surfaces have nanoscale irregularities in the form of elevations and depressions, which contribute to increased current per unit area that exceeds previously reported microelectrode designs. The ceramic substrate on the back face of the MEA was characterized by low nanoscale texture and the ceramic sides consisted of an extended network of ridges and cavities. Thus, individual recording sites have a unique Pt° composition and surface profile that has

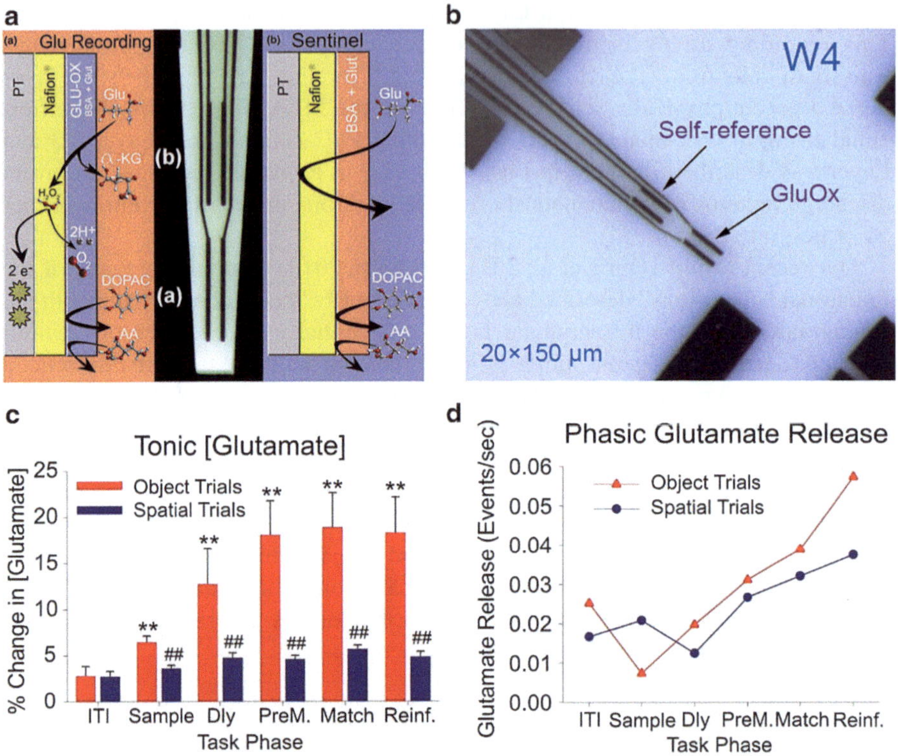

Fig. 17.5 Glutamate Recordings in Prefrontal Cortex of Awake, Behaving Monkey. (**a**) MEA for glutamate recording. (**b**) Glutamate recording with MEA in prefrontal cortical layer 2/3. (**c**) Electrochemical recording of tonic glutamate neurotransmitter concentrations in PFC Layer 2/3. Mean (±S.E.M.) glutamate concentration ([Glutamate]) measured as a percentage increase over baseline (average 8.69 ± 0.77 μM) glutamate concentration. Horizontal axis indicates phase of DMS task: intertrial interval (ITI), Sample phase, Delay phase (Dly), end of delay phase 5 s prior to Match (PreM), Match phase and reinforcement (Reinf.). Asterisks: $*p < 0.01$, $**p < 0.001$, *Object* vs. *Spatial* trials; $^{\#}p < 0.01$, $^{\#\#}p < 0.001$ DMS task phases vs. ITI. (**d**) Frequency of phasic glutamate release events measured as transient increase (<2.0 s duration) of at least 5 % in [Glutamate] for the same trials shown E. Frequency normalized to # events per second per DMS trial (Adapted with permission from Opris et al. 2012b)

not been previously observed for Pt-based MEAs. These features likely impact the physical chemistry of the device, which may influence adhesion of biological molecules and tissue as well as electrochemical recording. This is likely contributing as well to the unique electrophysiological recording capabilities of the MEAs for single and multi-unit neuronal activity (see above).

17.5.2 Glutamate Recording

17.5.2.1 In Rodents

A major issue that has surrounded *in vivo* measures of glutamate has surrounded the origin of the signal (see (Timmerman and Westerink 1997). Both neuronal and glial processes are involved in the regulation of tonic glutamate levels and many prior microdialysis studies support that resting levels sampled by microdialysis methods may not be derived from neurons. Recent pharmacological studies in awake rats using glutamate MEAs have shed light on the origins of tonic and phasic glutamate signals (Rutherford et al. 2007; Hascup et al. 2010). Modulation of resting glutamate levels in rat frontal cortex is shown in Fig. 17.5.

Resting glutamate levels were significantly lower (~40–50 %) than vehicle with local application of the Na^+ channel blocker TTX or the Ca^{2+} channel blocker ω-conotoxin. These results support that the resting levels of glutamate measured by the MEAs in the rat frontal cortex are derived from neurons because the inhibitory effects of both a sodium channel blocker and a calcium channel blocker are known to directly affect the release of neurotransmitters from neurons. Second, the contribution of Group II metabotropic glutamate receptors ($mGluR_{2/3}$), which act as inhibitory auto-receptors and reside primarily on presynaptic neurons in the rat frontal cortex, was examined. An ~35 % increase in resting levels was observed after delivery of the $mGluR_{2/3}$ antagonist LY341495 while an ~20 % decrease was found with the $mGluR_{2/3}$ agonist LY379268. These results support the idea that there is a significant neuronal glutamate contribution to resting levels as well as due to the modulation of resting levels detected following local administration of these drugs that can directly affect output of glutamate from nerve terminals. Third, when the non-selective glutamate transporter inhibitor D,L-*threo*-β-benzyloxyaspartate (TBOA) was locally applied there was an ~120 % increase in resting levels, supporting the idea that extracellular glutamate levels are regulated by glutamate transporters located largely on glia (see Danbolt 2001) and indicating that there is significant spontaneous efflux of glutamate that is detected following glutamate uptake inhibition by the NEAs in vivo.

17.5.2.2 In Nonhuman Primates

Biomorphic ceramic MEAs utilized in prior recordings of nonhuman primate PFC cortical layer 2/3 and layer 5 neurons were used in this study to record tonic glutamate levels and transient release in layer 2/3 PFC (Hampson et al. 2013b). Tonic glutamate levels were seen to increase in the Match (decision) phase of a visual delayed-match-to-sample (DMS) task, while increased transient glutamate release occurred in the Sample (encoding) phase of the task. Further, spatial vs. -object-oriented DMS trials evoked differential changes in glutamate levels. Thus, the same biomorphic MEAs were capable of electrophysiological and in vivo electrochemical recording, and revealed similar evidence of neural processing in layers 2/3 and layer 5 during cognitive processing in a behavioral task in awake nonhuman primates.

17.6 Future Developments of MEAs

17.6.1 Measuring Multiple Neurotransmitters Simultaneously

A major advantage of microdialysis is the ability to measure multiple neurotransmitters simultaneously (assuming good separation of peaks during offline detection by high performance liquid chromatographic-based methods). Our laboratory has successfully demonstrated the ability to measure acetylcholine and choline simultaneously (Burmeister et al. 2008), which is an essential element of the enzymes involved. Ideally, the long-term goal is to design an MEA to measure multiple neurotransmitters from the same brain structure. As previously mentioned, photolithographic methods allow us to design MEAs with well-defined, highly reproducible geometric configurations that can be patterned onto the front and back of a ceramic wafer. By using both sides of the ceramic wafer we can greatly enhance the recording site density of an MEA. We have designed several 16 site MEAs, one of which is shown in Fig. 17.2b. By increasing the recording site density we can selectively coat recording sites for self-referencing measurements of different analytes in the same brain region.

17.6.2 Simultaneous Electrophysiological and Electrochemical Recordings

The ability to simultaneously measure local field potentials (LFPs) and neurochemical dynamics would greatly enhance our understanding of neuromodulation. This would have profound implications for studying behavior or disorder related changes in the CNS. Until recently, few studies were capable of achieving simultaneous electrophysiological and electrochemical measures from the same recording preparation. And, those that did either switched between neuronal spike recording and the in vivo electrochemical recordings (Williams and Millar 1990; Stamford et al. 1993; Cheer et al. 2005) or required the use of two separate recording systems (Zhang et al. 2009, 2010). This made the experimental setups cumbersome and expensive while making data analysis and interpretation difficult. However, Zhang and colleagues (2009, 2010) are the first to demonstrate simultaneous LFPs and local neurochemical measurements using a single MEA coupled with constant potential amperometry. They successfully demonstrated that the high frequency component of the amperometric signal resembled LFPs. Using a single set-up and biosensor for the simultaneous acquisition of electrophysiological and neurochemical information can greatly enhance the field of *in vivo* electrochemistry (Zhang et al. 2009, 2010).

17.6.3 Optogenetics

Due to great diversity of cell types in neural circuits, it is critical to be able to analyze how these different kinds of cells work together. An emerging approach with broad implications for basic and clinical neuroscience is based on optogenetic stimulation (Gradinaru et al. 2007; Tye and Deisseroth 2012). Recent developments in optogenetics based on optical manipulation of activity in neural circuits with light-sensitive rhodopsins, such as the *Chlamydomonas* channelrhodopsin-2 (ChR2) are now capable to illuminate the inter-laminar microcircuits at the millisecond-scale, with cell type-specific optical perturbations in nonhuman primates (Diester et al. 2011; Han 2012), opening up new possibilities for repair and augmentation. In recent years, MEAs with integrated optical fibers use light-assisted perturbation and recording of local neural circuits during animal behavior (Royer et al. 2010; Marshel and Deisseroth 2013; Bernsterin and Boyden 2011).

17.6.4 Prosthetics

This aspect will be described in another chapter, here we just briefly touch on the relevance of MEAs in neural prosthetics.

17.6.4.1 Decision-Making in PFC

The cognitive approach to neural prostheses holds the promise of assisting individuals who are unable to move but who are capable of making decisions and movement plans. An executive prosthesis may even correct in real time a decision making process before a failed choice. A decision circuit is defined as a closed neural network that measures the probable value of a signal element and makes an output signal based on the value of the input signal and a predetermined criterion or threshold (Ratclif et al. 2003). The biomorphic MEAs provide the basis for applying a system specific model to control firing of cells via application of electrical stimulation (Opris et al. 2011, 2012a, b; Hampson et al. 2012a) to the same loci in which columnar firing has been detected (Opris et al. 2012b; Hampson et al. 2012a). This same model was implemented, tested and demonstrated to facilitate performance (Hampson et al. 2012a, b). This might very well serve as a blue print for a decision chip (Opris 2013).

17.6.4.2 Memory in Hippocampus

Berger and colleagues (Berger et al. 2007, 2008, 2011, 2012) demonstrated in rodents for the first time that a neural prosthesis is capable of real-time identification and manipulation of the encoding process that can restore and even enhance cognitive mnemonic processes. The idea was to block the ability to form long-term memories by using pharmacological agents that disrupt the neural circuitry between two fields of the hippocampus, CA1 and CA3, which interact to create long-term memory. By employing an artificial hippocampal system based on the multiple-input, multiple-output (MIMO) model (Song et al. 2009; Kim et al. 2006) they could duplicate the pattern of interaction between CA3-CA1 by monitoring the neural spikes in cells recorded by the electrode array, and then playing back the same pattern on the same array. Long-term memory capability returned to the pharmacologically blocked rats following activation of the electronic device, programmed to duplicate the memory-encoding function for that specific memory (which lever to pull). This suggests that if a prosthetic device and its associated electrodes were implanted in animals with a normal, functioning hippocampus, the device could actually be used to strengthen the memory being generated internally in the brain and enhance the memory capability of normal animals.

17.6.5 Nanotechnology

The US Presidential initiative for mapping the brain relies on the vast potential of nanotechnology. Recent developments in nanotechnological tools and in the design and synthesis of nano-materials have generated optical, electrical, and chemical methods that can readily be adapted for use in neuroscience. Nanotechnology was instrumental for development of the nanofabricated planar electrode array for high-density neuronal voltage recording (Du et al. 2011; Suyatin et al. 2013). Leveraging micro- and nanofabrication technology raises the prospect for less invasive implantable devices by creating smaller and "vastly" greater numbers of electrodes. A promising category for brain microcircuits is the planar electrode array (Viventi et al. 2011; Alivisatos et al. 2013), which is patterned on a crystalline, ceramic, or polymer support structure. The recording of neuronal activity with three-dimensional (3D) microelectrode arrays (Zorzos et al. 2012) represents a major advance in brain activity mapping techniques, by providing a tool to probe how intra and inter-laminar/regional neural circuits cooperate to process information.

17.7 Conclusions

Biomorphic MEAs are a crucial milestone in the advancement of our understanding of cortical modularity and have been used to simultaneously record neural activity from supra- and infra-granular layers along in adjacent cortical minicolumns. Currently used MEAs whether linear (Moxon et al. 2004; Mo et al. 2011) or bilinear (Hampson et al. 2004) are biomorphic to the laminar and/or columnar structure of the cortex, providing unparalleled insights into the function of cortical microcircuits through the use of ceramic-based MEAs. They hold great promise for unlocking other unique signaling features of the CNS through electrophysiological and/or neurochemical recordings.

References

Adams RN (1990) In vivo electrochemical measurements in the CNS. Prog Neurobiol 35:297–311

Alivisatos AP et al (2013) Nanotools for neuroscience and brain activity mapping. ACS Nano 7 (3):1850–1866

Bargon J (ed) (1984) Methods and materials in microelectronic technology, IBM research symposia series. Plenum Publishing Corp, New York

Bement SL, Wise KD, Anderson DJ, Najafi K, Drake KL (1986) Solid-state electrodes for multichannel multiplexed intracortical neuronal recording. IEEE Trans Biomed Eng 33 (2):230–241

Berger TW, Chapin JK, Gerhardt GA, McFarland DJ, Principe JC, Soussou WV (2007) WTEC panel report on international assessment of research and development in brain–computer interfaces. WTEC, Baltimore

Berger TW, Chapin JK, Gerhardt GA, McFarland DJ, Principe JC, Soussou WV, Taylor DM, Tresco PA (2008) The biotic-abiotic interface brain-computer interfaces: an international assessment of research and development trends, 1st edn. Springer, Berlin, pp 31–46

Berger TW, Hampson RE, Song D, Goonawardena A, Marmarelis VZ, Deadwyler SA (2011) A cortical neural prosthesis for restoring and enhancing memory. J Neural Eng 8(4):046017

Berger TW, Song D, Chan RHM, Shin D, Marmarelis VZ, Hampson RE, Sweatt AJ (2012) Role of the hippocampus in memory formation: restorative encoding memory integration neural device as a cognitive neural prosthesis. Pulse IEEE 3(5):17–22

Bernsterin JG, Boyden ES (2011) Optogenetic tools for analyzing the neural circuits of behavior. Trends Cogn Sci 15(12):592–600

Borland LM, Shi GY, Yang H et al (2005) Voltammetric study of extracellular dopamine near microdialysis probes acutely implanted in the striatum of the anesthetized rat. J Neurosci Methods 146:149–158

Bungay PM, Newton-Vinson P, Isele W et al (2003) Microdialysis of dopamine interpreted with quantitative model incorporating probe implantation trauma. J Neurochem 86:932–946

Burmeister JJ, Gerhardt GA (2001) Self-referencing ceramic-based multisite microelectrodes for the detection and elimination of interferences from the measurement of L-glutamate and other analytes. Anal Chem 73(5):1037–1042

Burmeister JJ, Gerhardt GA (2003) Ceramic-based multisite microelectrode arrays for in vivo electrochemical recordings of glutamate and other neurochemicals. TrAC-Trends Anal Chem 22:498–502

Burmeister JJ, Moxon K, Gerhardt G (2000) Ceramic-based multisite microelectrodes for electrochemical recordings. Anal Chem 72:187–192

Burmeister J, Pomerleau F, Palmer M et al (2002) Improved ceramic-based multisite microelectrode for rapid measurements of L-glutamate in the CNS. J Neurosci Methods 119:163–171
Burmeister JJ, Pomerleau F, Huettl P et al (2008) Ceramic-based multisite microelectrode arrays for simultaneous measures of choline and acetylcholine in CNS. Biosens Bioelectron 23:1382–1389
Buxhoeveden DP, Casanova MF (2002) The minicolumn hypothesis in neuroscience. Brain 125 (5):935–951
Campbell PK, Jones KE, Huber RJ, Horch KW, Normann RA (1991) A silicon-based, three-dimensional neural interface: manufacturing processes for an intracortical electrode array. IEEE Trans Biomed Eng 38(8):758–768
Casanova MF, Buxhoeveden DP, Brown C (2002a) Clinical and macroscopic correlates of minicolumnar pathology in autism. J Child Neurol 17(9):692–695
Casanova MF, Buxhoeveden DP, Switala AE, Roy E (2002b) Minicolumnar pathology in autism. Neurology 58:428–432
Casanova MF, Buxhoeveden D, Gomez J (2003) Disruption in the inhibitory architecture of the cell minicolumn: implications for autism. Neuroscientist 9(6):496–507
Casanova MF, El-Baz A, Switala AE (2011) Laws of conservation as related to brain growth, aging, and evolution: symmetry of the minicolumn. Front Neuroanat 5:66
Chapin JK, Loeb GE, Woodward DJ (1980) A simple technique for determination of footfall patterns of animals during treadmill locomotion. J Neurosci Methods 2(1):97–102
Cheer JF, Heien ML, Garris PA et al (2005) Simultaneous dopamine and single-unit recordings reveal accumbens GABAergic responses: implications for intracranial self-stimulation. Proc Natl Acad Sci U S A 102:19150–19155
Chen X, Matsumoto N, Hu Y et al (2002) Electrochemically mediated electrodeposition/electropolymerization to yield a glucose microbiosensor with improved characteristics. Anal Chem 74:368–372
Choi HB, Gordon GRJ, Zhou N, Tai C, Rungta RL, Martinez J, Milner TA, Ryu JK, McLarnon JG, Tresguerres M, Levin Lonny R, Buck J, MacVicar BA (2012) Metabolic communication between astrocytes and neurons via bicarbonate-responsive soluble adenylyl cyclase. Neuron 75:1094–1104
Clapp-Lilly KL, Roberts RC, Duffy LK et al (1999) An Ultrastructural analysis of tissue surrounding a microdialysis probe. J Neurosci Methods 90:129–142
Constantinople CM, Bruno RM (2013) Deep cortical layers are activated directly from thalamus. Science 340(6140):1591–1594
Dale N, Pearson T, Frenguelli BG (2000) Direct measurement of adenosine release during hypoxia in the CA1 region of the rat hippocampal slice. J Physiol 526(Pt 1):143–155
Danbolt NC (2001) Glutamate uptake. Prog Neurobiol 65:1–105
Dash MB, Douglas CL, Vyazovskiy VV, Cirelli C, Tononi G (2009) Long-term homeostasis of extracellular glutamate in the rat cerebral cortex across sleep and waking states. J Neurosci 29:620–629
Day BK, Pomerleau F, Burmeister JJ, Huettl P, Gerhardt GA (2006) Microelectrode array studies of basal and potassium-evoked release of L-glutamate in the anesthetized rat brain. J Neurochem 96:1626–1635
de Kock CP, Bruno RM, Spors H, Sakmann B (2007) Layer- and cell-type-specific suprathreshold stimulus representation in rat primary somatosensory cortex. J Physiol 581(Pt 1):139–154
Deadwyler SA, Hampson RE, Song D, Chan RHM, Opris I, Gerhardt GA, Marmaarelis V, Berger TW (2013) Donor/recipient enhancement of memory in rat hippocampus. Research Topic: "Augmentation of brain function: facts, fiction and controversy", Lebedev MA, Opris I, Casanova MF (eds). Front Syst Neurosci. **7**:120. doi:10.3389/fnsys.2013.00120
Delgado JM (1961) Cerebral and behavioral effects on the monkey of CAPP. Arch Int Pharmacodyn Ther 1961(133):163–172

Derdikman D, Yu C, Haidarliu S, Bagdasarian K, Arieli A, Ahissar E (2006) Layer-specific touch-dependent facilitation and depression in the somatosensory cortex during active whisking. J Neurosci 26(37):9538–9547

Diester I, Kaufman MT, Mogri M, Pashaie R, Goo W, Yizhar O, Ramakrishnan C, Deisseroth K, Shenoy KV (2011) An optogenetic toolbox designed for primates. Nat Neurosci 14(3):387–397

Dixon BM, Lowry JP, O'Neill RD (2002) Characterization in vitro and in vivo of the oxygen dependence of an enzyme/polymer biosensor for monitoring brain glucose. J Neurosci Methods 119:135–142

Drew KL, Pehek EA, Rasley BT et al (2004) Sampling glutamate and GABA with microdialysis: suggestions on how to get the dialysis membrane closer to the synapse. J Neurosci Methods 140:127–131

Du J, Blanche TJ, Harrison RR, Lester HA, Masmanidis SC (2011) Multiplexed high density electrophysiology with nanofabricated neural probes. PLoS One 6:e26204

Fischer EK (1957) Dehydration therapy of cerebral decompensation phenomena in cerebral injuries. Med Klin (Munich) 52(36):1592–1593

Friedemann MN, Robinson SW, Gerhardt GA (1996) O-phenylenediamine-modified carbon-fiber electrodes for the detection of nitric oxide. Anal Chem 68:2621–2628

Gerhardt GA, Burmeister JJ (2000) Voltammetry *in vivo* for chemical analysis of the nervous system. In: Meyers RA (ed) Encyclopedia of analytical chemistry. Wiley, Chichester

Gerhardt GA, Burmeister JJ (2006) Neurochemical arrays. In: Grimes CA (ed) Encyclopedia of sensors. American Scientific Publishers, California

Gerhardt GA, Oke AF, Nagy G et al (1984) Nafion-coated electrodes with high selectivity for CNS electrochemistry. Brain Res 290:390–395

González-Burgos G, Barrionuevo G, Lewis DA (2000) Horizontal synaptic connections in monkey prefrontal cortex: an in vitro electrophysiological study. Cereb Cortex 10(1):82–92

Gradinaru V, Thompson KR, Zhang F, Mogri M, Kay K, Schneider MB, Deisseroth K (2007) Targeting and readout strategies for fast optical neural control in vitro and in vivo. J Neurosci 27(52):14231–14238

Hampson RE, Coates TD Jr, Gerhardt G,A, Deadwyler SA (2004) Ceramic-based micro-electrode neuronal recordings in the rat and monkey. Proc Annu Int Conf IEEE Eng Med Biol Soc (EMBS) 25:3700–3703

Hampson RE, Song D, Chan RH, Sweatt AJ, Riley MR, Gerhardt GA, Shin DC, Marmarelis VZ (2012a) A nonlinear model for hippocampal cognitive prosthesis: memory facilitation by hippocampal ensemble stimulation. IEEE Trans Neural Syst Rehabil Eng 20(23):184–197

Hampson RE, Gerhardt GA, Marmarelis V, Song D, Opris I, Santos L, Berger TW, Deadwyler SA (2012b) Facilitation and restoration of cognitive function in primate prefrontal cortex by a neuroprosthesis that utilizes minicolumn-specific neural firing. J Neural Eng 9(5):056012

Hampson RE, Song D, Opris I, Santos LM, Shin DC, Gerhardt GA, Marmarelis VZ (2013a) Facilitation of memory encoding in primate hippocampus by a neuroprosthesis that promotes task-specific neural firing. J Neural Eng 10(6):066013

Hampson RE, Fuqua JL, Huettl P, Opris I, Song D, Shin D, Marmaarelis V, Berger TW, Gerhardt GA, Deadwyler SA (2013b) Conformal ceramic electrodes that record glutamate release and corresponding neural activity in primate prefrontal cortex. In: Engineering in medicine and biology society (EMBC), 2013 35th annual international conference of the IEEE, Osaka, Japan, pp 5954–5957

Han X (2012) In vivo application of optogenetics for neural circuit analysis. ACS Chem Neurosci 3(8):577–584

Hansen BJ, Chelaru MI, Dragoi V (2012) Correlated variability in laminar cortical circuits. Neuron 76(3):590–602

Hascup KN, Rutherford EC, Quintero JE et al (2006) Second-by-second measures of L-glutamate and other neurotransmitters using enzyme-based microelectrode arrays. In: Michael AC, Borland LM (eds) Electrochemical methods for neuroscience. CRC Press, Florida

Hascup KN, Hascup ER, Pomerleau F et al (2008) Second-by-second measures of L-glutamate in the prefrontal cortex and striatum of freely moving mice. J Pharmacol Exp Ther 324:725–731
Hascup ER, af Bjerken S, Hascup KN et al (2009) Histological studies of the effects of chronic implantation of ceramic-based microelectrode arrays and microdialysis probes in rat prefrontal cortex. Brain Res 1291:12–20
Hascup ER, Hascup KN, Stephens M et al (2010) Rapid microelectrode measurements and the origin and regulation of extracellular glutamate in rat prefrontal cortex. J Neurochem 115:1608–1620
Hebert MA, van Horne CG, Hoffer BJ et al (1996) Functional effects of GDNF in normal rat striatum: presynaptic studies using in vivo electrochemistry and microdialysis. J Pharmacol Exp Ther 279:1181–1190
Heien ML, Khan AS, Ariansen JL, Cheer JF, Phillips PE, Wassum KM, Wightman RM (2005) Real-time measurement of dopamine fluctuations after cocaine in the brain of behaving rats. Proc Natl Acad Sci U S A 102(29):10023–10028
Hess WR (1932) Beiträge zur Physiologie des Hirnstammes. I. Die Methodik der lokalisierten Reizung und Ausschaltung subkortikaler Hirnabschnitte. Thieme, Leipzig
Hoffman AF, Lupica CR, Gerhardt GA (1998) Dopamine transporter activity in the substantia nigra and striatum assessed by high-speed chronoamperometric recordings in brain slices. J Pharmacol Exp Ther 287:487–496
Howe WM, Berry AS, Francois J, Gilmour G, Carp JM, Tricklebank M, Lustig C, Sarter M (2013) Prefrontal cholinergic mechanisms instigating shifts from monitoring for cues to cue-guided performance: converging electrochemical and fMRI evidence from rats and humans. J Neurosci 33:8742–8752
Hu Y, Mitchell KM, Albahadily FN et al (1994) Direct measurement of glutamate release in the brain using a dual enzyme-based electrochemical sensor. Brain Res 659:117–125
Jung MW, Qin Y, McNaughton BL, Barnes CA (1998) Firing characteristics of deep layer neurons in prefrontal cortex in rats performing spatial working memory tasks. Cereb Cortex 8 (5):437–450
Kim SP, Sanchez JC, Rao YN, Erdogmus D, Carmena JM, Lebedev MA, Nicolelis MA, Principe JC (2006) A comparison of optimal MIMO linear and nonlinear models for brain-machine interfaces. J Neural Eng 3(2):145–161
Kinney GA, Overstreet LS, Slater NT (1997) Prolonged physiological entrapment of glutamate in the synaptic cleft of cerebellar unipolar brush cells. J Neurophysiol 78:1320–1333
Kissinger PT, Hart JB, Adams RN (1973) Voltammetry in brain tissue–a new neurophysiological measurement. Brain Res 55(1):209–213
Konradsson-Geuken A, Wu HQ, Gash CR, Alexander KS, Campbell A, Sozeri Y, Pellicciari R, Schwarcz R, Bruno JP (2010) Cortical kynurenic acid bi-directionally modulates prefrontal glutamate levels as assessed by microdialysis and rapid electrochemistry. Neuroscience 169:1848–1859
Kritzer MF, Goldman-Rakic PS (1995) Intrinsic circuit organization of the major layers and sublayers of the dorsolateral prefrontal cortex in the rhesus monkey. J Comp Neurol 359:131–143
Krupa DJ, Wiest MC, Shuler MG, Laubach M, Nicolelis MA (2004) Layer-specific somatosensory cortical activation during active tactile discrimination. Science 304(5679):1989–1992
Kulagina NV, Shankar L, Michael AC (1999) Monitoring glutamate and ascorbate in the extracellular space of grain tissue with electrochemical microsensors. Anal Chem 71:5093–5100
Lowry JP, Miele M, O'Neill RD et al (1998) An amperometric glucose-oxidase/poly (o-phenylenediamine) biosensor for monitoring brain extracellular glucose: in vivo characterisation in the striatum of freely-moving rats. J Neurosci Methods 79:65–74
Mahan MY, Georgopoulos AP (2013) Motor directional tuning across brain areas: directional resonance and the role of inhibition for directional accuracy. Front Neural Circuits 7:92
Marsden CA, Joseph MH, Kruk ZL et al (1988) In vivo voltammetry–present electrodes and methods. Neuroscience 25:389–400

Marshel JH, Deisseroth K (2013) Genetically encoded voltage sensor goes live. Nat Biotechnol 31 (11):994–995
Martin KF, Marsden CA (1987) In vivo electrochemistry – principles and applications. Life Sci 41:865–868
Matsumoto N, Chen X, Wilson GS (2002) Fundamental studies of glucose oxidase deposition on a Pt Electrode. Anal Chem 74:362–367
McCreery RL, Dreiling R, Adams RN (1974) Voltammetry in brain tissue: the fate of injected 6-hydroxydopamine. Brain Res 73(1):15–21
Michael DJ, Wightman RM (1999) Electrochemical monitoring of biogenic amine neurotransmission in real time. J Pharm Biomed Anal 19:33–46
Mo J, Schroeder CE, Ding M (2011) Attentional modulation of alpha oscillations in macaque inferotemporal cortex. J Neurosci 31(3):878–882
Mountcastle VB (1957) Modality and topographic properties of single neurons of cat's somatic sensory cortex. J Neurophysiol 20:408–434
Mountcastle VB (1997) The columnar organization of the neocortex. Brain 120:701–722
Moxon KA, Leiser SC, Gerhardt GA, Barbee KA, Chapin JK (2004) Ceramic-based multisite electrode arrays for chronic single-neuron recording. IEEE Trans Biomed Eng 51(4):647–656
Nickell J, Pomerleau F, Allen J, Gerhardt GA (2005) Age-related changes in the dynamics of potassium-evoked L-glutamate release in the striatum of Fischer 344 rats. J Neural Transm 112:87–96
Oldenziel WH, Dijkstra G, Cremers TI, Westerink BH (2006a) In vivo monitoring of extracellular glutamate in the brain with a microsensor. Brain Res 1118:34–42
Oldenziel WH, Dijkstra G, Cremers TI, Westerink BH (2006b) Evaluation of hydrogel-coated glutamate microsensors. Anal Chem 78:3366–3378
Onifer SM, Quintero JE, Gerhardt GA (2012) Cutaneous and electrically evoked glutamate signaling in the adult rat somatosensory system. J Neurosci Methods 208:146–154
Opris I (2013) Inter-laminar microcircuits across the neocortex: repair and augmentation. Research topic: "Augmentation of brain function: facts, fiction and controversy", Lebedev MA, Opris I and Casanova MF (eds), Front Syst Neurosci 7:80. doi:10.3389/fnsys.2013.00080
Opris I, Casanova MF (2014) Prefrontal cortical minicolumn: from executive control to disrupted cognitive processing. Brain 137(Pt 7):1863–1875. doi:10.1093/brain/awt359
Opris I, Barborica A, Ferrera VP (2005) Microstimulation of dorsolateral prefrontal cortex biases saccade target selection. J Cogn Neurosci 17(6):893–904
Opris I, Hampson RE, Stanford TR, Gerhardt GA, Deadwyler SA (2011) Neural activity in frontal cortical cell layers: evidence for columnar sensorimotor processing. J Cogn Neurosci 23 (6):1507–1521
Opris I, Hampson RE, Gerhardt GA, Berger TW, Deadwyler SA (2012a) Columnar processing in primate pFC: evidence for executive control microcircuits. J Cogn Neurosci 24(12):2334–2347
Opris I, Fuqua JL, Huettl PF, Gerhardt GA, Berger TW, Hampson RE, Deadwyler SA (2012b) Closing the loop in primate prefrontal cortex: inter-laminar processing. Front Neural Circuits 6:88
Opris I, Santos LM, Song D, Gerhardt GA, Berger TW, Hampson RE, Deadwyler SA (2013) Prefrontal cortical microcircuits bind perception to executive control. Sci Rep 3:2285
Parikh V, Sarter M (2006) Cortical choline transporter function measured in vivo using choline-sensitive microelectrodes: clearance of endogenous and exogenous choline and effects of removal of cholinergic terminals. J Neurochem 97:488–503
Parikh V, Sarter M (2008) Cholinergic mediation of attention: contributions of phasic and tonic increases in prefrontal cholinergic activity. Ann N Y Acad Sci 1129:225–235
Parikh V, Pomerleau F, Huettl P et al (2004) Rapid assessment of in vivo cholinergic transmission by amperometric detection of changes in extracellular choline levels. Eur J Neurosci 20:1545–1554
Parikh V, Kozak R, Martinez V et al (2007) Prefrontal acetylcholine release controls cue detection on multiple timescales. Neuron 56:141–154

Philips PE, Robinson DL, Stuber GD, Carelli RM, Wightman RM (2003) Real-time measurements of phasic changes in extracellular dopamine concentration in freely moving rats by fast-scan cyclic voltammetry. Methods Mol Med 79:443–464

Phillips PE, Wightman RM (2004) Extrasynaptic dopamine and phasic neuronal activity. Nat Neurosci 7(3):199

Photolithography (2001) Texas engineering extension service (ed), Texas

Pomerleau F, Day BK, Huettl P et al (2003) Real time in vivo measures of L-glutamate in the rat central nervous system using ceramic-based multisite microelectrode arrays. Ann N Y Acad Sci 1003:454–457

Quintero JE, Day BK, Zhang Z, Grondin R, Stephens ML, Huettl P, Pomerleau F, Gash DM, Gerhardt GA (2007) Amperometric measures of age-related changes in glutamate regulation in the cortex of rhesus monkeys. Exp Neurol 208:238–246

Ratclif R, Cherian A, Segraves M (2003) A comparison of macaque behavior and superior colliculus neuronal activity to predictions from models of two-choice decisions. J Neurophysiol 90(3):1392–1407

Robinson TW, Johnson EO (1967) Further studies on reference electrodes. Electroencephalogr Clin Neurophysiol 23(3):292

Royer S, Zemelman BV, Barbic M, Losonczy A, Buzsáki G, Magee JC (2010) Multi-array silicon probes with integrated optical fibers: light-assisted perturbation and recording of local neural circuits in the behaving animal. Eur J Neurosci 31(12):2279–2291

Rutherford EC, Pomerleau F, Huettl P et al (2007) Chronic second-by-second measures of L-glutamate in the central nervous system of freely moving rats. J Neurochem 102:712–722

Sakurai Y, Takahashi S (2006) Dynamic synchrony of firing in the monkey prefrontal cortex during working memory tasks. J Neurosci 26:10141–10153

Salcman M, Bak MJ (1973) Design, fabrication, and in vivo behavior of chronic recording intracortical microelectrodes. IEEE Trans Biomed Eng 20(4):253–260

Shepherd G, Grillner S (2010) Handbook of brain microcircuits. Oxford University Press, Oxford

Song D, Chan RHM, Marmarelis VZ, Hampson RE, Deadwyler SA, Berger TW (2009) Nonlinear modeling of neural population dynamics for hippocampal prostheses. Neural Netw 22 (9):1340–1351

Stamford JA, Palij P, Davidson C et al (1993) Simultaneous "real-time" electrochemical and electrophysiological recording in brain slices with a single carbon-fibre microelectrode. J Neurosci Methods 50:279–290

Stephens ML, Pomerleau F, Huettl P et al (2010) Real-time glutamate measurements in the putamen of awake rhesus monkeys using an enzyme-based human microelectrode array prototype. J Neurosci Methods 185:264–272

Stephens ML, Quintero JE, Pomerleau F, Huettl P, Gerhardt GA (2011) Age-related changes in glutamate release in the CA3 and dentate gyrus of the rat hippocampus. Neurobiol Aging 32:811–820

Suaud-Chagny MF, Cespuglio R, Rivot JP, Buda M, Gonon F (1993) High sensitivity measurement of brain catechols and indoles in vivo using electrochemically treated carbon-fiber electrodes. J Neurosci Methods 48:241–250

Suyatin DB, Wallman L, Thelin J, Prinz CN, Jorntell H et al (2013) Nanowire-based electrode for acute in vivo neural recordings in the brain. PLoS One 8:e56673

Takeuchi D, Hirabayashi T, Tamura K, Miyashita Y (2011) Reversal of interlaminar signal between sensory and memory processing in monkey temporal cortex. Science 331:1443–1447

Talauliker PM, Price DA, Burmeister JJ, Nagari S, Quintero JE, Pomerleau F, Huettl P, Hastings JT, Gerhardt GA (2011) Ceramic-based microelectrode arrays: recording surface characteristics and topographical analysis. J Neurosci Methods 198:222–229

Tian FM, Gourine AV, Huckstepp RTR, Dale N (2009) A microelectrode biosensor for real time monitoring of L-glutamate release. Anal Chim Acta 645:86–91

Timmerman W, Westerink BH (1997) Brain microdialysis of GABA and glutamate: what does it signify? Synapse 27:242–261

Tye KM, Deisseroth K (2012) Optogenetic investigation of neural circuits underlying brain disease in animal models. Nat Rev Neurosci 13:251–266

van Horne C, Hoffer BJ, Stromberg I et al (1992) Clearance and diffusion of locally applied dopamine in normal and 6-hydroxydopamine-lesioned rat striatum. J Pharmacol Exp Ther 263:1285–1292

Viventi J, Kim D-H, Vigeland L, Frechette ES, Blanco JA, Kim Y-S, Avrin AE, Tiruvadi VR, Hwang SW, Vanleer AC et al (2011) Flexible, foldable, actively multiplexed, high-density electrode array for mapping brain activity in vivo. Nat Neurosci 2011(14):1599–1605

Wassum KM, Tolosa VM, Wang J, Walker E, Monbouquette HG, Maidment NT (2008) Silicon wafer-based platinum microelectrode array biosensor for near real-time measurement of glutamate in vivo. Sensors (Basel) 8(8):5023–5036

Westerink RH (2004) Exocytosis: using amperometry to study presynaptic mechanisms of neurotoxicity. Neurotoxicology 25:461–470

Wightman RM, Strope E, Plotsky PM, Adams RN (1976) Monitoring of transmitter metabolites by voltammetry in cerebrospinal fluid following neural pathway stimulation. Nature 262 (5564):145–146

Williams GV, Millar J (1990) Concentration-dependent actions of stimulated dopamine release on neuronal activity in Rat striatum. Neuroscience 39:1–16

Wilson GS, Gifford R (2005) Biosensors for real-time in vivo measurements. Biosens Bioelectron 20:2388–2403

Wire Bond (1998) Texas engineering extension service (ed), Texas

Wise RA (1976) Moveable electrode for chronic brain stimulation in the rat. Physiol Behav 16(1):105–106

Wise KD, Angell JB (1975) A low-capacitance multielectrode probe for use in extracellular neurophysiology. IEEE Trans Biomed Eng 22(3):212–219

Wise KD, Angell JB, Starr A (1970) An integrated-circuit approach to extracellular microelectrodes. IEEE Trans Biomed Eng 17(3):238–247

Yang H, Peters JL, Michael AC (1998) Coupled effects of mass transfer and uptake kinetics on in vivo microdialysis of dopamine. J Neurochem 71:684–692

Zesiewicz T, Shaw J, Allison K, Staffetti J, Okun M, Sullivan K (2013) Update on treatment of essential tremor. Curr Treat Options Neurol 15:410–423

Zhang M, Alloway KD (2006) Intercolumnar synchronization of neuronal activity in rat barrel cortex during patterned airjet stimulation: a laminar analysis. Exp Brain Res 169:311–325

Zhang H, Lin SC, Nicolelis MA (2009) Acquiring local field potential information from amperometric neurochemical recordings. J Neurosci Methods 179:191–200

Zhang H, Lin SC, Nicolelis MA (2010) Spatiotemporal coupling between hippocampal acetycholine release and theta oscillations in vivo. J Neurosci 30:13431–13440

Zhou N, Rungta RL, Malik A, Han H, Wu DC, Macvicar BA (2013) Regenerative glutamate release by presynaptic NMDA receptors contributes to spreading depression. J Cereb Blood Flow Metab 33:1582–1594

Zorzos AN, Scholvin J, Boyden ES, Fonstad CG (2012) Three-dimensional multiwaveguide probe array for light delivery to distributed brain circuits. Opt Lett 37:4841–4843

Chapter 18
Uncovering Cortical Modularity by Nanotechnology

Marius Enachescu, Ruxandra Vidu, and Ioan Opris

Abstract Cortical modularity and nanotechnology might look like a strange pair of concepts taken together, but nevertheless they seem very much suited for each other. Indeed, cortical modularity is a fundamental microanatomic feature of the brain while nanotechnology with its nanometric precision provides nanoscale structures, namely nanowires and carbon nanotubes capable of interacting with the brain at the genetic, molecular, and microcircuit level. Research in neuroscience is essentially a combination of many interdisciplinary sciences where nanoscience and nanotechnology plays a pivotal role. In this chapter we examine carbon nanotubes (CNTs) and nanowires (NWs), and their potential to uncover the function of cortical microcircuits, as well as novel applications for diagnosis and treatment of brain diseases. For example, the simultaneous recording from cortical minicolumns with multi-electrode arrays (MEAs) consisting of CNTs or NWs is emerging for developing cognitive prostheses for a broad range of neurological and psychiatric dysfunctions.

Keywords Cortical microcircuits • Microelectrode arrays • Cortical minicolumn • Cortical layer • Nanowires • Carbon nanotubes • Nanotechnology • Neural prosthetics • Brain machine interface

M. Enachescu (✉)
Center for Surface Science and Nanotechnology, University "Politehnica" Bucharest, 313 Splaiul Independenţei, Bucharest 060042, Romania
e-mail: marius.enachescu@upb.ro

R. Vidu
Chemical Engineering and Materials Science, University of California Davis, Davis, CA 95621, USA
e-mail: rvidu@ucdavis.edu

I. Opris, Ph.D.
Department of Physiology and Pharmacology, Wake Forest University Health Sciences, Winston-Salem, NC 27103, USA
e-mail: ioanopris.phd@gmail.com

M.F. Casanova, I. Opris (eds.), *Recent Advances on the Modular Organization of the Cortex*, DOI 10.1007/978-94-017-9900-3_18

18.1 Introduction

Cortical modularity is a basic microanatomic feature of the brain relevant for the processing of sensory, motor, and cognitive information. A large body of literature provides supportive evidence for a modular micro-architecture of the neocortex, in which neurons form a minicolumnar-laminar arrangement similar to a "crystalline" structure (Mountcastle 1997; Casanova et al. 2011; Opris et al. 2013). The inter-laminar cortical microcircuits, with interconnected neuronal components, become analogous to integrated microprocessors that are used for neural processing (Opris et al. 2012a, b; Alivisatos et al. 2013; Opris 2013).

18.1.1 Neocortical Modules

Neocortical tissue is arrayed into three-dimensional mini-columnar and laminar arrangements of neurons that are linked by afferent and efferent connections distributed across many regions of the brain, with a broad functional significance. The laminar structure has six layers of cells grouped into supra-granular, granular and infra-granular layers, with the granular layer receiving sensory input from thalamus (Constantinople and Bruno 2013). Cortical minicolumns are chains of pyramidal neurons surrounded by inhibitory interneurons that regulate excitatory and inhibitory activity (Szentagothai and Arbib 1975).

18.1.2 Nanotechnology

The neocortical module mentioned above requires a set of tools with atomic/molecular precision that comes from nanotechnology. Neuro-nanotechnology is emerging at the interface between science and engineering as a tool for analyzing the unique features (properties, structure and function) of brain circuits as well as manipulating and repairing damaged neural circuits. Because the brain operates at the nanoscale level, we need to access the brain with tools and techniques that work at this level of detection.

The "nanoworld" in science was established in the 1980s when scientists were able to "see" the atom (i.e., the tiny "brick" of matter) in 3D real space for the first time. This was possible mainly due to the invention of the scanning tunneling microscope (Binnig et al. 1982, 1983), followed by additional techniques, such as atomic force microscopy (AFM) (Binnig et al. 1986). The "nanoworld" concept in science consists of several "nano"-fields such as *nanomaterials* (materials at nano-scale), *nanoarrays* (arrays of nano-objects), *nanotools* (tools needed to characterize the nanomaterials), *nanodevices* (new devices, many of them using quantum effects, built in by the synthesized nanomaterials), *et cetera*. More importantly, the ability to manipulate atoms and molecules (biomolecules) to induce unique properties and stability and to communicate signals has opened up incredible application opportunities. The design, characterization and synthesis of new materials with functional organization at nanoscale enable us to engineer and control functional bio-integrated systems. In addition, AFM offers the possibility of *in situ* imaging of growing surfaces under potential control (Ikemiya et al. 1996; Vidu and Hara 1999a, b; Vidu et al. 2001, 2006, 2007, 2012; Ku et al. 2004) or interacting soft surfaces such as lipids with proteins (Vidu et al. 2002; Zhang et al. 2002; Carmichael et al. 2004).

Neural probes like multi-electrode arrays (MEAs) and micro-devices are used for recording the activity of large neuronal assemblies (Wise 2007; Chang-Hsiao et al. 2010; Amaral et al. 2013) and to stimulate nerve cells for various clinical applications (Wise 2007). MEAs are produced for various applications in electrophysiology, chemical sensing, neurostimulation and optogenetics (Du et al. 2011; Tye and Deisseroth 2012; Alivisatos et al. 2013; Opris et al. 2013). Because MEAs are associated with minimally invasive methods, many research efforts are concentrated towards increasing the number and density of the extracellular electrodes while decreasing the size of the device. In this regard, nanoscale structures, namely the carbon nanotubes (CNTs), are capable of interacting with the brain at the genetic, molecular, and microcircuit level. The trend toward miniaturization and the high output of integrated circuits has stimulated the development of various nanostructured materials. In this chapter we examine CNTs and nanowires (NWs), and their potential to uncover the function of these microcircuits, together with the novel applications for diagnosis and treatment of brain diseases. An example is the nanofabrication of extracellular electrode array with high density electrical leads such as the low noise multi-channel silicon system (Du et al. 2011) presented in Sect. 18.4.

18.2 Carbon Nanotubes and Nanowires

Nanotubes and nanowires are nanomaterials that basically represent quasi-one-dimensional (1D) conductors and semiconductors. In recent years, a broad platform for electronic interfaces with cells and tissue using CNTs and NWs devices has been implemented. Compared to standard techniques that are used to measure, record and observe extracellular signals from individual tissues and cells, the CNTs and NWs devices have several orders of magnitude smaller recording area. The millivolt range signals of CNTs/NWs platforms device are significantly larger than those measured using planar devices or multiple electrode arrays, likely due to enhanced coupling between cell membrane and nanoscale device. CNTs and NWs represent the natural building blocks for a nanoscale interface and the base for fundamental studies of the biological-nanomaterial interface, studies that are creating hybrid nanoelectronic–neuronal devices, opening unique directions in neuronal research and applications. The possibility of tuning at the time of synthesis their material composition and corresponding properties, offers to CNTs and NWs a remarkable property, opening up the design of ultra-high sensitivity devices at nanoscale to future opportunities.

18.2.1 Nanotechnology of Carbon Nanotubes and Nanowires

Carbon nanotubes (CNTs) exhibit outstanding mechanical, thermal, and conductive properties. They were discovered by Iijima (1991). CNTs are formed by rolling-up one or more graphene sheets bestowing excellent chemical and thermal stability, extreme electronic properties, large surface area, and high mechanical strength while carrying ultralight weight (Ajayan 1999).

Two forms of CNTs can be efficiently prepared under well-defined conditions of synthesis: (a) single-wall carbon nanotubes (SWCNTs) or (b) nested multiwall carbon nanotubes (MWCNTs) (see Fig. 18.1). Because of its relation to graphene, the CNTs are usually close to perfection at the atomic scale (Enachescu et al. 1999; Bota et al. 2014a, b) making them chemically inert. Although the CNTs have one-sixth the weight of steel, similar to graphene, nanotubes are two orders of magnitude stronger than steel. As it was shown via computer simulations, the estimation of melting point of nanotubes is around 3,700 °C, i.e. higher than for any metal, but close to that of graphite. SWCNTs can act as very good conductors of electrons or can show semiconducting behavior, depending on their diameter and the atomic structure of nanotubes. CNTs are excellent conductors of heat (with a very high thermal conductivity that exceeds the same of isotopically pure diamond), perfectly positioned in devices to dissipate the heat from the PC chips.

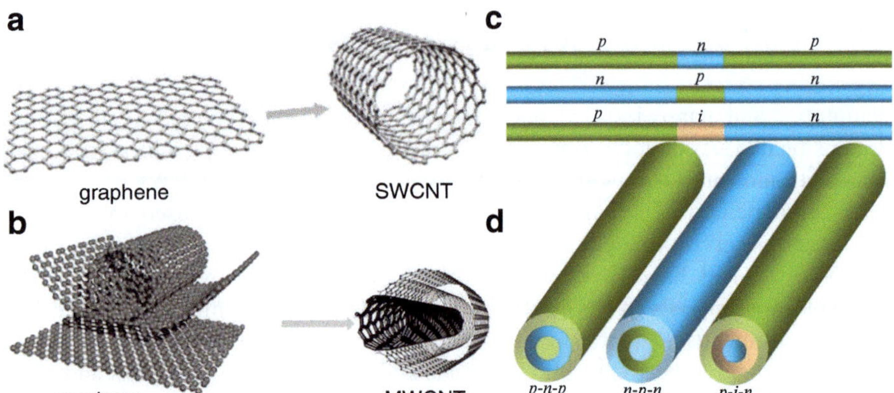

Fig. 18.1 NanoTechnology elements for NanoNeuroscience. (**a**) From graphene to single wall carbon nanotubes. (**b**) From graphene to multiple wall carbon nanotubes. (**c–d**) Axial and radial nanowires based heterojunctions, e.g., *p-n-p, n-p-n, p-i-n*

Different technological fields are witnessing huge promises from CNTs based on their unique properties, such as being used in probes and interconnects as conductive composites, nanometer-sized semiconductor devices, field emission displays and radiation sources, hydrogen storage media, sensors, energy storage and energy conversion devices (Sharma and Ahuja 2008). Functionalization of the CNTs surface was performed using many different approaches. CNTs have been proposed either by themselves or as components for biosensors (Wenrong et al. 2010), ion channel blockers (Park et al. 2003), biocatalysts (Feng and Ji 2011), photo-thermal probes in cancer therapy (Moon et al. 2009), nanovectors (Klumpp et al. 2006) and imaging applications (Kam et al. 2005a; Wu et al. 2005; Klumpp et al. 2006).

Imaging applications of CNTs to living cells and neural tissues bring promising advantages for neurobiological applications based on the optical properties of nanotubes. Due to the high photostability of SWCNTs photoluminescence, a longer excitation time is attainable in CNTs at higher laser fluency compared to quantum dots or organic fluorophores. An attenuated absorption, combined with autofluorescence and scattering characteristics makes opaque tissue visible in the range of 700–1,400 nm allowing imaging of the whole blood or thick tissue (Heller et al. 2006). This is mainly because the fluorescence profiles of many semiconducting nanotubes overlap with the wavelength range. The imaging SWCNTs in tissue sections and the nanotubes concentration measured in blood is based on nanotube fluorescence (Cherukuri et al. 2006). On the other hand, CNTs can be detected due to their large resonance-enhanced Raman scattering (Heller et al. 2005). In addition, SWCNTs show confined heating upon near infrared (NIR) absorption because SWCNTs absorb strongly in NIR wavelengths range (Endo et al. 1990). SWCNTs released DNA due to exposure at NIR radiation permits its translocation into the cell nucleus (Kam et al. 2005b). Cell death demonstrated in the same study was due to the fact that SWCNTs were causing a local heating upon exposure at NIR radiation, triggering the killing of the cells.

The transport of nanotubes into cells is of fundamental importance for the biomedical applications mentioned. As yet, the way in which CNTs enter cells is still under hot debate, generating both controversy and confusion about the mechanism by which they enter cells. Work by Bianco and colleagues (Bianco et al. 2011) suggests that ammonium-functionalized SWCNTs and MWCNTs are formed via a passive, endocytosis-independent mechanism (Pantarotto et al. 2004b), while Dai and colleagues (Dai et al. 2001) conclude that the mechanism of the acid-functionalized SWCNTs entering the cell involves an endocytosis pathway (Kam et al. 2004). Alternate mechanism is proposed for MWCNTs, which cannot use the endocytosis pathway due to their size. The mechanism proposed for MWCNTs takes into consideration the flipping of lipid molecules of the membrane to allow CNTs to enter the cell (Kam et al. 2004; Lopez et al. 2004; Lu et al. 2004; Pantarotto et al. 2004a, b; Cai et al. 2005; Kam and Dai 2005).

The communication between cell and nanotube is ruled by the type of coating on the nanotube surface. By transferring CNTs into cells, proteins are adsorbed to the nanotube surface, coating the nanotube with serum containing proteins, such as albumin and fibronectin. However, it was suggested that CNTs transfer into cells have a natural switching mechanism of lipids in the membrane (Lopez et al. 2004; Pantarotto et al. 2004a) and do not have an endocytotic pathway for the MWCNTs that are 200 nm in length with 10 nm radius. Open questions are what happens to CNTs once they have entered the cell and also when or how CNTs would be exocytosed by the cells (Sakhtianchi et al. 2013). The possibility for nanotubes to be subsequently expelled from the cell seems advantageous for most biological applications; although, as yet, this has not been reported in the literature. There is still much work necessary to understand the CNTs cellular transport to be able to control the CNTs placement inside cells (Endo et al. 1990).

The therapeutic effect of drugs is steadily improving through the development of new delivery vehicles. CNTs have a higher surface area to volume ratio than spheres have, giving nanotubes the potential to be conjugated with more functional agents (Wu et al. 2005) and to accommodate large loads of therapeutic agents (Kam et al. 2005a). Previously, these vehicles included viral vectors, liposomes, cationic lipids, polymers, and nanoparticles. Although nonviral vehicles have versatility of shape, size, and materials, one issue of concern is the poor penetration of some therapeutic agents into cells. CNTs are readily internalized by cells, and after surface modification, they exhibit low cytotoxicity over the period of a few days (Kam et al. 2004, 2005a; Lu et al. 2004; Pantarotto et al. 2004a, b; Cai et al. 2005; Kam and Dai 2005; Wu et al. 2005).

18.2.2 Nanowire Junctions

The range of length scale in biology varies by several orders of magnitudes—from nanometer sized nucleic or amino acids to several centimeters for organs and neuronal circuits. There is a need for interfaces with nanoscale spatial resolution in order to investigate processes at the subcellular level. Besides carbon nanotubes, these interfaces can be achieved through the use of other nanostructures, such as semiconductor nanowires having dimensions as small as a protein molecule. CNTs and NWs represent the building blocks for nanoscale electronics (McEuen et al. 2002; Lieber 2003). The critical feature sizes (atomic scale) of these building blocks can be well controlled during synthesis, in contrast with the nanostructures fabricated by the "top-down" process. Even for isolated CNTs transistors that have shown exceptional properties (Javey et al. 2003), large scale integration challenges remain due to difficulties in preparing pure semiconductor nanotubes. The issues faced by CNTs could be overcome by nanowires because of the reproducible control over size and electronic properties that current growth methods enable (Cui et al. 2001b, 2003; Cui and Lieber 2001; Wu et al. 2004). A good class of NWs has been developed, ranging from NWs based on classic semiconductors, such as silicon NWs (Chen et al. 2006; Goncher et al. 2006; Yajie et al. 2008), GaP (Dujavova-Laurencikova et al. 2013), GaN (Lee et al. 2007), CdS and ZnS (Barrelet et al. 2003), heterostructures as Ge-Si (Xiang et al. 2006a, b), InAs-InP (Jiang et al. 2007), oxide nanowires MgO (Yin et al. 2002), Cu_2O (Jiang et al. 2002), SiO_2

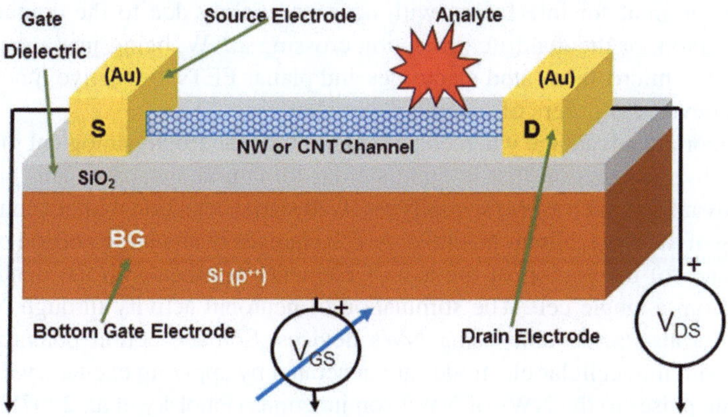

Fig. 18.2 Schematics of a field-effect transistor (FET) using as a gate channel a NW or a CNT. Its main elements are: Source electrode (*S*), Drain electrode (*D*), Bottom/back gate electrode (*BG*), Gate dielectric (SiO_2 or high-κ dielectric). An analyte binding to the FET channel is suggested

(Yu et al. 1998; Liu et al. 2001; Zheng et al. 2002), Ga_2O_3 (Wu et al. 2000; Sharma and Sunkara 2002), Al_2O_3 (Valcarcel et al. 1998; Xiao et al. 2002), In_2O_3 (Li et al. 2003), SnO_2 (Dai et al. 2001), MnO_2 (Wang and Li 2002), Sb_2O_3 (Guo et al. 2000), TiO_2 (Seraji et al. 2000; Miao et al. 2002), ZnO (Tian et al. 2003; Vayssieres 2003), $LiNiO_2$ (Zhou et al. 2002), and others.

The field-effect transistors (FETs) configuration of semiconductor NWs is one of the most appropriate detection schemes for biological events (Cui and Lieber 2001; Cui et al. 2001a, 2003; Zheng et al. 2004; Xiang et al. 2006b). An illustration of a FET schematic is depicted in Fig. 18.2. The binding to the dielectric gate of a polar/charged species appears analogous to applying a voltage to a gate electrode. For example, accumulation of positive carriers (holes) together with an increase/variation in device conductance can be generated by binding a protein with a negative charge to the surface of a *p*-type FET. Silicon based NWs (or other types of semiconductors) also may function as FET devices (Cui and Lieber 2001; Cui et al. 2003; Lieber 2003; Zheng et al. 2004; Li et al. 2006; Xiang et al. 2006b). One-dimensional morphology of NWs is the main feature that determines overcoming sensitivity limitations for planar FET devices. Thus, a more substantial change in device conductance for the NWs versus a planar FET will take place if any analyte binding event will happen (this event leads to accumulation or depletion of carriers).

18.2.3 Nanowires-Based Neural Devices

One of the most powerful and versatile platforms for building functional interfaces to biological systems (including neurons) is based on NWs devices. NWs are highly local probes for neuronal projections because individual NWs devices are becoming optimal for interfacing with neurons, mainly due to the contact length along the axon, or the dendrite projection crossing a NW, being just about 20 nm. Compared to micro-fabricated electrodes and planar FETs, the active junction area for NWs devices is orders of magnitude smaller.

An important advantage when compared to other electrophysiological methods is the small hybrid junction that is quite similar to natural synapses. This small size creates advantages such as: (a) spatially resolved signal detection without complicated averaging of the extracellular potential, which changes over a large portion of a given neuron, and (b) integration of the axon's elements together with the dendrite projections from a single cell. The stimulation of neuronal activity through NW/axon junctions is also achievable using NWs devices. Somatic action potential spikes detected with intracellular electrodes are generated by applying excitatory sequences of biphasic pulses to the NWs of NW/axon junctions (Patolsky et al. 2007).

Additionally, a NW-based FET device array can be designed to promote neuron growth (Patolsky et al. 2007). Thus, interfacing ensembles of NWs inputs and outputs to different neuron ensembles enables the implementation of stimulation, inhibition, or reversibly blocking signal propagation through specific pathways allowing the signal flow to be simultaneously mapped throughout the network. Besides single or multiple arrays of NWs-based FET devices used for investigating neuronal activity, the NWs were also used to design and build NWs-based electrodes for neural recordings in the brain.

The potential to revolutionize neuroscience research and clinical therapy (Benabid 2007; Kipke et al. 2008; Vaadia and Birbaumer 2009; Suyatin et al. 2013) is represented by implantable neural interfaces (Rutten 2002; Fromherz 2003; Cogan 2008). However, the recorded neurons and tissue reactions that encapsulate and insulate the implant are still presenting instability results (Schouenborg 2011). The nanostructured electrodes are considered as a promising alternative to conventional neuronal interfaces because the recording properties depend mainly on electrode surface properties and tissue reactions to the surface (Kotov et al. 2009; Timko et al. 2010; Dvir et al. 2011; Suyatin et al. 2013). Nanostructured electrodes provide additional advantages such as improved electrical properties (Keefer et al. 2008; Cellot et al. 2009; Martin et al. 2010; Ansaldo et al. 2011; Duan et al. 2012), a shorter cell-to-electrode distance (Tian et al. 2010; Duan et al. 2012; Xie et al. 2012), and better spatial resolution. They also have a potential for less tissue damage (Almquist and Melosh 2010; Martin et al. 2010; Tian et al. 2010; Duan et al. 2012), better biocompatibility (Hallstrom et al. 2007; Kim et al. 2007; Martin et al. 2010; Berthing et al. 2011) and new functionalities, such as selective guidance of neuronal fibers (Hallstrom et al. 2009). Importantly, recordings of cell signals with different nanowire-based electrodes have been achieved *in vitro* (Kotov et al. 2009; Tian et al. 2010; Timko et al. 2010; Brueggemann et al. 2011; Dvir et al. 2011; Duan et al. 2012; Robinson et al. 2012; Xie et al. 2012), thus demonstrating that epitaxially grown wires of small diameter may provide a minimally invasive tissue penetration (Kawano et al. 2010; Takei et al. 2010; Tian et al. 2010; Duan et al. 2012; Xie et al. 2012). Until now, using mainly carbon nanotubes without structural feature control and in combination with rather big surfaces, has shown improved cell survival for neuronal interfaces (Hallstrom et al. 2007) and improved cell adhesion with focal adhesions forming specifically on the nanowires epitaxially grown, e.g., gallium phosphide (GaP) NWs have beneficial properties (Prinz et al. 2008). Compared to other material NWs, GaP NWs can be synthesized with very little tapering and exceptional control over their position and geometry, and with a high aspect ratio (the ratio between their length and their diameter) is over 50 (Suyatin et al. 2009). Also, the design and fabrication of a first generation of GaP NW-based electrode with a controllable nanomorphology was reported (Suyatin et al. 2013). The first functional testing *in vivo* of a NWs based device was performed during acute recordings in the rat cerebral cortex, where the NWs were used as a backbone for metal nanostructured electrode with a three-dimensional (3D) structure. This electrode design opened the development of a new model system, with the prospect of enabling a more reliable tissue anchoring as well as a more intimate contact between the electrode and the neurons (Xie et al. 2010). Further research on the functionality of nanostructure-based neuronal interfaces *in vivo* is providing better electrode-cell electrical coupling (Hai et al. 2010; Robinson et al. 2012; Xie et al. 2012).

18.3 Carbon Nanotubes in NeuroNanoTechnology

Carbon nanotubes have a multitude of useful properties (electrical, mechanical and chemical) that make them very promising materials for applications in neuroscience. The ease with which carbon nanotubes can be patterned permits use for studying the organization of neural networks while the electrical conductivity of nanotubes can provide a vital mechanism to monitor or stimulate the neurons. Carbon nanotubes can interact with neurons and affect neuronal function, especially at the level of ion channels (Malarkey and Parpura 2007, 2010). Both SWCNTs and MWCNTs have been increasingly used as "scaffolds for neuronal growth". Recently, CNTs have been used in neural stem cell growth and differentiation research. Also, CNTs have been used as interface materials with neurons, where they can deliver electrical stimulation to these cells and detect electrical activity. Here we mention just few applications of the CNTs:

18.3.1 Interfacing Cultured Neurons with Carbon Nanotubes

To demonstrate induced neuronal signaling by electrical simulation involving single-wall carbon nanotubes (SWCNTs), Mazzatenta and colleagues developed an integrated SWCNTs neuronal system (Mazzatenta et al. 2007) showing that hippocampal cells can be grown on pure SWCNTs substrates. These experimental results point to the fact that SWCNTs can be directly used to stimulate brain circuit activity. This result may have an impact in the future developments and architectural design of microsystems for neural prosthetics (Mazzatenta et al. 2007).

18.3.2 Intracellular Neural Recording with Pure Carbon Nanotubes Probes

A novel millimeter-long electrode can be used to produce extracellular and intracellular recordings from vertebrate neurons *in vitro* and *in vivo* experiments, when it is terminated with a tip fabricated from self-entangled pure CNTs with sub-micron dimensions (Yoon et al. 2013). Assembling intracellular electrodes from CNTs using self-entangled CNTs fabrication technology is opening the way to "harness" nanotechnology for neuroscience applications.

18.3.3 Analog Neuromorphic Modules Based on Carbon Nanotube Synapses

Shen et al. (2013) recently reported an analog neuromorphic module consisting of an integrate-and-fire circuit and *p*-type carbon nanotubes synapses. The CNTs synapse resembles a FET structure using as its gate an aluminum oxide dielectric layer implanted with indium ions and as its channel a random CNTs network. The electrons are attracted into the defect sites of the gate aluminum oxide layer by a positive voltage pulse applied to the gate, followed by a gradual release of the trapped electrons after the pulse is removed. Thus, the electrons induce a dynamic postsynaptic current in the CNTs channel by modifying the holes concentration. The excitatory or inhibitory postsynaptic currents generated by multiple input pulses via excitatory or inhibitory CNTs synapses, flow toward an integrate-and-fire circuit that triggers output pulses. Further, the analysis of the dynamic transfer functions between the input and output pulses of the neuromorphic module can be performed. An emulation of biological neural networks and their functions could be implemented by scaling up such a neuromorphic module.

18.3.4 Nanotechnology and Nanocomputing

The last decade witnessed an increasing use of artificial intelligence tools in nanotechnology research (Sacha and Varona 2013). Convergence of the two sciences, nanocomputing and nanotechnology, has the potential to reshape current research directions and technological developments in virtually all sciences. Thus, nanocomputing hardware development can boost the field of artificial-intelligence-based applications. Combining nanotechnology and nanocomputing has also shown great potential of hybrid technologies (i.e., nano-device and biological components) in neuroscience, in bioengineering, which combines new data representations and computer architectures, and in a large variety of related disciplines.

18.4 Role of Carbon Nanotubes in Deciphering Cortical Microcircuit Function

The tools used to study the functions of the neocortex need to operate at the nano-level, i.e. the same scale as brain functions operate. Nanoscience together with nanotechnology bring together an arsenal of unique methods to examine the complexity of brain function by allowing concurrent recording of thousands of neurons and manipulating the activity of millions of cells. One major reason why we are interested in examining them relates to the fact that they are ideal elements to uncover the micro-connectivity between cortical neurons intra-layer and within cortical minicolumns (Wade and Katzman 1975; Gandhi et al. 2010; Qiu et al. 2010; Choi et al. 2012; Jiang et al. 2013). Huge effort is now being devoted to the decoding of specific neural interactions and circuits, a goal that has emerged as the Brain Activity Mapping Project (Alivisatos et al. 2013). Several examples of the rich synergy between nanotechnology and neuroscience are given as follows:

18.4.1 Multi-electrode Array Technologies for Brain Circuits

The "substrate-integrated" microelectrode array (MEA) is the finest approach to study brain circuitry and connectivity, neurophysiology, or pathology, both *in vivo* and *in vitro*. MEAs add real-time versatility to long-term recording of neurophysiological activity in the extra-cellular micro-environment along with biochemical fluctuations, while being minimally invasive (Wise 2007; Chang-Hsiao et al. 2010; Amaral et al. 2013). The micro-anatomy of a neural ensemble and its synaptic interconnectivity, excitability, and plasticity may be better monitored by MEAs.

18.4.1.1 Carbon Nanotube MEAs

Suzuki et al. (2013) developed MEA chips of planar CNTs that can measure both the electrophysiological responses (such as action potentials and field postsynaptic potentials) and the release of dopamine neurotransmitter. These CNTs-MEA chips have been fabricated directly on the microelectrode surfaces by electroplating an indium-tin oxide (Fig. 18.3a). Chronoamperometric measurements based on such CNTs-MEA chips detected dopamine concentration at nanomolar level with high

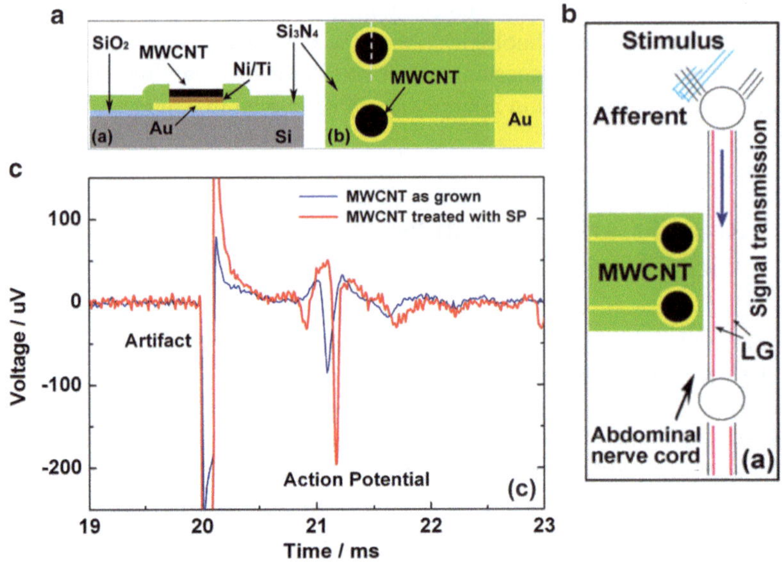

Fig. 18.3 Recording neural activity with MEA based on MWCNTs. (**a**) Schematic of multi electrode array based MWCNTs. (*Left*) cross-sectional view *right*. (*Right*) *top view* of the electrode scheme. (**b**) Schematic view of recording system. (**c**) Extracellular signal with large signal-to-noise ratio was obtained from lateral giant nerve fiber of an American crayfish using MEA based on MWCNTs and treated with steam plasma, compared with MWCNTs as grown (With permission from Chen et al. 2010)

temporal resolution and a 100-fold better signal to noise ratio. MEA chips may be useful for various applications such as drug screening and toxicity, *in vitro* stem cell differentiation, synaptic plasticity, or pathogenic processes associated with stroke, epilepsy, Alzheimer's disease and other neurodegenerative conditions.

18.4.1.2 Multi-walled Carbon Nanotube MEAs

MEAs using MWCNTs have the advantage of using a small size microelectrode with increased impedance and decreased charge-transfer capability (Gacem et al. 2013). To decrease impedance, the effective surface area for recording of the electrode needs to be increased. With a steam-plasma treatment the surface of MWCNTs becomes converted from super-hydrophobic to super-hydrophilic. This hydrophilic property is attributed to the OH bonding on the surface of MWCNTs. This electrode type allows the separation of neural signals with their distinct shapes for long-term recordings and improved recording performance.

Long-term recordings of modular features of the brain may be applied by this MEA type. In extracellular recording, action potentials were recorded with the treated MWCNTs; the signal-to-noise ratio (SNR) was up to 40 dB when used to record neural signals of a lateral giant cell from an American crayfish (Fig. 18.3b).

Figure 18.3c shows the extracellular signal with large SNR obtained from the nerve fiber using MEA based on MWCNTs treated with steam plasma, compared with MWCNTs as grown.

18.4.1.3 Multiplexed High Density MEAs

Recent neural probes based on silicon (Du et al. 2011) have employed nanofabricated, high-density electrical leads that can read out multichannel data. MEA uses "application-specific integrated circuit" (ASIC) which intensifies signals, multiplexing functions and band-pass filtering. A multiplex high density device with fully integrated, low noise, 64-channel system can perform high spatial resolution extracellular measurements and weighs just 330 mg (Du et al. 2011). These on-chip multiplexers allow recordings with "substantially fewer external wires than the number of input channels". The combination of ASICs and nanofabricated probes that is both "minimally invasive and highly scalable" (Du et al. 2011) was employed for carrying out large-scale, high-density electrophysiology in small animals.

Similarly, Viventi et al. (2011) integrated "ultrathin and flexible silicon nanomembrane transistors" into a MEA, enabling "dense arrays" of thousands of amplified and multiplexed sensors to be connected with fewer wires. This system was employed to record in cat the spatial properties of brain activity *in vivo*, including patterns of activity such as sleep spindles, single-trial visual evoked responses, or electrographic seizures. Such developments promise to provide diagnostic and therapeutic brain-machine interface devices.

The MEA system containing a 64-site array (Fig. 18.4a) was implanted into the mouse hippocampus (Fig. 18.4a, b). The animal was then connected to a flexible 12 wire tethered cable and allowed to explore a 30 cm × 50 cm enclosure during acquisition of neurophysiological data. Current source density (CSD) analysis of local field potential signals with a vertical resolution of ~28 mm revealed relatively uniform theta oscillations in layers between the CA1 and dentate gyrus (Fig. 18.4b). However, marked shifts in amplitude and phase were observed on sites around the suprapyramidal and infrapyramidal blades of the dentate gyrus. Theta phase dependent oscillatory firing of hippocampal neurons were observed during exploratory behavior (Fig. 18.4b). As expected, hippocampal neurons were found to be preferentially active in the vicinity of the negative peak of the theta potential.

Figure 18.5 displays a sample of spontaneous firing activity in the ventral posteromedial thalamic nucleus of an anesthetized mouse, recorded by Du and colleagues (Du et al. 2011). The data shows the majority of functional (63/64) electrodes in the 1.3 mm long MEA have spiking activity, indicating that more than one electrode often records from the same neuron. The close electrode spacing is thus well suited for high-density electrophysiology with extensive coverage of extracellular fields along the length of the array. High spatial resolution

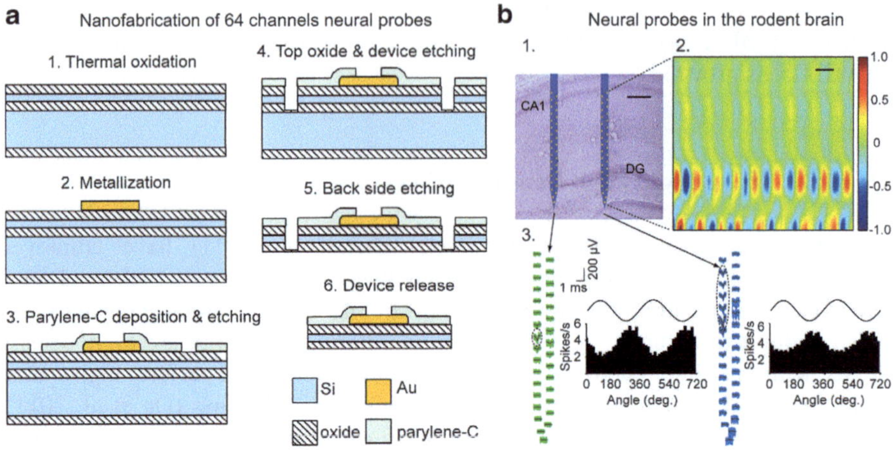

Fig. 18.4 Recording with nanofabricated multi-electrode-arrays in awake behaving mice. (**a**) Nanofabrication of 64 channels neural probes. (**b**) Neuron recordings with nanofabricated probes in awake behaving mice. (1) Nissl-stained brain section overlaid with a schematic of the probe at its stereotaxically implanted location. Each silicon shaft is 60 mm wide. Scale bar, 200 mm. (2) Current source density analysis of local field potentials across the hippocampus with a vertical resolution of ~28 mm. Measurements correspond to the rightmost shaft, which is co-localized with both the supra- and infra-pyramidal blades of the dentate gyrus (DG). The data were gathered during home cage exploratory behavior. The CSD is normalized to ±1. Scale bar, 100 ms. (3) Waveforms of two putative single units recorded from this probe across all sites on their entire respective shafts, with histograms showing theta phase locking of spikes. *Dashed ellipses* indicate the sites exhibiting the highest extracellular action potentials for these units. Measured theta oscillations on top of the histograms are for reference. Theta oscillations were measured from the upper rightmost electrode near the CA1 pyramidal cell layer (Du et al. 2011)

extracellular measurements with a fully integrated, low noise 64-channel system enabled reliable spike detection with peak amplitudes as low as 40–50 μV (Du et al. 2011).

18.4.1.4 Nanowire-Based Electrode for Acute Neural Recordings in the Brain

A new kind of electrode is based on "structurally controlled nanowires," for neurophysiological measurements *in vivo* (Suyatin et al. 2009). This electrode (Fig. 18.6) has a sensing part made of a thin metal layer deposited on epitaxially grown GaP nanowires. Suyatin and colleagues realized the first functional NWs-based electrode (Suyatin et al. 2009). The team also has successfully tested the electrode by *in vivo* recordings in the cortex of rat in multiple brain implantations. This kind of electrode with a controllable geometry of the nanowires can be further used for the investigation of many *in vivo* functional properties in nanostructured neuronal interfaces.

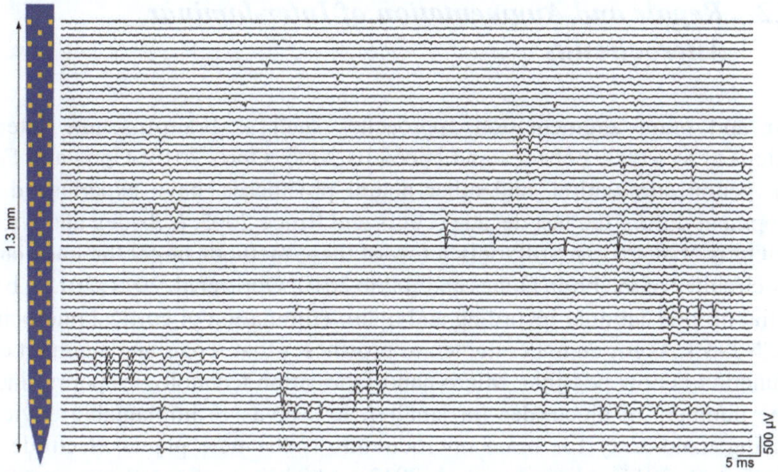

Fig. 18.5 Parallel recording capabilities of the multiplexed neurophysiological system. Traces are filtered from 0.3–5 kHz to highlight spiking activity, and are plotted near their corresponding location on the shaft of the probe displayed on the left (for each horizontal left-right pair of pads, the trace for the left-hand pad appears just above the trace for the right-hand probe). Measurements were made in the mouse thalamus with a probe with post-plating impedances of 0.6 MΩ ± 0.4 MΩ (mean ± 1 s.d.). The majority of the 64 sites report action potentials (Du et al. 2011)

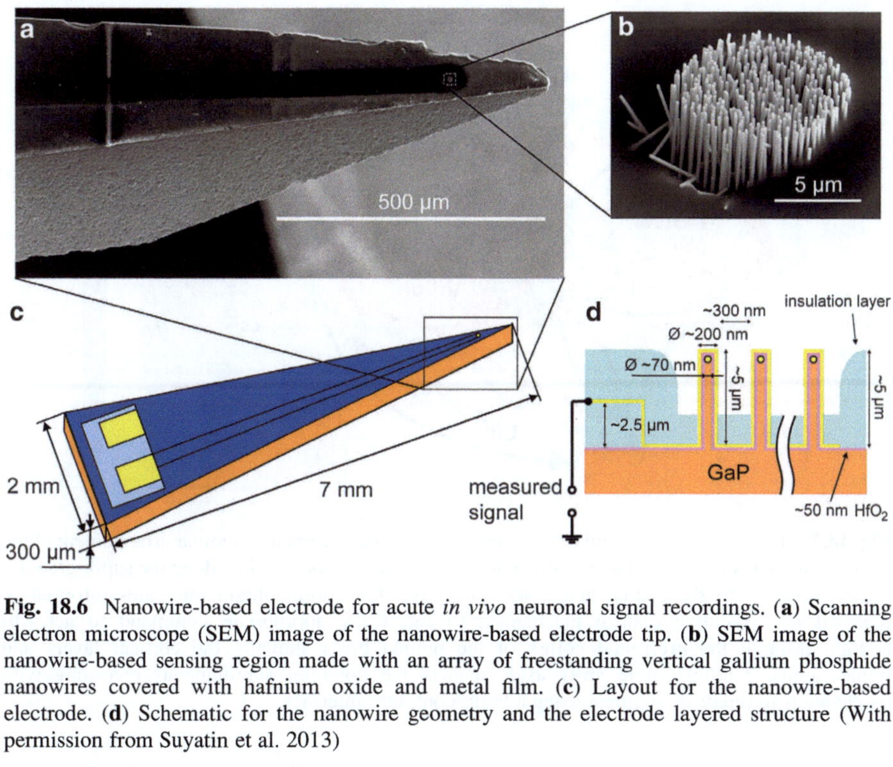

Fig. 18.6 Nanowire-based electrode for acute *in vivo* neuronal signal recordings. (**a**) Scanning electron microscope (SEM) image of the nanowire-based electrode tip. (**b**) SEM image of the nanowire-based sensing region made with an array of freestanding vertical gallium phosphide nanowires covered with hafnium oxide and metal film. (**c**) Layout for the nanowire-based electrode. (**d**) Schematic for the nanowire geometry and the electrode layered structure (With permission from Suyatin et al. 2013)

18.4.2 Repair and Augmentation of Inter-laminar Microcircuits

Repair and brain augmentation approaches such as brain-machine interfaces (Nicolelis et al. 2003; Lebedev and Nicolelis 2006; Opris 2013; Opris and Ferrera 2014), neural stimulation, and other neural prostheses, have experienced rapid development during the last decade (Opris and Bruce 2005; Lebedev and Nicolelis 2006; Opris et al. 2013). Still, only a few of these methods target the inter-laminar micro-circuitry of the brain (Jones and Rakic 2010; Opris et al. 2011, 2012a, b). The potential of inter-laminar recording and micro-stimulation of cortical microcircuits with CNT-MEAs for engineering neural prostheses that repair and augment cognitive function is now possible. Micro/nano-fabrication technologies permit increasing the numbers of electrodes for smaller, less invasive implantable devices. A promising nano-array for brain microcircuits is the new planar electrode array (Viventi et al. 2011; Alivisatos et al. 2013), which is configured on a crystalline, ceramic, or polymer support structure (Fig. 18.7). Additionally, recording neuronal firing with three-dimensional microelectrode arrays (Zorzos et al. 2012) represents a major advance in brain activity mapping techniques, by providing a tool to demonstrate how intra and inter-laminar/regional neural circuits cooperate together to process information. Building prosthetic minicolumns (Mountcastle 1957, 1997;

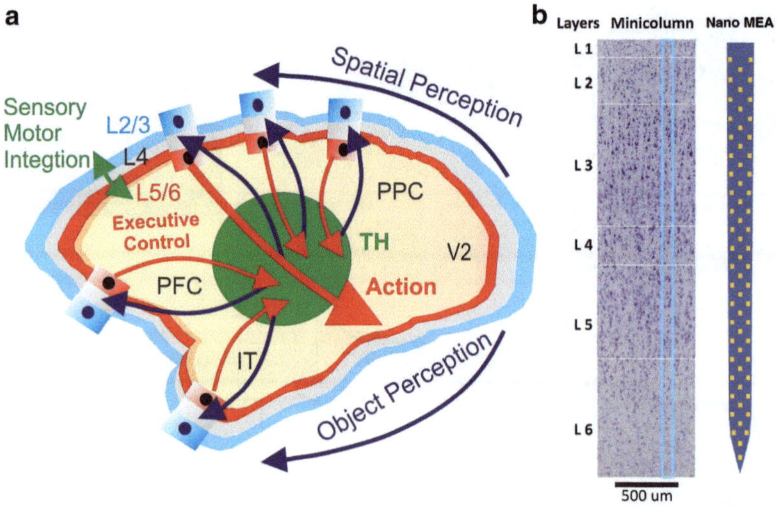

Fig. 18.7 Implications for modular NanoNeuroscience. (**a**) Laminar-columnar arrangement of the neurons in the neocortex of the primate brain. Visual information flowing along the supra-granular layers is split into the spatial and object pathways. Each minicolumn integrates information required for executive control mechanism to select the appropriate command to act. (**b**) Section through the prefrontal cortex of the human brain showing the cortical layers and minicolumns. Multielectrode array aside depicts the potential simultaneous recording with 64 channels to repair or augment using a brain-computer interface

Buxhoeveden and Casanova 2002; Mahan and Georgopoulos 2013; Opris and Casanova 2014) as basic modules to repair damaged cortical tissue is becoming a valuable approach for cognitive neuroprosthetics. This may be accomplished by designing artificial minicolumns that can be surgically inserted into the human brain, or using nanowire contacts to place a device with minicolumn function within the damaged circuitry (Lebedev and Nicolelis 2006; Bokara et al. 2013; Marmarelis et al. 2014). In the future, such microcircuit based prostheses will provide efficient therapies for patients with neurological and psychiatric disorders (Casanova 2007, 2013; Casanova et al. 2008; Chance et al. 2011). Moreover, neural enhancement approaches may be applied to inter-laminar microcircuits across the entire cortex (Opris 2013).

18.4.3 Associative Memories with Enhanced Storage Capacity

The modeling of associative memories with non-monotonic neural networks (Nishimori and Opris 1993) was demonstrated by Monte Carlo computer simulations to yield an enhanced storage capacity. Recent experiments employing MIMO (Multi-Input/Multi-Output) micro-stimulation of prefrontal cortical inter-laminar microcircuits (based on multiplexing principle) was shown to enhance cognitive performance and memory in nonhuman primates performing a behavioral task (Hampson et al. 2012; Opris et al. 2012a, b). Future use of multiplexed high density MEAs (Du et al. 2011) holds the promise to provide unprecedented memory enhancement for both artificial intelligence and humans with implanted chips.

18.5 Regeneration and Repair of Brain Microcircuits

18.5.1 Carbon Nanotubes as Interfaces for Neural Prosthetics

The neural prostheses that successfully help patients to increase their daily living activities are quite simple implants that yield some definite tissue response and are well recognized as foreign body (Stieglitz 2007). Based on the latest developments in materials sciences new avenues towards highly advanced systems to interface the human brain have emerged. Nanotechnology is opening the door to macromolecular approaches to implants that mimic the "biologic topology" and take into account the surface interaction of biologic cells. Combinations of neural cells with micro-implants become the platform of stable bio-hybrid interfaces. Furthermore, "converging technologies" that exploit the synergies between computer science, engineering, neuroscience, and psychology are envisioned to completely change the understanding of the entire science.

Artificial synapses in neuromorphic circuits based on nanoscale memory devices have been recently accepted as a promising route for creating novel circuit architectures that tolerate variability and/or defects (Gacem et al. 2013). Still, the implementations of neural network-type circuits that are based on non-CMOS (complementary metal–oxide–semiconductor) memory devices with learning capabilities are rare. Gacem et al. (2013) showed that memory elements based on CNTs may be used as "artificial synapses," combined with "conventional neurons" to be further trained to perform several functions (by applying a supervised learning algorithm). This is possible because the same device ensemble can be trained many times to code successively any type of 3-input combination of Boolean logic functions despite the variability among devices. This approach has huge potential for application to parallel learning of several devices with more complex functions.

Carbon nanowires used as interface material in contact with neurons can both deliver electrical stimulation to these cells and detect neuronal electrical activity. Carbon nanowires or nanotubes have emerged as materials that do not have recognizable adverse effects. Consequently, they can be successfully used in brain-machine interfaces (Malarkey and Parpura 2010). In the last years, carbon nanotube research on growing CNTs substrates have been used to examine *in vivo* formation of neurons and neuronal networks during guided growth by artificial nano-scaled cues. Additionally, prostheses for monitoring brain activity were developed using interfaces based on nanotube architecture (Stankova et al. 2014). Fabbro et al. (2012) recently demonstrated the alteration of various hippocampal neurons responses by the CNTs substrates in cultures. This observation highlighted the exceptional ability of the CNTs substrate to interfere with nerve tissue growth; they confirmed the hypothesis that CNT scaffolds promote the development of immature neurons isolated from the neonatal rat spinal cord and maintained *in vitro*. Results from electrophysiological studies associated with gene expression analysis indicated that spinal neurons plated on electro-conductive CNTs show an assisted expansion suggesting that CNT platforms activate healable activities involving microglia, in the absence of reactive gliosis.

18.5.2 Neuroregeneration of Brain Circuits

CNT technology can now be applied to develop new devices that are capable to drive repair of nerve tissue, neuronal differentiation and growth, and network reconstruction. Development of future strategies for tissue repair in order to promote functional recovery after brain damage is one of the main aims of nanotech studies (Fabbro et al. 2012). CNTs based technologies are emerging as particularly innovative tools due to their ability to interface neuronal circuits, synapses and membranes (Sakhtianchi et al. 2013), as well as due to the outstanding physical properties of these nanomaterials.

A class of ideal biomaterials for a wide range of regenerative medicine applications is MWCNTs polymer composites consisting of hybrid materials. Using a composite as a substrate to increase electronic interfacing between neurons and micro-machined electrodes, Antoniadou et al. (2011) reported the synthesis and characterization of a novel biomaterial for the development of nerve guidance channels in order to promote nerve regeneration, opening up potential applications for prosthetic devices, neural probes, and brain implants. CNTs used as scaffolds in brain tissues and neural cells have shown promising results, supporting treatment strategies based on transferring stem cells containing scaffolds to the damaged regions. Protection of neurons and enhancement of recovery of behavioral functions in rats with induced stroke were observed when rats were pretreated with amine-modified SWCNTs (Lee et al. 2011) as indicated by the low levels of apoptotic, angiogenic and inflammation markers. In another study, it was shown that CNTs promote the recovery from stroke when they are impregnated with neural progenitor cells in subventricular zones to improve stem cell differentiation (Moon et al. 2012).

18.5.3 Biocompatability of Carbon Nanotubes with Stem Cells to Treat Brain Injuries

Regenerative medicine, especially for central nervous system, (CNS) has extensively looked into the possibility to use stem cell therapy to replace lost cells during CNS injuries (Bokara et al. 2013). However, the survival rate of the transplanted stem cells that affects tissue restoration may be limited by the toxic byproducts and the complexity of the CNS injuries. Nanotechnologies are emerging as useful platforms for understanding, measuring, and manipulating stem cells (Ferreira et al. 2008). Such examples are, (i) magnetic nanoparticles and quantum dots for labelling stem cell and *in vivo* tracking, (ii) nanoparticles, carbon nanotubes, and polyplexes used for the intracellular delivery of genes (i.e., oligonucleotides) and protein (i.e., peptides), (iii) nanometer-scale engineered scaffolds for stem cell differentiation and transplantation, and, (iv) nanotechnological support for tracking, differentiation and transplant of stem cells.

18.6 Conclusion

Nanoscience is intersecting neuroscience by opening the neuro-world at a scale never imagined before. Tiny nanostructures that are many thousand times smaller than a human hair and hundred times smaller than neurons are on their development road to become decisive components and/or devices in repairing brain. In this regard, nano-neuroscience has the potential for diagnosing and repairing the smallest neuronal circuits both from inside and from outside, i.e., bringing internal and external “nano-surgeons” to interface the neuronal circuits. Carbon nanotubes and nanowires, due to their exceptional electrical and chemical properties, represent some of the most promising nanomaterials for the nervous system. They can be used to control the patterns of neuronal processes or designed to monitor or stimulate neurons or used as effective material to promote axonal repair and regeneration. The nanostructures’ ability to conduct electricity on contact with different chemicals induces rapid changes brain chemistry. Acting as “on-site” laboratories, nano-sized multi-electrode arrays can be integrated into sensing, monitoring, recording, and stimulating devices. Properly integrated with optimal human brain operation, the use of nanomaterials is capable of recovering or even improving performance in complex tasks, as demonstrated via closed-loop facilitation of columnar processing in primate brain. In fact, the promise of NanoNeuroTechnology is to provide chips that will interface with the brain and allow online detection and correction of malfunctioning brain microcircuits. The nanostructures at the interface between nanotechnology and neuroscience will play a pivotal role in addressing the multitude of brain disorders as well as in repair/augmentation of brain functions (Vidu et al. 2014).

References

Ajayan PM (1999) Nanotubes from carbon. Chem Rev 99:1787–1799

Alivisatos AP, Andrews AM, Boyden ES, Chun M, Church GM, Deisseroth K, Donoghue JP, Fraser SE, Lippincott-Schwartz J, Looger LL, Masmanidis S, McEuen PL, Nurmikko AV, Park H, Peterka DS, Reid C, Roukes ML, Scherer A, Schnitzer M, Sejnowski TJ, Shepard KL, Tsao D, Turrigiano G, Weiss PS, Xu C, Yuste R, Zhuang X (2013) Nanotools for neuroscience and brain activity mapping. ACS Nano 7:1850–1866

Almquist BD, Melosh NA (2010) Fusion of biomimetic stealth probes into lipid bilayer cores. Proc Natl Acad Sci U S A 107:5815–5820

Amaral J, Pinto V, Costa T, Gaspar J, Ferreira R, Paz E, Cardoso S, Freitas PP (2013) Integration of TMR sensors in silicon microneedles for magnetic measurements of neurons. IEEE Trans Magn 49:3512–3515

Ansaldo A, Castagnola E, Maggiolini E, Fadiga L, Ricci D (2011) Superior electrochemical performance of carbon nanotubes directly grown on sharp microelectrodes. ACS Nano 5:2206–2214

Antoniadou EV, Ahmad RK, Jackman RB, Seifalian AM, IEEE (2011) Next generation brain implant coatings and nerve regeneration via novel conductive nanocomposite development. In: 2011 annual international conference of the IEEE Engineering in Medicine and Biology Society, Boston, 31 August–3 September 2011, pp 3253–3257

Barrelet CJ, Wu Y, Bell DC, Lieber CM (2003) Synthesis of CdS and ZnS nanowires using single-source molecular precursors. J Am Chem Soc 125:11498–11499

Benabid AL (2007) What the future holds for deep brain stimulation. Expert Rev Med Devices 4:895–903

Berthing T, Bonde S, Sorensen CB, Utko P, Nygard J, Martinez KL (2011) Intact mammalian cell function on semiconductor nanowire arrays: new perspectives for cell-based biosensing. Small 7:640–647

Bianco A, Kostarelos K, Prato M (2011) Making carbon nanotubes biocompatible and biodegradable. Chem Commun 47:10182–10188

Binnig G, Rohrer H, Gerber C, Weibel E (1982) Surface studies by scanning tunneling microscopy. Phys Rev Lett 49:57–61
Binnig G, Rohrer H, Gerber C, Weibel E (1983) 7 × 7 reconstruction on SI(111) resolved in real space. Phys Rev Lett 50:120–123
Binnig G, Quate CF, Gerber C (1986) Atomic force microscope. Phys Rev Lett 56:930–933
Bokara KK, Kim JY, Lee YI, Yun K, Webster TJ, Lee JE (2013) Biocompatability of carbon nanotubes with stem cells to treat CNS injuries. Anat Cell Biol 46:85–92
Bota PM, Dorobantu D, Boerasu I, Bojin D, Enachescu M (2014a) Synthesis of single wall carbon nanotubes by excimer laser ablation. Surf Eng Appl Electrochem 50:294–299
Bota PM, Dorobantu D, Boerasu I, Bojin D, Enachescu M (2014b) New laser ablation chamber for producing carbon nanomaterials using excimer laser. Mater Res Innov 19:33–39
Brueggemann D, Wolfrum B, Maybeck V, Mourzina Y, Jansen M, Offenhaeusser A (2011) Nanostructured gold microelectrodes for extracellular recording from electrogenic cells. Nanotechnology 22(26):265104
Buxhoeveden DP, Casanova MF (2002) The minicolumn hypothesis in neuroscience. Brain 125:935–951
Cai D, Mataraza JM, Qin ZH, Huang ZP, Huang JY, Chiles TC, Carnahan D, Kempa K, Ren ZF (2005) Highly efficient molecular delivery into mammalian cells using carbon nanotube spearing. Nat Methods 2:449–454
Carmichael M, Vidu R, Maksumov A, Palazoglu A, Stroeve P (2004) Using wavelets to analyze AFM images of thin films: surface micelles and supported lipid bilayers. Langmuir 20:11557–11568
Casanova MF (2007) Schizophrenia seen as a deficit in the modulation of cortical minicolumns by monoaminergic systems. Int Rev Psychiatry 19:361–372
Casanova MF (2013) Neural mechanisms in autism. In: Encyclopedia of autism spectrum disorders: 1994–2007. Springer, New York
Casanova MF, Kreczmanski P, Trippe J II (2008) Neuronal distribution in the neocortex of schizophrenic patients. Psychiatry Res 158:267–277
Casanova MF, El-Baz A, Switala A (2011) Laws of conservation as related to brain growth, aging, and evolution: symmetry of the minicolumn. Front Neuroanat 5:66
Cellot G, Cilia E, Cipollone S, Rancic V, Sucapane A, Giordani S, Gambazzi L, Markram H, Grandolfo M, Scaini D, Gelain F, Casalis L, Prato M, Giugliano M, Ballerini L (2009) Carbon nanotubes might improve neuronal performance by favouring electrical shortcuts. Nat Nanotechnol 4:126–133
Chance SA, Clover L, Cousijn H, Currah L, Pettingill R, Esiri MM (2011) Microanatomical correlates of cognitive ability and decline: normal ageing, MCI, and Alzheimer's disease. Cereb Cortex 21:1870–1878
Chang-Hsiao C, Shih-Chang C, Yu-Tao L, Yen-Chung C, Shih-Rung Y, Da-Jeng Y (2010) Three-dimensional flexible microprobe for recording the neural signal. J Micro/Nanolithograph MEMS MOEMS 9:031007 (031006 pp)–031007 (031006 pp)
Chen Y, Barrelet CJ, Capasso F, Lieber CM (2006) Single p-type/intrinsic/n-type silicon nanowires as nanoscale avalanche photodetectors. Nano Lett 6:6pp–6pp
Chen C-H, Su H-C, Chuang S-C, Yen S-J, Chen Y-C, Lee Y-T, Chen H, Yew T-R, Chang Y-C, Yeh S-R, Yao D-J (2010) Hydrophilic modification of neural microelectrode arrays based on multi-walled carbon nanotubes. Nanotechnology 21(48):485501
Cherukuri P, Gannon CJ, Leeuw TK, Schmidt HK, Smalley RE, Curley SA, Weisman RB (2006) Mammalian pharmacokinetics of carbon nanotubes using intrinsic near-infrared fluorescence. Proc Natl Acad Sci U S A 103:18882–18886
Choi J, Kim HY, Ju EJ, Jung J, Park J, Chung HK, Lee JS, Lee JS, Park HJ, Song SY, Jeong SY, Choi EK (2012) Use of macrophages to deliver therapeutic and imaging contrast agents to tumors. Biomaterials 33:4195–4203
Cogan SF (2008) Neural stimulation and recording electrodes. Annu Rev Biomed Eng 10:275–309

Constantinople CM, Bruno RM (2013) Deep cortical layers are activated directly by thalamus. Science 340:1591–1594

Cui Y, Lieber CM (2001) Functional nanoscale electronic devices assembled using silicon nanowire building blocks. Science 291:851–853

Cui Y, Wei QQ, Park HK, Lieber CM (2001a) Nanowire nanosensors for highly sensitive and selective detection of biological and chemical species. Science 293:1289–1292

Cui Y, Lauhon LJ, Gudiksen MS, Wang JF, Lieber CM (2001b) Diameter-controlled synthesis of single-crystal silicon nanowires. Appl Phys Lett 78:2214–2216

Cui Y, Zhong ZH, Wang DL, Wang WU, Lieber CM (2003) High performance silicon nanowire field effect transistors. Nano Lett 3:149–152

Dai ZR, Pan ZW, Wang ZL (2001) Ultra-long single crystalline nanoribbons of tin oxide. Solid State Commun 118:351–354

Du J, Blanche TJ, Harrison RR, Lester HA, Masmanidis S (2011) Multiplexed, high density electrophysiology with nanofabricated neural probes. PLoS One 6(10):e26204

Duan X, Gao R, Xie P, Cohen-Karni T, Qing Q, Choe HS, Tian B, Jiang X, Lieber CM (2012) Intracellular recordings of action potentials by an extracellular nanoscale field-effect transistor. Nat Nanotechnol 7:174–179

Dujavova-Laurencikova A, Hasenoehrl S, Elias P, Stoklas R, Blaho M, Novotny I, Krizanova Z, Novak J (2013) Ohmic contacts to p-GaP/n-ZnO core/shell nanowires based on Au metallization. Appl Surf Sci 269:60–64

Dvir T, Timko BP, Kohane DS, Langer R (2011) Nanotechnological strategies for engineering complex tissues. Nat Nanotechnol 6:13–22

Enachescu M, Schleef D, Ogletree DF (1999) Integration of point-contact microscopy and atomic-force microscopy: application to characterization of graphite/Pt(111). Phys Rev B 60:16913–16919

Endo M, Strano MS, Ajayan PM (1990) Potential applications of carbon nanotubes. American Electroplaters and Surface Finishers Society, Orlando

Fabbro A, Bosi S, Ballerini L, Prato M (2012) Carbon nanotubes: artificial nanomaterials to engineer single neurons and neuronal networks. ACS Chem Neurosci 3:611–618

Feng W, Ji P (2011) Enzymes immobilized on carbon nanotubes. Biotechnol Adv 29:889–895

Ferreira L, Squier T, Park H, Choe H, Kohane DS, Langer R (2008) Human embryoid bodies containing nano- and microparticulate delivery vehicles. Adv Mater 20:2285–+

Fromherz P (2003) Semiconductor chips with ion channels, nerve cells and brain slices. In: Wolf LJ, Strock JL (eds) 1st international IEEE Embs conference on neural engineering 2003, conference proceedings, pp 1–1

Gacem K, Retrouvey J-M, Chabi D, Filoramo A, Zhao W, Klein J-O, Derycke V (2013) Neuromorphic function learning with carbon nanotube based synapses. Nanotechnology 24 (38):384013

Gandhi N, Saiyed ZM, Napuri J, Samikkannu T, Reddy PV, Agudelo M, Khatavkar P, Saxena SK, Nair MP (2010) Interactive role of human immunodeficiency virus type 1 (HIV-1) clade-specific tat protein and cocaine in blood-brain barrier dysfunction: implications for HIV-1-associated neurocognitive disorder. J Neurovirol 16:294–305

Goncher G, Solanki R, Carruthers JR, Conley J Jr, Ono Y (2006) p-n junctions in silicon nanowires. J Electron Mater 35:1509–1512

Guo L, Wu ZH, Liu T, Wang WD, Zhu HS (2000) Synthesis of novel Sb_2O_3 and Sb_2O_5 nanorods. Chem Phys Lett 318:49–52

Hai A, Shappir J, Spira ME (2010) In-cell recordings by extracellular microelectrodes. Nat Methods 7:200–U250

Hallstrom W, Martensson T, Prinz C, Gustavsson P, Montelius L, Samuelson L, Kanje M (2007) Gallium phosphide nanowires as a substrate for cultured neurons. Nano Lett 7:2960–2965

Hallstrom W, Prinz CN, Suyatin D, Samuelson L, Montelius L, Kanje M (2009) Rectifying and sorting of regenerating axons by free-standing nanowire patterns: a highway for nerve fibers. Langmuir 25:4343–4346

Hampson RE, Gerhardt GA, Marmarelis V, Song D, Opris I, Santos L, Berger TW, Deadwyler SA (2012) Facilitation and restoration of cognitive function in primate prefrontal cortex by a neuroprosthesis that utilizes minicolumn-specific neural firing. J Neural Eng 9(5):056012

Heller DA, Baik S, Eurell TE, Strano MS (2005) Single-walled carbon nanotube spectroscopy in live cells: towards long-term labels and optical sensors. Adv Mater 17:2793–2799

Heller DA, Jeng ES, Yeung TK, Martinez BM, Moll AE, Gastala JB, Strano MS (2006) Optical detection of DNA conformational polymorphism on single-walled carbon nanotubes. Science 311:508–511

Iijima S (1991) Helical microtubules of graphitic carbon. Nature 354:56–58

Ikemiya N, Iwai D, Yamada K, Vidu R, Hara S (1996) Atomic structures and growth morphologies of electrodeposited Te film on Au(100) and Au(111) observed by in situ atomic force microscopy. Surf Sci 369:199–208

Javey A, Guo J, Wang Q, Lundstrom M, Dai HJ (2003) Ballistic carbon nanotube field-effect transistors. Nature 424:654–657

Jiang XC, Herricks T, Xia YN (2002) CuO nanowires can be synthesized by heating copper substrates in air. Nano Lett 2:1333–1338

Jiang X, Xiong Q, Nam S, Qian F, Li Y, Lieber CM (2007) InAs/InP radial nanowire heterostructures as high electron mobility devices. Nano Lett 7:3214–3218

Jiang DP, Li J, Zhang J, Xu SL, Kuang F, Lang HY, Wang YF, An GZ, Li JH, Guo GZ (2013) Electromagnetic pulse exposure induces overexpression of beta amyloid protein in rats. Arch Med Res 44:178–184

Jones EG, Rakic P (2010) Radial columns in cortical architecture: it is the composition that counts. Cereb Cortex 20:2261–2264

Kam NWS, Dai HJ (2005) Carbon nanotubes as intracellular protein transporters: generality and biological functionality. J Am Chem Soc 127:6021–6026

Kam NWS, Jessop TC, Wender PA, Dai HJ (2004) Nanotube molecular transporters: internalization of carbon nanotube-protein conjugates into mammalian cells. J Am Chem Soc 126:6850–6851

Kam NWS, Liu Z, Dai HJ (2005a) Functionalization of carbon nanotubes via cleavable disulfide bonds for efficient intracellular delivery of siRNA and potent gene silencing. J Am Chem Soc 127:12492–12493

Kam NWS, O'Connell M, Wisdom JA, Dai HJ (2005b) Carbon nanotubes as multifunctional biological transporters and near-infrared agents for selective cancer cell destruction. Proc Natl Acad Sci U S A 102:11600–11605

Kawano T, Harimoto T, Ishihara A, Takei K, Kawashima T, Usui S, Ishida M (2010) Electrical interfacing between neurons and electronics via vertically integrated sub-4 mu m-diameter silicon probe arrays fabricated by vapor-liquid-solid growth. Biosens Bioelectron 25:1809–1815

Keefer EW, Botterman BR, Romero MI, Rossi AF, Gross GW (2008) Carbon nanotube coating improves neuronal recordings. Nat Nanotechnol 3:434–439

Kim W, Ng JK, Kunitake ME, Conklin BR, Yang P (2007) Interfacing silicon nanowires with mammalian cells. J Am Chem Soc 129:7228–+

Kipke DR, Shain W, Buzsaki G, Fetz E, Henderson JM, Hetke JF, Schalk G (2008) Advanced neurotechnologies for chronic neural interfaces: new horizons and clinical opportunities. J Neurosci 28:11830–11838

Klumpp C, Kostarelos K, Prato M, Bianco A (2006) Functionalized carbon nanotubes as emerging nanovectors for the delivery of therapeutics. Biochimica Et Biophysica Acta-Biomembranes 1758:404–412

Kotov NA, Winter JO, Clements IP, Jan E, Timko BP, Campidelli S, Pathak S, Mazzatenta A, Lieber CM, Prato M, Bellamkonda RV, Silva GA, Kam NWS, Patolsky F, Ballerini L (2009) Nanomaterials for neural interfaces. Adv Mater 21:3970–4004

Ku JR, Vidu R, Talroze R, Stroeve P (2004) Fabrication of nanocables by electrochemical deposition inside metal nanotubes. J Am Chem Soc 126:15022–15023

Lebedev MA, Nicolelis MAL (2006) Brain-machine interfaces: past, present and future. Trends Neurosci 29:536–546
Lee SY, Kim TH, Suh DI, Park JE, Kim JH, Youn CJ, Ahn BK, Lee SK (2007) An electrical characterization of a hetero-junction nanowire (NW) PN diode (n-GaN NW/p-Si) formed by dielectrophoresis alignment. Physica E-Low-Dimens Syst Nanostruct 36:194–198
Lee G-J, Choi SK, Choi S, Park JH, Park H-K (2011) Enzyme-immobilized CNT network probe for in vivo neurotransmitter detection. In: Wang P (ed) Nanoscale biocatalysis: methods and orotocols. Humana Press, New York, pp 65–75
Li C, Zhang DH, Han S, Liu XL, Tang T, Zhou CW (2003) Diameter-controlled growth of single-crystalline In_2O_3 nanowires and their electronic properties. Adv Mater 15:143–+
Li Y, Qian F, Xiang J, Lieber CM (2006) Nanowire electronic and optoelectronic devices. Mater Today 9:18–27
Lieber CM (2003) Nanoscale science and technology: building a big future from small things. MRS Bull 28:486–491
Liu ZQ, Xie SS, Sun LF, Tang DS, Zhou WY, Wang CY, Liu W, Li YB, Zou XP, Wang G (2001) Synthesis of alpha-SiO_2 nanowires using Au nanoparticle catalysts on a silicon substrate. J Mater Res 16:683–686
Lopez CF, Nielsen SO, Moore PB, Klein ML (2004) Understanding nature's design for a nanosyringe. Proc Natl Acad Sci U S A 101:4431–4434
Lu Q, Moore JM, Huang G, Mount AS, Rao AM, Larcom LL, Ke PC (2004) RNA polymer translocation with single-walled carbon nanotubes. Nano Lett 4:2473–2477
Mahan MY, Georgopoulos AP (2013) Motor directional tuning across brain areas: directional resonance and the role of inhibition for directional accuracy. Front Neural Circuits 7:92
Malarkey EB, Parpura V (2007) Applications of carbon nanotubes in neurobiology. Neurodegener Dis 4:292–299
Malarkey EB, Parpura V (2010) Carbon nanotubes in neuroscience. In: Czernicki Z, Baethmann A, Ito U, Katayama Y, Kuroiwa T, Mendelow D (eds) Brain edema XIV. Springer, London, pp 337–341
Marmarelis VZ, Shin DC, Song D, Hampson RE, Deadwyler SA, Berger TW (2014) On parsing the neural code in the prefrontal cortex of primates using principal dynamic modes. J Comput Neurosci 36:321–337
Martin DC, Wu J, Shaw CM, King Z, Spanninga SA, Richardson-Burns S, Hendricks J, Yang J (2010) The morphology of poly (3,4-ethylenedioxythiophene). Polym Rev 50:340–384
Mazzatenta A, Giugliano M, Campidelli S, Gambazzi L, Businaro L, Markram H, Prato M, Ballerini L (2007) Interfacing neurons with carbon nanotubes: electrical signal transfer and synaptic stimulation in cultured brain circuits. J Neurosci 27:6931–6936
McEuen PL, Fuhrer MS, Park HK (2002) Single-walled carbon nanotube electronics. IEEE Trans Nanotechnol 1:78–85
Miao Z, Xu DS, Ouyang JH, Guo GL, Zhao XS, Tang YQ (2002) Electrochemically induced sol-gel preparation of single-crystalline TiO_2 nanowires. Nano Lett 2:717–720
Moon HK, Lee SH, Choi HC (2009) In vivo near-infrared mediated tumor destruction by photothermal effect of carbon nanotubes. ACS Nano 3:3707–3713
Moon SU, Kim J, Bokara KK, Kim JY, Khang D, Webster TJ, Lee JE (2012) Carbon nanotubes impregnated with subventricular zone neural progenitor cells promotes recovery from stroke. Int J Nanomedicine 7:2751–2765
Mountcastle VB (1957) Modality and topographic properties of single neurons of cats somatic sensory cortex. J Neurophysiol 20:408–434
Mountcastle VB (1997) The columnar organization of the neocortex. Brain 120:701–722
Nicolelis MA, Dimitrov D, Carmena JM, Crist R, Lehew G, Kralik JD (2003) Chronic, multisite, multi-electrode recordings in macaque monkeys. Proc Natl Acad Sci U S A 100:11041–11046
Nishimori H, Opris I (1993) Retrieval process of an associative memory with a general input-output function. Neural Netw 6:1061–1067

Opris I (2013) Inter-laminar microcircuits across the neocortex: repair and augmentation. In: Lebedev MA, Opris I, Casanova MF (eds) Frontiers in systems neuroscience research topic: "augmentation of brain function: facts, fiction and controversy"
Opris I, Bruce CJ (2005) Neural circuitry of judgment and decision mechanisms. Brain Res Rev 48:509–526
Opris I, Casanova MF (2014) Prefrontal cortical minicolumn: from executive control to disrupted cognitive processing. Brain 137:1863–1875
Opris I, Ferrera VP (2014) Manipulating cognition and behavior with microstimulation: implications for cognitive prostheses. Neurosci Biobehav Rev 42:303
Opris I, Hampson RE, Stanford TR, Gerhardt GA, Deadwyler SA (2011) Neural activity in frontal cortical cell layers: evidence for columnar sensorimotor processing. J Cogn Neurosci 23:1507–1521
Opris I, Hampson RE, Gerhardt GA, Berger TW, Deadwyler SA (2012a) Columnar processing in primate pFC: evidence for executive control microcircuits. J Cogn Neurosci 24:2334–2347
Opris I, Fuqua JL, Huettl PF, Gerhardt GA, Berger TW, Hampson RE, Deadwyler SA (2012b) Closing the loop in primate prefrontal cortex: inter-laminar processing. Front Neural Circuits 6:88
Opris I, Santos L, Gerhardt GA, Song D, Berger TW, Hampson RE, Deadwyler SA (2013) Prefrontal cortical microcircuits bind perception to executive control. Sci Rep 3:2285
Pantarotto D, Briand JP, Prato M, Bianco A (2004a) Translocation of bioactive peptides across cell membranes by carbon nanotubes. Chem Commun 1:16–17
Pantarotto D, Singh R, McCarthy D, Erhardt M, Briand JP, Prato M, Kostarelos K, Bianco A (2004b) Functionalized carbon nanotubes for plasmid DNA gene delivery. Angew Chem Int Ed 43:5242–5246
Park KH, Chhowalla M, Iqbal Z, Sesti F (2003) Single-walled carbon nanotubes are a new class of ion channel blockers. J Biol Chem 278:50212–50216
Patolsky F, Timko BP, Zheng G, Lieber CM (2007) Nanowire-based nanoelectronic devices in the life sciences. MRS Bull 32:142–149
Prinz C, Haellstroem W, Martensson T, Samuelson L, Montelius L, Kanje M (2008) Axonal guidance on patterned free-standing nanowire surfaces. Nanotechnology 19(34):345101
Qiu LB, Ding GR, Li KC, Wang XW, Zhou Y, Zhou YC, Li YR, Guo GZ (2010) The role of protein kinase C in the opening of blood-brain barrier induced by electromagnetic pulse. Toxicology 273:29–34
Robinson JT, Jorgolli M, Shalek AK, Yoon M-H, Gertner RS, Park H (2012) Vertical nanowire electrode arrays as a scalable platform for intracellular interfacing to neuronal circuits. Nat Nanotechnol 7:180–184
Rutten WLC (2002) Selective electrical interfaces with the nervous system. Annu Rev Biomed Eng 4:407–452
Sacha GM, Varona P (2013) Artificial intelligence in nanotechnology. Nanotechnology 24 (45):452002
Sakhtianchi R, Minchin RF, Lee K-B, Alkilany AM, Serpooshan V, Mahmoudi M (2013) Exocytosis of nanoparticles from cells: role in cellular retention and toxicity. Adv Colloid Interf Sci 201:18–29
Schouenborg J (2011) Biocompatible multichannel electrodes for long-term neurophysiological studies and clinical therapy-novel concepts and design. Brain Mach Interfaces Implic SciClin Pract Soc 194:61–70
Seraji S, Wu Y, Jewell-Larson NE, Forbess MJ, Limmer SJ, Chou TP, Cao GZ (2000) Patterned microstructure of sol-gel derived complex oxides using soft lithography. Adv Mater 12:1421–1424
Sharma P, Ahuja P (2008) Recent advances in carbon nanotube-based electronics. Mater Res Bull 43:2517–2526

Sharma S, Sunkara MK (2002) Direct synthesis of gallium oxide tubes, nanowires, and nanopaintbrushes. J Am Chem Soc 124:12288–12293

Shen AM, Chen C-L, Kim K, Cho B, Tudor A, Chen Y (2013) Analog neuromorphic module based on carbon nanotube synapses. ACS Nano 7:6117–6122

Stankova L, Fraczek-Szczypta A, Blazewicz M, Filova E, Blazewicz S, Lisa V, Bacakova L (2014) Human osteoblast-like MG 63 cells on polysulfone modified with carbon nanotubes or carbon nanohorns. Carbon 67:578–591

Stieglitz T (2007) Restoration of neurological functions by neuroprosthetic technologies: future prospects and trends towards micro-, nano-, and biohybrid systems. Operative neuromodulation, vol 1: Functional neuroprosthetic surgery, an Introduction 97, pp 435–442

Suyatin DB, Hallstram W, Samuelson L, Montelius L, Prinz CN, Kanje M (2009) Gallium phosphide nanowire arrays and their possible application in cellular force investigations. J Vac Sci Technol B 27:3092–3094

Suyatin DB, Wallman L, Thelin J, Prinz CN, Jorntell H, Samuelson L, Montelius L, Schouenborg J (2013) Nanowire-based electrode for acute in vivo neural recordings in the brain. Plos One 27:3092–3094

Suzuki I, Fukuda M, Shirakawa K, Jiko H, Gotoh M (2013) Carbon nanotube multi-electrode array chips for noninvasive real-time measurement of dopamine, action potentials, and postsynaptic potentials. Biosens Bioelectron 49:270–275

Szentagothai J, Arbib MA (1975) Conceptual models of neural organization. MIT Press, Cambridge, MA

Takei K, Kawano T, Kawashima T, Sawada K, Kaneko H, Ishida M (2010) Microtube-based electrode arrays for low invasive extracellular recording with a high signal-to-noise ratio. Biomed Microdevices 12:41–48

Tian ZRR, Voigt JA, Liu J, McKenzie B, McDermott MJ, Rodriguez MA, Konishi H, Xu HF (2003) Complex and oriented ZnO nanostructures. Nat Mater 2:821–826

Tian B, Cohen-Karni T, Qing Q, Duan X, Xie P, Lieber CM (2010) Three-dimensional, flexible nanoscale field-effect transistors as localized bioprobes. Science 329:830–834

Timko BP, Cohen-Karni T, Qing Q, Tian B, Lieber CM (2010) Design and implementation of functional nanoelectronic interfaces with biomolecules, cells, and tissue using nanowire device arrays. IEEE Trans Nanotechnol 9:269–280

Tye KM, Deisseroth K (2012) Optogenetic investigation of neural circuits underlying brain disease in animal models. Nat Rev Neurosci 13:251–266

Vaadia E, Birbaumer N (2009) Grand challenges of brain computer interfaces in the years to come. Front Neurosci 3:151–154

Valcarcel V, Souto A, Guitian F (1998) Development of single-crystal alpha-Al_2O_3 fibers by vapor-liquid-solid deposition (VLS) from aluminum and powdered silica. Adv Mater 10:138–+

Vayssieres L (2003) Growth of arrayed nanorods and nanowires of ZnO from aqueous solutions. Adv Mater 15:464–466

Vidu R, Hara S (1999a) Surface alloying at the Cd vertical bar Au(100) interface in the upd region: electrochemical studies and in situ EC-AFM observation. J Electroanal Chem 475:171–180

Vidu R, Hara S (1999b) In situ EC-AFM observation of Cd electrodeposition on Au(100). Scr Mater 41:617–624

Vidu R, Hirai N, Hara S (2001) Comparative kinetic study of Cd diffusion into Au(100) and Ag (100) during electrodeposition. Phys Chem Chem Phys 3:3320–3324

Vidu R, Zhang LQ, Waring AJ, Lehrer RI, Longo ML, Stroeve P (2002) Phospholipid bilayers on a polyion-alkylthiol layer pair: microprobe imaging, electrochemical properties and peptide association. Mater Sci Eng B-Solid State Mater Adv Technol 96:199–208

Vidu R, Ku J-R, Stroeve P (2006) Growth of ultrathin films of cadmium telluride and tellurium as studied by electrochemical atomic force microscopy. J Colloid Interface Sci 300:404–412

Vidu R, Ku J-R, Stroeve P (2007) Fabrication of multiscale nanostructures from polymeric membrane templates. In: Nalwa HS (ed) Polymeric nanostructures and their applications. American Scientific Publishers, Stevenson Ranch, pp 124–149

Vidu R, Li S, Quach DV, Stroeve P (2012) Electrochemical deposition of Co-Sb thin films on nanostructured gold. J Appl Electrochem 42:333–339

Vidu R, Rahman M, Mahmoudi M, Enachescu M, Poteca TD, Opris I (2014) Nanostructures: a platform for brain repair and augmentation. Front Syst Neurosci 8:91

Viventi J, Kim D-H, Vigeland L, Frechette ES, Blanco JA, Kim Y-S, Avrin AE, Tiruvadi VR, Hwang S-W, Vanleer AC, Wulsin DF, Davis K, Gelber CE, Palmer L, Van der Spiegel J, Wu J, Xiao J, Huang Y, Contreras D, Rogers JA, Litt B (2011) Flexible, foldable, actively multiplexed, high-density electrode array for mapping brain activity in vivo. Nat Neurosci 14:1599–U1138

Wade LA, Katzman R (1975) 3-0-Methyldopa uptake and inhibition of L-dopa at the blood-brain barrier. Life Sci 17:131–136

Wang X, Li YD (2002) Selected-control hydrothermal synthesis of alpha- and beta-MnO_2 single crystal nanowires. J Am Chem Soc 124:2880–2881

Wenrong Y, Ratinac KR, Ringer SP, Thordarson P, Gooding JJ, Braet F (2010) Carbon nanomaterials in biosensors: should you use nanotubes or graphene? Angew Chem Int Ed 49:2114–2138

Wise KD (2007) Integrated sensors, MEMS, and microsystems: reflections on a fantastic voyage. Sens Actuat Phys 136:39–50

Wu XC, Song WH, Huang WD, Pu MH, Zhao B, Sun YP, Du JJ (2000) Crystalline gallium oxide nanowires: intensive blue light emitters. Chem Phys Lett 328:5–9

Wu Y, Cui Y, Huynh L, Barrelet CJ, Bell DC, Lieber CM (2004) Controlled growth and structures of molecular-scale silicon nanowires. Nano Lett 4:433–436

Wu W, Wieckowski S, Pastorin G, Benincasa M, Klumpp C, Briand JP, Gennaro R, Prato M, Bianco A (2005) Targeted delivery of amphotericin B to cells by using functionalized carbon nanotubes. Angew Chem Int Ed 44:6358–6362

Xiang J, Vidan A, Tinkham M, Westervelt RM, Lieber CM (2006a) Ge/Si nanowire mesoscopic Josephson junctions. Nat Nanotechnol 1:208–213

Xiang J, Lu W, Hu YJ, Wu Y, Yan H, Lieber CM (2006b) Ge/Si nanowire heterostructures as high-performance field-effect transistors. Nature 441:489–493

Xiao ZL, Han CY, Welp U, Wang HH, Kwok WK, Willing GA, Hiller JM, Cook RE, Miller DJ, Crabtree GW (2002) Fabrication of alumina nanotubes and nanowires by etching porous alumina membranes. Nano Lett 2:1293–1297

Xie C, Hanson L, Xie W, Lin Z, Cui B, Cui Y (2010) Noninvasive neuron pinning with nanopillar arrays. Nano Lett 10:4020–4024

Xie C, Lin Z, Hanson L, Cui Y, Cui B (2012) Intracellular recording of action potentials by nanopillar electroporation. Nat Nanotechnol 7:185–190

Yajie D, Guihua Y, McAlpine MC, Wei L, Lieber CM (2008) Si/a-Si core/shell nanowires as nonvolatile crossbar switches. Nano Lett 8:386–391

Yin YD, Zhang GT, Xia YN (2002) Synthesis and characterization of MgO nanowires through a vapor-phase precursor method. Adv Funct Mater 12:293–298

Yoon I, Hamaguchi K, Borzenets IV, Finkelstein G, Mooney R, Donald BR (2013) Intracellular neural recording with pure carbon nanotube probes. Plos One 8(6):e65715

Yu DP, Hang QL, Ding Y, Zhang HZ, Bai ZG, Wang JJ, Zou YH, Qian W, Xiong GC, Feng SQ (1998) Amorphous silica nanowires: intensive blue light emitters. Appl Phys Lett 73:3076–3078

Zhang L, Vidu R, Waring AJ, Lehrer RI, Longo ML, Stroeve P (2002) Electrochemical and surface properties of solid-supported, mobile phospholipid bilayers on a polyion/alkylthiol layer pair used for detection of antimicrobial peptide insertion. Langmuir 18:1318–1331

Zheng B, Wu YY, Yang PD, Liu J (2002) Synthesis of ultra-long and highly oriented silicon oxide nanowires from liquid alloys. Adv Mater 14:122–+

Zheng GF, Lu W, Jin S, Lieber CM (2004) Synthesis and fabrication of high-performance n-type silicon nanowire transistors. Adv Mater 16:1890–+

Zhou YK, Huang J, Shen CM, Li HL (2002) Synthesis of highly ordered $LiNiO_2$ nanowire arrays in AAO templates and their structural properties. Mater Sci Eng Struct Mater Prop Microstruct Process 335:260–267

Zorzos AN, Scholvin J, Boyden ES, Fonstad CG (2012) Three-dimensional multiwaveguide probe array for light delivery to distributed brain circuits. Opt Lett 37:4841–4843

Chapter 19
Extraction of Cortical Modularity Patterns for Neural Prosthetics

Sam A. Deadwyler, Ioan Opris, Lucas M. Santos, Robert E. Hampson, Greg A. Gerhardt, Dong Song, Vasilis Z. Marmarelis, and Theodore W. Berger

Abstract Cortical modularity is a fundamental microanatomic feature of the brain with direct implications for the cognitive function of cortical microcircuits. Neural activity recorded simultaneously with multi-electrode arrays (MEAs) from supra- and infra-granular layers along adjacent cortical minicolumns in PFC was shown to extract microanatomic codes relevant for successful behavioral performance. In addition, it is shown that pharmacologic agents disrupt the micro-anatomic processing and cognitive performance, but that recovery from cognitive impairment is produced by application of a cognitive prosthesis, via nonlinear multi-input multi-output (MIMO) model stimulation of microanatomic outputs with successful MIMO codes. The functional basis of this approach provides the potential for applying cognitive prostheses to a broad range of neurological and psychiatric dysfunctions involving cortical processes.

Keywords Cortical microcircuits • Microelectode arrays • Cortical minicolumn • Cortical layer • Recording • Prosthetics • Brain machine interface

---

S.A. Deadwyler (✉) • L.M. Santos • R.E. Hampson
Department of Physiology and Pharmacology, Wake Forest University Health Sciences, Medical Center Boulevard, Winston-Salem, NC 27157, USA
e-mail: sdeadwyl@wakehealth.edu; santos@wakehealth.edu; rhampson@wakehealth.edu

I. Opris, Ph.D.
Department of Physiology and Pharmacology, Wake Forest University Health Sciences, Winston-Salem, NC 27103, USA
e-mail: ioanopris.phd@gmail.com

G.A. Gerhardt
Department of Anatomy and Neurobiology, Parkinson's Disease Translational Research Center of Excellence, Center for Microelectrode Technology, University of Kentucky, Lexington, KY 40506, USA
e-mail: gregg@uky.edu

D. Song • V.Z. Marmarelis • T.W. Berger
Department of Biomedical Engineering, University of Southern California, 403 Hedco Neuroscience Building, Los Angeles, CA 90089, California
e-mail: dsong@usc.edu; marmarelis@hotmail.com; berger@usc.edu

M.F. Casanova, I. Opris (eds.), *Recent Advances on the Modular Organization of the Cortex*, DOI 10.1007/978-94-017-9900-3_19

19.1 Introduction

Most of our existence depends on our brain's ability to perceive environmental stimuli and to decide about the most appropriate actions to be taken. The primate neocortex, whether sensorimotor, cognitive or emotionally driven, provokes a broad spectrum of behavioral routines. Cortical modularity is a fundamental microanatomic feature of the brain with direct implications for the cognitive function of cortical microcircuits (Bugbee and Goldman-Rakic 1983; Favorov et al. 1987; Kritzer and Goldman-Rakic 1995; Mountcastle 1957, 1978; Shepherd and Grillner 2010; Wiesel and Hubel 1974). Neocortical micro-architecture is modular, with neurons forming a minicolumnar-laminar arrangement analogous to a "crystaline" structure (Casanova et al. 2011; Bastos 2012), with inter-laminar cortical microcircuits, similar to integrated circuits, with permanent inter-connected neuron components used for dedicated processing.

19.1.1 Microanatomy of Cortical Modules

Neocortical microanatomy consisting of minicolumnar and laminar arrangements of neurons, linked into a unified network by afferent and efferent connections, is distributed across many regions of the brain with broad functional significance. Cortical minicolumns consist of chains of pyramidal neurons surrounded by a "curtain of inhibition" to regulate excitatory and inhibitory roles (Szentágothai and Arbib 1975). Cortical mantle, on the other hand, has a laminar structure with six layers grouped into supra-granular, granular and infra-granular laminae, with the granular layer 4 that receives sensory input from thalamus (Constantinople and Bruno 2013). Granular layer 4 separates the functionality of cortical minicolumn in perceptual and executive/behavior related actions. In essence, the supra-granular layers consisting of small pyramidal neurons generate a complex network of intra-cortical connections between cortical areas. Ramón y Cajal S (1906) was the first to emphasize the possibility that this network is primarily responsible for high level functions like cognition, learning, and memory. The supra-granular layers also provide a major input to the infra-granular layers of relatively large pyramidal neurons that generate most of the output of cerebral cortex to other parts of the brain (Mountcastle 1997) and, in fact, infra-granular layers are essentially the "executive" part of the cerebral cortex (Opris et al. 2011, 2013). According to this three layer microstructure of a functional cortical module, infra-granular layers execute the "cognitive" computations elaborated in associated supra-granular layers. In the frontal cortex, minicolumnar microcircuits can be regarded as modules that integrate perceptual stimuli of various sensory modalities (visual, auditory or somatosensory) in supra-granular layers and select behavioral relevant signals in the infra-granular layers.

A canonical microcircuit consists of interconnected neurons between cortical layers of the same minicolumn or with different neighboring minicolumns (Bastos 2012; Lorente de No 1938). One prevalent view regarding functional diversity (sensory, motor, cognitive or emotional) of columnar/laminar microcircuitry across the entire neocortex is their "canonical" similarity from frontal to occipital and from parietal to temporal regions (Hubel and Wiesel 1974; Jones and Rakic 2010; Kaas 2012; Buxhoeveden and Casanova 2002; Defelipe 2011; DeFelipe et al. 2012; Douglas et al. 1989; Jones 2000; Mahan and Georgopoulos 2013; Opris 2013; Takeuchi et al. 2011). An element of variability in these inter-laminar circuits is granular layer 4 that is thicker in sensory areas than in the cognitive regions, and thinner in the motor cortex (DeFelipe et al. 2012). However, it is intriguing how the key aspects of cortical microcircuits emerge on the basis of specific inputs, outputs and inter-laminar interactions within the cortical hierarchy, from simple to more complex functionality. Furthermore, disruption of inter-laminar microcircuits within cortical minicolumns is a signature of a broad spectrum of neurologic and psychiatric disorders e.g. autism, schizophrenia, Alzheimer's and drug addiction (Casanova 2007, 2008; Casanova et al. 2007; Chance et al. 2006, 2011; Opris et al. 2012a). Recent technological advances provided an unprecedented insight into cortical modularity and function of primate brain suggesting that targeting cortical microcircuitry is key to the development of successful medical treatments and enhancement methods including neural prosthetics.

19.2 Recording Modularity in Primate Prefrontal Cortex

19.2.1 Neuromorphic Assessment of Prefrontal Cortical Neural Processing During Target Selection

The relevance of prefrontal cortical (PFC) neural activity to decision making has been investigated under several conditions in the past (Rao et al. 1999; Goldman-Rakic 1996; Opris and Bruce 2005; Pesaran et al. 2008; Resulaj et al. 2009; Heekeren et al. 2008; Opris et al. 2011, 2012a, 2013). Many of the prior reports of neural correlations during executive function and decision making in a sensorimotor hierarchy (Miller and Cohen 2001; Pesaran et al. 2008; Opris and Bruce 2005; Opris et al. 2012a, b; Sugrue et al. 2005) describe recordings from the area of the dorsolateral PFC (DLPFC) illustrated in Fig. 19.1a, c in nonhuman primates trained to perform a delayed match to sample (DMS) task in Fig. 19.1b (Hampson et al. 2004a, 2010, 2011; Opris et al. 2012a; Porrino et al. 2013; Moxon et al. 2004). In this study, interlaminar connectivity was sensed by conformal-configured MEAs (Fig. 19.1c, d) positioned to simultaneously record PFC L2/3 and L5 neurons in adjacent 'minicolumns' during the performance of the DMS task (Hampson et al. 2004b) as shown in Fig. 19.1b.

The DMS task required hand-tracking movements by NHPs to place a cursor into different positions on a screen to obtain a juice reward for selection of the correct (sample) image, the location of which varied randomly in one of seven spatial positions on the screen on each trial (Hampson et al. 2004b). Key variables in the task were the number of images (2–7) presented on the screen in the match

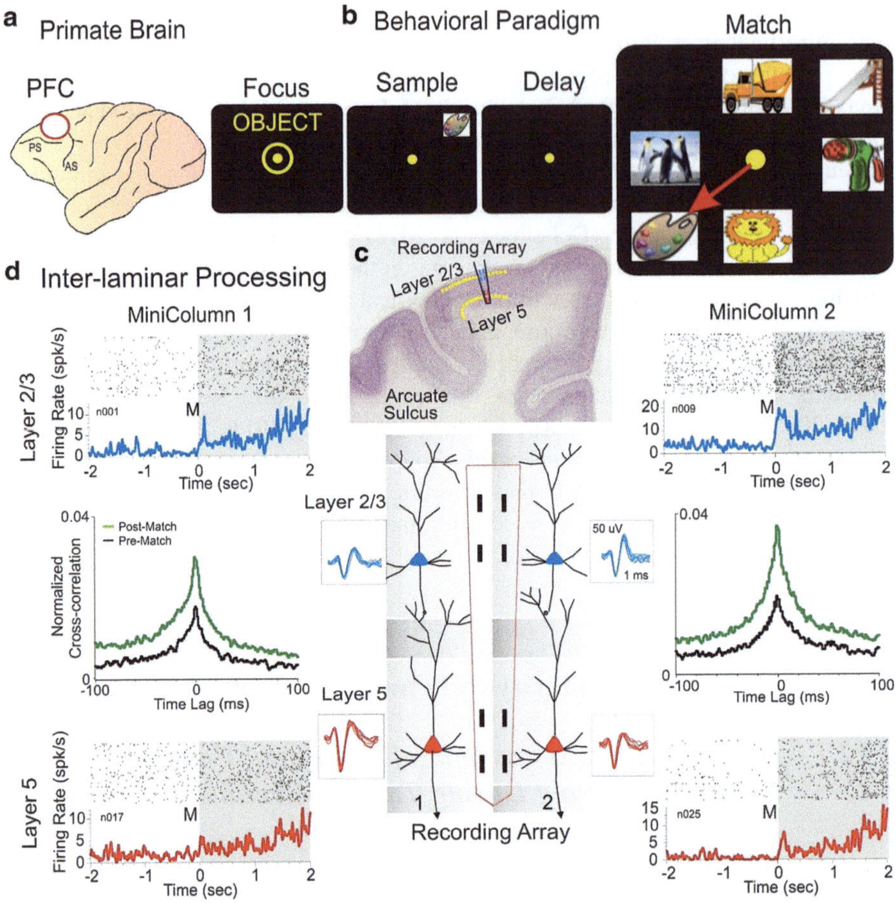

Fig. 19.1 Behavioral paradigm and recording of inter-laminar activity in PFC. Behavioral task and Prefrontal columnar recording. (**a**) Diagram of NHP brain with PFC recording locations in areas 46, 8, 6. (**b**) Behavioral paradigm showing sequence of events in DMS task. (1) 'Focus ring' presentation and response to initiate the trial commencing with (2) presentation of the 'sample target' image, followed by (3) 'sample response' by cursor movement into the image which initiated (4) a variable 'delay' period of 1–90 s prior to (5) the presentation of the 'match' target (sample image) accompanied by 1–6 non-match (distracter) images on the same screen. *Cursor* movement into correct (match target) image for 0.5 s was rewarded by juice reward (0.5 ml) via a sipper tube next to the animal's mouth. Placement of the *cursor* into a non-match image for 0.5 s caused the screen to blank without reward delivery. Intertrial interval (ITI): 10.0 s. (**c**) Histologic section of DLPFC brain showing relative location of supra-granular layer 2/3 (L2/3, *blue*) and infra-granular layer 5 (L5, *red*) with tract used for placement of conformal multielectrode recording array (MEA) probe (W3). (**d**) Correlated inter-laminar activity within two adjacent prefrontal minicolumns (1&2) during DMS task performance. Center panel illustrates conformal MEA positioned for simultaneous inter-laminar columnar recording from adjacent minicolumns 1 and 2 (40 μm separation) from L2/3 and L5 cell pairs (blue and red extracellular spike waveforms). Associated cell activity is shown as individual trial rasters and average PEHs obtained from both cell pairs recorded simultaneously from L2/3 (*blue*) and L5 (*red*) within the respective minicolumns (1&2). Rasters and PEHs depict ±2.0 s relative to onset (0.0 s) of Match phase screen on each trial (Fig. 19.1b) within a single DMS session. In the middle of the raster-PEH displays are shown cross-correlation histograms (CCHs) for the same cell pair minicolumns

phase of the trial, the duration of the interposed delay (1.0–90.0 s), as well as the random placement of the sample image in the match phase (after the delay interval) that differed from the position responded to in the sample phase. In addition, to incorporating key cognitive features such as attention, short-term memory, cognitive workload and reward expectancy, subjects were also executing a 'decision process' in the match phase (Fig. 19.1b) which involved 'target selection', i.e. a process that was related directly to columnar related neuron firing in the PFC (Hampson et al. 2011).

A demonstration of this type of columnar processing is shown in Fig. 19.1d for two adjacent PFC minicolumns recorded simultaneously with the MEA during the match phase of the task. Raster/PEHs for the two simultaneously recorded cell pairs (L2/3 and L5) at the indicated locations on the conformal MEA array (Fig. 19.1d minicolumns 1&3) positioned via pad separation dimensions of 1300u are shown in Fig. 19.1c. There were significant increases in overall mean firing rates at the onset of the match phase ('M' onset = 0.0 s; pre vs. post, $p < 0.001$) over the time (from 0.0 to + 2.0 s) in which arm movements involved in target selection were initiated (Hampson et al. 2011, 2012b; Opris et al. 2012a; Porrino et al. 2013). In addition, the L2/3 neurons shown in Fig. 19.1d (upper raster/PEHs) exhibited significantly higher overall mean firing rates ($p < 0.001$) than the L5 neurons (lower raster/PEHs) recorded in minicolumns 1 and 2 on the MEA during the same phase of the task. Most important was the determination that the firing of cell pairs recorded in the same vertical positions on the MEA reflected columnar-based interlaminar communication via cross-correlations between cell pairs (minicolumns 1 and 2, Fig. 19.1d) which showed increased firing synchrony at the time of target presentation and response selection in the match phase.

The normalized CCHs in Fig. 19.1d depict differences in the correlation of L2/3 and L5 firing of the same cell pairs with respect to (1) the 'pre' match phase onset period ($M = 0.0$) baseline (−2.0–0.0 s) versus after presentation of images on the screen in the 'post'-match phase (0.0 s to +2.0 s) onset period (Fig. 19.1b). The specificity of the cross-correlations between both L2/3 and L5 cell pairs (minicolumns 1 and 2) is shown in Fig. 19.1d by the significant increase in CCHs ($p < 0.001$) during performance of the Match response (green, Post-Match) relative to the 'pre' match phase period when the animal was waiting for the Delay interval to time out prior to image presentation (black, Pre-Match). Results revealed reduced correlated firing between cell pairs within single minicolumns on error trials with inappropriate target selection (Opris et al. 2011, 2012a, b). This was a direct demonstration of task-specific, real-time columnar processing in prefrontal cortex indicating the role of inter-laminar microcircuits in executive control of decision-making in primate brain.

Fig. 19.1 (continued) which show display increased inter-laminar synchronization during target selection (*green*) in the match phase (0.0 + 2, 0 s, post) relative to similar but reduced correlations between the same cell pairs constructed 2.0 s prior to (*black*, pre) Match phase onset (−2.0 to 0.0 s in PEHs) (Compiled from Opris et al. 2012a, b)

19.3 Derivation by MIMO Model of PFC Columnar Processing

19.3.1 Application of the MIMO Model to PFC Columnar Processing

In prior studies (Berger et al. 2011; Hampson et al. 2012b, Hampson et al. 2013; Opris et al. 2013), it has been shown that a multi-input multi-output model (MIMO) nonlinear dynamic model applied to spatiotemporal patterns of multiple recordings from synaptically connected neurons is capable of extracting patterns of firing related to successful task performance of this same task shown in Fig. 19.1a. It was shown that the MIMO model could be used to facilitate and recover performance when administered as patterns of electrical pulses to the same recording locations from which output signals were detected on correct trials (Hampson et al. 2012a, 2013; Berger et al. 2011). The MIMO model applied to minicolumnar data recorded by the conformal MEA probes in the animals performing the DMS task shown in Fig. 19.1 as a modeling strategy for the nonlinear dynamics underlying spike-train transformations between L2/3 and L5. This was then used to predict L5 output firing patterns from input patterns of L2/3 neural activity, as a representation of multicolumnar firing patterns (Berger et al. 2005, 2011, 2012; Hampson et al. 2012a; Song et al. 2007, 2009). In this application, the identification of spatio-temporal pattern transformation from the PFC layer 2/3 to layer 5 in MEA identified columns was formulated by the MIMO model and analyses included extraction of the first-, second- and third-order temporal firing within at least two defined minicolumns on MEAs inserted repetitively on multiple recording sessions in order to extract relevant patterns of minicolumnar activity related to successful image selection during the match phase of the task.

The same MIMO model was applied in this study to multicolumnar PFC L2/3 and L5 cell pairs recorded during the Match phase of the DMS task. Spatiotemporal patterns associated with trials of correct performance were constructed from 2–4 L2/3 and L5 confirmed MEA cell pairs recorded in five different NHPs ($n = 5$) performing the same task. MIMO model derived L5 output firing associated with L2/3 input firing, was used to predict performance on individual trials. Figure 19.2 shows the relationship of L5 firing to task performance with respect to the firing patterns extracted by the MIMO model. Performance of the task shown in Fig. 19.2a was differential with respect to the number of distracter images with an increase in image number associated with poorer performance. However if performance on trials was segregated and the type of MIMO output associated with performance identified, a clear distinction between the types of patterns present on successful (strong code) vs. poorer (weak code) performed trials for each number of images was clearly present as shown in Fig. 19.2b, c. The MIMO model clearly can derive appropriate firing patterns but what is more important detection of patterns that are less appropriate for successful performance can be

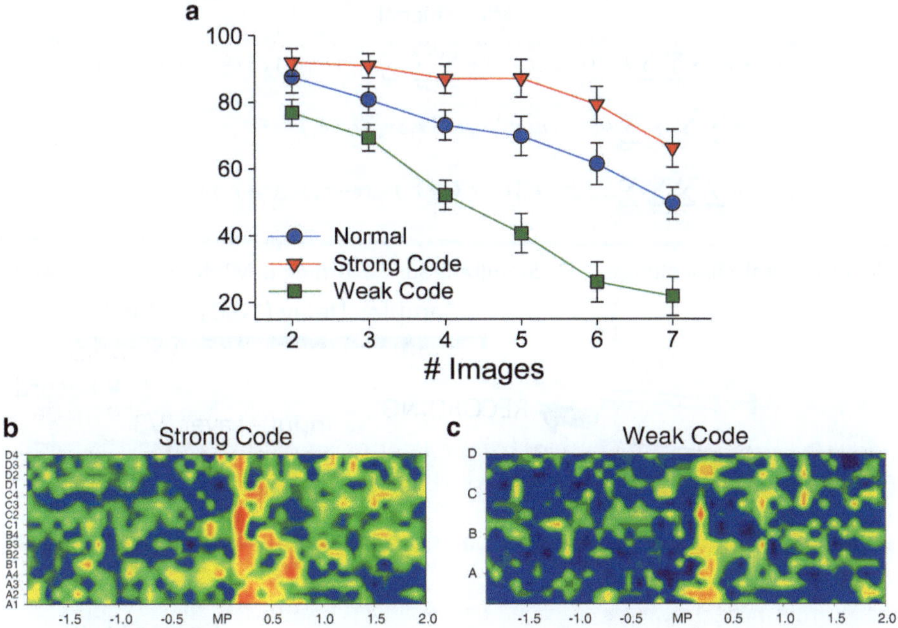

Fig. 19.2 Prefrontal cortical module code strength vs. DMS performance. (**a**) Behavioral performance, quantified as percent correct, segregated according to normal average performance over all trials (*blue*), above average performance on trials in which spatiotemporal strong code patterns (*red*) were detected, or trials in which performance was below average and different "weak code" (*green*) patterns detected. *Heat maps* show the time course of PFC layer 2/3 cell firing ($n = 20$ cells) on MIMO model derived strong code (**b**), vs. weak code (**c**) trials, shown in (a) prior (−2.0 to 0 s) to and following (0 to +2.0 s) Match Phase presentation (MP)

predicted by monitoring the associated L2/3 input patterns for wrong code trials online that provides time for modifying the L5 firing pattern on the same trial.

19.3.2 Facilitation of PFC Columnar Processing by MIMO Microstimulation

Prior investigations applying MIMO model-derived stimulation patterns to hippocampus in the rodent provided the means to enhance performance and overcome deficits induced by pharmacological treatments (Berger et al. 2011). In this manner, predictions of the L5 output related to the successful performance were monitored online during the task to define when successful or non-successful trials were about to be completed prior to target selection in the Match phase as shown in Fig. 19.3. This online monitoring by the MIMO model provided the basis for interposing the correct (strong code) L5 firing pattern as electrical pulses delivered to the same L5 recording pads of the MEAs at the same time as normal occurrence after Match

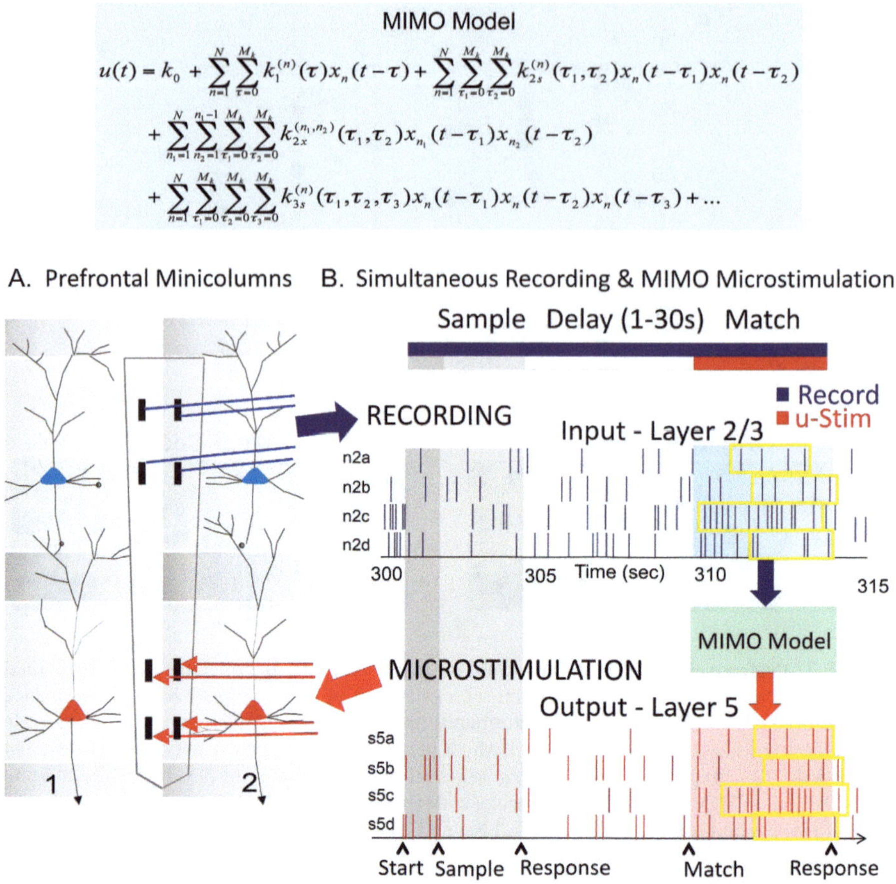

Fig. 19.3 Columnar microstimulation delivered via MIMO model. Online application of MIMO model for calculating Match phase response codes from L2/3 recordings and delivering electrical stimulation pulses to L5 recording pads that mimic prior associated L5 strong codes during the same trial. Schematic at *left* shows prefrontal cortical (PFC L2/3) recording and the NHP MIMO model with feedback stimulation applied to PFC L5. Neural recordings from layer 2/3 are analyzed to predict layer 5 neural activity from multilevel temporal analyses, which is used to generate stimulation patterns applied to layer 5 recording sites after assessment of L2/3 patterns on the same trial. MIMO model coefficients applicable to PFC recordings distinguish different features of the DMS task. This has provided the means to test the specificity of the MIMO codes recorded in L2/3 that occur on different types of trials (with different cognitive load) when applied as stimulation patterns to L5

phase onset and completion of target selection (Fig. 19.3). Stimulation pulses consisted of biphasic constant current square waves, 0.5 ms per phase, adjusted in intensity (10–50 μA) to produce local field potentials monitored on adjacent L5 recoding pads on the same MEA. Model-derived stimulation was applied on 50 % of trials in each session to compare stimulated and non-stimulated trials in terms of correct performance. Effective stimulation patterns that were determined to facilitate performance under normal conditions were also applied at the same time during the trial, irrespective of prior L2/3 activity, when employed for the recovery of function as a neural prosthesis as described below (Fig. 19.5). In this manner, it was demonstrated that the MIMO model served as a prostheses for recovering decision making that required appropriate PFC columnar processing related to successful target selection in the DMS task.

The effects of MIMO generated electrical stimulation patterns averaged over all NHPs and trials (Fig. 19.4) were highly significant ($p < 0.001$, ANOVA). It is clear that MIMO stimulation facilitated performance above normal levels but was not consistent for the same types of trials in all animals (Fig. 19.4a, individual NHPs); however, Fig. 19.4b shows that the overall effect of MIMO stimulation was to significantly facilitate performance in all animals across trials with increased number of images ($p < 0.001$). Figure 19.4c shows an additional facilitatory effect of the MIMO-derived stimulation on trials with longer delays (30–60 s, 61–90 s) where performance was less impaired on stimulated versus nonstimulated trials ($p < 0.001$). Finally, the facilitatory effects of MIMO stimulation on target selection and execution were strongly supported by a significant decrease ($F(4{,}401) = 9.14$, $p < 0.01$) in the latency to make the match response (Fig. 19.4d) as a function of trial difficulty (number of images) on MIMO stimulation versus nonstimulation trials. The controls associated with these results are found in Hampson et al. (2012b).

19.4 Drug Induced Disruption of Modularity and Cognition

19.4.1 Pharmacologic Agents Disrupt Microanatomic Processing and Cognitive Performance

Extensive prior investigation of features that affect cognitive processing in DMS tasks have shown that Match phase activation of PFC is altered by the modulation of dopamine influences on task-related PFC cell firing by cocaine and other agents that alter dopamine uptake (Hampson et al. 2011; Porrino et al. 2013; Arnsten and Robbins 2009). To determine similar actions on PFC columnar activity in this task firing was assessed in the same minicolumns before and after systemic injection of cocaine (0.40 mg kg^{-1}) midway through the DMS session. Figure 19.5a shows raster/PEHs for a PFC inter-laminar cell pair (L2/3 upper, L5 lower) recorded: (1) in the first 60 trials of a DMS session (IV Saline control) followed by, (2) the un-signaled administration (IV) of cocaine on trial 61 in the same session of 120 trials. It is clear that a significant reduction in match phase firing occurred in L2/3 ($p < 0.001$) and L5 ($p < 0.001$) in the cocaine versus control half of sessions (Fig. 19.5a), and that these changes in firing resembled closely the type of reduced firing that occurs on error trials (Opris et al. 2012a). The generality of this effect on

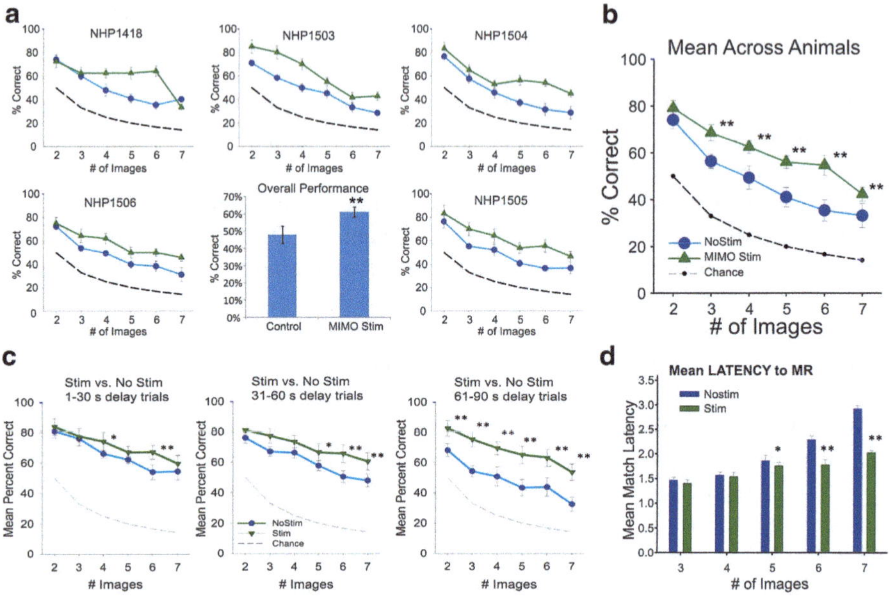

Fig. 19.4 Facilitation of DMS performance by MIMO microstimulation. Facilitation of DMS performance by MIMO stimulation (**a**). MIMO stimulation applied to prefrontal cortex in five different NHPs (indicated by *number*) performing the DMS task under normal conditions. MIMO model utilized L2/3 input to predict L5 output patterns delivered as electrical stimulation (Fig. 19.3) during target image selection in the Match phase of the task (Fig. 19.1 Behavioral graphs for each animal show mean DMS performance (±SEM) as a function of trial complexity indicated by number of images (2–7) for 3–5 sessions consisting of 120 trials each). Performance is shown on normal trials in which stimulation was not delivered (No Stim) and trials in the same session in which the MIMO model stimulation was delivered to PFC L5 as shown in Fig. 19.3. *Dashed line* indicates performance that could be achieved by random 'chance' selection at each degree of match difficulty related to the number of images to select from. *Inset*: average performance summed across trials, animals and sessions for Control (nonStim) vs. MIMO Stim trials. $^{**}F(5{,}239) > 42.16$, $p < 0.001$ increase compared to no stimulation. (**b**) Mean DMS performance as a function of images across all animals shown in (**a**) for stimulation versus no stimulation trials. $^{**}F(1{,}239) > 18.34$, $p < 0.001$ increase compared to no stimulation. (**c**) Effect of MIMO stimulation on DMS trials with differing delays as a function of number of images. Same results shown in (**b**) sorted by duration of trial delay prior to match phase onset (Fig. 19.1a) for no stimulation versus stimulation trials. $^{*}F(1{,}239) > 8.22$, $p < 0.01$; $^{**}F(1{,}239) > 13.40$, $p < 0.001$ increase compared to no stimulation. (**d**) Effect of MIMO stimulation on mean Match response (MR) latency to respond to the match image on stimulation versus no stimulation trials. Mean (±SEM) latencies (in seconds) across animals and sessions as a function of number of images presented in the match phase. $^{*}F(1{,}239) = 9.51$, $p < 0.01$; $^{**}F(1{,}239) > 21.29$, $p < 0.001$ versus no stimulation (With permission Hampson et al. 2012a)

match phase firing over a large population of cell pairs ($n = 30$) is shown in Fig. 19.5b as a significant decrease in the L2/3 cell activity ($p < 0.001$) relative to the saline half of the session. The reduction in L5 average firing rates in the cocaine half of the session approached but did not reach significance ($p > 0.10$) which perhaps reflects decreased dopamine sensitive processes (Gulledge and Jaffe

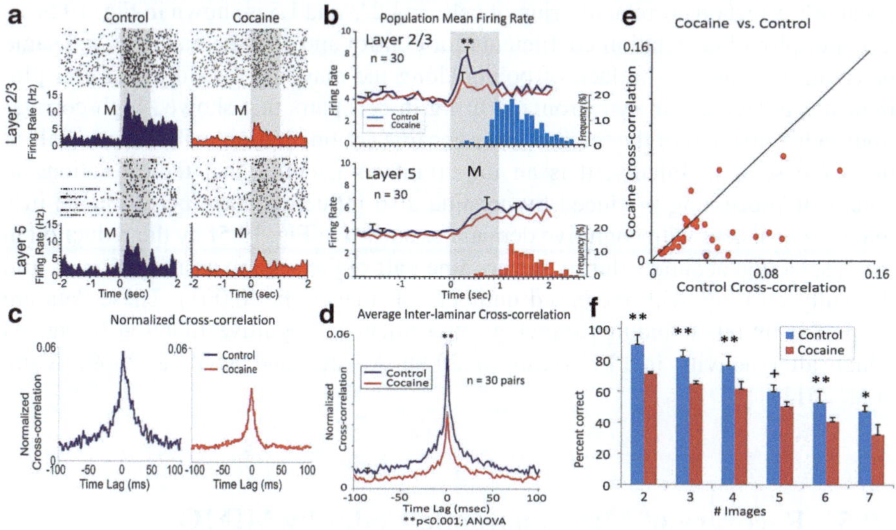

Fig. 19.5 Disruption of cortical modularity and cognition. Pharmacological interruption of DMS-dependent inter-laminar processing. Effects of administration cocaine (0.40 mg kg^{-1} IV) at midsession on MEA L2/3-L5 cell pair firing during performance of DMS task. (**a**): Rasters and PEHs show Match phase L2/3 and L5 inter-laminar activity during the initial control (*saline*, *blue*) portion of the session and after cocaine administration (*cocaine*, *red*) midway through the same session. $^{**}F(1{,}958) > 19.72$, $p < 0.001$ versus Control (*saline*). (**b**) Average PEHs for control (*upper*) versus cocaine trials (*lower*) summed across animals over all inter-laminar L2/3 (*blue*) & L5 (*red*) PFC cell pairs ($n = 30$) recorded in the same sessions with cocaine administered at the midpoint (trial 62) of the session. *Blue* (control) and *red* (cocaine) histograms show mean frequency distributions of associated MR latencies relative to Match phase onset (M, 0.0 s) for the same trials. $^{*}F(1{,}958) = 13.43, p < 0.001$. (**c**) Cross-correlograms (CCHs) of firing between the L2/3 and L5 cell pair shown in (**a**) constructed from control trials ($n = 62$) in the first half of the session (*blue-left*), and after cocaine administration during the second half of the same session (*red-right*). $^{**}F(1{,}401) = 17.22, p < 0.001$ versus Control. (**d**) Mean cross-correlation histograms (CCHs) for the same inter-laminar cell pairs ($n = 30$) shown in (**b**) constructed from trials in the control (*blue*) versus cocaine (*red*) halves of the session. $^{**}F(1{,}401) = 11.22, p < 0.001$, versus Control. (**e**) Scatter plot of normalized cross-correlation coefficients from cell pairs shown in (**d**) for control (*horizontal axis*) and cocaine (*vertical axis*) halves of the same DMS sessions. Distribution of coefficients along the *diagonal line* would represent no change in correlation coefficients between the two halves of the session, whereas the demonstrated asymmetry of coefficient distribution reflects a significant change in inter-laminar correlated cell firing after cocaine administration. (**f**) Reduction in DMS (% correct) performance for all animals on trials with varying number of images (Fig. 19.4) for control versus cocaine segments of the same sessions ($n = 19$). $^{**}F(1{,}239) > 16.01, p < 0.001$ Cocaine versus Control (With permission Hampson et al. 2012a)

1998) relative to L2/3 cells as has been indicated in prior studies of dopamine receptor actions in these PFC layers in NHPs (Bordelon-Glausier et al. 2008; Arnsten and Robbins 2009). However, the more specific columnar firing indicator, CCHs for the single cell pair in Fig. 19.5a and for all cell pairs in Fig. 19.5b ($n = 30$), showed mark clear decreases in correlated firing ($p < 0.001$) after cocaine administration (Fig. 19.5c, d). The significance of this change with respect to cocaine's effect on correlated firing of cells in L2/3 and L5 is shown in Fig. 19.5e as a scatter plot of correlation coefficients for control and cocaine halves of the same behavioral sessions. The lack of points along the diagonal line in the scatter plot indicates a decrease in synchronized firing in cell pairs that showed high correlations in the first half of the session, after cocaine administration in the second half of the same session. Finally, it is an important coincidence that the alterations in columnar processing produced by cocaine also influenced task performance in a manner consistent with cognitive demand as shown in Fig. 19.5f by the reduction in mean selection accuracy during the cocaine half of the session as a function of trial difficulty on trials with increased numbers of images ($p < 0.001$). These data are supported by other findings showing impairment in cognitive function by agents which alter activity in PFC (Arnsten 2000; Arnsten and Robbins 2009; Wang et al. 2011).

19.5 Recovery of Microanatomic Codes by MIMO Microstimulation

19.5.1 Cognitive Impairment Recovered via Stimulation of Microanatomic Outputs with Successful MIMO Codes

Figures 19.3 and 19.4 show that MIMO L5 stimulation was exceedingly effective in facilitating DMS-task performance under normal conditions and this provided the basis for testing the effects of the MIMO model as a neuroprosthetic by delivering MIOMO stimulation to L5 cells on trials disrupted by cocaine injection during the session. Since, as shown in Fig. 19.5e, cocaine's overall effect was to alter PFC columnar firing in the same minicolumns that processed information effectively in the first half of the same session ($p < 0.001$), application of the MIMO model under these conditions (Fig. 19.6a) provided the means to detect this non-effective trial specific firing in L2/3 cells. This provided the online signal for delivery of L5 stimulation patterns associated with columnar firing shown to be facilitatory when delivered under normal (nondrug) conditions (Fig. 19.4).

Figure 19.6b shows that online delivery of MIMO model controlled stimulation on trials during which cocaine disrupted performance in the second half of the session, reversed the detrimental effects on trials with 3–7 images (cocaine vs. cocaine + MIMO Stim; $p < 0.001$; $F(1, 239) > 16.82$). MIMO stimulation not only re-established control performance but also increased it to the level achieved on stimulated trials (control versus drug + MIMO Stim, $p < 0.001$) under control (nondrug) conditions (Fig. 19.6b). These results provide the first evidence that a neural prostheses in the MIMO model format can effectively reverse externally induced impairments in cortical function that directly influence cognitive processing in primate brain. The findings clearly demonstrate that the application

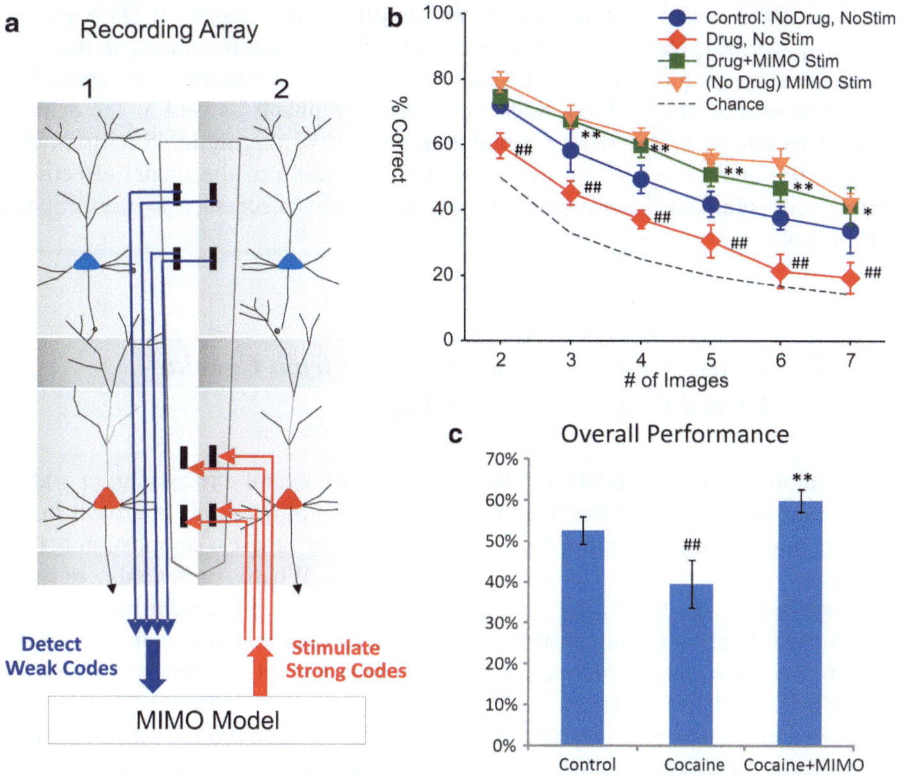

Fig. 19.6 Recovery of microanatomic columnar codes by MIMO stimulation. MIMO-based neural prosthetic recovery of PFC-dependent DMS performance. (**a**) Application of MIMO model detects increased 'weak code' L2/3 firing associated with error trials in DMS performance following cocaine exposure. Output of the MIMO model is then utilized to stimulate L5 (Stim) with a previously determined 'strong code' L5 pattern associated with correct (control) performance at the time of target selection in the match phase. (**b**) DMS performance resulting from MIMO stimulation applied to prefrontal cortex in five NHPs receiving split sessions in which each animal received saline injection prior to start of the session, and then was administered cocaine (0.4 mg kg–1 IV) at the midpoint of the session. DMS mean (±SEM) performance during (1) control (no drug, No Stim) half of the session compared to (2) nonstimulated trials (drug, No Stim) in the cocaine half of the session and (3) MIMO stimulated (drug + MIMO Stim) trials in the cocaine half of the same session. Performance of MIMO stimulation trials in the absence of drug (Fig. 19.4) is also shown for comparison (no drug MIMO Stim). ##$F(1,239) > 16.82$, $p < 0.001$ decrease versus Control. *$F(1,239) = 7.22$, $p < 0.01$; **$F(1,239) > 10.63$, $p < 0.001$ increase versus Control. (**c**) Overall performance (mean ± SEM) shown for all animals on trials in (1) - non-drug half of session (control), (2) cocaine half of session on trials with no stimulation (cocaine) and (3) cocaine half of session on trials with MIMO stimulation (cocaine + MIMO). ##$F(5,239) = 42.53$, $p < 0.001$; performance decrease versus control; **$F(5,239) > 15.05$, $p < 0.001$ performance increase versus control (With permission Hampson et al. 2012a)

of the MIMO model reversed the disruptive effects of cocaine on DMS performance ($p < 0.001$) resulting from reduced PFC inter-laminar processing in the same minicolumns to which successful MIMO L5 stimulation patterns were applied in the same session (Fig. 19.6a). Such inherent determinants as well as the demonstrated specificity of the MIMO stimulation patterns in altering DMS performance (Figs. 19.3 and 19.4) provide evidence that application of the model effectively mimicked columnar information processing in PFC necessary to perform the DMS task.

19.5.2 MIMO Model Induced Recovery from Cocaine Altered Cognitive Processing

Prior applications of MIMO models to disrupted neural processing in rodent hippocampus established the functional basis for employing this approach in the design and implementation of the cortical neuroprostheses demonstrated here (Berger et al. 2011, 2012; Hampson et al. 2012a, b). What is presented here is the first application of the MIMO model to primate brain via a conformal electrode MEA capable of extracting spatiotemporal neural firing patterns related to known underlying columnar microcircuitry in PFC, which not only extends application of the MIMO model to other brain areas but also to the performance of human-like cognitive tasks. The recovery from cocaine-induced disruption shown in Fig. 19.6 utilized the MIMO model to (1) extract, characterize and predict spatiotemporal patterns critical for effective performance and (2) interpose those same patterns into layer 5 using multichannel electrical stimulation.

19.6 Conclusion

These unique results show that columnar interactions between prefrontal neurons that encode and process information relevant to executive function and decision making (Goldman-Rakic 1996; Opris and Bruce 2005; Heekeren et al. 2008; Opris et al. 2005) necessary for the successful performance of this DMS task (Fig. 19.1), are capable of being simulated and interposed to facilitate and recover performance via application of a MIMO model-based neuroprostheses (Berger and Deadwyler 2012; Deadwyler and Granacki 2012; Deadwyler et al. 2007). Since the possible neural basis for effective performance in this task relates to significantly increased transmission within PFC minicolumns to select relevant cues during target presentation in the behavioral task (Figs. 19.2, 19.3, and 19.4), interposing a MIMO model to control this type of processing provides a means of reducing random fluctuations in performance under normal conditions. It was also possible to re-establish appropriate task-dependent processing by recovering columnar processing when performance was impaired by cocaine administration that decreased columnar firing (Figs. 19.5 and 19.6). In addition, to provide potential insight into other types of cognitive impairments involving decision making and executive function as a result of disease or injuries in human brain (Brennan and Arnsten 2008; Dobbs 2010; Duncan et al. 1997; Shallice and Burgess 1991; Wang et al. 2011; Casanova et al. 2010), a MIMO based functional device, if properly integrated with normal brain operation, is capable of recovering or even improving performance in complex tasks.via closed-loop facilitation of columnar processing as demonstrated here in primate brain.

References

Arnsten AF (2000) Stress impairs prefrontal cortical function in rats and monkeys: role of dopamine D1 and norepinephrine alpha-1 receptor mechanisms. Prog Brain Res 126:183–192

Arnsten AF, Robbins TW (2009) The neuropsychopharmacology of fronto-executive function: monoaminergic modulation. Annu Rev Neurosci 32:267–287

Bastos AM et al (2012) Canonical microcircuits for predictive coding. Neuron 76:695–711

Berger TW, Deadwyler SA (2012) A nonlinear model for hippocampal cognitive prostheses: memory facilitation by hippocampal ensemble stimulation. IEEE Trans Neural Syst Rehabil Eng 20:184–197

Berger TW et al (2005) Restoring lost cognitive function. IEEE Eng Med Biol Mag 24:30–44

Berger TW, Hampson RE, Song D, Goonawardena A, Marmarelis VZ, Deadwyler SA (2011) A cortical neural prosthesis for restoring and enhancing memory. J Neural Eng 8:046017

Berger TW, Song D, Chan RHM, Marmarelis VZ, LaCoss J, Wills J, Hampson RE, Deadwyler SA, Granaki JJ (2012) A hippocampal cognitive prosthesis: multi-input, multi-output nonlinear modeling and VLSI implementation. IEEE Trans Neural Syst Rehabil Eng 20(2):198–211

Bordelon-Glausier JR, Khan ZU, Muly EC (2008) Quantification of D1 and D5 dopamine receptor localization in layers I, III, and V of Macaca mulatta prefrontal cortical area 9: coexpression in dendritic spines and axon terminals. J Comp Neurol 508(6):893–905

Brennan AR, Arnsten AF (2008) Neuronal mechanisms underlying attention deficit hyperactivity disorder: the influence of arousal on prefrontal cortical function. Ann N Y Acad Sci 1129:236–245

Bugbee NM, Goldman-Rakic PS (1983) Columnar organization of corticocortical projections in squirrel and rhesus monkeys: similarity of column width in species differing in cortical volume. J Comp Neurol 220:355–364

Buxhoeveden DP, Casanova MF (2002) The minicolumn hypothesis in neuroscience. Brain 125 (5):935–951

Casanova MF (2007) Schizophrenia seen as a deficit in the modulation of cortical minicolumns by monoaminergic systems. Int Rev Psychiatry 19(4):361–372

Casanova MF (2008) The significance of minicolumnar size variability in autism: a perspective from comparative anatomy (ch. 16). In: Zimmerman A (ed) Autism current theories and evidence. Current clinical neurology. The Human Press, Totowa, pp 349–360

Casanova MF, Trippe JT II, Switala AE (2007) A temporal continuity to the vertical organization of the human neocortex. Cereb Cortex 17(1):130–137

Casanova MF, Ayman El-Baz A, Vanbogaert E, Narahari P, Switala A (2010) A topographic study of minicolumnar core width by lamina comparison between autistic subjects and controls: possible minicolumnar disruption due to an anatomical element in-common to multiple laminae. Brain Pathol 20:451–458

Casanova MF, El-Baz AS, Switala AE (2011) Laws of conservation as related to brain growth, aging, and evolution: symmetry of the minicolumn. Front Neuroanat 5:66

Chance SA, Casanova MF, Switala AE, Crow TJ, Esiri MM (2006) Minicolumn thinning in temporal lobe association cortex but not primary auditory cortexin normal human ageing. Acta Neuropathol 111(5):459–464

Chance SA, Clover L, Cousijn H, Currah L, Pettingill R, Esiri MM (2011) Microanatomical correlates of cognitive ability and decline: normal ageing, MCI, and Alzheimer's disease. Cereb Cortex 21(8):1870–1878

Constantinople CM, Bruno RM (2013) Deep cortical layers are activated directly from thalamus. Science 340(6140):1591–1594

Deadwyler SA, Granacki JJ (2012) A hippocampal cognitive prosthesis: multi-input, multi-output nonlinear modeling and VLSI instrumentation. IEEE Trans Neural Syst Rehabil Eng 20:198–211

Deadwyler SA, Porrino L, Siegel JM, Hampson RE (2007) Systemic and nasal delivery of orexin-A (hypocretin-1) reduces the effects of sleep deprivation on cognitive performance in nonhuman primates. J Neurosci 27:14239–14247

Defelipe J (2011) The evolution of the brain, the human nature of cortical circuits, and intellectual creativity. Front Neuroanat 5:29. doi:10.3389/fnana.2011.00029. eCollection 2011

DeFelipe J, Markram H, Rockland KS (2012) The neocortical column. Front Neuroanat 6:22

Dobbs D (2010) Schizophrenia: the making of a troubled mind. Nature 468:154–156

Douglas RJ, Martin KAC, Whitteridge D (1989) A canonical microcircuit for neocortex. Neural Comput 1:480–488

Duncan J, Johnson R, Swales M, Freer C (1997) Frontal lobe deficits after head injury: unity and diversity of function. Cogn Neuropsychol 14:713–741

Favorov OV, Diamond ME, Whitsel BL (1987) Evidence for a mosaic representation of the body surface in area 3b of the somatic cortex of cat. Proc Natl Acad Sci U S A 84(18):6606–6610

Goldman-Rakic PS (1996) The prefrontal landscape: implications of functional architecture for understanding human mentation and the central executive. Philos Trans R Soc Lond B Biol Sci 351:1445–1453

Gulledge AT, Jaffe DB (1998) Dopamine decreases the excitability of layer V pyramidal cells in the rat prefrontal cortex. J Neurosci 18(21):9139–9151

Hampson RE, Coates TD Jr, Gerhardt GA, Deadwyler SA (2004a) Ceramic-based microelectrode neuronal recordings in the rat and monkey. EMBS Proc Annu Int Conf IEEE Eng Med Biol Soc 25:3700–3703

Hampson RE, Pons TP, Stanford TR, Deadwyler SA (2004b) Categorization in the monkey hippocampus: a possible mechanism for encoding information into memory. Proc Natl Acad Sci U S A 101:3184–3189

Hampson RE, Opris I, Deadwyler SA (2010) Neural correlates of fast pupil dilation in nonhuman primates: relation to behavioral performance and cognitive workload. Behav Brain Res 212:1–11

Hampson RE, Porrino LJ, Opris I, Stanford T, Deadwyler SA (2011) Effects of cocaine rewards on neural representations of cognitive demand in nonhuman primates. Psychopharmacology (Berl) 213:105–118

Hampson RE, Gerhardt GA, Marmarelis V, Song D, Opris I, Santos L, Berger TW, Deadwyler SA (2012a) Facilitation and restoration of cognitive function in primate prefrontal cortex by a neuroprosthesis that utilizes minicolumn-specific neural firing. J Neural Eng 9(5):056012

Hampson RE, Song D, Chan RHM, Sweatt AJ, Riley MR, Goonawardena AV, Marmarelis VZ, Gerhardt GA, Berger TW, Deadwyler SA (2012b) Closing the loop for memory prosthesis: detecting the role of hippocampal neural ensembles using nonlinear models. IEEE Trans Neural Syst Rehabil Eng 20:510–525

Hampson RE, Song D, Opris I, Santos LM, Shin D, Gerhardt GA, Marmaarelis V, Berger TW, Deadwyler SA (2013) Facilitation of memory encoding in primate hippocampus by a neuroprosthesis that promotes task specific neural firing. J Neural Eng 10(6):066013

Heekeren HR, Marrett S, Ungerleider LG (2008) The neural systems that mediate human perceptual decision making. Nat Rev Neurosci 9(6):467–479
Hubel DH, Wiesel TN (1974) Sequence regularity and geometry of orientation columns in the monkey striate cortex. J Comp Neurol 158(3):267–293
Jones EG (2000) Microcolumns in the cerebral cortex. Proc Natl Acad Sci U S A 97(10): 5019–5021. A historical overview of the concept of micro- or minicolumns
Jones EG, Rakic P (2010) Radial columns in cortical architecture: it is the composition that counts. Cereb Cortex 20(10):2261–2264
Kaas JH (2012) Evolution of columns, modules, and domains in the neocortex of primates. Proc Natl Acad Sci U S A 109(Suppl 1): 10655–10660
Kritzer MF, Goldman-Rakic PS (1995) Intrinsic circuit organization of the major layers and sublayers of the dorsolateral prefrontal cortex in the rhesus monkey. J Comp Neurol 359:131–143
Lorente de Nó R (1938) Analysis of the activity of the chains of internuncial neurons. J Neurophysiol 1:207–244
Mahan MY, Georgopoulos AP (2013) Motor directional tuning across brain areas: directional resonance and the role of inhibition for directional accuracy. Front Neural Circuits 7:92
Miller EK, Cohen JD (2001) An integrative theory of prefrontal cortex function. Annu Rev Neurosci 24:167–202
Mountcastle VB (1957) Modality and topographic properties of single neurons of cat's somatic sensory cortex. J Neurophysiol 20:408–434
Mountcastle V (1978) An organizing principle for cerebral function: the unit module and the distributed system. In: Edelman GM, Mountcastle VB (eds) The mindful brain. MIT Press, Cambridge, MA, pp 7–50
Mountcastle VB (1997) The columnar organization of the neocortex. Brain 120(4): 701–722. A comprehensive review of the literature indicating the modular architecture of the cortex
Moxon KA, Leiser SC, Gerhardt GA, Barbee KA, Chapin JK (2004) Ceramic-based multisite electrode arrays for chronic single-neuron recording. IEEE Trans Biomed Eng 51:647–656
Opris I (2013) Inter-laminar microcircuits across the neocortex: repair and augmentation. research topic: "Augmentation of brain function: facts, fiction and controversy", Lebedev MA, Opris I, Casanova MF (eds). Front Syst Neurosci 7:80
Opris I, Bruce CJ (2005) Neural circuitry of judgment and decision mechanisms. Brain Res Rev 48:509–526
Opris I, Barborica A, Ferrera VP (2005) Microstimulation of dorsolateral prefrontal cortex biases saccade target selection. J Cogn Neurosci 17(6):893–904
Opris I, Hampson RE, Stanford TR, Gerhardt GA, Deadwyler SA (2011) Neural activity in frontal cortical cell layers: evidence for columnar sensorimotor processing. J Cogn Neurosci 23 (6):1507–1521
Opris I, Hampson RE, Gerhardt GA, Berger TW, Deadwyler SA (2012a) Columnar processing in primate pFC: evidence for executive control microcircuits. J Cogn Neurosci 24(12):2334–2347
Opris I, Fuqua JL, Huettl PF, Gerhardt GA, Berger TW, Hampson RE, Deadwyler SA (2012b) Closing the loop in primate prefrontal cortex: inter-laminar processing. Front Neural Circuits 6:88
Opris I, Santos LM, Song D, Gerhardt GA, Berger TW, Hampson RE, Deadwyler SA (2013) Prefrontal cortical microcircuits bind perception to executive control. Sci Rep 3:2285
Pesaran B, Nelson MJ, Andersen RA (2008) Free choice activates a decision circuit between frontal and parietal cortex. Nature 453:406–409
Porrino LJ, Hampson RE, Opris I, Deadwyler SA (2013) Acute cocaine induced deficits in cognitive performance in rhesus macaque monkeys treated with baclofen. Psychopharmacology 225(1):105–114
Ramón y Cajal S (1906) The structure and connexions of neurons. Nobelprize.org [online], http://nobelprize.org/ medicine/laureates/1906/cajal-lecture.pdf

Rao SG, Williams GV, Goldman-Rakic PS (1999) Isodirectional tuning of adjacent interneurons and pyramidal cells during working memory: evidence for microcolumnar organization in PFC. J Neurophysiol 81:1903–1916

Resulaj A, Kiani R, Wolpert DM, Shadlen MN (2009) Changes of mind in decision-making. Nature 461:263–266

Shallice T, Burgess PW (1991) Deficits in strategy application following frontal lobe damage in man. Brain 114(2):727–741

Shepherd G, Grillner S (2010) Handbook of brain microcircuits. Oxford University Press, Oxford

Song D, Chan RH, Marmarelis VZ, Hampson RE, Deadwyler SA, Berger TW (2007) Nonlinear dynamic modeling of spike train transformations for hippocampal-cortical prostheses. IEEE Trans Biomed Eng 54(6 Pt 1):1053–1066

Song D, Chan RH, Marmarelis VZ, Hampson RE, Deadwyler SA, Berger TW (2009) Nonlinear modeling of neural population dynamics for hippocampal prostheses. Neural Netw 22(9):1340–1351. doi:10.1016/j.neunet.2009.05.004

Sugrue LP, Corrado GS, Newsome WT (2005) Choosing the greater of two goods: neural currencies for valuation and decision making. Nat Rev Neurosci 6:363–375

Szentágothai J, Arbib MA (1975) Conceptual models of neural organization. MIT Press, Cambridge, MA

Takeuchi D, Hirabayashi T, Tamura K, Miyashita Y (2011) Reversal of interlaminar signal between sensory and memory processing in monkey temporal cortex. Science 331 (6023):1443–1447

Wang M et al (2011) Neuronal basis of age-related working memory decline. Nature 476:210–213

Wiesel TN, Hubel DH (1974) Ordered arrangement of orientation columns in monkeys lacking visual experience. J Comp Neurol 158(3):307–318

Chapter 20
Hippocampal Microcircuits, Functional Connectivity, and Prostheses

Dong Song, Min-Chi Hsiao, Ioan Opris, Robert E. Hampson, Vasilis Z. Marmarelis, Greg A. Gerhardt, Sam A. Deadwyler, and Theodore W. Berger

Abstract Hippocampus is a brain region critical for the formation of new long-term declarative memories. It transmits and processes memory information with its distinct feedforward trisynaptic pathway. Identifying functional properties of the hippocampal circuits is important for understanding the mechanisms of memory formation and building hippocampal prostheses for restoring memory functions lost in diseases or injuries. In hippocampal slices, trisynaptic responses can be elicited and recorded using conformal multi-electrode arrays. A proof-of-principle hippocampal prosthetic system has been successfully developed based on a computational model that accurately describes the input-output properties of the hippocampal circuit. In behaving animals, hippocampal functional connectivities are analyzed with a nonlinear dynamical multi-input, multi-output (MIMO) model using behaviorally-driven spiking data. Results show that the hippocampal CA3-CA1 functional connection is diffusive along the septo-temporal axis, as opposed to strictly laminar. There are strong causal relations between the CA3 and CA1 spiking activities. The MIMO model can accurately predict the spatio-temporal patterns of the CA1 output spikes based on the ongoing spatio-temporal

D. Song (✉) • M.-C. Hsiao • V.Z. Marmarelis • T.W. Berger
Department of Biomedical Engineering, University of Southern California, 403 Hedco Neuroscience Building, Los Angeles, CA 90089, California
e-mail: dsong@usc.edu; mhsiao@usc.edu; marmarelis@hotmail.com; berger@bmsr.usc.edu

I. Opris, Ph.D.
Department of Physiology and Pharmacology, Wake Forest University Health Sciences, Winston-Salem, NC 27103, USA
e-mail: ioanopris.phd@gmail.com

R.E. Hampson • S.A. Deadwyler
Department of Physiology and Pharmacology, Wake Forest University Health Sciences, Medical Center Boulevard, Winston-Salem, NC 27157, USA
e-mail: rhampson@wfubmc.edu; sdeadwyl@wfubmc.edu

G.A. Gerhardt
Department of Anatomy and Neurobiology, Parkinson's Disease Translational Research Center of Excellence, Center for Microelectrode Technology, University of Kentucky, Lexington, KY 40506, USA
e-mail: gregg@uky.edu

M.F. Casanova, I. Opris (eds.), *Recent Advances on the Modular Organization of the Cortex*, DOI 10.1007/978-94-017-9900-3_20

patterns of the CA3 input spikes. MIMO model-based electrical stimulation to the CA1 region effectively restores the hippocampal memory function by reinstating the CA1 activities. The recording component, the nonlinear dynamical MIMO model, and the stimulation component essentially constitute a closed-loop prosthetic system that bypasses the impaired hippocampal region.

Keywords Cortical microcircuit • Multi-electrode array • Cortical layer • Recording • Stimulation • Prosthesis • Brain machine interface

20.1 Introduction

The brain processes information with neurons and neuronal circuits (Golgi 1967). Understanding how neurons are organized anatomically and functionally to perform high-level cognitive functions is one of the ultimate goals of neuroscience. Such knowledge will allow us to build cortical prostheses that can restore cognitive functions lost in diseases or injuries (Berger et al. 2005), and develop biologically-inspired cognitive architectures to achieve brain-like intelligence (Goertzel et al. 2010).

The hippocampal formation is a brain region responsible for the formation of new long-term declarative memories (Squire and Zola-Morgan 1991; Eichenbaum 1999): the "fact-based" or episodic (autobiographical) memories. If the hippocampus is damaged, the short-term memory function and the long-term memories before the damage remain intact. However, the animals/patients will no longer be able to form new long-term memories (Milner 1970).

The hippocampal formation consists of three sub-regions (Fig. 20.1): the dentate gyrus (DG), the hippocampus proper (CA3, CA2, and CA1), and the subiculum (Sub). Different from the neocortices that typically have six layers, the hippocampal formation is a three-layered cortex (Fig. 20.1). The most prominent and well-known feature of the hippocampal formations is perhaps the trisynaptic pathway: first, granule cells in the DG receive synaptic inputs from the entorhinal cortex (EC) through the perforant path; second, CA3 pyramidal neurons are innervated by the mossy fibers from the DG granule cells; third, CA3 pyramidal neurons make synaptic connections to the CA1 pyramidal neurons via the Schaffer collaterals. More recent studies show the existence of two additional pathways, i.e., the EC to CA3 pathway and the EC to CA1 pathway (Yeckel and Berger 1990, 1995), and thus the hippocampal circuits should be more accurately described as a combination of monsynaptic, disynaptic, and trisynaptic pathways. Despite these additional pathways, however, the hippocampus remains largely a feedforward neuronal network with mostly local feedbacks within each subregions (Amaral et al. 1990).

The trisynaptic or feedforward circuit can be clearly identified in the transverse planes of the hippocampus, along almost the whole range of its longitudinal axis (Fig. 20.2). This unusual anatomical organization has inspired the so-called "laminar hypothesis", as described by Andersen: "by means of this lamellar organization, small strips of the hippocampal cortex may operate as independent functional

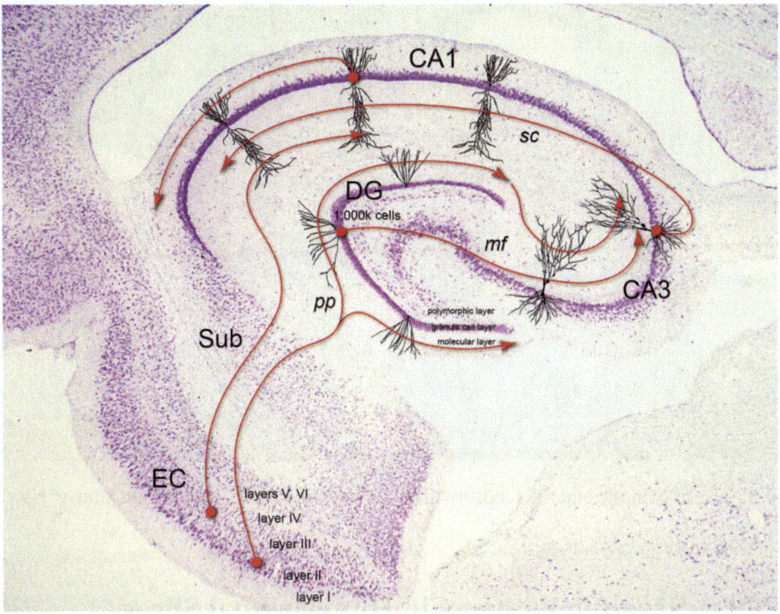

Fig. 20.1 The internal circuit of the hippocampus. *EC* entorhinal cortex, *DG* dentate gyrus, *Sub* subiculum, *pp* perforant path, *mf* mossy fiber, *sc* Schaffer collateral

units, although excitatory and inhibitory transverse connections may modify the behavior of neighboring lamellae." The laminar hypothesis has been supported by electrophysiological studies on the in vitro hippocampus (Andersen et al. 1971). However, more recent anatomical studies have shown that most of the excitatory synaptic projects within the hippocampus are more spread out compared with what have been suggested by the laminar hypothesis, with the exception of DG to CA3 connections being the only "true" laminar pathway (Freund and Buzsaki 1996).

Compared with the neocortex, the hippocampus has a more regular anatomical structure and a relatively simple circuit. Given also its pivotal roles in memory functions, it has been one of the most studied regions in the brain, and the first target of neural engineers in building cortical prostheses that can restore cognitive functions (Berger et al. 2005, 2010, 2011, 2012; Hampson et al. 2012, 2013). In this chapter, we will describe the hippocampal microcircuit with an emphasis on its functional consequences in the context of building hippocampal memory prostheses for restoring memory functions. We will show evidence of the hippocampal trisynaptic circuit with extracellular recordings from hippocampal slices, functional connectivities revealed by nonlinear dynamical modeling of spiking activities recorded in behaving animals. We will also describe how we build hippocampal memory prostheses by taking advantages of the characteristics of the hippocampal microcircuit.

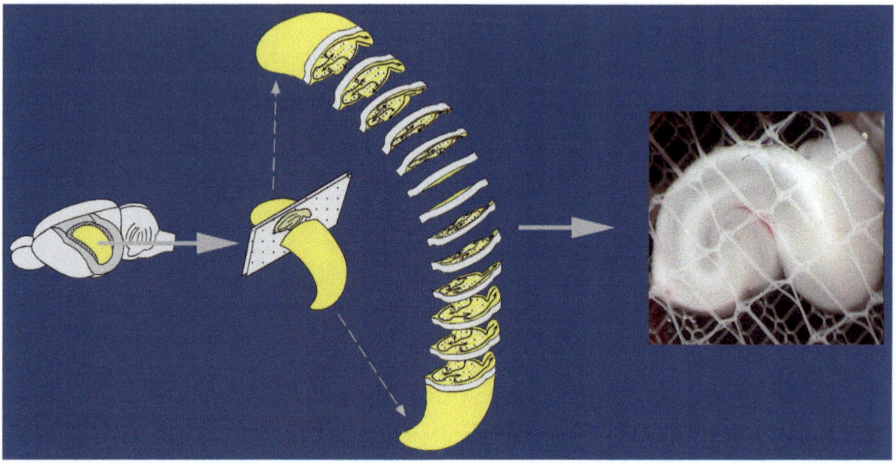

Fig. 20.2 The trisynaptic and the feedforward pathways can be observed in the transverse slices of the hippocampus

20.2 Tri-Synaptic Pathway in Hippocampal Slices

The signal propagation along the hippocampal trisynaptic circuit can be observed from the transverse hippocampal slices using field potential recordings and electrical stimulation (Fig. 20.3). We have designed a conformal planary multi-electrode array (MEA) with three sets of electrode arrays in the DG, CA3 and CA1, respectively. These electrodes allow simultaneous recordings from multiple locations of each region. Field potentials with different waveforms can be recorded along the main axis of the neurons in multiple recording sites, reflecting the distribution of the current sources and current sinks during the formation of EPSPs and population spikes. Figure 20.3 shows simultaneously recordings in DG, CA3, and CA1 regions cell body layers. In this experiment, the perforant path is stimulated with a bipolar stimulating electrode. Three arrays of electrodes are placed in the cell body layers of the three regions. Extracellular field potentials, i.e., excitatory post-synaptic potentials (EPSPs) and population spikes, are elicited in all three regions by the stimulation. These field potentials show progressively longer delays from the stimulation to the onsets of the responses, indicating the existence of the trisynaptic circuit in the transverse slices of the hippocampus.

The hippocampal slice has allowed us to build the first proof-of-principal hippocampal "prosthesis" to bypass a damaged brain region (Hsiao et al. 2013). In this prosthetic system, random-interval electrical pulses are delivered to the perforant path to mimic the input action potentials to the hippocampus (Fig. 20.4). The experiments are performed in three steps (Fig. 20.5). In the first step, in the intact slice, DG, CA3 and CA1 field potentials driven by the perforant path stimulations are simultaneously recorded (Fig. 20.4). The CA3 population spikes and the CA1 population spikes are taken as the input and output signals,

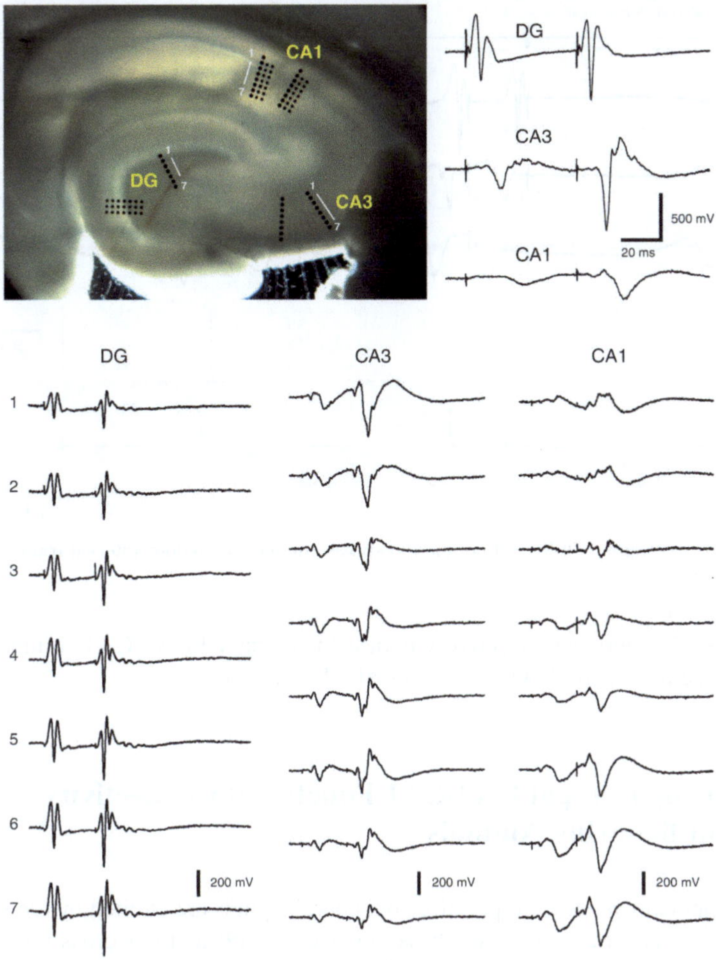

Fig. 20.3 Field potential recordings from a hippocampal slice using a conformal multi-electrode array (MEA)

respectively. A single-input, single-output (SISO) Volterra model is built to describe the nonlinear dynamical transformation from the amplitudes of the CA3 population spikes to the amplitudes of the CA1 field EPSPs (Hsiao et al. 2013). In the second step, a cut is made between CA3 and CA1 regions. The CA1 region becomes silent while the DG and CA3 signals remain intact, showing the disruption of the trisynaptic pathway. In the third step, a biomimetic FPGA device containing the SISO model is connected to the system. The device receives CA3 population spikes as its input signal, calculates the corresponding amplitudes of CA1 EPSPs as it output signal, and stimulates the CA3 with appropriate current intensities to restore the CA1 field potentials. Results show that this closed-loop, model-driven,

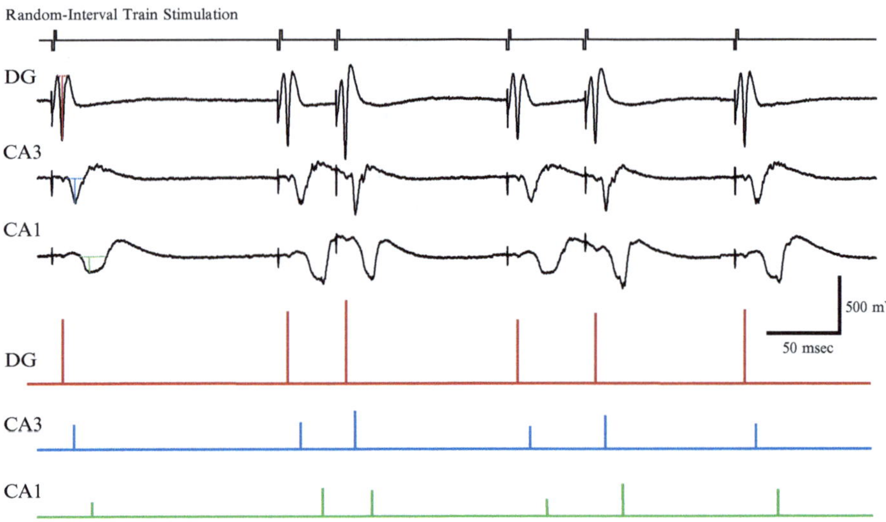

Fig. 20.4 Trisynaptic EPSPs and population spikes elicited by random-interval train stimulation in a hippocampal slice

"prosthesis" system can effectively bypass the damaged CA3-CA1 connection and recover accurately the CA1 output signals (Fig. 20.6).

20.3 Hippocampal CA3-CA1 Functional Connectivity in Behaving Animals

In the hippocampal slice preparation, the neural signals are elicited by the electrical stimulation. The waveforms of these signals, as well as their cross-region transformations, provide valuable insights into the anatomical and electrophysiological properties of the hippocampal circuits. In behaving animals, the neural signals are the spatio-temporal patterns of spikes driven by the cognitive activities of the animal. These signals carry the information of short-tem memories for the formation of long-term memories. Identification of the functional connectivities of the hippocampus using spiking activities provides the most functionally relevant analysis of how the hippocampal microcircuit transmits and process memory information.

We have developed a multiple-input, multiple-output (MIMO), point-process, nonlinear dynamical modeling approach for such a task (Song et al. 2006, 2009a, b, 2013, 2014; Song and Berger 2010). In this approach, the functional connectivity between neurons is represented as the causal relationship between their spiking activities. A MIMO model consists of a series of physiologically plausible multi-input, single-output (MISO) spiking neuron models that each contains five components (Fig. 20.7, right): (1) a feedforward MISO Volterra model transforming the

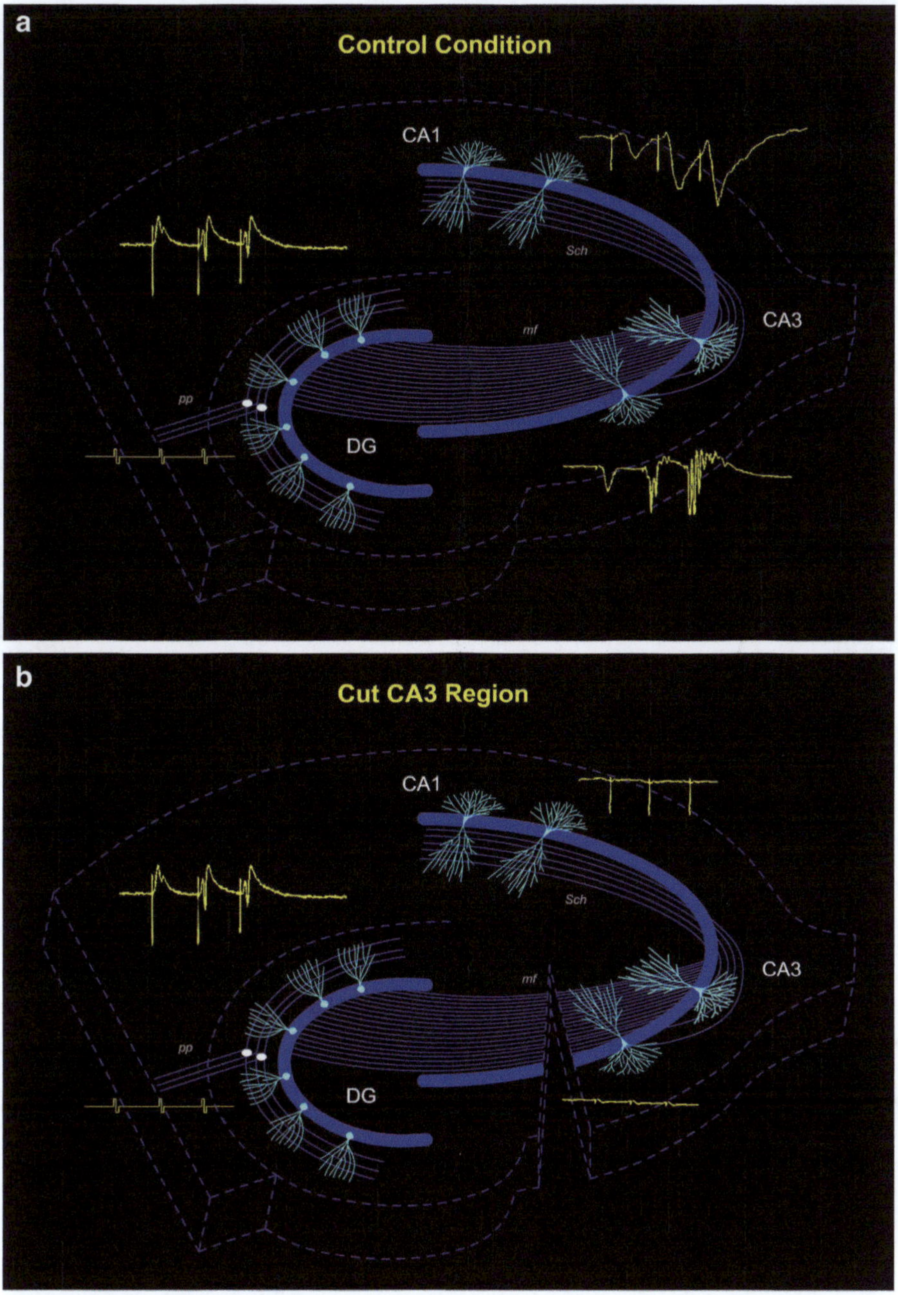

Fig. 20.5 Proof-of-principle hippocampal FPGA "prosthesis" on a hippocampal slice

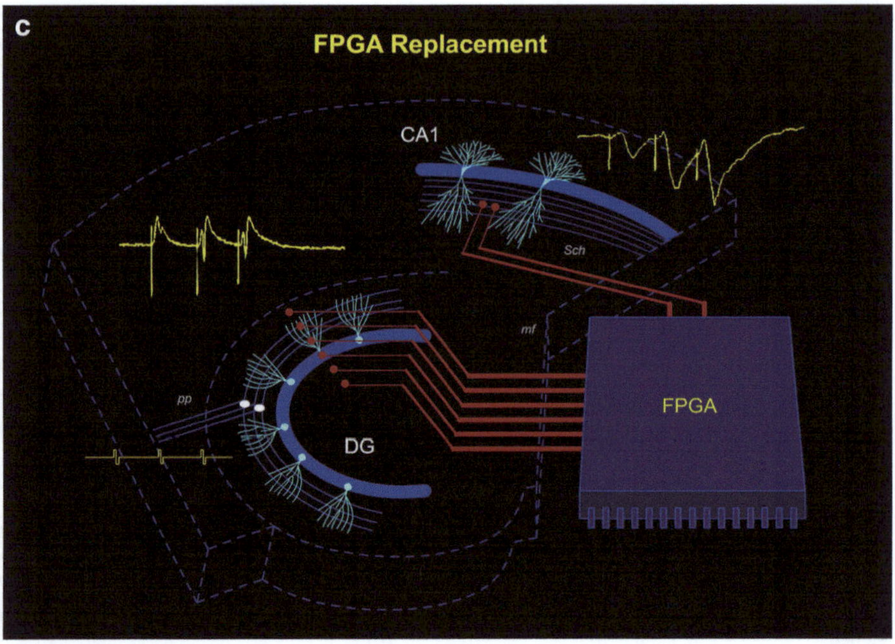

Fig. 20.5 (continued)

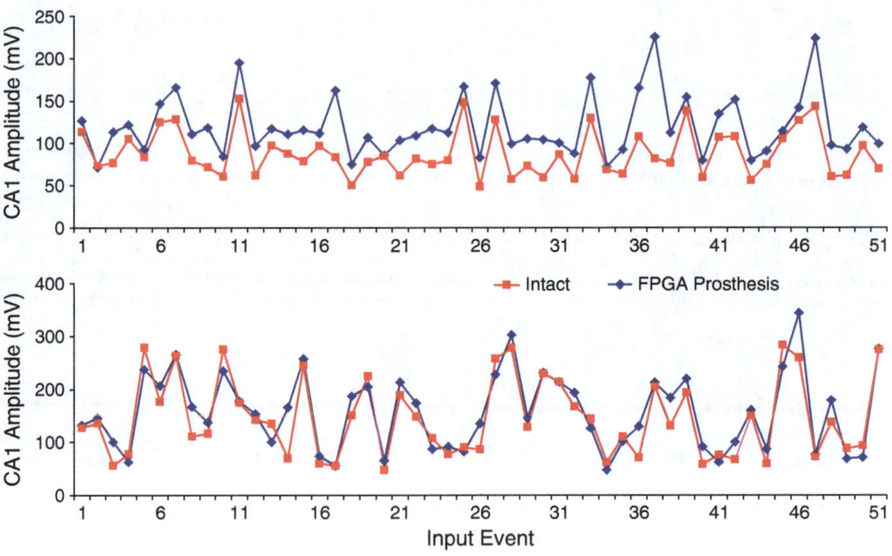

Fig. 20.6 Hippocampal CA1 output signals restored by the hippocampal prosthesis

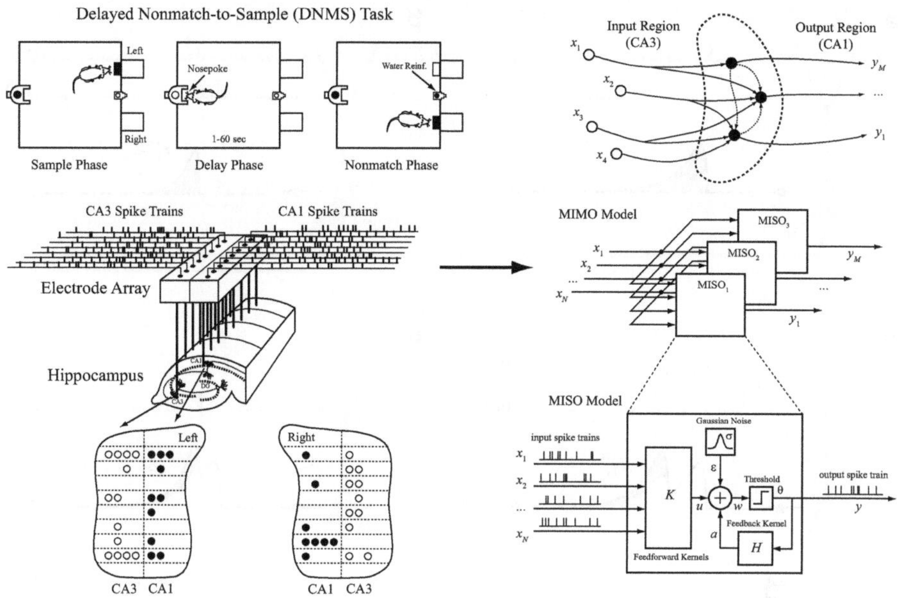

Fig. 20.7 Identification of hippocampal CA3 and CA1 functional connectivity using a multi-input, multi-output (MIMO) nonlinear dynamical model with spike trains recorded from animals performing a memory-dependent delayed nonmatch-to-sample (DNMS) task

input spike trains into the synaptic potential of the output neuron, (2) a feedback Volterra kernel transforming the output spikes into the spike-triggered after-potential, (3) a noise term capturing the stochastic properties of firing, (4) an adder generating the pre-threshold potential, and (5) a threshold function generating output spikes.

The model can be expressed by the following equations:

$$w = u(k, x) + a(h, y) + \varepsilon(\sigma) \tag{20.1}$$

$$y = \begin{cases} 0 & \text{when } w < \theta \\ 1 & \text{when } w \geq \theta \end{cases} \tag{20.2}$$

$$\begin{aligned} u(t) = k_0 &+ \sum_{n=1}^{N} \sum_{\tau=0}^{M_k} k_1^{(n)}(\tau) x_n(t-\tau) \\ &+ \sum_{n=1}^{N} \sum_{\tau_1=0}^{M_k} \sum_{\tau_2=0}^{M_k} k_{2s}^{(n)}(\tau_1, \tau_2) x_n(t-\tau_1) x_n(t-\tau_2) \end{aligned} \tag{20.3}$$

$$a(t) = \sum_{\tau=1}^{M_h} h(\tau) y(t-\tau) \tag{20.4}$$

The zeroth-order kernel, k_0, is the value of u when the input is absent. First-order kernels $k_1^{(n)}$ describe the first-order linear relation between the nth input x_n and u, as

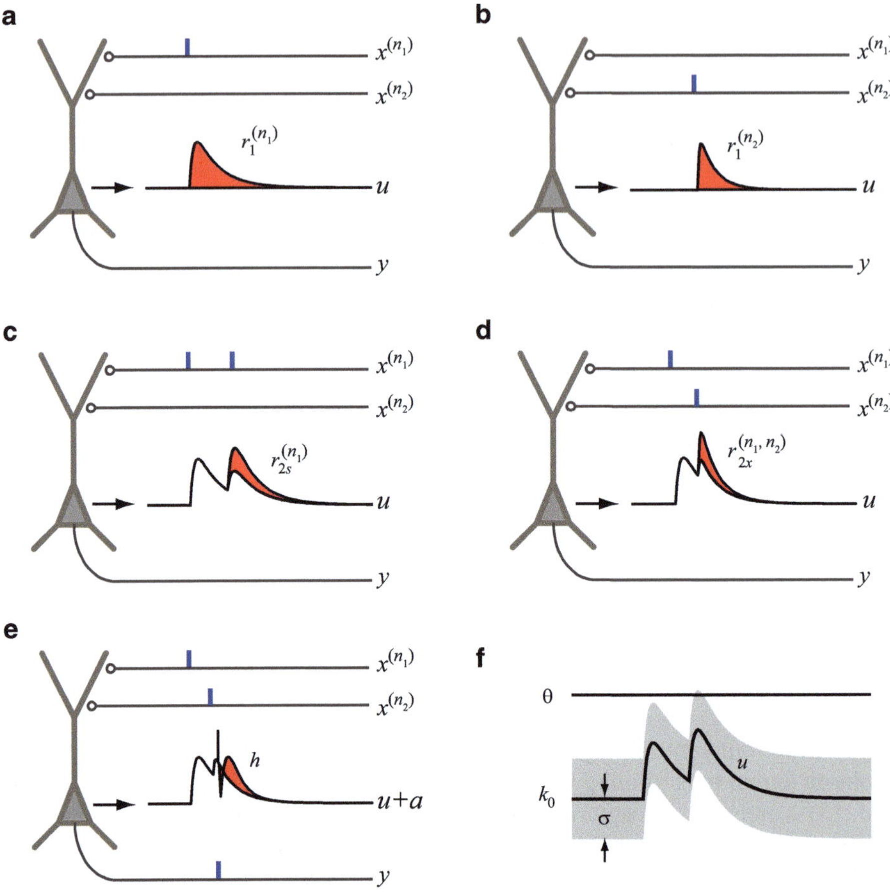

Fig. 20.8 Interpretation of the model variables in the nonlinear dynamical spiking neuron model

functions of the time intervals τ between the present time and the past time. Second-order self kernels $k_{2s}^{(n)}$ describe the second-order nonlinear interaction between pairs of spikes in the nth input x_n as they affect u. N is the number of inputs. M_k and M_h denote the memory lengths of the feedforward process and feedback process, respectively. Additional terms such as second-order cross kernels and higher-order (e.g., third-order) kernels are not included for simplicity.

Single-pulse and paired-pulse response functions (r_1 and r_2) can be derived from the first-order and second-order self kernels (Song et al. 2009c). r_1 and r_2 can be interpreted as the post-synaptic potential (PSP) and the paired-pulse facilitation/depression function, respectively (Fig. 20.8).

$$r_1^{(n)}(\tau) = k_1^{(n)}(\tau) + k_{2s}^{(n)}(\tau, \tau) \tag{20.5}$$

$$r_2^{(n)}(\tau_1, \tau_2) = 2k_{2s}^{(n)}(\tau_1, \tau_2) \tag{20.6}$$

To facilitate model estimation and avoid overfitting, the Volterra kernels are expanded with Laguerre basis functions b as in:

$$u(t) = c_0 + \sum_{n=1}^{N}\sum_{j=1}^{J} c_1^{(n)}(j)v_j^{(n)}(t) + \sum_{n=1}^{N}\sum_{j_1=1}^{J}\sum_{j_2=1}^{j_1} c_{2s}^{(n)}(j_1, j_2)v_{j_1}^{(n)}(t)v_{j_2}^{(n)}(t) \quad (20.7)$$

$$a(t) = \sum_{j=1}^{L} c_h(j)v_j^{(h)}(t) \quad (20.8)$$

where

$$v_j^{(n)}(t) = \sum_{\tau=0}^{M_k} b_j(\tau)x_n(t-\tau) \quad (20.9)$$

$$v_j^{(h)}(t) = \sum_{\tau=1}^{M_h} b_j(\tau)y(t-\tau) \quad (20.10)$$

$c_1^{(n)}$, $c_{2s}^{(n)}$, and c_h are the sought Laguerre expansion coefficients of $k_1^{(n)}$, $k_{2s}^{(n)}$, and h, respectively (c_0 is equal to k_0). J is the number of basis functions.

To determine whether neurons have significant functional connections, model coefficients are estimated with a penalized likelihood estimation method, i.e., group LASSO (Song et al. 2013), to yield group-level sparse model representations. Zero-valued coefficients mean that there is no functional connection (for example, if both $c_1^{(n)}$ and $c_{2s}^{(n)}$ are equal to zero, the nth input neuron is not connected to the output neuron), while non-zero coefficients indicate a significant functional connection. The estimated kernels between an input neuron and an output neuron collectively describe the functional connectivity between them.

In group LASSO, the composite penalized criterion is written as

$$\begin{aligned} S(c) &= -l(c) + \lambda\left(\sum_{n=1}^{N}\|c_1^{(n)}(j)\|_2^{\frac{1}{2}} + \sum_{n=1}^{N}\|c_{2s}^{(n)}(j_1, j_2)\|_2^{\frac{1}{2}}\right) \\ &= -l(c) + \lambda\left(\sum_{n=1}^{N}\left(\sum_{j=1}^{J} c_1^{(n)}(j)^2\right)^{\frac{1}{2}} + \sum_{n=1}^{N}\left(\sum_{j_1=1}^{J}\sum_{j_2=1}^{j_1} c^{(n)}(j_1, j_2)^2\right)^{\frac{1}{2}}\right) \end{aligned} \quad (20.11)$$

-$l(c)$ is the negative log likelihood function in maximum likelihood estimation (MLE). Tuning parameter λ controls the relative importance of the likelihood and the penalty term. Model coefficients c are estimated by minimizing S. When λ becomes larger, the estimation yields sparser result of the coefficients. λ is optimized with cross-validation methods. Due to its sparse representation, utilization of Laguerre basis function and Volterra kernels, as well as the noisy threshold which is equivalent to a *probit* link function in the generalized linear model, the resulting model is formally termed as sparse generalized Laguerre-Volterra model (sGLVM).

To identify the functional connectivity between hippocampal CA3 and CA1, we first estimate MISO models for every CA1 neuron using the CA3 and CA1 spike trains recorded from behaving animals. Specifically, rats are trained to perform a memory-dependent, delayed nonmatch-to-sample (DNMS) task with random delay intervals (Deadwyler et al. 1996; Hampson et al. 1999) (Fig. 20.7, top-left). Animals perform the task by pressing a single lever presented in one of the two positions in the sample phase (left or right). This event is called the "sample response". The lever is then retracted and the delay phase initiates; for the variable durations (0–30 s) of the delay phase, the animal is required to nose-poke into a lighted device on the opposite wall. When the delay is ended, nose-poke light is extinguished, both levers are extended, and the animal is required to press the lever *opposite* to the sample lever. This event is called the "nonmatch response". If the correct lever is pressed, the animal is rewarded. A session includes approximately 100 successful DNMS tasks that each consists of two of the four behavioral events, i.e., right sample (RS) and left nonmatch (LN), or left sample (LS) and right nonmatch (RN). Spike trains are obtained with MEAs from different septo-temporal regions of the hippocampus (Fig. 20.7, bottom-left). For each hemisphere of the brain, a microwire MEA is surgically implanted into the hippocampus, with 8 electrodes in the CA3 (input) region and 8 electrodes in the CA1 (output) region. All arrays are implanted in the same anatomical location of the hippocampus, and have a fixed geometry with 200 μm between CA3 and CA1 pairs (septo-temporal axis) and 800 μm between rows (medial-lateral axis). From each electrode, up to 4 neurons can be identified and recorded. The maximal number of neurons that can possibly be recorded thus is 128, with 64 in CA3 and 64 in CA1 (Fig. 20.9). In reality, a dataset typically contains approximately 10 CA3 neurons and 10 CA1 neurons.

Figure 20.10 shows an example of a MISO hippocampal CA3-CA1 model. In this dataset, there are 24 CA3 neurons and 18 output neurons. The fifth CA1 neuron is chosen to be the output neuron of the MISO model. The non-zero, first-order and second-order self kernels of the 24 inputs and the single output are plotted. The result indicates that, out of 24 inputs, there are 6 inputs having significant functional connections to this specific output neuron. The MISO connectivity is represented by a 24-by-1 binary vector (Fig. 20.10, left). The kernels further quantitatively describe the input-output relations in means of PSPs (i.e., r_1) and paired-pulse facilitation/depression functions (i.e., r_{2s}) of each input neuron to the output neuron (Fig. 20.10, right).

The MIMO connectivity matrix is then obtained by concatenating the MISO connectivity vectors of each outputs. In this specific case, the MIMO matrix is a 24-by-18 binary matrix representing the functional connectivity between the 24 CA3 neurons and the 18 output neurons (Fig. 20.11a). To reveal the topographical organization of the functional connectivity, the MIMO matrix is then reorganized with respect to the 16 unique CA3 and CA1 electrodes into a 16-by-16 matrix. This matrix contains the counts of identified connections between one

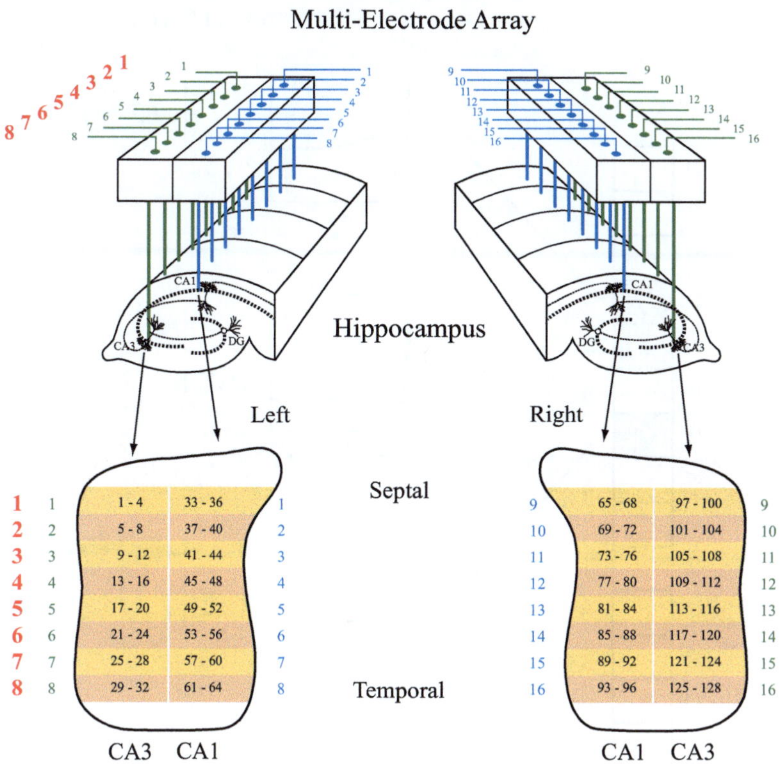

Fig. 20.9 MEA recording sites in the hippocampal CA3 and CA1 regions

CA3 electrode and one CA1 electrode (Fig. 20.11b). This procedure is applied to a total number of 61 MIMO datasets from 61 different animals. The summation of the 61 16-by-16 matrices yields a population-level MIMO matrix containing the counts of the identified connections between CA3 and CA1 electrodes in all animals (Fig. 20.11c). To further calculate the probability (as opposed to the counts) of having functional connections, all possible connections constrained by the availabilities of neuronal recordings in the datasets are calculated. In the specific case shown above, the possible connection matrix starts with a 24-by-18 matrix with values all equal to 1 (Fig. 20.11a′). It is then reorganized with respect to the electrodes (Fig. 20.11b′), and summed with matrices from other datasets to produce the population-level possible connection matrix (Fig. 20.11c′). The population-level MIMO matrix (C) is element-wise divided by the population-level possible connection matrix (C′) to yield the population-level connection probability matrix (Fig. 20.11d). In this matrix, each element is the probability of having functional connection between one CA3 electrode to one CA1 electrode in all 61 animals.

In total, there are 661 (10.8 per animal) CA3 neurons and 697 (11.4 per animal) CA1 neurons recorded from the 61 animals. The sparseness of CA3-CA1

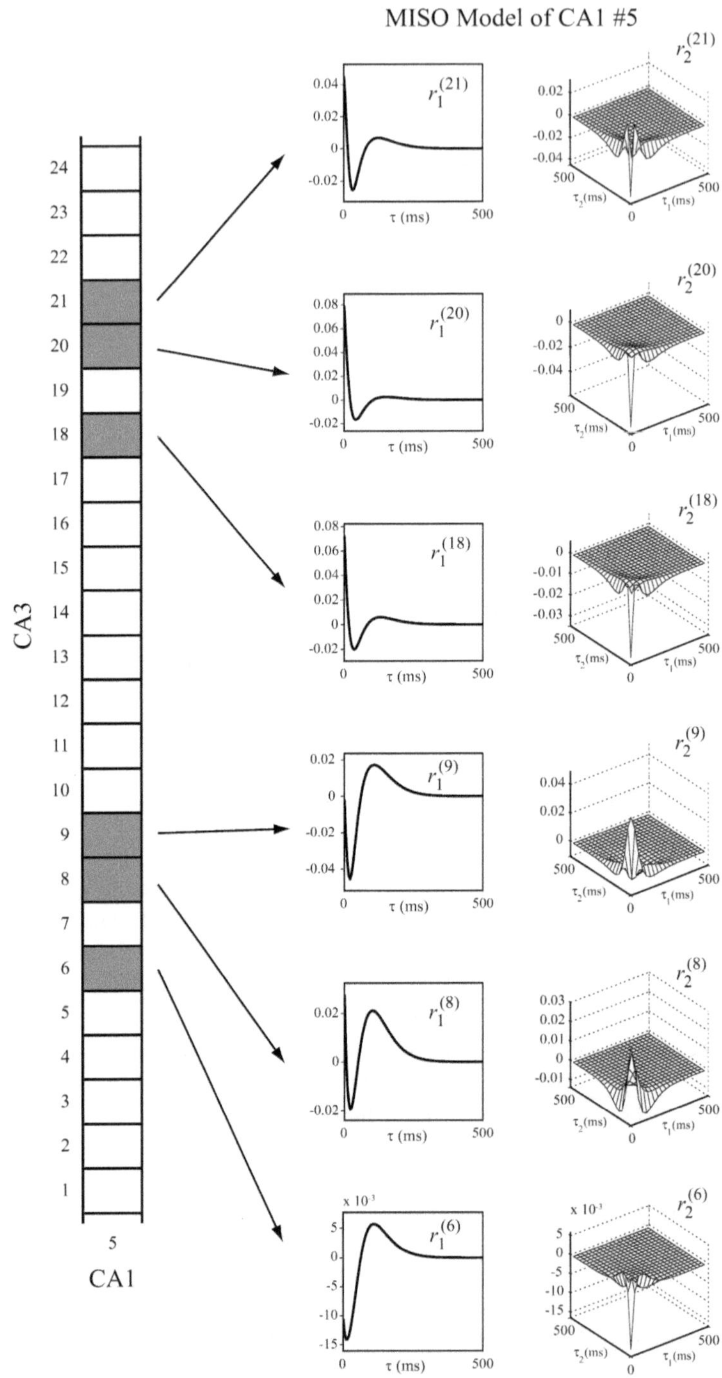

Fig. 20.10 A sparse multi-input, single-output (MISO) nonlinear dynamical model representing the functional connectivity between 24 CA3 neurons and one CA1 neuron

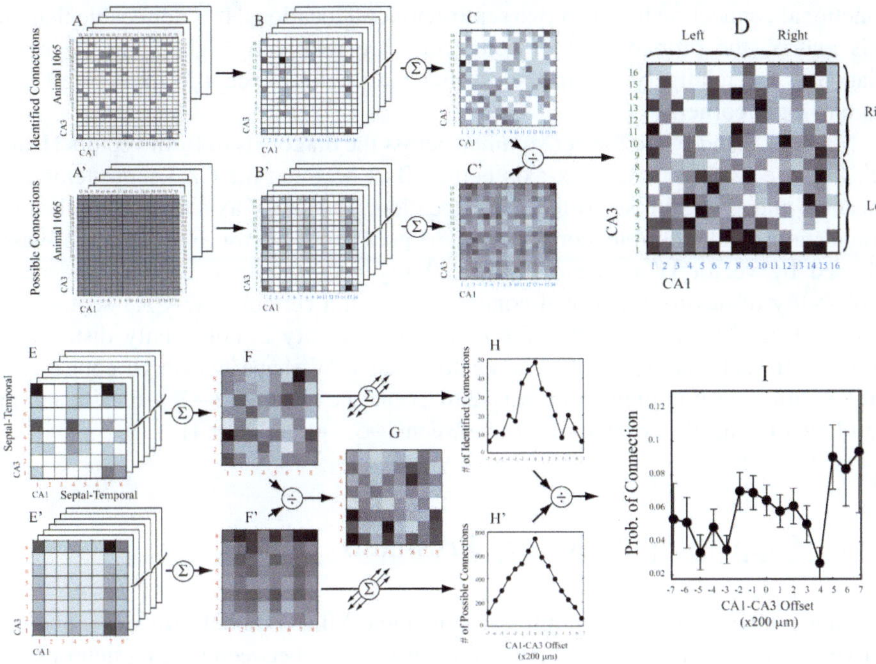

Fig. 20.11 Hippocampal CA3–CA1 functional connectivity revealed by sparse MIMO nonlinear dynamical models ($n = 61$)

connections, defined as the ratio between the total number of estimated functional connections (676) and the total number of possible connections (8,856), is estimated to be 7.63 ± 0.28 %. In the probability matrix D, it is discernible that the left and right halves of the hippocampus have unevenly distributed functional CA3-CA1 connections. The left-to-left and right-to-right quarters of the matrix has higher densities of connections compared with the left-to-right and right-to-left quarters of the matrix. The sparseness of ipsilateral (same side) and contralateral (different side) CA3-CA1 connections are estimated to be 8.32 ± 0.41 % and 6.94 ± 0.38 %, respectively. This result shows that although hippocampal CA3 neurons have strong connections to both the ipsilateral and contralateral CA1 neurons, the density of the ipsilateral connections is significantly higher than that of the contralateral connections.

To further investigate the laminar organization of the CA3-CA1 connections, the functional connections are analyzed with respect to the septo-temporal locations of the neurons. The MIMO matrix B of each dataset is divided into four 8-by-8 quarters and summed to form a 8-by-8 matrix E. Ipsilateral and contralateral connections are not differentiated in this matrix. Matrices E of all animals are summed to form the population-level septo-temporal MIMO matrix F. F is then element-wise divided by its corresponding possible connection matrix F′ to form the probability matrix G. In G, each element is the probability of having CA3-CA1 functional connection between two septo-temporal locations. It is noticeable that, in this matrix, the probabilities of connection are relatively high along the main diagonal, and relatively low in the off-diagonal regions except in the top-left and bottom-right corners.

Matrices F and F′ are further summed across the diagonals to form vectors H and H′. In these two vectors, the x-axis is the offset between the CA3 septo-temporal location and the CA1 septo-temporal location, and the y-axis is the number of connections and possible connections, respectively. Vector H is element-wise divided by vector H′ to yield the probability vector I. In I, each value is the probability of having functional connection given a certain CA3-CA1 septo-temporal offset. Results show that the connection density is not evenly distributed across different CA3-CA1 offsets. Vector I has a W-shape with high connection probabilities existing in the middle (−2 to 2) and the two ends (−7, −6, 5, 6, and 7) and lower connection probability in between (−5, −4, −3 and 4).

20.4 Hippocampal Memory Prosthesis

As shown above, the sparse nonlinear dynamical MIMO model provides a quantitative way of identifying the functional connectivities between spiking neurons. In addition, it also has been used for predicting the spatio-temporal patterns of output spike trains based on the spatio-temporal patterns of input spike trains, and served as the computational basis of the hippocampal memory prostheses.

Our hippocampal memory prosthesis experiments are conducted in rats performing the DNMS task (Fig. 20.7). The experiment consists of three steps (Fig. 20.12). First, in the control step, a mini-pump implanted into the hippocampus continuously introduces saline during the DNMS session. The memory function of animals is quantified with a forgetting curve describing the percentage of correct nonmatch responses as the function of the delay duration. It shows that animal can successfully maintain memory of the sample locations (left or right) for delays as long as 30 s, with the percentage of correct responses monotonically decreases as the delay is prolonged (Fig. 20.12, green line). During the same period, CA3 and CA1 spike trains are recorded simultaneously with a MEA. Nonlinear dynamical MIMO models reflecting the normal CA3-CA1 spatio-temporal pattern transformations are estimated using these normal condition spike train input-output data. The estimated MIMO models can accurately predict the CA1 spike trains based on the ongoing CA3 spike trains in real time (Fig. 20.13). Both CA3 and CA1 show different spatio-temporal patterns for the two lever positions (Song et al. 2014). Second, in the blockade step, glutamatergic NMDA channel blocker MK801, instead of saline, is introduced to the hippocampal CA1 region. MK801 drastically suppresses the activities of CA1 pyramidal neurons, while leaves the CA3 pyramidal neuron activities relatively intact. Consequently at the behavioral level, animal performances during the DNMS task are significantly decreased. The forgetting curve shifts down to the percentage of correct responses at 70–80 % for delays

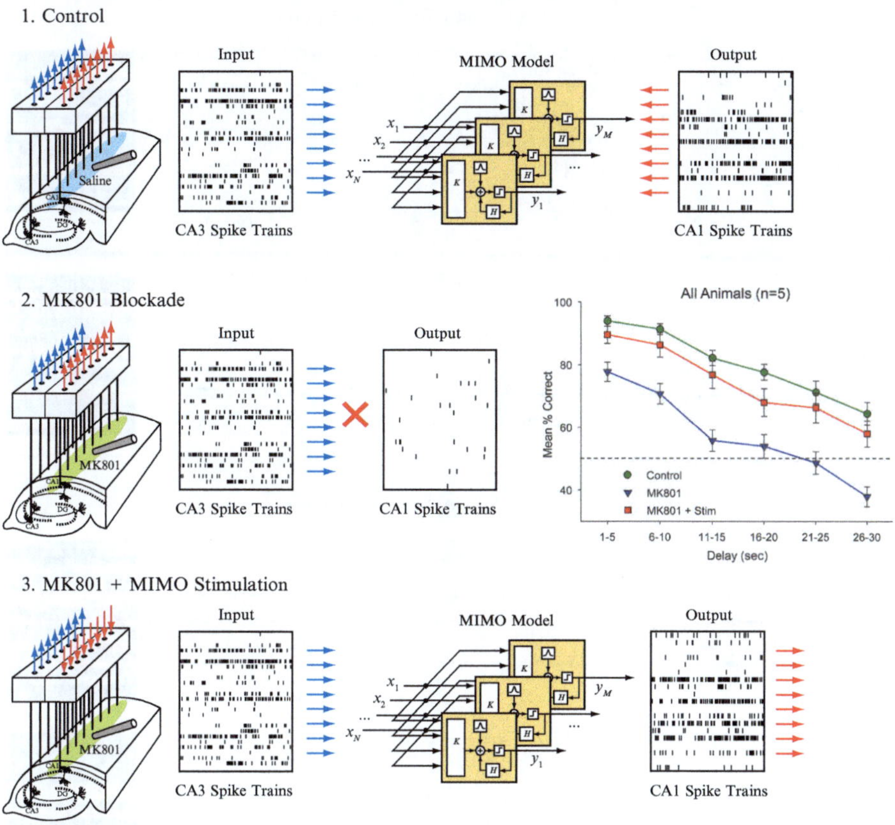

Fig. 20.12 Hippocampal memory prosthesis for behaving rat

shorter than 10 s and near or below chance level (50 %) for delays longer than 10 s (Fig. 20.12, blue line). These results show that MK801 impairs the hippocampal memory function by disrupting the CA3-CA1 signal transmission. Finally in the MIMO stimulation step, the CA1 regions are electrically stimulated with CA1 output patterns predicted by MIMO models during the sample phase, with the infusion of MK801. The CA1 stimulations are driven by the relatively intact CA3 activities based on the MIMO models estimated during the control step. At the behavioral level, the DNMS performances are enhanced from those of the blockade step. The forgetting curve shifts up to the percentage of correct responses at 90–85 % for delays shorter than 10 s and above chance level (50 %) for delays longer than 10 s (Fig. 20.12, red line). These results show that the MIMO model-based CA1 stimulation effectively restores the hippocampal memory function by reinstating the CA1 activities. The recording component, the nonlinear dynamical MIMO model, and the stimulation component essentially constitute a closed-loop prosthetic system that bypasses the impaired hippocampal region.

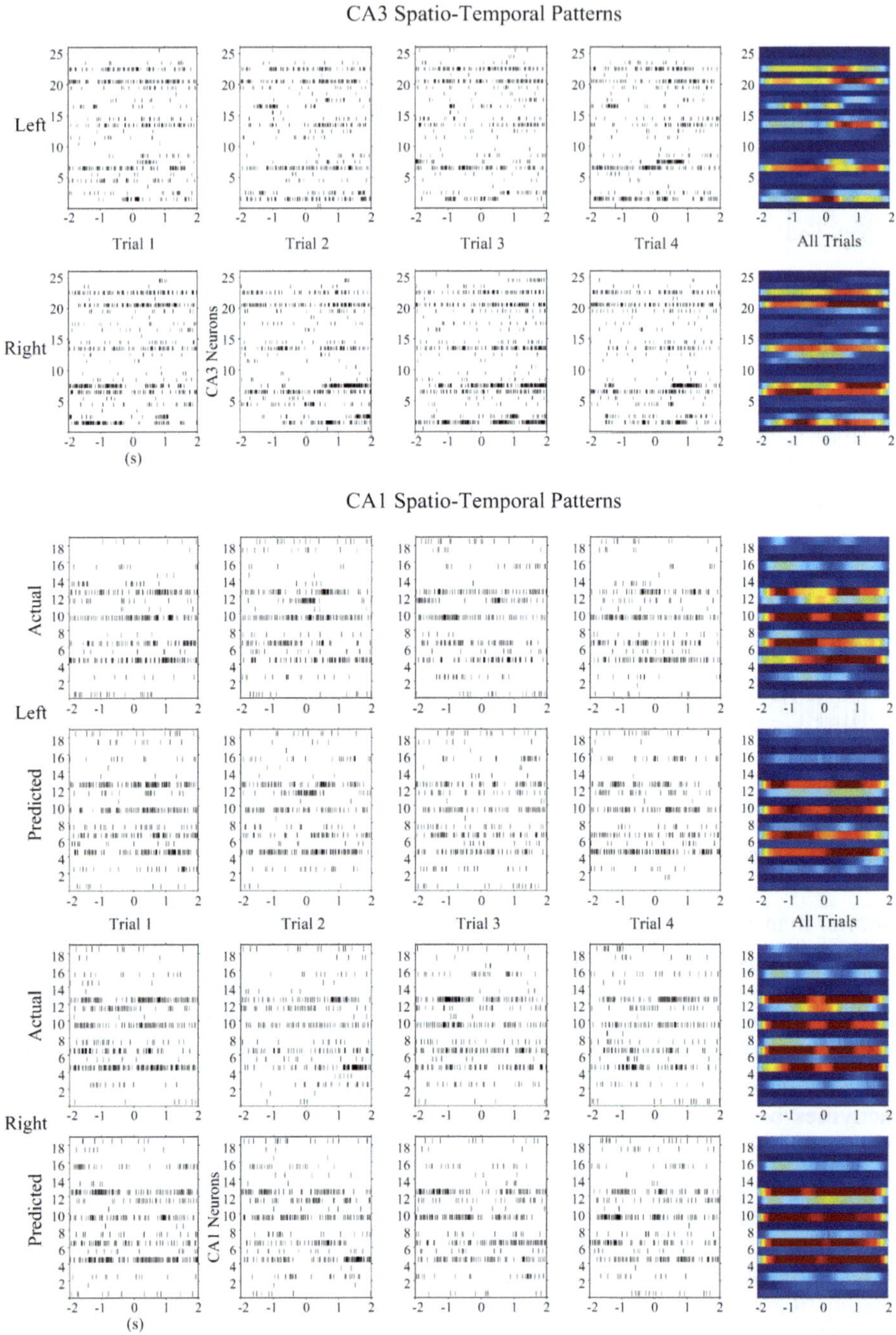

Fig. 20.13 CA1 spatio-temporal patterns of spikes are predicted from the CA3 spatio-temporal patterns of spikes using nonlinear dynamical MIMO models

20.5 Conclusions and Discussions

This book chapter describes a series of electrophysiological and computational studies on the hippocampus. Different from anatomical approach, these studies focus on the functional properties of the hippocampal circuits, i.e., how the hippocampal neuronal networks process and transmit memory information. In slice studies, we show the physiological evidence of the hippocampal trisynaptic pathway: electrical stimulation to the perforant path can elicit a cascade of signals along the whole pathway; disruption of the pathway abolishes the downstream signals. We also show that the input-output signal transformation within the hippocampal can be modeled computationally. A prosthetic device with the appropriate computational model can restore the hippocampal signal transmission. In behaving animal studies, we develop and apply a sophisticated nonlinear dynamical MIMO model for the identification of hippocampal functional connectivity using spiking data. The hippocampal CA3-CA1 connectivity is identified from a large number of animals performing a memory-dependent behavioral task. Consistent with anatomical studies (Freund and Buzsaki 1996), our results show that the hippocampal CA3-CA1 connectivity is diffusive along the septo-temporal axis, as opposed to strictly laminar. There are strong causal relations between the CA3 and CA1 spiking activities. These causal relations allow us to build MIMO models that accurately predict the CA1 spike trains based on the CA3 spike trains. Such MIMO models serve as the computational basis of the hippocampal memory prosthesis.

The hippocampal functional connections described here are likely to be caused by the monosynaptic connections between CA3 and CA1 neurons. Indeed, the Schaffer collateral to CA1 pyramidal neuron synapses are among the most studied and best-known synapses in the brain. In addition to anatomical evidences, physiological studies have repeatedly shown that stimulation of the CA3 pyramidal neuron can reliably excite the CA1 pyramidal neurons. The results shown in this chapter provide further evidence of such connections in behaving animals. However, it must be noted that functional connectivity is not equivalent to synaptic connectivity. In this chapter, functional connectivity is defined as the causal relations between the input and output neurons. Besides actual synaptic connection, other factors such as common input can also cause functional connectivity. In the hippocampus, both CA3 and CA1 regions receive direct inputs from the EC through the perforant path (Fig. 20.1). This common input may contribute to the functional connectivity observed between CA3 and CA1 neurons. Despite such a difference and the consequent complication in interpretations, the functional connectivity described here provides a quantitative characterization of the hippocampal signal transformation that is essential for understanding the formation of long-term memories and developing hippocampal memory prostheses.

Acknowledgements This work was supported by the Defense Advanced Research Projects Agency (DARPA) through the Restorative Encoding Memory Integration Neural Device (REMIND) Program and N66001-14-C-4016.

References

Amaral DG, Ishizuka N, Claiborne B (1990) Neurons, numbers and the hippocampal network. In: Storm-Mathisen J, Zimmer J, Ottersen OP (eds) Progress in brain research, Elsevier, Amsterdam, pp 1–11

Andersen P, Bliss TVP, Skrede KK (1971) Lamellar organization of hippocampal excitatory pathways. Exp Brain Res 13(2):222–238

Berger TW, Ahuja A, Courellis SH, Deadwyler SA, Erinjippurath G, Gerhardt GA, Gholmieh G, Granacki JJ, Hampson R, Hsiao M-C, LaCoss J, Marmarelis VZ, Nasiatka P, Srinivasan V, Song S, Tanguay AR Jr, Wills J (2005) Restoring lost cognitive function: hippocampal-cortical neural prostheses. IEEE EMBS Spec Issue: Toward Biomim Microelectron Neural Prosthes 24:30–44

Berger TW, Song D, Chan RHM, Marmarelis VZ (2010) The neurobiological basis of cognition: identification by multi-input, multioutput nonlinear dynamic modeling. Proc IEEE 98:356–374

Berger TW, Hampson RE, Song D, Goonawardena A, Marmarelis VZ, Deadwyler SA (2011) A cortical neural prosthesis for restoring and enhancing memory. J Neural Eng 8:046017

Berger TW, Song D, Chan RHM, Marmarelis VZ, LaCoss J, Wills J, Hampson RE, Deadwyler SA, Granacki JJ (2012) A hippocampal cognitive prosthesis: multi-input, multi-output nonlinear modeling and VLSI implementation. IEEE Trans Neural Syst Rehabil Eng 20:198–211

Deadwyler SA, Bunn T, Hampson RE (1996) Hippocampal ensemble activity during spatial delayed-nonmatch-to-sample performance in rats. J Neurosci 16:354–372

Eichenbaum H (1999) The hippocampus and mechanisms of declarative memory. Behav Brain Res 103:123–133

Freund TF, Buzsaki G (1996) Interneurons of the hippocampus. Hippocampus 6:345–470

Goertzel B, Lian R, Arel I, de Garis H, Chen S (2010) A world survey of artificial brain projects, Part II: biologically-inspired cognitive architectures. Neurocomputing 74(1–3):30–49

Golgi C. [1906] (1967) The neuron doctrine: theory and facts. In: Nobel lectures: physiology or medicine, 1901–1921. Elsevier, Amsterdam, pp 189–217

Hampson RE, Simeral JD, Deadwyler SA (1999) Distribution of spatial and nonspatial information in dorsal hippocampus. Nature 402:610–614

Hampson RE, Song D, Chan RHM, Sweatt AJ, Fuqua J, Gerhardt GA, Marmarelis VZ, Berger TW, Deadwyler SA (2012) A nonlinear model for cortical prosthetics: memory facilitation by hippocampal ensemble stimulation. IEEE Trans Neural Syst Rehabil Eng 20:184–197

Hampson RE, Song D, Opris I, Santos L, Shin D, Gerhardt GA, Marmarelis VZ, Berger TW, Deadwyler SA (2013) Facilitation of memory encoding in primate hippocampus by a neuroprosthesis that promotes task specific neural firing. J Neural Eng 9:056012

Hsiao M-C, Song D, Berger TW (2013) Nonlinear dynamical model based control of in vitro hippocampal output. Front Neural Circuits 7:1–14

Milner B (1970) Memory and the medial temporal regions of the brain. In: Pribram KH, Broadbent DE (eds) Biology of memory. Academic, New York, pp 29–50

Song D, Berger TW (2010) Identification of nonlinear dynamics in neural population activity. In: Oweiss KG (ed) Statistical signal processing for neuroscience and neurotechnology. Elsevier, Amsterdam, pp 103–128

Song D, Chan RHM, Marmarelis VZ, Hampson RE, Deadwyler SA, Berger TW (2006) Nonlinear dynamic modeling of spike train transformations for hippocampal-cortical prostheses. IEEE Tran Biomed Eng 54:1053–1066

Song D, Chan RHM, Marmarelis VZ, Hampson RE, Deadwyler SA, Berger TW (2009a) Nonlinear modeling of neural population dynamics for hippocampal prostheses. Neural Netw 22:1340–1351

Song D, Chan RHM, Marmarelis VZ, Hampson RE, Deadwyler SA, Berger TW (2009b) Sparse generalized Laguerre-Volterra model of neural population dynamics. In: Proceedings of the IEEE EMBS conference, pp 4555–4558

Song D, Wang Z, Marmarelis VZ, Berger TW (2009c) Parametric and non-parametric modeling of short-term synaptic plasticity. Part I: Computational study. J Comput Neurosci 26:1–19

Song D, Wang H, Tu CY, Marmarelis VZ, Hampson RE, Deadwyler SA, Berger TW (2013) Identification of sparse neural functional connectivity using penalized likelihood estimation and basis functions. J Comput Neurosci 35:335–357

Song D, Harway M, Marmarelis VZ, Hampson RE, Deadwyler SA, Berger TW (2014) Extraction and restoration of hippocampal spatial memories with nonlinear dynamical modeling. Front Syst Neurosci. doi:10.3389/fnsys.2014.00097

Squire LR, Zola-Morgan S (1991) The medial temporal lobe memory system. Science 253:1380–6

Yeckel MF, Berger TW (1990) Feedforward excitation of the hippocampus by afferents from the entorhinal cortex: redefinition of the role of the trisynaptic pathway. Proc Natl Acad Sci 87:5832–5836

Yeckel MF, Berger TW (1995) Monosynaptic excitation of hippocampal CA1 pyramidal cells by afferents from the entorhinal cortex. Hippocampus 5:108–114

Chapter 21
Brain-Machine Interfaces: From Macro- to Microcircuits

Mikhail Lebedev and Ioan Opris

Abstract Brain-machine interfaces (BMIs) establish unidirectional and bidirectional communication channels between the brain and assistive devices, such as wheelchairs, limb prostheses and computers. BMI technologies can also link areas within the brain and even individual brains. BMI approach holds promise to provide effective treatment for a range of neurological conditions. Moreover, BMIs can be utilized to augment brain function in healthy individuals. Here we consider three broadly defined BMI types: motor, sensory and cognitive. For these BMI types, both noninvasive and invasive methods have been employed for neural recordings and stimulation. While original BMIs were implemented at the level of brain macrocircuits, a recent trend was to develop BMIs that utilize neuronal microcircuits.

Keywords Brain-machine interface • Brain circuits • Neuroprosthesis • Multichannel neural recordings

21.1 Brain Repair with BMIs

We usually take for granted our ability to produce voluntary movements, experience sensations, and perform mental operations – all driven by what seems to be our free will. In reality, we are consciously unaware of the vast complexity of neural mechanisms that subserve our motor, sensory and cognitive capacities. These mechanisms involve continuous processing of information streams by sophisticated brain circuits composed of billions of neurons.

Unfortunately, well organized operation of the normal brain can go wrong. This may happen after a relatively minor, measured in millimeters, damage to neural structures. Millions of people worldwide suffer from severe neurological

M. Lebedev (✉)
Department of Neurobiology, Duke University, Durham, NC, USA
e-mail: lebedev@neuro.duke.edu

I. Opris, Ph.D.
Department of Physiology and Pharmacology, Wake Forest University Health Sciences, Winston-Salem, NC 27103, USA
e-mail: ioanopris.phd@gmail.com

M.F. Casanova, I. Opris (eds.), *Recent Advances on the Modular Organization of the Cortex*, DOI 10.1007/978-94-017-9900-3_21

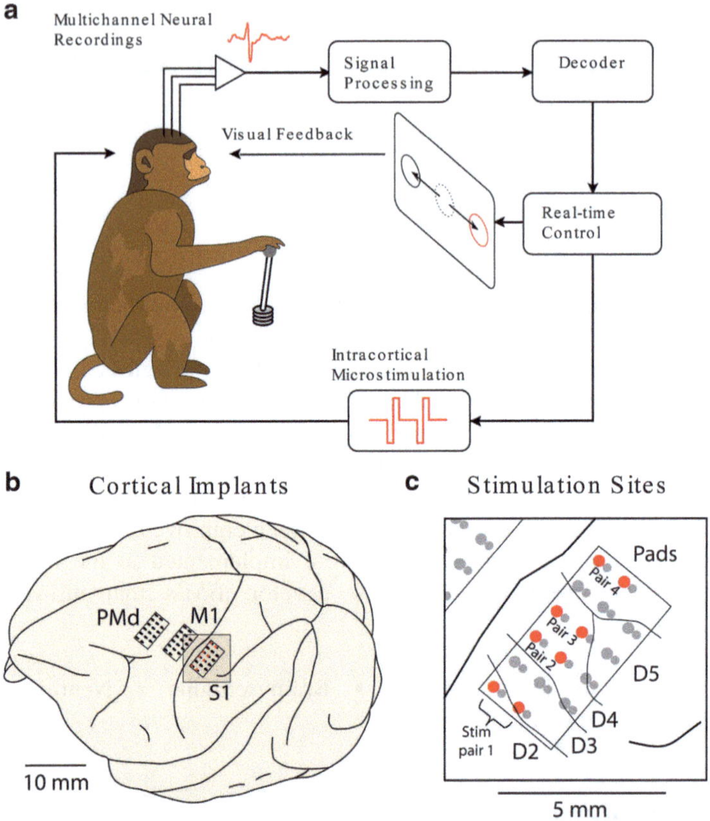

Fig. 21.1 Bidirectional brain-machine interface. (**a**) A monkey is engaged in a motor task that requires reaching toward screen targets with a cursor. The animal performs the task either with a hand held joystick or using a BMI that converts cortical activity into the cursor position. Two targets are shown, and the correct one is signaled by joystick vibration or intracortical microstimulation applied to somatosensory cortex. (**b**) Locations of cortical implants. (**c**) Cortical locations to which microstimulation was applied. Neurons in this area had receptive fields on digits D2–D5

conditions, such as spinal cord injury (SCI), stroke, autism, Parkinson's disease, Alzheimer's disease, and amyotrophic lateral sclerosis (ALS). No efficient treatments exist for many of these disabilities.

The modern fields of neuroengineering and BMIs strive to repair neurological damage by establishing functional links between the intact neural areas and external devices that perform motor, sensory and even cognitive operations (Lebedev and Nicolelis 2006; Nicolelis and Lebedev 2009; Schwartz et al. 2006; McFarland et al. 2006; Hatsopoulos and Donoghue 2009) (Fig. 21.1). For example, a BMI that connects sensorimotor cortex directly to a prosthetic body part could restore some motor and sensory functions in SCI patients.

Furthermore, BMIs could be used to augment brain function in healthy people. Examples include BMIs that aid computer gaming (Tangermann et al. 2009) and BMIs that detect drowsiness in drivers (Lin et al. 2010).

We will consider three broadly defined classes of BMIs: (1) motor, (2) sensory, and (3) cognitive. Brain circuits are often classified the same way. While this classification is useful, one should bear in mind that there are no separate motor, sensory and cognitive systems in the brain. Rather, it has been long known that these functions are represented in distributed way by many brain areas, and each individual area multiplexes numerous functions (Lilly 1956; Evarts 1973). Notably, a recent trend in BMIs was to combine sensory and motor functions in one design (O'Doherty et al. 2011).

Facilitated by the improvement of neurophysiological recording methods and the advances in computer systems, BMIs have experienced an explosive development during the last two decades. The number of BMI publications has been growing exponentially. Because of this impressive progress, many ideas previously entertained only in science fiction (e.g., cyborgs, reading-out thoughts and memories) have become feasible today.

Cochlear implants have been by far the most successful neuroprosthetic application, with hundreds of thousands of patients with auditory deficits receiving these devices (Shannon 2012; Wilson and Dorman 2008). Many other BMI types are still at the stage of laboratory research, and much work will be needed to translate them to clinical practice and consumer applications.

As BMIs develop, they bring out a number of ethical issues related to the possibility of BMIs interfering with our thoughts, intruding the mind and taking away our free will (Farah 2002; Vlek et al. 2012). These questions are especially relevant to BMIs that work in the cognitive domain, for example BMIs that mediate decision making (Andersen et al. 2010) and memory processing (Berger et al. 2005).

21.2 Brief History

Multichannel neural recordings is the key technology of modern BMIs. Recording devices that include many electrodes date back to the 1950s, when Lilly implanted several hundreds of electrodes in the monkey cortex and evoked body part movements by the application of electrical stimulation through those electrodes (Lilly 1956).

In the 1960s, research started on neurofeedback, another essential component of modern BMIs. Nowlis, Kamiya, Black and Sterman experimented with electroencephalograms (EEGs) as a source of such feedback (Lebedev and Nicolelis 2006). They converted EEG parameters into auditory or visual signals. Aided with this neurofeedback, subjects (both humans and animals) learned to modulate their brain potentials voluntarily.

In 1963, Walter conducted the first demonstration of an interface between the brain and a mechanical device. He recorded from the motor cortex of patients undergoing neural surgery (Dennett 1992). Walter instructed his patients to advance a slide projector by pushing a button with their hands. The motor cortex electrodes registered readiness potentials, which occurred shortly before the movements were initiated. The potentials were prominent enough to serve as a reliable trigger for the projector instead of the button. Accordingly, Walter disconnected the button and made human brain potentials control the projector directly. The patients continued to perform button presses, and were surprised that the brain interface often completed the job before they moved.

Several years later, Fetz and his colleagues utilized single-unit recordings from monkey cortex as the source of neurofeedback (Fetz 1969). They operantly conditioned monkeys to control the firing of single cortical neurons. Neuronal rates were fed back to the monkeys as visual or auditory signals. The monkeys successfully learned this volitional control, and were even able to modulate single neurons without producing any movements of their own limbs.

Similar research started at National Institutes of Health (NIH) under the leadership of Frank (Frank 1968). Frank proclaimed: "We will be engaged in the development of principles and techniques by which information from the nervous system can be used to control external devices such as prosthetic devices, communications equipment, teleoperators . . . and ultimately perhaps even computers". The NIH group recorded from several neurons in monkey motor cortex while their monkeys performed wrist movements (Humphrey et al. 1970). In an offline analysis, wrist kinematics were decoded from the neuronal activity using multiple linear regression (Humphrey et al. 1970). Eventually the decoding was performed online (Schmidt 1980). In these experiments monkeys learned to control a cursor on a LED screen with their cortical modulations.

Research on sensory BMIs (i.e. BMIs that deliver sensory information to the nervous system) started in the late 1950s. In 1957, Djourno and Eyriès implanted deaf patients with single-channel cochlear implants that electrically stimulated the auditory nerve to produce auditory sensations (Wilson and Dorman 2008). Simmons enhanced the cochlear implant with multiple stimulation channels in 1964, and House and Urban obtained an FDA approval for their design of the implant in the 1970s.

Research on prosthetic vision was pioneered by the laboratories led by Brindley and Lewin (1968) and Dobelle et al. (1974). Electrical stimulated of the primary visual cortex was employed to produce artificial visual sensations in blind humans. The subjects experienced sensations of light spots in their visual field, called phosphens. Multichannel stimulation of the visual cortex successfully conveyed simple visual patterns composed of several phosphens.

The late 1990s were marked by a renewed interest to BMI research and its considerable acceleration. Nicolelis and Chapin pioneered invasive BMIs that controlled robotic devices (Chapin et al. 1999). In their experiments, motor commands were decoded from neuronal population activity recorded in the cortex and thalamus of awake, behaving rats. The animals learned to operate a one-dimensional robot with their brain activity alone.

Following this successful study in rodents, Nicolelis moved on to implement a number of motor and sensorimotor BMIs in nonhuman primates, namely cortical BMIs that enacted reaching movements of robotic arms (Wessberg et al. 2000; Carmena et al. 2003; Lebedev et al. 2005), delivered artificial tactile sensations to the brain (O'Doherty et al. 2011), subserved bimanual movements (Ifft et al. 2013), and decoded locomotion kinematics (Fitzsimmons et al. 2009).

Kennedy and his colleagues invented a neurotrophic electrode where myelinated fibers grew into the recording tip (Kennedy and Bakay 1998). An ALS patient implanted with this electrode was able to perform neural on/off control.

A series of BMI studies were conducted by Donoghue and his colleagues. In particular, they implanted paralyzed patients with motor cortical implants. The patients were able to control a computer cursor (Hochberg et al. 2006) and a robotic arm (Hochberg et al. 2012).

Schwartz and his colleagues developed a BMI that enabled three dimensional (3D) movements of a cursor (Taylor et al. 2002) and a robotic arm (Velliste et al. 2008). They also implanted human patients and achieved cortical control of complex movements performed by an antropomorphic robotic arm (Collinger et al. 2013).

Significant contributions to the development of invasive BMIs were made by the laboratories of Andersen, Shenoy and Vaadia. A number of laboratories have been developing noninvasive BMIs, which were extensively tested in humans. Here, valuable contributions came from the groups of Birbaumer, Millan, Müller, Neuper, Kübler, Pfurtscheller, Schalk, Walpaw, and others.

21.3 Neuronal Tuning and Decoding

BMI researchers are to a certain extent lucky, as they can decode behavioral variables from the brain activity even though they very often do not have a clear understanding of what this activity means. Such decoding is possible because neuronal rates are correlated with behaviors. This correlation is fairly consistent, but noisy.

The relationship between neuronal rates and behavioral parameters is often called neuronal tuning. Pioneering studies of neuronal tuning were conducted in the late 1950s–1960s by Mouncastle in the somatosensory system (Mountcastle 2005), Hubel and Wiesel in the visual system (Hubel and Wiesel 2005), and Evarts in the motor system (Evarts 1973). In the 1980s. Wise (1985), Kalaska et al. (1997) and others studied neuronal tuning that represented motor preparation, spatial attention and visuomotor transitions.

An important milestone was achieved by Georgopoulos and his colleagues who studied directional tuning of motor cortical neurons and explored the possibility of decoding motor parameters from neuronal populations. They described the relationship between neuronal rates and movement direction as cosine tuning curves (Georgopoulos et al. 1982). Furthermore, the concept of population vector was introduced to describe encoding at the population level. Population vector proved to be a powerful method not only for decoding ongoing movements, but also for decoding mental transformations, such as mental rotation required to transform stimulus spatial location into movement direction (Georgopoulos et al. 1989). Georgopoulus suggested decoding accuracy can be improved by the recordings from large neuronal populations.

21.4 Decoding from Neuronal Ensembles

Georgopoulos's population approach was extended by other laboratories and eventually became the mainstream in BMI research (Nicolelis and Lebedev 2009). Decoding improves with the size of neuronal population because noisy fluctuations in the firing of individual neurons cancel each other when the signals of many neurons are combined by the decoding algorithm (Lebedev and Nicolelis 2006; Nicolelis and Lebedev 2009).

Although large neuronal ensembles are generally preferred for BMI decoding, small neuronal samples can be useful in certain cases (Moritz et al. 2008; Taylor et al. 2002). The best known of highly informative neurons, which can be useful for such decoding, are neurons specialized to represent a particular person – the so-called grandmother neurons or Jennifer Aniston neurons (Quiroga et al. 2005).

21.5 Decoding Algorithms

Many BMI decoding algorithms have been developed over the years. The choice of algorithm in each particular case depends on the parameters that have to be decoded from neuronal activity, characteristics of neuronal signals, number of recording channels, and other factors.

Historically, population vector was for a long time the mainstream method for decoding of movement kinematics from cortical activity (Georgopoulos et al. 1989). In this method, each neuron in the population is characterized by an individual vector that points in the preferred direction, i.e. movement direction for which the neuron exhibits the highest firing. The population vector is calculated as a weighted sum of the individual-neuron vectors. The weights are equal to the firing rates of the corresponding neurons. Notwithstanding the historical significance of this method, it is suboptimal for decoding because it does not minimize decoding error.

The Wiener filter is an optimal linear method which, similar to the population vector, represents decoded parameters as a weighted sum of neuronal rates (Haykin

2001). The weights are assigned to the measurements of neuronal rates at several time points in the past, called lags or taps. The weights are optimized to minimize mean-square error.

The Kalman offers additional advantages compared to the Wiener filter, particularly for the paradigms that involve stereotyped behaviors. The Kalman filter handles two groups of variables: state variables, such as limb position and velocity, and observable variables, such as neuronal firing rates. During the decoding, the states are updated in discrete steps, typically every 10 ms–100 ms. Each update incorporates two computations. First, an estimation of a new state is derived from the current state. Second, this estimation is adjusted based on the values of neuronal signals. This latter adjustment utilizes a model of the relationship between neuronal activity and the state variables, i.e. neuronal tuning model.

The unscented Kalman filter (UKF) adds several improvements (Li et al. 2009). It introduces nonlinearity to the neuronal tuning model and also adds a more detailed time history of neuronal rates.

Artificial neural networks (ANNs) are often utilized for BMI decoding (Chapin et al. 1999; Wessberg et al. 2000). Popular ANN methods include Gaussian classifier, multilayer perceptron, Bayesian logistic regression network, adaptive logic network and learning vector quantization network. A recently developed dynamical ANN, called recurrent neural network (RNN), improves decoding by treating neuronal activity as a function of its history (Sussillo et al. 2012).

In addition to continuous decoding BMIs can employ discrete classifiers. For example, BMI spellers extract discrete selections of font characters from EEG activity (Birbaumer et al. 1999, 2008). The list of discrete classifiers used in BMIs includes linear discriminant analysis, support vector machine, hidden Markov models, k nearest neighbors algorithm, and fuzzy logic decoder.

21.6 Theoretical Basis for Motor Decoding

A clear relationship can be seen between BMI approaches to motor decoding and the theories of motor control, which prevailed in the past or currently prevail.

Motor circuitry of the nervous system is commonly described as a hierarchical structure. Cortical areas are typically attributed advanced motor functions, and placed at the top of motor hierarchy. Lower-order areas, such as the brainstem and spinal cord, are attributed simpler, automated motor functions, for example spinal reflexes (Sherrington 1906) and locomotion patterns produced by central pattern generators (CPGs) (Guertin 2009).

The concept of reflex arc (Sherrington 1906) has prevailed for a long time in the motor control field. Reflexes are highly automated motor responses generated in response to sensory inputs. Many simple reflex arcs reside in the spinal cord. Voluntary movements are more complexly organized. They are controlled by higher-order motor areas, including a network of cortical areas. Motor activities usually incorporate both voluntary and reflex components (Cordo and Gurfinkel 2004).

A distinction between voluntary and reflex components can be found in BMI designs. For example, in a shared control BMI the user has control over higher-order parameters, while a robotic controller handles reflex-like tasks.

The concept of internal model is essential for several modern theories of motor control. This concept was first introduced a century ago by Head and Holmes (1911). They proposed that the brain maintains an internal representation of the body, called the "body schema". A more recent theory, the internal model theory (Kawato 1999), defines two parts of the motor system: the plant and the controller. The controller programs future motor states based on a model of the plant. If there is discrepancy between the programmed state and the sensory feedback from the plant, a correction command is issued by the controller. One possible implementation of such control mechanism is proposed by the equilibrium point hypothesis (Feldman et al. 1998). Here, higher-order motor centers generate a goal for the plant, called an equilibrium point. The plant is then placed to the equilibrium point by a servo mechanism that resides in the spinal cord and utilizes spinal reflexes.

21.7 Arm Movements Enabled by BMIs

A considerable number of BMI studies focused on the control of reaching and grasping movements. In one implementation of such a BMI, monkey cortical activity was converted into the reaching and grasping performed by a robotic arm (Carmena et al. 2003; Lebedev et al. 2005). Monkeys did not see the actual robot, but rather viewed a computer screen. The robot position was depicted by the position of a screen cursor, and the gripping force was indicated by the cursor diameter. Cortical activity was recorded with multielectrode implants placed in multiple cortical areas. Discharges of several hundreds of neurons were sampled simultaneously. The behavioral task consisted of placing the cursor on a screen target and then adjusting the cursor diameter to match the target size. The animals first learned to perform this task with a joystick. The joystick movements positioned the robot, and the animals squeezing the joystick handle to exert a gripping force by. Multiple Wiener filters were trained to extract the robot kinematics and gripping force from cortical activity. During BMI control, the joystick was electrically disconnected from the robot and in some experiments physically removed from the apparatus, and the BMI output controlled the robot instead.

In a similar experiment, Schwartz and his colleagues decoded monkey cortical activity with a population vector decoder (Taylor et al. 2002). This BMI enacted cursor reaching movements in three dimensions (3D). A co-adaptive algorithm was introduced to improve the performance on this task. The algorithm adjusted the population vector weights to minimize the discrepancy between the BMI-generated trajectories and the ideal trajectories connecting the starting point and the target. Following this study, Schwartz and his colleagues developed a BMI driven robotic arm that grasped pieces of food and brought them to the monkey's mouth (Velliste et al. 2008).

In the next development, the groups of Donoghue (Hochberg et al. 2012) and Schwartz (Collinger et al. 2013) constructed BMI controlled robotic arms for paralyzed patients. The patients were implanted with invasive cortical arrays. The BMIs executed real-time control of reaching, grasping, and object manipulations with the robotic arms.

The researchers at the Nicolelis laboratory advanced this research further by developing a bimanual BMI that operated a pair of virtual arms (Ifft et al. 2013).

21.8 Functional Electrical Stimulation

Several groups explored the possibility of connecting cortical output to the subjects' own muscles. In these studies, muscle contractions were induced by functional electrical stimulation (FES).

Pfurtscheller and his colleagues powered the paralyzed hand of a tetraplegic patient with an FES device attached to the forearm (Pfurtscheller et al. 2003). The FES was driven by cortical commands extracted from EEG beta rhythms, which the subject modulated by imagining foot movements.

Fetz and his colleagues developed a BMI that decoded activity of motor cortical neurons and drove temporarily paralyzed monkey hands with FES (Moritz et al. 2008).

Miller and his colleagues developed a more advanced cortically controlled FES device that gave monkeys an ability to grasp objects with temporarily paralyzed hands (Ethier et al. 2012; Pohlmeyer et al. 2009).

21.9 BMIs for Locomotion

Recently there has been an increased interest to BMIs that enable locomotion. The Nicolelis laboratory pioneered this research by showing that kinematics of bipedal locomotion can be decoded from neuronal ensembles recorded in monkey cortex (Fitzsimmons et al. 2009). Following this proof of concept demonstration, Nicolelis and his colleagues founded the Walk Again Project, an international consortium for the advancement of BMI controlled exoskeletons (Nicolelis and Lebedev 2009). An European project, called Mindwalker, declared similar goals (Cheron et al. 2012). Yet another group, led by Contreras Vidal, focused on EEG controlled BMIs that enable human walking (Presacco et al. 2012).

In addition to restoration of locomotion by cortical signals, there is an alternative strategy, which is based on the idea of reactivating spinal CPGs. This approach was tested in rats with spinal cord transactions (Courtine et al. 2009). Spinal cord circuitry was pharmacologically conditioned with serotonergic agonists, after which epidural electrical stimulation reactivated the CPG and induced locomotion.

21.10 BMIs and Brain Plasticity

BMI control of an external actuator has much in common with normal usage of tool, the behavior known to provoke brain plasticity. Plasticity associated by tool usage was first shown by Iriki and his colleagues (Iriki et al. 1996). Their monkeys learned to reach toward distant objects with a rake. This learning was accompanied by cortical plasticity: neurons in posterior parietal cortex acquired visual receptive fields that extended along the length of the rake.

Several studies sought for brain plasticity that would result from learning to control a BMI. Overall, the results of these studies support the suggestion that the brain can plastically adapt to represent a BMI-controlled prosthesis, and effectively incorporate the prosthesis into the body schema (Lebedev and Nicolelis 2006; Nicolelis 2011). Several manifestations of such plasticity have been reported: changes in directional tuning of the neurons involved in BMI control (Lebedev et al. 2005), novel modulation patterns associated with learning to operate a BMI (Zacksenhouse et al. 2007), changes in pairwise correlation between neurons (Carmena et al. 2003; Ifft et al. 2013), and adaptions to rotational transformation applied to a subset of neurons used for BMI decoding (Chase et al. 2012).

21.11 Noninvasive BMIs

The safest, nonivasive BMIs, sample neural signals without penetrating into the body. Because of the simplicity of their implementation, noninvasive BMIs have been extensively tested in humans. These BMIs have been utilized for communication tasks, control of limb prostheses, and wheelchair navigation (Galán et al. 2008; Muller-Putz and Pfurtscheller 2008; Nicolas-Alonso and Gomez-Gil 2012; Sellers et al. 2010). Aided with noninvasive BMIs, severely impaired "locked in" patients were able to communicate with the outside world (Birbaumer et al. 1999, 2008).

EEGs are used most commonly in noninvasive systems. There are two major classes of EEG-based BMIs: independent (endogenous) and dependent (exogenous). In an independent BMI, subjects volitionally modulate their brain activity, for example by imagining movements of their body parts. All known EEG rhythms have been utilized in BMIs (McFarland et al. 2006). EEG-based BMIs often employ adaptive decoders (Wolpaw and McFarland 2004).

A dependent BMI relies on an external stimulus that paces EEG responses. Such stimuli are usually presented on a computer screen. A dependent BMI then reads out subjects' intentions by detecting the differences in EEG patterns associated with attended versus unattended stimuli. For example, visual evoked potential (VEP) based BMIs detect enhanced VEPs over the visual cortex when the subject attends to a particular stimulus on the screen. Steady state visual evoked potentials (SSVEPs) can be also used for the same purpose (Vialatte et al. 2010). SSVEP based BMIs employ/register visual responses to several screen objects, each flickering at its own frequency. P300 evoked potentials can be utilized in a very similar way (Farwell and Donchin 1988).

Several clinically relevant implementations of EEG-based BMIs been developed, such as BMIs that control robots (Millan et al. 2004), wheelchair navigation (Galán et al. 2008), and limb orthoses (Tavella et al. 2010; Pfurtscheller et al. 2003).

EEG artifacts present a considerable problem for EEG-based BMIs. Such artifacts are often not handled adequately (Fatourechi et al. 2007). Consequently, artifacts could be confused with neural signals and could be even used as a control signal for a BMI.

Electrocorticographic (ECoG) BMIs work similarly to the ones based on EEGs while offering a neural signal of better quality. However, implantation of ECoG electrodes is an invasive procedure that requires piercing the skull and dura mater.

Magnetoencephalography (MEG) has found use in BMIs (Mellinger et al. 2007). Brain magnetic fields are detected by magnetometers, superconducting quantum interference devices (SQUIDs). MEG recordings have better spatial and temporal resolution compared to EEGs, but require a specialized, magnetically shielded facility.

Another recording method, Near infrared spectroscopy (NIRS), utilizes an infrared light that penetrates through the skull and monitors oxyhemoglobin and deoxyhemoglobin concentration in the brain blood supply (Sitaram et al. 2009). The spatial resolution of this method is approximately 1 cm, and temporal resolution is on the order of 100 ms. A drawback of this approach is a considerable delay of the hemodynamic response (several seconds).

Functional magnetic resonance imaging (fMRI) has been used in BMIs, as well (Sitaram et al. 2009). The spatial resolution is superb throughout the brain. The temporal resolution is 1 s–2 s, and response lag is several seconds.

21.12 Sensory BMIs

Sensory BMIs strive to repair neural circuitry that processes sensory information. Here, there are many sensory modalities that could be recreated with a BMI: hearing, sight, touch, smell, taste, vestibular sensation, and proprioception.

Normal sensation involves processing of the signals originating from peripheral receptors by a hierarchically organized neural network. Sensory receptors send their information to the spinal cord and brainstem. From there, information ascends to the higher-order areas: thalamus, cerebellum, cortex, and basal ganglia. Neural areas that encode sensory information are often called somatotopic maps. These somatotopic representations resemble a homunculus – a disproportional body with large hands and face, a small trunk and small legs (Schott 1993).

Sensory impairments can be related to problems with sensory periphery, or they can result from damage to higher-order areas. For example, patients with lesions to visual cortex do not perceive visual stimuli, but they may have blindsight, an ability to utilize visual information subconsciously (Barton 2011).

Current sensory BMIs mostly focus on neurological damage to peripheral sensors. BMI systems strive to replace damaged sensory components by artificial sensors. An artificial sensor is interfaced to an intact sensory area, such as cortex or thalamus, and electrical stimulation is usually employed to evoke neural responses (Romo et al. 2000; Fitzsimmons et al. 2007; O'Doherty et al. 2011). Recently invented optogenetic stimulation methods can be used for this purpose, as well (Zhang et al. 2007).

The approach adopted by sensory BMIs bears similarity to the method called sensory substitution. Sensory substitution implies that a compromised sensory channel is substituted by an intact physiological sensation, such as tactile sensation from an unaffected body part. The modality of the substitution sensation can be different from the sensory modality being repaired. For example, vision can be substituted by a skin tactile display connected to a video camera (Jones 2011). As long as intact physiological receptors are used for information delivery, the approach is called sensory substitution. If the method to produce sensation is artificial (e.g., electrical stimulation of the cortex), the approach is called sensory BMI. In some cases, it is difficult to decide if the stimulation method is natural or artificial. For instance, vision can be recreated by electrical stimulation of the tongue (Bach-y-Rita and Kercel 2003; Sampaio et al. 2001).

21.12.1 Auditory Prosthesis

Cochlear implant is the best success story in the field of sensory BMIs (Shannon 2012; Wilson and Dorman 2008). The implant recipients can recognize speech, distinguish between voices and even listen to melodies. Bilateral implants restore directional hearing.

The implant converts sounds recorded with an external microphone into a multichannel electrical stimulation applied to the intact parts of the auditory nerve. Different stimulation patterns are delivered to different parts of the nerve. Several types of stimulation patterns are currently used. In the continuous interleaved sampling strategy, the auditory signal is split into frequency bands, which are converted into a narrow range of stimuli using a nonlinear transform function. There are also methods that select a different electrode for each frame of stimulation.

A brainstem implant has been proposed for people with severe cochlea damage (Shannon 2012). This device stimulates the cochlear nucleus.

21.12.2 Visual Prosthesis

Although not as successful yet as the cochlear implants, a number of clinical visual prostheses have been developed (Fernandes et al. 2012). Such prostheses can be retinal, or non-retinal. Retinal prostheses are applicable to cases where parts of the eye and optic nerve are spared. Non-retinal prostheses are employed when the optic nerve is a severely damaged.

There are several types of retinal prostheses, each intended for a particular state of retinal degeneration. In the epiretinal implant, an intraocular electrode array stimulates the layer of nerve fibers of retinal ganglion cells. The array conveys images from a video camera, which is mounted outside of the eye in the current systems, and will be implanted inside the eye in the future. This implant restores perception of visual objects shape, brightness and motion.

The subretinal prosthesis stimulates ganglion cells and bipolar cells. In this implant, an array of microphotodiodes is connected to an array of stimulation electrodes. Testing has only started for this system.

The transchoroidal prosthesis employs electrodes that are placed in the suprachoroidal space. The implantation surgery is relatively simple for this device. The implant evokes perception of phosphens and simple images.

In non-retinal prostheses, stimulation is applied to the visual cortex. Dobelle's visual prosthesis utilized an array of 64 surface electrodes placed over the visual cortex (Dobelle et al. 1974). This system restored rudimentary vision to blind individuals.

21.13 Bidirectional BMIs

Bidirectional BMIs combine extraction of information from brain activity with the delivery of sensory feedback to the brain. A pioneering study of such an interface was conducted by O'Doherty and his colleagues (O'Doherty et al. 2009, 2011). The experiments were conducted in monkeys chronically implanted with microelectrode arrays in the sensorimotor cortex. Motor commands were extracted from the activity of motor cortex neurons. To deliver artificial somatosensory feedback, patterns of intracortical microstimulation (ICMS) were delivered to the somatosensory cortex. Monkeys used this bidirectional BMI to actively explore virtual objects, which looked identical, but were associated with different ICMS patterns. The exploration was conducted by a realistic image of the monkey arm controlled by motor cortical activity through the BMI motor loop.

21.14 Cognitive BMIs

Cognitive BMIs strive to read out high-level neural representations, such as percepts, thoughts, or decisions (Andersen et al. 2010; Nicolelis 2001). According to Nicolelis (2001), "hybrid brain–machine interfaces have the potential to enhance our perceptual, motor and cognitive capabilities."

Assistive technologies based on cognitive BMIs can in the future restore function to people with neurological deficits. Such BMIs could process cortical and subcortical signals that represent such cognitive processes as executive control, attention, and working memory. If successful, cognitive BMIs may be useful to a large population of patients, including those suffering from Parkinson's disease, schizophrenia, autism, and drug addiction (Opris and Casanova 2014).

21.15 BMIs to Restore Memory

Berger and colleagues (2011) demonstrated in rodents that a BMI is capable of real-time identification and manipulation of the encoding process that can restore and even enhance cognitive mnemonic processes. To produce a model of memory deficit, the ability to form long-term memories was blocked by pharmacological agents that disrupted the neural circuitry between two fields of the hippocampus that interact to create long-term memory: CA1 and CA3.

Berger's artificial hippocampus was based on the multiple-input, multiple-output (MIMO) model (Song et al. 2009; Kim et al. 2006). The MIMO model duplicated the pattern of interaction between CA3-CA1 by monitoring the neural activity in cells recorded by the electrode array, and then playing back that pattern on the same array. Berger and his colleagues succeeded to restore long-term memory to the pharmacologically blocked rats. Their electronic device was programmed to duplicate the memory-encoding function for a memory of what lever the rat had to pull.

Building on this result, Berger and Deadwyler started a series of experiments in monkeys, with the aim of eventually developing a BMI that could restore memory and other cognitive functions to humans suffering from Alzheimer's disease, stroke, or brain injury. To this end, the MIMO model (Hampson et al. 2012; Opris 2013) was applied in the primate prefrontal cortex (Casanova 2013; Opris et al. 2011, 2012a, b, 2013, 2014). Ensemble firing patterns of up to 16 prefrontal neurons were recorded using a wireless recording system. MIMO algorithm successfully predicted of cortical layer 5 output from layer 2/3 input. Furthermore, an 8-channel wireless stimulator delivered that prediction to layer 5 as patterns of electrical pulses. This study showed that behavioral performance on a memory task was significantly improved when the stimulation scheme was employed (Hampson et al. 2012, 2013; Opris 2013).

21.16 BMIs for Executive Control

An executive BMI could enhance decision making, for example by correcting an incorrect decision. One possible design (Fig. 21.2) for such a decision-making BMI involves collecting sensory signals recorded from the primary sensory areas (for visual modality V1; for tactile modality S1), reward signals from ventral tegmental area (VTA) or nucleus accumbens (NAc), working memory signals from PFC and FEF, and decision bias signals from caudate nucleus (CN). This BMI processed the critical variables for the decision process: (i) visual signals encoding the sensory evidence (dorsal stream nodes in V1, MT and LIP and/or S1), (ii) dopaminergic signals (VTA and SNc) to weigh the options based on reward or gain, and (iii) signals in prefrontal cortex to monitor cognitive workload and potential failures in

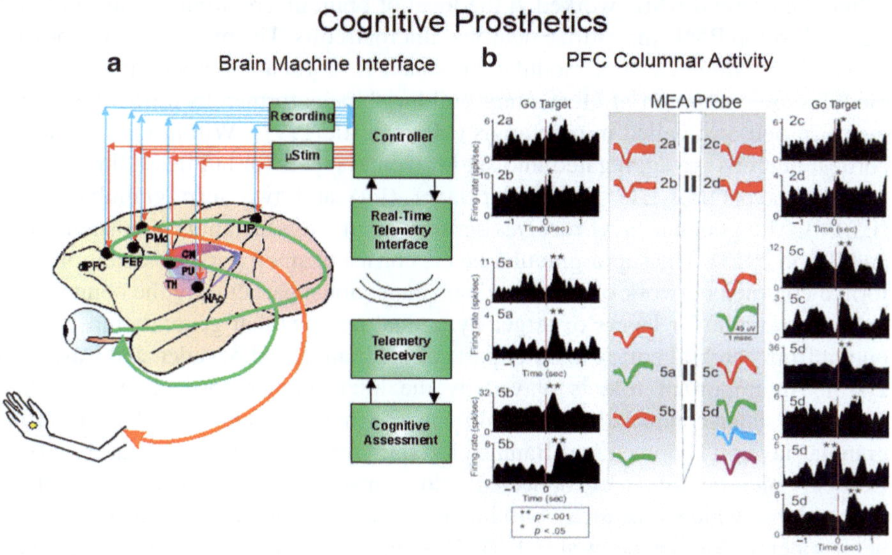

Fig. 21.2 BMI for Executive Control. (**a**) The diagram shows the recording of sensory (visual spatial) signals from lateral intra-parietal area (LIP), reward signals from nucleus accumbens (NAc), working memory signals from prefrontal cortex (PFC), the signals from dorsolateral PFC carrying cognitive workload and a decision bias signal from caudate nucleus (CN). Signals are recorded, decoded and interpreted so that the optimal selection is made. An error signal instructs the microstimulator to apply a microcurrent in the right brain region and the decision signal should be corrected in real time. (**b**) Frontal Cortex neurons simultaneously recorded with a W3 ceramic multi-electrode array (MEA) in primates during a behavioral Go-Nogo task. Display shows the activity recorded from 14 frontal cortical neurons (superimposed waveforms, calibration 1 msec, 150 µV) in Layer 2/3 and Layer 5 from an animal performing the Go No-go task. Neurons were detected and electronically isolated at the individual recording sites on the ceramic MEA probe diagram with more than one neuron (indicated by differences in waveform) isolated from some recording sites. PEHs next to waveforms show mean firing rate (Hz) of the same neurons during go target presentations (*red line* 0.0 s) within the go–no-go session (With permission from Opris and Ferrera 2014, Biobehavioral and Brain Reviews)

the cognitive process. The normal decision making signal follows a temporal pattern characterized by a rising to threshold when the preferred option is selected or a temporal decay for non-preferred option. Signals are recorded, decoded and interpreted so that the optimal selection is made (Nurmikko et al 2010). When the decision signal is not optimal, predicting that a failed choice will occur soon, an error signal instructs the microstimulator to apply a microcurrent in the appropriate brain region (for example, prefrontal cortical layer 2/3 if the accumulation of sensory evidence is weak, layer 5 if the selection signal is weak, or caudate nucleus if the decision bias is weak) and the decision signal will be corrected in real time.

21.17 Microcircuit-Based BMIs

While the original BMIs worked at the level of brain macrocircuits, a recent trend was to develop BMIs that utilize neuronal microcircuits. The microcircuit approach is based on Mountcastle's modular approach to cortical microcircuits. Cortical modules are composed of elementary building blocks formed by vertical arrangements of neurons, called minicolumns (Mountcastle 1997). Within mini-columns, cortical neurons are aggregated into six horizontal layers (or laminae): three supra-granular layers (L1-L3), a granular layer (L4) and two infra-granular layers (L5/L6). The granular layer receives sensory input from thalamus (Constantinople and Bruno 2013). The supra-granular layers consist of small pyramidal neurons that form a complex network of intra-cortical connections, particularly the connections to the infra-granular layers of larger pyramidal neurons that generate most of the output from cerebral cortex to other parts of the brain (Buxhoeveden and Casanova 2002). An important role is played by the inter-laminar cortical microcircuits formed by inter-connected pyramidal neurons from the supra-granular and infra-granular layers (Thomson and Bannister 2003; Opris et al. 2011, 2012a, b, 2013). Cortical microcircuits are connected into a macro-network by cortico-cortical connections, which link areas within the same hemisphere, as well as between hemi-spheres (Van Essen et al. 1982). This super network sub-serves the "perception-to-action" cycle – a group of processes that handle environmental stimuli and convert them into actions (Romo et al. 2002; Fuster and Bressler 2012; Opris et al. 2013). Recording such microcircuits requires a special type of microelectrode arrays (MEAs) shown in Fig. 21.3.

With recent technological advancement, cortical microcircuits are becoming instrumental to build BMIs for repair and augmentation of cognitive function (Opris 2013). This implies that increasing the number of channels in a multi-input / multi-output (MIMO) device that use cortical microcircuits may improve performance. MIMO model is based on the principle of multiplexing, where a high rate signal is split into several low rate signals, which are then sent to multiple cells via multiple channels. The MIMO model provides a reliable communication method. To perform cognitive augmentation, inter-laminar recordings are analyzed via a non-linear MIMO model, whose output is then converted into patterns of

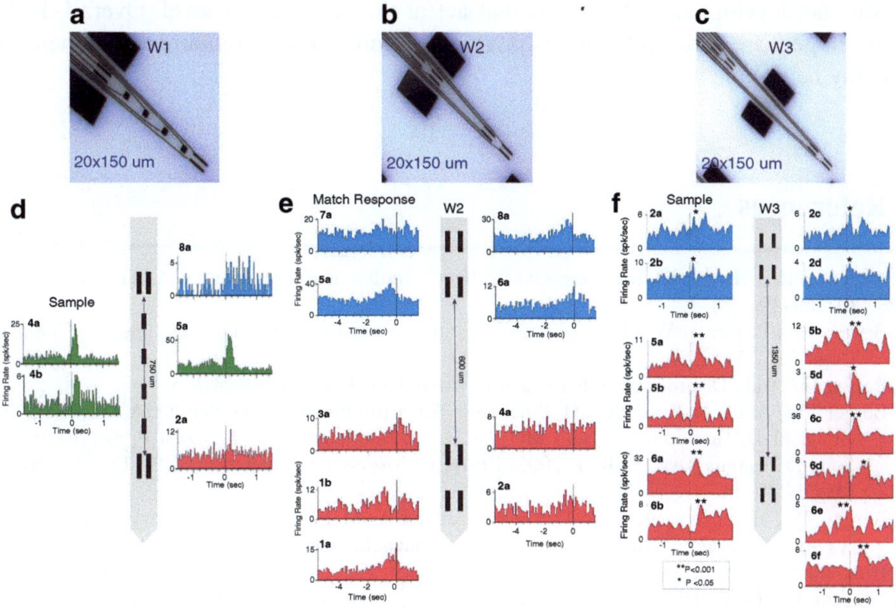

Fig. 21.3 Example of prefrontal neurons recorded simultaneous with neuromorphic multi-electrode arrays. (**a–c**) Illustration of configuration for three different types of neuromorphic probes W1, W2, W3 used in columnar recordings. (**d–f**) Example of simultaneous recorded cells in the prefrontal cortex. Code color for the neural activity in cortical layers is: *blue* (layer 2/3), *green* (layer 4) and *red* (layer 5. Peri-event histograms (PEHs) show neural activity simultaneously recorded with neuromorphic probes during single session. Separation distance of the MEA pads is shown for each diagram with cells recorded from those locations indicated by different markers (With permission from Opris et al. 2014, Journal of Neuroscience Methods)

microstimulation (Berger et al. 2011). In these studies, MIMO models used a precise topographically matched stimulation by extracting the patterns of firing that relate to the successful behavioral performance. This allowed the substitution of task-related laminar L5 neuron firing patterns with electrical stimulation in the same recording regions during columnar transmission from lamina L2/3 at the time of target selection. Such stimulation improved normal task performance, but more importantly, recovered performance impaired by pharmacological disruption of decision making (Hampson et al. 2012).

21.18 Conclusion

BMIs have experienced a rapid development in recent years. BMI approach has been successfully applied to arm movements and bipedal locomotion, somatosensory sensation, hearing and vision. Moreover, BMIs have been applied to cognitive functions, such as decision making and memory. A recent trend in BMI research was the development of systems that act at the microcircuit level. Overall, BMI developers are confident that this is a viable approach to revolutionize treatment of many neurological conditions.

References

Andersen RA, Hwang EJ, Mulliken GH (2010) Cognitive neural prosthetics. Annu Rev Psychol 61:169–190, C1-3

Bach-y-Rita P, Kercel W (2003) Sensory substitution and the human-machine interface. Trends Cogn Sci 7:541–546

Barton JJ (2011) Disorder of higher visual function. Curr Opin Neurol 24:1–5

Berger TW, Ahuja A, Courellis SH, Deadwyler SA, Erinjippurath G, Gerhardt GA, Gholmieh G, Granacki JJ, Hampson R, Hsaio MC, LaCoss J, Marmarelis VZ, Nasiatka P, Srinivasan V, Song D, Tanguay AR, Wills J (2005) Restoring lost cognitive function. IEEE Eng Med Biol Mag 24:30–44

Berger TW, Hampson RE, Song D, Goonawardena A, Marmarelis VZ, Deadwyler SA (2011) A cortical neural prosthesis for restoring and enhancing memory. J Neural Eng 8(4):046017

Birbaumer N, Ghanayim N, Hinterberger T, Iversen I, Kotchoubey B, Kübler A, Perelmouter J, Taub E, Flor H (1999) A spelling device for the paralysed. Nature 398:297–298

Birbaumer N, Murguialday AR, Cohen L (2008) Brain-computer interface in paralysis. Curr Opin Neurol 21:634–638

Brindley GS, Lewin WS (1968) The sensations produced by electrical stimulation of the visual cortex. J Physiol 196:479–493

Buxhoeveden DP, Casanova MF (2002) The minicolumn hypothesis in neuroscience. Brain 125(5):935–951

Carmena JM, Lebedev MA, Crist RE, O'Doherty JE, Santucci DM, Dimitrov DF, Patil PG, Henriquez CS, Nicolelis MA (2003) Learning to control a brain-machine interface for reaching and grasping by primates. PLoS Biol 1(2), E42

Casanova MF (2013) Canonical circuits of the cerebral cortex as enablers of neuroprosthetics. Research Topic: "Augmentation of brain function: facts, fiction and controversy", Lebedev MA, Opris I and Casanova MF (eds). Front Syst Neurosci 7.77. doi:10.3389/fnsys.2013.00080

Chapin JK, Moxon KA, Markowitz RS, Nicolelis MA (1999) Real-time control of a robot arm using simultaneously recorded neurons in the motor cortex. Nat Neurosci 2:664–670

Chase SM, Kass RE, Schwartz AB (2012) Behavioral and neural correlates of visuomotor adaptation observed through a brain-computer interface in primary motor cortex. J Neurophysiol 108:624–644

Cheron G, Duvinage M, De Saedeleer C, Castermans T, Bengoetxea A, Petieau M, Seetharaman K, Hoellinger T, Dan B, Dutoit T, Sylos LF, Lacquaniti F, Ivanenko Y (2012) From spinal central pattern generators to cortical network: integrated BCI for walking rehabilitation. Neural Plast 2012:375148

Collinger JL, Wodlinger B, Downey JE, Wang W, Tyler-Kabara EC, Weber DJ, McMorland AJ, Velliste M, Boninger ML, Schwartz AB (2013) High-performance neuroprosthetic control by an individual with tetraplegia. Lancet 381:557–564

Constantinople CM, Bruno RM (2013b) Deep cortical layers are activated directly from thalamus. Science 340(6140):1591–1594

Cordo PJ, Gurfinkel VS (2004) Motor coordination can be fully understood only by studying complex movements. Prog Brain Res 143:29–38

Courtine G, Gerasimenko Y, van den Brand R, Yew A, Musienko P, Zhong H, Song B, Ao Y, Ichiyama RM, Lavrov I, Roy RR, Sofroniew MV, Edgerton VR (2009) Transformation of

nonfunctional spinal circuits into functional states after the loss of brain input. Nat Neurosci 12:1333–1342
Dennett DC (1992) Consciousness explained. Back Bay Books, New York, 528 pp. [This book contains a description of the pioneering demonstration of a brain-machine interface by Grey Walter]
Dobelle WH, Mladejovsky MG, Girvin JP (1974) Artificial vision for the blind: electrical stimulation of visual cortex offers hope for a functional prosthesis. Science 183:440–444
Ethier C, Oby ER, Bauman MJ, Miller LE (2012) Restoration of grasp following paralysis through brain-controlled stimulation of muscles. Nature 485:368–371
Evarts EV (1973) Motor cortex reflexes associated with learned movement. Science 179:501–503
Farah MJ (2002) Emerging ethical issues in neuroscience. Nat Neurosci 5:1123–1129
Farwell LA, Donchin E (1988) Talking off the top of your head: toward a mental prosthesis utilizing event-related brain potentials. Electroencephalogr Clin Neurophysiol 70:510–523
Fatourechi M, Bashashati A, Ward RK, Birch GE (2007) EMG and EOG artifacts in brain computer interface systems: a survey. Clin Neurophysiol 118:480–494
Feldman AG, Ostry DJ, Levin MF, Gribble PL, Mitnitski AB (1998) Recent tests of the equilibrium-point hypothesis (lambda model). Mot Control 2:189–205
Fernandes RA, Diniz B, Ribeiro R, Humayun M (2012) Artificial vision through neuronal stimulation. Neurosci Lett 519:122–128
Fetz EE (1969) Operant conditioning of cortical unit activity. Science 163:955–958
Fitzsimmons NA, Drake W, Hanson TL, Lebedev MA, Nicolelis MA (2007) Primate reaching cued by multichannel spatiotemporal cortical microstimulation. J Neurosci 27:5593–5602
Fitzsimmons NA, Lebedev MA, Peikon ID, Nicolelis MA (2009) Extracting kinematic parameters for monkey bipedal walking from cortical neuronal ensemble activity. Front Integr Neurosci 3:3
Frank K (1968) Some approaches to the technical problem of chronic excitation of peripheral nerve. Ann Otol Rhinol Laryngol 77:761–771
Fuster JM, Bressler SL (2012) Cognit activation: a mechanism enabling temporal integration in working memory. Trends Cogn Sci 16(4):207–218
Galán F, Nuttin M, Lew E, Ferrez PW, Vanacker G, Philips J, Millán JR (2008) A brain-actuated wheelchair: asynchronous and non-invasive brain-computer interfaces for continuous control of robots. Clin Neurophysiol 119:2159–2169
Georgopoulos AP, Kalaska JF, Caminiti R, Massey JT (1982) On the relations between the direction of two-dimensional arm movements and cell discharge in primate motor cortex. J Neurosci 2:1527–1537
Georgopoulos AP, Lurito JT, Petrides M, Schwartz AB, Massey JT (1989) Mental rotation of the neuronal population vector. Science 243:234–236
Guertin PA (2009) The mammalian central pattern generator for locomotion. Brain Res Rev 62:45–56
Hampson RE, Gerhardt GA, Marmarelis V, Song D, Opris I, Santos L, Berger TW, Deadwyler SA (2012) Facilitation and restoration of cognitive function in primate prefrontal cortex by a neuroprosthesis that utilizes minicolumn-specific neural firing. J Neural Eng 9(5):056012
Hampson RE, Song D, Opris I, Santos LM, Shin DC, Gerhardt GA, Marmarelis VZ, Berger TW, Deadwyler SA (2013) Facilitation of memory encoding in primate hippocampus by a neuroprosthesis that promotes task-specific neural firing. J Neural Eng 10(6):066013. doi:10.1088/1741-2560/10/6/066013
Hatsopoulos NG, Donoghue JP (2009) The science of neural interface systems. Annu Rev Neurosci 32:249–266
Haykin S (2001) Adaptive filter theory, 4th edn. Prentice Hall, Upper Saddle River, 936 pp
Head H, Holmes G (1911) Sensory disturbances from cerebral lesions. Brain 34:102–254
Hochberg LR, Serruya MD, Friehs GM, Mukand JA, Saleh M, Caplan AH, Branner A, Chen D, Penn RD, Donoghue JP (2006) Neuronal ensemble control of prosthetic devices by a human with tetraplegia. Nature 442:164–171

Hochberg LR, Bacher D, Jarosiewicz B, Masse NY, Simeral JD, Vogel J, Haddadin S, Liu J, Cash SS, van der Smagt P, Donoghue JP (2012) Reach and grasp by people with tetraplegia using a neurally controlled robotic arm. Nature 485:372–375

Hubel DH, Wiesel TN (2005) Brain and visual perception: the story of a 25-year collaboration. Oxford University Press, Oxford, 744 pp

Humphrey DR, Schmidt EM, Thompson WD (1970) Predicting measures of motor performance from multiple cortical spike trains. Science 170:758–762

Ifft PJ, Shokur S, Li Z, Lebedev MA, Nicolelis MA (2013) A brain-machine interface enables bimanual arm movements in monkeys. Sci Transl Med 5:210ra154

Iriki A, Tanaka M, Iwamura Y (1996) Coding of modified body schema during tool use by macaque postcentral neurones. Neuroreport 7:2325–2330

Jones LA (2011) Tactile communication systems optimizing the display of information. Prog Brain Res 192:113–128

Kalaska JF, Scott SH, Cisek P, Sergio LE (1997) Cortical control of reaching movements. Curr Opin Neurobiol 7:849–859

Kawato M (1999) Internal models for motor control and trajectory planning. Curr Opin Neurobiol 9:718–727

Kennedy PR, Bakay RA (1998) Restoration of neural output from a paralyzed patient by a direct brain connection. Neuroreport 9:1707–1711

Kim SP, Sanchez JC, Rao YN, Erdogmus D, Carmena JM, Lebedev MA, Nicolelis MA, Principe JC (2006) A comparison of optimal MIMO linear and nonlinear models for brain-machine interfaces. J Neural Eng 3(2):145–161

Lebedev MA, Nicolelis MA (2006) Brain-machine interfaces: past, present and future. Trends Neurosci 29:536–546

Lebedev MA, Carmena JM, O'Doherty JE, Zacksenhouse M, Henriquez CS, Principe JC, Nicolelis MA (2005) Cortical ensemble adaptation to represent velocity of an artificial actuator controlled by a brain-machine interface. J Neurosci 25:4681–4693

Li Z, O'Doherty JE, Hanson TL, Lebedev MA, Henriquez CS, Nicolelis MA (2009) Unscented Kalman filter for brain-machine interfaces. PLoS ONE 4:e6243

Lilly JC (1956) Distribution of 'motor' functions in the cerebral cortex in the conscious, intact monkey. Science 124:937

Lin CT, Chang CJ, Lin BS, Hung SH, Chao CF, Wang IJ (2010) A real-time wireless brain–computer interface system for drowsiness detection. IEEE Trans Biomed Circuits Syst 4 (4):214–222

McFarland DJ, Krusienski DJ, Wolpaw JR (2006) Brain-computer interface signal processing at the Wadsworth Center: mu and sensorimotor beta rhythms. Prog Brain Res 159:411–419

Mellinger J, Schalk G, Braun C, Preissl H, Rosenstiel W, Birbaumer N, Kübler A (2007) An MEG-based brain–computer interface (BCI). Neuroimage 36:581–593

Millan JR, Renkens F, Mouriño J, Gerstner W (2004) Noninvasive brain-actuated control of a mobile robot by human EEG. IEEE Trans Biomed Eng 51:1026–1033

Moritz CT, Perlmutter SI, Fetz EE (2008) Direct control of paralysed muscles by cortical neurons. Nature 456:639–642

Mountcastle VB (1997) The columnar organization of the neocortex. A comprehensive review of the literature indicating the modular architecture of the cortex. Brain 120(4):701–722

Mountcastle VB (2005) The sensory hand: neural mechanisms of somatic sensation. Harvard University Press, Cambridge, MA, 640 pp

Muller-Putz GR, Pfurtscheller G (2008) Control of an electrical prosthesis with an SSVEP-based BCI. IEEE Trans Biomed Eng 55:361–364

Nicolas-Alonso LF, Gomez-Gil J (2012) Brain computer interfaces, a review. Sensors (Basel) 12:1211–1279

Nicolelis MAL (2001) Actions from thoughts. Nature 409:403–407

Nicolelis MA (2011) Beyond boundaries: the new neuroscience of connecting brains with machines – and how It will change our lives. Times Books, New York, 354 pp

Nicolelis MA, Lebedev MA (2009) Principles of neural ensemble physiology underlying the operation of brain-machine interfaces. Nat Rev Neurosci 10:530–540

Nurmikko AV, Donoghue JP, Hochberg LR, Patterson WR, Song YK, Bull CW, Borton DA, Laiwalla F, Park S, Ming Y, Aceros J (2010) Listening to brain microcircuits for interfacing with external world-progress in wireless implantable microelectronic neuroengineering devices. Proc IEEE Inst Electr Electron Eng 98(3):375–388

O'Doherty JE, Lebedev MA, Hanson TL, Fitzsimmons NA, Nicolelis MA (2009) A brain-machine interface instructed by direct intracortical microstimulation. Front Integr Neurosci 3:20. doi:10.3389/neuro.07.020.2009

O'Doherty JE, Lebedev MA, Ifft PJ, Zhuang KZ, Shokur S, Bleuler H, Nicolelis MA (2011) Active tactile exploration using a brain-machine-brain interface. Nature 479(7372):228–231. doi:10.1038/nature10489

Opris I (2013) Inter-laminar microcircuits across the neocortex: repair and augmentation. Research topic: "Augmentation of brain function: facts, fiction and controversy", Lebedev MA, Opris I and Casanova MF (eds). Front Syst Neurosci 7:80. doi:10.3389/fnsys.2013.00080

Opris I, Casanova MF (2014) Prefrontal cortical minicolumn: from executive control to disrupted cognitive processing. Brain 137(7):1863–1875

Opris I, Hampson RE, Stanford TR, Gerhard GA, Deadwyler SA (2011) Neural activity in frontal cortical cell layers: evidence for columnar sensorimotor processing. J Cogn Neurosci 23 (6):1507–1521

Opris I, Hampson RE, Gerhardt GA, Berger TW, Deadwyler SA (2012a) Columnar processing in primate pFC: evidence for executive control microcircuits. J Cogn Neurosci 24(12):2334–2347

Opris I, Fuqua JL, Huettl PF, Gerhardt GA, Berger TW, Hampson RE, Deadwyler SA (2012b) Closing the loop in primate prefrontal cortex: inter-laminar processing. Front Neural Circuits 6:88

Opris I, Santos LM, Song D, Gerhardt GA, Berger TW, Hampson RE, Deadwyler SA (2013) Prefrontal cortical microcircuits bind perception to executive control. Sci Rep 3:2285

Opris I, Fuqua JL, Gerhardt GA, Hampson RE, Deadwyler SA (2014) Prefrontal cortical record ings with biomorphic MEAs reveal complex columnar-laminar microcircuits for BCI/BMI implementation. J Neurosci Methods. S0165-0270(14)00197-6

Pfurtscheller G, Müller GR, Pfurtscheller J, Gerner HJ, Rupp R (2003) 'Thought'–control of functional electrical stimulation to restore hand grasp in a patient with tetraplegia. Neurosci Lett 351:33–36

Pohlmeyer EA, Oby ER, Perreault EJ, Solla SA, Kilgore KL, Kirsch RF, Miller LE (2009) Toward the restoration of hand use to a paralyzed monkey: brain-controlled functional electrical stimulation of forearm muscles. PLoS ONE 4:e5924

Presacco A, Forrester LW, Contreras-Vidal JL (2012) Decoding intra-limb and inter-limb kinematics during treadmill walking from scalp electroencephalographic (EEG) signals. IEEE Trans Neural Syst Rehabil Eng 20:212–219

Quiroga RQ, Reddy L, Kreiman G, Koch C, Fried I (2005) Invariant visual representation by single neurons in the human brain. Nature 435:1102–1107

Romo R, Hernández A, Zainos A, Brody CD, Lemus L (2000) Sensing without touching: psychophysical performance based on cortical microstimulation. Neuron 26:273–278

Romo R, Hernández A, Salinas E, Brody CD, Zainos A, Lemus L, de Lafuente V, Luna R (2002) From sensation to action. Behav Brain Res 135(1–2):105–118

Sampaio E, Maris S, Bach-y-Rita P (2001) Brain plasticity: 'visual' acuity of blind persons via the tongue. Brain Res 908:204–207

Schmidt EM (1980) Single neuron recording from motor cortex as a possible source of signals for control of external devices. Ann Biomed Eng 8:339–349

Schott GD (1993) Penfield's homunculus: a note on cerebral cartography. J Neurol Neurosurg Psychiatry 56:329–333

Schwartz AB, Cui XT, Weber DJ, Moran DW (2006) Brain-controlled interfaces: movement restoration with neural prosthetics. Neuron 52:205–220

Sellers EW, Vaughan TM, Wolpaw JR (2010) A brain-computer interface for long-term independent home use. Amyotroph Lateral Scler 11:449–455

Shannon RV (2012) Advances in auditory prostheses. Curr Opin Neurol 25:61–66

Sherrington CS (1906) The integrative action of the nervous system. Charles Scribner's Sons, Liverpool, 411 pp

Sitaram R, Caria A, Birbaumer N (2009) Hemodynamic brain-computer interfaces for communication and rehabilitation. Neural Netw 22:1320–1328

Song D, Chan RH, Marmarelis VZ, Hampson RE, Deadwyler SA, Berger TW (2009b) Nonlinear modeling of neural population dynamics for hippocampal prostheses. Neural Netw 22(9):1340–1351. doi:10.1016/j.neunet.2009.05.004

Sussillo D, Nuyujukian P, Fan JM, Kao JC, Stavisky SD, Ryu S, Shenoy K (2012) A recurrent neural network for closed-loop intracortical brain-machine interface decoders. J Neural Eng 9:026027

Tangermann M, Krauledat M, Grzeska K, Sagebaum M, Blankertz B, Vidaurre C, Müller KR (2009) Playing pinball with non-invasive BCI. Adv Neural Inf Process Syst 21:1641–1648

Tavella M, Leeb R, Rupp R, Millán JdR (2010) Towards natural non-invasive hand neuroprostheses for daily living. In: Proceedings of the 32nd annual international conference of the IEEE engineering in medicine and biology society, Buenos Aires

Taylor DM, Tillery SI, Schwartz AB (2002) Direct cortical control of 3D neuroprosthetic devices. Science 296:1829–1832

Thomson AM, Bannister AP (2003) Interlaminar connections in the neocortex. Cereb Cortex 13(1):5–14

Van Essen DC, Newsome WT, Bixby JL (1982) The pattern of interhemispheric connections and its relationship to extrastriate visual areas in the macaque monkey. J Neurosci 2(3):265–283

Velliste M, Perel S, Spalding MC, Whitford AS, Schwartz AB (2008) Cortical control of a prosthetic arm for self-feeding. Nature 453(7198):1098–1101

Vialatte FB, Maurice M, Dauwels J, Cichocki A (2010) Steady-state visually evoked potentials: focus on essential paradigms and future perspectives. Prog Neurobiol 90:418–438

Vlek RJ, Steines D, Szibbo D, Kübler A, Schneider MJ, Haselager P, Nijboer F (2012) Ethical issues in brain-computer interface research, development, and dissemination. J Neurol Phys Ther 36:94–99

Wessberg J, Stambaugh CR, Kralik JD, Beck PD, Laubach M, Chapin JK, Kim J, Biggs SJ, Srinivasan MA, Nicolelis MA (2000) Real-time prediction of hand trajectory by ensembles of cortical neurons in primates. Nature 408:361–365

Wilson BS, Dorman MF (2008) Cochlear implants: a remarkable past and a brilliant future. Hear Res 242:3–21

Wise SP (1985) The primate premotor cortex: past, present, and preparatory. Annu Rev Neurosci 8:1–19

Wolpaw JR, McFarland DJ (2004) Control of a two-dimensional movement signal by a noninvasive brain-computer interface in humans. Proc Natl Acad Sci U S A 101:17849–17854

Zacksenhouse M, Lebedev MA, Carmena JM, O'Doherty JE, Henriquez C, Nicolelis MA (2007) Cortical modulations increase in early sessions with brain-machine interface. PLoS ONE 2: e619

Zhang F, Aravanis AM, Adamantidis A, de Lecea L, Deisseroth K (2007) Circuit-breakers: optical technologies for probing neural signals and systems. Nat Rev Neurosci 8:577–581

Index

M.F. Casanova, I. Opris (eds.), *Recent Advances on the Modular Organization of the Cortex*, DOI 10.1007/978-94-017-9900-3

MIX
Papier aus verantwortungsvollen Quellen
Paper from responsible sources
FSC® C105338
FSC
www.fsc.org

If you have any concerns about our products,
you can contact us on
ProductSafety@springernature.com

In case Publisher is established outside the EU,
the EU authorized representative is:
Springer Nature Customer Service Center GmbH
Europaplatz 3, 69115 Heidelberg, Germany

Printed by Libri Plureos GmbH
in Hamburg, Germany